Computer-assisted Bacterial Systematics

Edited by

M. Goodfellow
Department of Microbiology
The Medical School
Newcastle upon Tyne, UK

D. Jones
Department of Microbiology
University of Leicester
Leicester, UK

F. G. Priest
Department of Brewing and Biological Sciences
Heriot–Watt University
Edinburgh, UK

1985

Published for the
Society for General Microbiology
by
ACADEMIC PRESS
(Harcourt Brace Jovanovich, Publishers)
London Orlando San Diego New York
Toronto Montreal Sydney Tokyo

ACADEMIC PRESS INC. (LONDON) LTD.
24–28 Oval Road
LONDON NW1 7DX

United States Edition published by
ACADEMIC PRESS, INC.
Orlando, Florida 32887

BRITISH LIBRARY CATALOGUING IN PUBLICATION DATA
Computer-assisted bacterial systematics.—(Special
publications of the Society for General
Microbiology)
1. Microbiology—Classification
I. Goodfellow, M. II. Jones, D. III. Priest,
Fergus, G. IV. Series
576'.012 QR12

LIBRARY OF CONGRESS CATALOGING-IN-PUBLICATION DATA
Main entry under title:
Computer-assisted bacterial systematics.
(Special publications of the Society for General
Microbiology)
Papers presented at a symposium entitled "Twenty-five
years of numerical taxonomy" held at the University of
Warwick in April 1983.
Includes index.
1. Bacteriology—Classification—Data processing—
Congresses. 2. Numerical taxonomy—Data processing—
Congresses. I. Goodfellow, M. II. Jones, D.
III. Priest, F. G. IV. Society for General Microbiology.
V. Series.
QR81.C66 1985 589.9'0012 85-5987
ISBN 0-12-289665-3 (alk. paper)

PRINTED IN THE UNITED STATES OF AMERICA

85 86 87 88 9 8 7 6 5 4 3 2 1

Computer-assisted Bacterial Systematics

Special Publications of the Society for General Microbiology

A complete list of books in this series appears at the end of the volume.

This book is based on a Symposium of the SGM held at Warwick, April 1983.

Dedicated to
Peter Sneath and Robert Sokal
The Fathers of Numerical Taxonomy

Contents

Contributors

Numbers in parentheses indicate the pages on which the authors' contributions begin.

G. ALDERSON (227), *School of Medical Sciences, University of Bradford, Bradford BD7 1DP, UK*

R. P. AMBLER (307), *Department of Molecular Biology, University of Edinburgh, Edinburgh EH9 3JR, UK*

E. A. BARBOUR (137), *Technical Centre, Scottish and Newcastle Breweries PLC, Edinburgh EH8 8DD, UK*

S. BASCOMB (37), *Department of Medical Microbiology, St. Mary's Hospital Medical School, London W2 1PG, UK*

R. R. COLWELL (107), *Department of Microbiology, The University of Maryland, College Park, Maryland 20742, USA*

C. H. DICKINSON (165), *Department of Plant Biology, The University, Newcastle upon Tyne NE1 7RU, UK*

M. GOODFELLOW (165, 289), *Department of Microbiology, The Medical School, Newcastle upon Tyne NE2 4HH, UK*

C. S. GUTTERIDGE (369), *Cadbury Schweppes PLC, Lord Zuckerman Research Centre, University of Reading, Reading RG6 2LA, UK*

L. R. HILL (265), *National Collection of Type Cultures, Central Public Health Laboratory, London NW9 5HT, UK*

B. HOLMES (265), *National Collection of Type Cultures, Central Public Health Laboratory, London NW9 5HT, UK*

K. KERSTERS (337), *Laboratorium voor Microbiologie en Microbiële Genetica, Faculteit Wetenschappen, Rijksuniversiteit, B-9000 Ghent, Belgium*

M. T. MACDONELL (107), *Department of Microbiology, The University of Maryland, College Park, Maryland 20742, USA*

H. J. H. MACFIE (369), *Agricultural and Food Research Council, Food Research Institute (Bristol), Bristol BS18 7DY, UK*

A. G. O'DONNELL (403), *Department of Soil Science, The University, Newcastle upon Tyne NE1 7RU, UK*

F. G. PRIEST (137), *Department of Brewing and Biological Sciences, Heriot–Watt University, Edinburgh EH1 1HX, UK*

M. J. SACKIN (21), *Department of Microbiology, University of Leicester, Leicester LE1 7RH, UK*

P. H. A. SNEATH (415), *Department of Microbiology, University of Leicester, Leicester LE1 7RH, UK*

R. R. SOKAL (1), *Department of Ecology and Evolution, State University of New York at Stony Brook, Stony Brook, New York 11794, USA*

L. VALLIS (369), *Agricultural and Food Research Council, Food Research Institute (Bristol), Bristol BS18 7DY, UK*

J. C. VICKERS (289), *Department of Microbial Biochemistry, Glaxo Group Research Ltd., Greenford, Middlesex UB6 0HE, UK*

L. G. WAYNE (91), *Tuberculosis Research Laboratory, Veterans Administration Medical Center, Long Beach, California 90822, USA, and Department of Microbiology and Immunology, California College of Medicine, University of California, Irvine, California, USA*

J. WILLIAMS (61), *Department of Biochemistry, University of Bristol, Bristol BS8 1TD, UK*

S. T. WILLIAMS (289), *Department of Botany, University of Liverpool, Liverpool L69 3BX, UK*

Preface

Numerical taxonomy, the grouping by numerical methods of taxonomic units based on their character states, was introduced by Peter Sneath and Robert Sokal more than 25 years ago. Their collaboration has resulted in several joint publications, including two books, *Principles of Numerical Taxonomy,* published in 1963, and *Numerical Taxonomy,* which appeared in 1973. The 1963 book was the first to enunciate the principles and to detail the methodology of numerical taxonomy, and both are the main reference books in the field. Numerical taxonomy seeks to develop methods that are objective and reproducible both in evaluation of taxonomic relationships and in the construction of taxa. The rapid developments that have taken place in numerical taxonomy since its inception owe much to the simultaneous development of computer techniques. The latter are being increasingly applied in every aspect of bacterial systematics.

Numerical taxonomic methods were first applied to bacteriology more than 25 years ago by Peter Sneath. His two papers, 'Some Thoughts on Bacterial Classification' and the 'Application of Computers to Taxonomy', published in 1957, revolutionised microbial classification, placing it on a new, firmer scientific basis by allowing large amounts of data to be analysed and assessed more objectively than before. A survey of the literature now reveals that numerical taxonomy has been applied not only to more than 100 bacterial genera but has also been used in other biological disciplines, including botany, mycology, and zoology. The purpose of the present book is to give readers a balanced survey of the theoretical basis of numerical taxonomy and its impact on microbial classification and identification.

All of the chapters have been written by specialists so that the book as a whole forms a unique collection of papers on a fundamental scientific discipline. There are valuable and detailed reviews of the principles of numerical taxonomy, stability of classifications, cladistics and evolution of proteins, protein sequences and taxonomy, computer-assisted identification, and the impact of computer-assisted methods on the systematics of lactic acid bacteria, mycobacteria, and Gram-negative bacteria and on the description of microbial populations in natural habitats. Additional chapters deal with the numerical analysis of various types of chemical data using multivariate statistics and with the value of non-hierarchical methods in bacterial taxonomy. A final chapter considers the future of numerical taxonomy and the shape of things to come.

This book, number 15 in the series Special Publications of the Society for General Microbiology, arose as a result of a symposium entitled 'Twenty-five Years of Numerical Taxonomy', organised by the Microbial Systematics Group of the Society of General Microbiology and held at the University of Warwick in April 1983. The book contains contributions from all of those who presented papers at the symposium. Additional chapters were solicited from experts in fields not covered at the symposium. The majority of the chapters include work published up until about mid-1984.

We would like to express our thanks to all the contributors for the help they have given in the preparation of this book. We are also grateful to the staff of Academic Press for all of their help and encouragement during the preparation of this volume.

M. Goodfellow
D. Jones
F. G. Priest

1

The Principles of Numerical Taxonomy: Twenty-five Years Later

R. R. SOKAL

Department of Ecology and Evolution, State University of New York at Stony Brook, Stony Brook, New York, USA

Introduction

In this chapter, I shall first recount the origins of modern numerical taxonomy, with emphasis on the collaboration between P. H. A. Sneath and myself. Then I shall turn to the principles of this discipline as first enunciated and the changes these have undergone in the past 25 years. Next I shall give my views on the current controversy concerning optimality criteria in numerical taxonomic research. Because numerical taxonomy has been and continues to be a largely empirical field, principles and methodology have been closely intertwined over the years. I cannot within the confines of this chapter give a review of the development of taxometric methodology but necessarily I shall have to refer briefly to some of the methodological developments during the past 25 years.

Sneath and Sokal (1973) defined numerical taxonomy as 'the grouping by numerical methods of taxonomic units into taxa on the basis of their character states. The term includes the drawing of phylogenetic inferences from the data'. Because microbiologists have had little faith in phylogenetic interpretations of their data and have practiced phenetic taxonomy in the great majority of cases, I shall restrict myself to developments in numerical phenetics for the purposes of this chapter. There has been impressive progress in numerical methods for inferring phylogenies, and I can recommend the review by Felsenstein (1982) to the interested reader. Although I shall not address the problem of phylogenetic inference directly, I shall be comparing numerical phenetic procedures with numerical cladistic ones, with the cladistic methods considered principally as classificatory tools rather than as estimators of phylogenetic lineages.

The idea of quantifying relations between taxa (typically species) has occurred to various taxonomists since the beginning of this century and even before that (see Sneath and Sokal, 1973, p. 13, for an account; these authors cited 18

ISBN 0-12-289665-3

references in biological taxonomy alone, not counting those in cognate fields such as ecology, geology, or psychology). Why were none of the methods proposed in these studies adopted by other workers and further elaborated? Sneath and Sokal (1973) listed several reasons for this failure of the early developments to be exploited. The methods were generally not elaborated into a complete system of character coding, resemblance coefficient computation, clustering method, plus construction of a classification. Established taxonomic practice employed weighting of characters without an explicit rule for how these weights should be applied; yet taxonomists thought that equally weighted characters violated established taxonomic practice. The necessity to use many characters and the consequent tedium of computations made numerical taxonomy impractical in precomputer days. Finally, the prevailing theory of classification looked for a taxonomy that would reflect evolutionary lineages. It was obvious that these early numerical taxonomic attempts did not yield phylogenies, yet an explicit alternative set of principles, such as that of phenetic taxonomy, had not been established. Biologists in general, and taxonomists especially, were far less numerative than they are today. Thus they were simply not inclined to adopt such methods, and there is little evidence in the literature that the early proposals made became the subjects of controversy. All of these points clearly contributed to the failure of numerical taxonomy to develop much earlier than it did but, in my judgment, the two single most important considerations were contextual rather than conceptual. They were (1) the advent of computing machinery, which made it possible to carry out more than a pilot study, and (2) the development of a vigorous controversy which had the effect of drawing the attention of other systematists to the methods and of forcing the proponents of the new techniques to formulate principles defending their methods.

Modern numerical taxonomy originated in four separate centres—two in Britain and two in the United States—all working independently of each other. P. H. A. Sneath was attempting a revision of the genus *Chromobacterium* and was becoming dissatisfied by the subjective procedures he encountered in bacterial taxonomy. Faced with the usual binary data tables for different strains and species, he conceived of the need for a similarity coefficient and independently rediscovered Jaccard's coefficient and single-linkage clustering (Sneath, 1957b), being unaware then of the work of Florek *et al.* (1951a,b), published in Polish journals. Sneath used various aids, such as superimposed sheets of X-ray film bearing data matrices to carry out the considerable amount of computation involved in his large study of 45 OTUs and 105 characters, but also was the first biologist to employ a genuine electronic computer, the Elliott 405, for taxonomic purposes. In two publications (Sneath, 1957a,b) he presented his principles and methodology as well as his results.

While Sneath's work stemmed from a need to arrive at an efficient classification of his group of organisms, my own entry into the field came not as a practising taxonomist but as a biometrician filled with youthful enthusiasm to convert my colleagues to the joys and manyfold uses of biological statistics. Following graduation from the University of Chicago I had the good fortune at my first position to work in an active centre of taxonomic research at the University of Kansas, in a group headed by Professor Charles D. Michener, one of the leading American entomologists of this generation. An informal discussion over lunch in 1953 concerning the subjectivity of taxonomic principles and practice led to the rash boast on my part that I could devise a method for quantifying taxonomic relationships and for establishing classifications more objectively than was possible by the traditional methods. I was challenged on this point and before I knew it had entered into a wager (to set the historical record straight, it was for a six-pack of beer) that I could develop such a technique. Thus began my own rather accidental entry into numerical taxonomy, which over the following years went on to absorb more and more of my time and thinking. Some of my work at the time dealt with the factor analysis of character correlations, and I had become familiar with the psychometric literature. In trying to think of methods for solving the taxonomic problem, I turned to psychometrics for suggestions on how to tackle it. Various types of primitive cluster analyses had been suggested in the 1940s by Holzinger and Harman (1941) and Cattell (1944). Once the idea of cluster analysis was implanted in my mind, the subsequent development of the average-linkage methods took only a few weeks. Professor Michener had furnished me with a data matrix of 97 species in four genera of bees known as the *Hoplitis* complex. Each species was described by 122 mostly multistate ordered characters. The computations for this analysis, carried out by means of IBM tabulating machinery and electric mechanical desk calculators, took several months and a number of assistants but can now be carried out in less than 1 min on a computer. They resulted in a classification of the bees which was similar to the established classification but provided some useful new insights to Michener, the specialist on the group. Having succeeded in developing a method that seemed to work, Michener and I turned to analysing the theoretical implications of this approach to taxonomy. The problems of coding and scaling multistate characters, of parallelism and convergence, as well as of equal weighting of characters, were all faced in this study (Michener and Sokal, 1957; Sokal and Michener, 1958).

P. H. A. Sneath, who among his other accomplishments is an outstanding bibliographer, learned of my work through an abstract of a paper given at a meeting in 1956 and wrote to me, initiating a correspondence and eventual collaboration that has spanned many years. We first met in 1959 when Sneath was in the United States on a fellowship in bacterial genetics, and we met again later that year when I spent a year at University College, London. At that time, we decided to collaborate on a position paper stating both principles as well as methods of numerical taxonomy. The manuscript grew until it became the basis for our first book on the subject (Sokal and Sneath, 1963). Interest in numerical taxonomy continued at a high level through the late 1950s and early 1960s, fanned by a vigorous controversy involving the proprieties of quantifying taxonomy, the justification for phenetics, the supposedly baneful effects of convergence and parallelism, and the like. Another boost for the development of numerical methods was the simultaneous introduction of computers into universities. By 1956 most sizable universities had acquired their first computer, and tasks that had at one time taken months began to take days at first, hours somewhat later, and before long, only minutes. It is obvious that the great variety of techniques could not have been developed if there had not been computational means to execute them.

One of the other two groups developing NT, as the subject came to be called before long, comprised two British workers, the zoologist A. J. Cain and the anthropologist G. A. Harrison at Oxford. In several important papers, Cain clarified taxonomic principles and with Harrison (Cain and Harrison, 1958) was first to distinguish clearly between phenetic and cladistic relationships. He also computed average Manhattan distance for the first time in numerical taxonomy. After their initial, independent, and fundamental work in this area, the authors went on to the consideration of other problems and did not continue in numerical taxonomy. The fourth group was founded by the American botanist David Rogers, who, in association with the mathematician T. T. Tanimoto, developed an early approach to probabilistic distance coefficients and clustering (Rogers and Tanimoto, 1960). Out of Rogers' group came George Estabrook, whose contributions to numerical taxonomy and especially numerical phylogenetic inference have been outstanding.

Principles of Classification

Natural Classification

With the first successes in forming classifications by numerical methods, numerical taxonomists faced the challenge of integrating their work within the existing framework of classificatory theory or of creating their own body of principles. It had been generally accepted among biological taxonomists that classifications were to be natural rather than artificial and that 'natural' in some sense implied a true reflection of the arrangement of the organisms in nature. When numerical taxonomy came on the scene, natural taxa were generally considered to be those that reflected phylogenetic history, that is, were monophyletic. Rigorous definitions of monophyly were lacking (see discussions in Simpson, 1961, and Sokal and Sneath, 1963). Numerical taxonomists realized from the beginning (Michener and Sokal, 1957; Sneath, 1961) that the classifications they were producing based on unweighted phenetic similarity would be affected by parallelisms and convergence as well as by unequal evolutionary rates in diverging lineages and hence would not necessarily yield monophyletic taxa. They took the attitude that criteria for recognizing monophyletic taxa were imprecise and not operational, and turned for their principles to the writings of J. H. Woodger, M. Beckner, and especially, J. S. L. Gilmour, whose general theses might be described as follows: taxa are defined by the correspondences of many characters (Woodger, 1937); character states need not correspond for all members of a taxon—that is, taxa are polythetic (Beckner, 1959; Sneath, 1962); and a system of classification is the more natural, the more propositions that can be made regarding its constituent classes (Gilmour, 1937, 1940, 1951, 1961). Gilmour traced this concept to the nineteenth-century philosophers of science such as J. S. Mill. Sneath (1957a) traced analogous taxonomic principles back to the French botanist Adanson who worked in the eighteenth century. Gilmour's principle is essentially one of predictiveness: natural taxa are thought to be those that are most highly predictive overall—that is, not for any one special purpose but in terms of a number of logically independent statements that can be made concerning its members. Given that uniformity of character states is not required—that is, that the taxa are polythetic (Sneath, 1962)—overall predictiveness should be maximized for a given classification but predictiveness for a given variable in a classification would rarely if ever be perfect. The notions of Woodger, Beckner, and Gilmour also led to the concept of *degrees* of naturalness as distinct from designating a classification categorically as natural or not natural depending on its conformity with the monophyly criterion. As will be discussed later, it was assumed but not demonstrated that the numerical taxonomic methods that yielded taxa with mutually most similar membership would also yield taxa that were correspondingly high in terms of Gilmour naturalness.

With respect to natural classifications, the views of numerical taxonomists have not changed substantially during the past 25 years. They contrast markedly with the views of cladists (Eldredge and Cracraft, 1980; Wiley, 1981), who have redefined monophyletic taxa more rigorously than before, equating natural supraspecific taxa with holophyletic groups, which are groups of species that include a common ancestor and all of its descendants (Farris, 1974). Cladists consider supraspecific taxa as real entities, not as classes, and therefore to them a natural taxon is 'a taxon that exists in nature independent of man's ability to perceive it'. Such taxa must therefore be discovered rather than invented, and they will originate 'according to natural processes and thus must be consistent with these natural processes [Wiley, 1981, p. 72]'. By contrast, phenetic numerical taxonomists consider supraspecific taxa as classes and are therefore not especially concerned with their reality. Taxa are human constructs and natural taxa are those that are natural to humans. It has been shown that naive as well as professional taxonomists appear to base their classifications more on phenetic than on cladistic criteria irrespective of the principles which the experimental subjects profess to have followed (Sokal and Rohlf, 1980). The notion that a natural taxon should be a natural concept to humans rather than a real entity in nature finds support in studies in cognitive psychology, where psychologists and computer scientists have investigated the nature of human concepts and the processing of information for assigning membership of an object in a given class or concept (Smith and Medin, 1981). The currently predominant view of concept formation is the so-called probabilistic one which holds that the representation of a concept is a summary description of an entire class and cannot be restricted to a set of necessary and sufficient conditions, but is rather a measure of central tendency of the pattern of its members. This corresponds to the notion of polythetic classes in biological taxonomy. If humans process information and form concepts on the basis of polythetic classes, then it seems to me desirable to form such classes also for classifications constructed by computers for human use. Such an argument leads to considering a natural classification as a maximal predictive one.

Many and Equally Weighted Characters

From the above principles adopted for numerical taxonomy it followed almost naturally that taxa should be based on many characters, since statements on overall predictivity of classifications as well as those on overall phenetic similarity would be unreliable when made on few characters. In this, the position differed greatly from the then-established one in taxonomy, which selected a few phylogenetically 'meaningful' or 'important' characters. The next problem faced by numerical taxonomists was whether to weight characters according to some definable scheme or whether to give each character equal weight in the computation of a resemblance coefficient. Several lines of reasoning combined to lead to the latter decision. Sneath (1957a) reached this conclusion on epistomological grounds stemming from Gilmour's work. Michener and Sokal (1957) concluded that there was no rational way of allocating weight to characters even where the entire genomic constitution known. It was later pointed out by critics that equal weighting of characters, at least using the algorithms employed by the earlier numerical taxonomists, was, strictly considered, an impossibility. Not only were those characters or character suites omitted from the analysis effectively weighted zero, but characters with many states carry a heavier weight in determining similarity than those with few states. Thus, characters effectively weight themselves in terms of the number of differentiable states that can be considered in a given study. This is especially true when ordered multistate characters are recoded in binary form, although one could argue that they thereby have weights in proportion to their 'content of information', and thus this is in the spirit of Gilmour's concepts. Sneath and Sokal (1973) state that even unequal weighting when carried out by an explicit algorithm is permissible in phenetic taxonomy. However, most work in phenetic numerical taxonomy has eschewed explicit weighting, since no one has come up with a convincing algorithm for doing so.

There has been little work done in the intervening years to determine how important the effects of weighting characters are in numerical taxonomic studies. Perhaps this lack of interest reflects experience based on the few published reports (e.g., Moss, 1968) which show that even a 1000-fold weighting for 18 of 135 characters has only a negligible effect in a 17-OTU study. Although the differential evaluation of characters underlies much of numerical phylogenetic inference, in practice, as Felsenstein (1982) pointed out, most numerical phylogenetic work is also based on equal weighting of character states. Such unequal weighting as exists in this field is largely related to the assumptions made about character change in a given model and is based on the probabilities of character state change.

How Many and Which Characters?

In any numerical taxonomic study, the questions of how many and which types of characters to employ are of immediate importance. Although the two questions are closely interlinked, it is useful to consider them separately at first for simplicity's sake. In many cases a taxonomist is faced only with characters of a single type, such as external morphology or biochemical information. How many characters should be employed? Early theorizing on this issue has not stood the test of time. Unless each character studied would provide independent information on a large number of loci, one cannot argue that a moderate number of characters, say between 40 and 100, would sample a reasonable fraction of the genetic information in the taxon being described. The notion that the characters chosen lead to a genetically based estimate of an overall similarity (the *matches asymptote hypothesis* of Sokal and Sneath, 1963) is also not useful as the increasingly complex relations underlying molecular genetics are unravelled. It is not clear which part of the genome should in fact be counted and what the similarity metric should be, given the complexities of repeat sequences, silent DNA, transposons, and the like. In fact, various measures of genetic similarity based on DNA hybridization, restriction enzymes, and related techniques all assay somewhat different aspects of genetic similarity, and no one of these can be considered an absolute standard. But more fundamentally it could be argued that at this stage in our understanding of molecular genetics and development, the links between the molecular basis of the genome and the phenotypic characters being used in conventional taxonomy, at least in higher organisms, are only slightly understood. Therefore, it is premature if not quixotic to aim for a molecular rationale for justifying overall similarity, and to base a sampling strategy of phenotypic characters on molecular considerations.

Against these discouraging considerations, one can marshal considerable empirical evidence that numerical classifications become quite stable when a reasonable number, say 60 characters, is employed. By stable, I mean that when the number of characters is increased substantially, to say 100, the resulting taxonomic structure does not differ by much. It is as though there were an asymptote of similarity which once approached will not change much by the addition of a substantial number of characters. Sokal and Rohlf (1970) tested this in an experiment in which independent, naive taxonomists described new character sets on the same group of organisms. They found that independent individuals coding their own characters largely described the same variational patterns and that additional information provided by independently varying characters did not appreciably change the classification or the underlying character-correlational structure. Sneath and Sokal (1973) suggested a simulation experiment involving character suites of different correlational structure which might test the effect of the addition of the new characters by sampling from the respective suites. This experiment was never done. However, a more realistic approach to this question, which we are currently carrying out in my laboratory, is to compare classifications that share a given cladistic topology but are based on different character suites evolved along the cladogram. The character suites result from separate simulation runs, employing different assumptions about evolutionary rates and reversals, which in turn yields different classes of characters having different correlational structures. It should then be possible to study how stability would be reached by increasing the number of characters from a single class and furthermore whether and in what manner stability would be reached by mixtures of classes of characters.

Even though the results of this study are not yet known, it can be safely assumed that there will be few changes in a classification once an appreciable number of characters is employed in establishing it. However, the incorporation of different classes of characters into this simulation model leads us into the second question, namely, What kinds of characters should be used in a numerical taxonomic study? Once biologically meaningless and logically correlated characters are eliminated from consideration, is there any guide to preferring one type of character such as those based on external morphology, internal morphology, histology, biochemistry, behaviour, and the like? Numerical taxonomists have traditionally used all types of characters, and there seems no reason to prefer one type over the other. There is no current evidence that there are classes of genes coding for classes of characters. Rather, most genes affect various characters and most characters are affected by more than one gene. This *nexus hypothesis* postulated in the early days of numerical taxonomy (Sneath and Sokal, 1962) is still believed to be true. From it followed the *nonspecificity hypothesis,* which postulated that there are no large, distinct classes of genes affecting exclusively one class of characters such as morphological, physiological, or ethological characters, or affecting special regions of the organisms such as head, skeleton, or leaves. This hypothesis may be true at a genetic level.* The implications of this hypothesis are that any class of characters should provide information about a random sample of the genome, and by inference, about the taxonomic structure of the group. Another way of looking at the problem is to assume that because any taxonomic group has a single phylogeny, differential information about its taxonomic structure may come from characters that differ in their evolutionary parameters. But if such differences are not reflected in classes of phenotypic characters, then the only other reason there should be lack of congruence in classifications is the existence of different adaptive evolution for different classes of characters. We plan to test this hypothesis by simulation.

Nonspecificity in data sets of real organisms has been tested repeatedly. A summary of the findings until 1973 are furnished by Sneath and Sokal (1973, p. 100). Matrix correlations between resemblance matrices based on different types of characters range from a low of .29 to a high of .80 over a taxonomically very diverse group. Subsequent results fall well within this range (e.g., Rohlf *et al.*, 1983b; Neel *et al.*, 1974). Recent studies have frequently compared findings between biochemical and traditional morphological information, often with a view to finding discrepancies between genomic and morphological divergence. Matrix correlations in such studies still fall within the range observed in the earlier, purely morphological studies (Cherry *et al.*, 1978). Thus, the resemblance matrices based on different classes of characters resemble each other only partially. Other recent studies have compared congruence between phenetic and cladistic classifications in the hope of discovering whether the failure of nonspecificity to hold would more seriously affect phenetic or cladistic classifications. Studies by Mickevich (1978, 1980), based on some of the same groups of organisms earlier reported by Sneath and Sokal (1973), purported to demonstrate that phylogenetic classifications invariably were more congruent than phenetic ones. However, Rohlf and Sokal (1980, 1981) have criticized the design of her test, and Rohlf *et al.* (1983a,b) have shown that the numerical results obtained by her were largely not repeatable. When the computations were carried out correctly, the results on congruence for a phenetic method (UPGMA clustering) and a cladistic method, minimum-length Wagner trees, were very similar. The fact that nonspecificity is only partially true in real organisms has important consequences for phenetic taxonomy. It means that phenetic techniques will not reach perfect congruence of classifications when these are based on different sets of characters, even when a large number of characters is employed for each character class. This will be so because different classes of characters frequently

*Stated exactly in this way it may be difficult to prove. What exactly does it mean that there are no large classes of genes affecting one of these classes of characters? One would have to state that a defined set of genes located variously on the genome affected only one class of character.

reflect different adaptations. Rohlf and Sokal (1981) predicted that 'ideally, cladistic techniques should yield fully congruent cladograms since there is only one true cladogram for a given set of data regardless of the set of characters on which it is based'. But, since cladograms based on character sets are only estimates of the true cladogram and will furthermore be subject to errors due to the sampling of characters employed, errors in the determination of their states, and any shortcomings of the algorithms employed, it may be that cladistic solutions in practice will be no more congruent than phenetic classifications based on different classes of characters (as shown by Rohlf *et al.*, 1983a,b).

The consequences to be realized from the failure of nonspecificity to hold strictly is that characters from different character classes should be employed with some stratified sampling design. In fact, of course, such designs are frequently impossible and taxonomists must use whatever data they can obtain. I do not think that this situation is appreciably different in cladistic taxonomy, and therefore, if the findings of Rohlf *et al.* (1983a,b) are corroborated and there are indeed few differences between cladistic and phenetic analyses with respect to the nonspecificity hypothesis, the types of recommendations for each of the methods for using stratified samples will clearly be the same. A question that has not yet been tested is whether the consensus of separate classifications derived from separate classes of characters would be similar to the classification derived from a single analysis based on a mixture of the separate classes of characters and whether in this respect phenetic and cladistic classifications will perform differently. Experiments also need to be carried out which would weight characters in different amounts to simulate basing classifications on different numbers of characters in each class to see to what degree such a weighting of characters would affect the overall solutions. This is clearly the model that will be most likely to reflect actual taxonomic problems encountered in the real world.

Resemblance

Necessarily, it became an early consideration of numerical taxonomists to develop satisfactory measures of resemblance. The two types of resemblance measures, similarity and dissimilarity coefficients, both have their roots in work carried out before the development of modern numerical taxonomy, and essentially all commonly used coefficients were established during the early formulation of the subject. Among similarity coefficients, Jaccard's coefficient was employed in Sneath's pioneering paper (1957a), and the simple-matching coefficient was proposed by Sokal and Michener (1958), although it had an earlier history in other disciplines. These are the two association coefficients for binary data most commonly employed in biological taxonomy. For continuous variables, the product-moment correlation coefficient was proposed by Michener and Sokal (1957) and is still the most commonly used measure despite criticisms of its properties by Eades (1965), Minkoff (1965), Jardine and Sibson (1971), and Dunn and Everitt (1982). Both commonly used distances were used early on: the Manhattan distance by Cain and Harrison (1958), who termed it mean character difference, and the Euclidean distance by Sokal (1961), who termed it taxonomic distance.

To some degree the choice of a resemblance coefficient is determined by the nature of the original data matrix. With binary and unordered multistate data matrices one necessarily employs association coefficients; with continuous characters one employs distances and correlations. Yet another aspect of deciding which resemblance measure to employ is knowledge of what type of resemblance the taxonomist wishes to portray—presumably the type of resemblance that taxonomists have traditionally recognized. At least for organisms that differ markedly in size and shape, classifications based on correlation coefficients seem to correspond more to those based on intuitive notions of similarity than do distance-based classifications (Rohlf and Sokal, 1965; Boyce, 1964, 1965; Sokal and Rohlf, 1980; Sokal, 1983a). But correlations are not especially helpful with binary data (nor is standardization of binary data).

Despite the proposal and occasional application of other similarity measures, the vast majority of published studies has employed one of the above-named coefficients. A factor in their continued use has been the relative simplicity of each of the measures for a given type of data matrix and resemblance measure. This is true of all but the product-moment correlation coefficient, which could have been simplified further, perhaps to a rank correlation, which has so far been employed but once as far as I know (Daget and Hureau, 1968). I am currently working on a further simplification of a taxonomic resemblance coefficient in an effort to develop robust methods of numerical taxonomy. If such methods can gain general acceptance they should alleviate the criticisms of instability of methods as a function of variation in computational techniques.

Taxonomic Structure

Assuming that one can agree on a measure of resemblance, taxa constructed by numerical taxonomy must have members which resemble each other more than they do those of other taxa at the same or at other rank levels in the hierarchy. Because one's intuitive notion of within-taxon similarity encompasses an average measure rather than the minimum or maximum thresholds, single-linkage and complete-linkage clustering are used in special cases only, and average-linkage clustering (usually UPGMA) has become the method of choice (Sneath and Sokal, 1973). But while UPGMA clustering may indeed lead to classifications with high average within-group similarity, there is no assurance that classifications reflecting maximal resemblance have been attained. In this connection, mutually highest similarity for a taxon needs to be defined more rigorously. Is it to be expressed only by the level of the average similarity, or should scatter of pairwise values around this average—or what amounts indirectly to the same thing, scatter of OTUs in the phenetic space—be taken into consideration? Tests of hierarchic clustering methods so far have concentrated mainly on goodness of fit of the similarity values implied by the phenogram (the cophenetic values) to the original similarity matrix. It is well known (Rohlf and Sokal, 1981) that cophenetic correlation can be increased by iterative improvements of a phenogram.

However, the problem of arriving at an optimal classification with respect to the resemblances among taxa is more complex than maximizing the cophenetic correlation coefficient. Even if we limit ourselves only to bifurcating trees, a very large number of different-rooted topologies can be constructed. The two extreme forms of this range of topologies is the completely symmetrical bifurcating tree (for OTU numbers that are powers of two) and the 'comb', which is an ultimate pair of OTUs at the tips being successively joined by the other $t - 2$ OTUs, one at a time. An overall criterion of intra-taxon similarity needs to be established for a classification which will allow for different topologies along this range and for different rank levels of OTUs in a classification. Such an overall criterion should take into consideration the average level of similarity within a taxon, its variance, and the rank of a given taxonomic subset with respect to the entire study. Since such a criterion can be constructed in a number of ways and weighted as a function of taxon size, rank, variance of similarity values, and so forth, it may be difficult to construct a universally acceptable criterion. It is not obvious that the cophenetic correlation coefficient will be monotonically related to such a criterion, since the former measures the conformity of a hierarchic clustering of the resemblance matrix to the matrix itself, whereas the latter would presumably measure level and dispersion of similarity values.

An unfinished problem of phenetic numerical taxonomy (as pointed out by Farris, 1979) is that, although pheneticists have called on Gilmour naturalness, which leads to high predictivity, as a goal criterion in their work, they have in fact largely clustered by similarity in the hope that high similarity will also lead to high predictivity. There has not, so far, been any work showing how to maximize similarity except by relocation techniques in nonhierarchic classifications. Although there must obviously be a close relation between classifications in which objects within any given taxon resemble each other closely and predictivity of characters within that taxon, the relationship has not as yet been established. It is currently under investigation in at least two laboratories.

It is interesting that efforts to provide a theoretical foundation to numerical taxonomy were largely aimed at providing a rationale for classification by resemblance and not expressly at the resulting taxonomic structure. Thus, there was less justification for clustering methods than there was for resemblance coefficients. As long as a clustering method worked reasonably well, it was employed. The principal constraint in biological taxonomy was that the resulting classifications should be hierarchic and nonoverlapping in conformity with the principles of a Linnaean classification; that is, the classifications should have ultrametric properties (Rohlf and Sokal, 1981). Subsequently, numerical taxonomists attempted to have the resulting classification reflect as faithfully as possible the actual similarities among all pairs of OTUs expressed in the resemblance matrix. Measures such as the cophenetic correlation coefficient (Sokal and Rohlf, 1962) were used to express the goodness of a classification. It soon became evident that different clustering methods yielded different fits by cophenetic correlation coefficients to the same resemblance matrix. Sneath (1966) showed, for an artificial data set, that average-linkage clustering represented resemblances expressed as distances better than complete or single linkage, and I and others have repeatedly made the same observations on real data sets. It also became obvious rapidly that other, non-Linnaean modes of arranging data to show their taxonomic relationships would give better fits to the original resemblance matrix than dendrograms produced by hierarchic cluster analysis methods. Thus, many phenetic taxonomists employed various methods of ordination analysis to represent taxonomic structure. However, for classifications in the narrow sense, the hierarchic models have been normative.

Optimality Criteria

The development of numerical cladistic procedures and the proposal for a system of classification based entirely on Hennigian principles have inevitably led to competing claims for the superiority of three systems of classification. Pheneticists, cladists, and a third group, the so-called evolutionary systematists, also known as syncreticists, are engaged in a controversy about which of the three classificatory systems is best.

How can one decide what is the best classification? It seems reasonable that classifications be judged by the criteria which they attempt to achieve. This suggests criteria of optimality that are defined by the methods which postulate them. Three such criteria are postulated by pheneticists: mutually highest similarity, predictivity, and fit to the original resemblance matrix among OTUs. Others are postulated by cladists: closeness to the true cladogeny and minimum tree length or maximum character compatibility, depending on the phylogenetic method employed. For such criteria, it may therefore be possible, and protagonists in the various controversies have indeed claimed, that classifications established on the basis of methods propounded by school A might in fact be better by the criteria of school B than classifications established by the methods of school B. Thus, it is possible that cladistic classifications, by which is meant classifications established by methods generally held to be cladistic, are better by phenetic criteria of optimality than phenetic classifications of the same data; similarly, the converse could be true.

Phenetic classifications should group together the most similar OTUs and, in fact, various methods of cluster analysis have attempted to measure homogeneity of OTUs with respect to character states and have attempted to develop clustering procedures which maximize homogeneity (see Sneath and Sokal, 1973). For reasons already stated, such homogeneity criteria can be implemented easily only at one hierarchic level and there exists no algorithm for maximizing, or for that matter even unequivocally computing, global homogeneity of a classification. As we have seen, the linkage between similarity and predictivity, while necessarily close, has not been defined clearly. Since generally accepted measures of homogeneity or mutual similarity of OTUs within taxa have not been established, it is not surprising that there are no comparative studies of phenetic and cladistic classifications in this regard. Given the existence of homoplasy, one would expect that phenetic classifications should be more homogeneous than cladistic ones which would tend to discount homoplasy. The only comparative study of predictivity (Archie, 1980) has led to ambiguous results in which the decision as to which method is best depends on how one measures predictivity.

The problem of how to evaluate fit to a similarity matrix, which is essentially a phenetic criterion, is a complex one. Pheneticists have traditionally preferred classifications with better fits to the original similarity matrix over others with worse fits. It is evident that better fits will yield a higher cophenetic correlation, but such a high value will only obtain if the similarity matrix itself is of a nested nature so that one can indeed fit a hierarchic classification to it. Using the cophenetic correlation coefficient between a resemblance matrix and a hierarchic classification implies a belief that biological taxa are indeed hierarchic in their diversity. To the degree this is not so, resemblance matrices will not reflect the hierarchy and the results of cluster analyses will not yield high cophenetic correlations against these resemblance matrices. Ordinations may be necessary to summarize the taxonomic structure of such data. It should be stated clearly that phenetic criteria of goodness of classification, as currently employed, generally deal with the data set under the restrictions of hierarchic Linnaean classifications. When these restrictions are relaxed, as in a four-point metric or an ordination, the criterion of closeness of fit to an original similarity or distance matrix will be improved. At that point one no longer has a Linnaean classification (Rohlf and Sokal, 1981). Comparisons, at this time at least, have been made and should continue to be made in terms of Linnaean classifications.

In the few cases where the true cladogram is known or can be guessed at with near certainty, phenetic methods have been shown in many cases to be as good estimators of cladistic sequences as numerical cladistic methods (Tateno *et al.*, 1982; Fiala, 1983). I have shown that in the Caminalcules with a known phylogeny, cladistic methods have a better fit to a known cladistic relationship than phenograms for a large character set (85 characters), but this relationship is reversed when subsamples of characters are chosen (Sokal, 1983b,c). Thus, for small character samples (the situation typical of many cladistic analyses), a phenetic method would give a better estimate of the cladogram than a cladistic method (if the results from the Caminalcules can be generalized). With respect to minimum-length trees, there does not seem to have been any study fitting characters to phenograms and so estimating the length of a phenogram viewed as a phylogenetic tree. This might indeed be an interesting experiment.

Other optimality criteria are those that seem inherently desirable for any classification. In this category fall properties such as taxonomic stability, which is desirable for any classification. The first type of stability is character stability, which means robustness of a classification to the addition of new characters or to different selections of characters. There are two kinds of character stability: that of randomly chosen samples of characters as well as that of sampling from different classes of characters. This second sampling is the well-known test for congruence related to the nonspecificity hypothesis. The second type of stability is OTU stability, which is robustness of a classification to the addition (or subtraction) of OTUs. A third type of stability is invariance of classifications under methodological alternatives such as differences in character coding or in computing similarity coefficients. I shall take these up in turn.

In both phenetic and cladistic studies of character stability, one must be on guard that the asymptotic or standard dendrogram itself meets the desired criterion. Thus, take a cladistic study in which the cladogram based on the total character set is a poor estimate of the true cladogeny of the organisms. If one should demonstrate in such a study that the sampling of characters shows less variation around the estimate for the overall cladogram than in a comparable phenetic study, the claim for increased stability of cladistic methods would be rather hollow, since increased stability for a false estimate would be of little taxonomic advantage. If this argument is pushed to its limits, one obtains the point made by Janowitz (1979) that perfect stability can be obtained when the classification arrived at has no necessary relationship to the data. It is for this reason that the few studies, such as the Caminalcules, where a true cladogeny is known are of importance, since here the stability of classifications based on subsuites of characters can be tested against the true cladogram. The converse danger exists as well. If a standard phenogram obtained by a given phenetic method does not meet the desired criteria of phenetic taxonomy (e.g., resemblance, homogeneity of taxa, predictivity), then demonstration of the stability of classifications based on subsuites of characters will not prove of much value. A valid counterargument to these statements could be made if it could be shown that there is homogeneity of sampling error in either numerical cladistic or in numerical phenetic classifications, so that regardless of the nature of the parametric dendrogram from which samples are taken, its sampling error will be more or less the same. The above arguments apply to estimates of OTUs as well.

Character stability has been investigated on a comparative basis by Schuh and Polhemus (1980) and Schuh and Farris (1981) in the hemipteran group Leptopodomorpha. These authors found that character stability to random selection of characters is considerably higher in classifications based on Wagner trees than those on UPGMA phenograms. The first of these studies was fraught with numerous errors, some of which were corrected in the second study, which, however, has several quite inappropriate procedures which lead to built-in biases in favour of the phylogenetic method. These were discussed in detail by Sokal (1983c). When these biases are corrected for, there is no evidence in favour of either method. In the Caminalcules, character stability is very much higher in phenetic than in cladistic classifications (Sokal, 1983c).

Comparisons of congruence between classifications based on different classes of characters have been carried out by Mickevich (1978, 1980) whose work was severely criticized by Rohlf and Sokal (1980). In addition to pointing out several unconventional computational procedures in her paper, these authors concluded that it was difficult to evaluate the effect of failure of nonspecificity in these data, since there was no benchmark of random sampling of characters so that the additional effect of lack of nonspecificity could be evaluated. Thus, any incongruence observed could be due either to sampling of characters or to differences in the taxonomic relationships based on the different classes of characters. Furthermore, Rohlf *et al.* (1983b) have shown that many of Mickevich's results are not reproducible and that, in fact, when the computations are repeated in what is believed to be a correct mode, there is little difference in congruence between phenetic and cladistic classifications. Thus, the citation of her work by a number of cladist authors (Farris, 1979; Wiley, 1981; Schuh and Farris, 1981) in support of cladistic classifications may be unfounded.

In the only published OTU stability study so far, Schuh and Farris (1981) claimed that classifications based on the Wagner tree method (applied to the Leptopodomorpha) were invariably more stable (measured as congruence of the subsets based on OTUs with the classification based on the entire set of OTUs) than were similar phenetic classifications. Again, these authors achieved these results by building in a computational bias in favour of the cladistic method. Once this bias is removed, there is a slight but nonsignificant difference in favour of the phenetic method, which would have been predicted on theoretical grounds. Further studies of OTU stability are currently under way in my laboratory.

Outlook

The period of reexamination of phenetic principles will continue for a while longer until the equivalence of overall similarity within a taxon to overall predictivity of that taxon can be established. If such a correspondence can be demonstrated, then attempts to maximize one or the other by improved algorithms will be undertaken. Should it turn out, surprisingly, that these two criteria are not essentially identical, then pheneticists will have to make the decision which of the criteria should be the primary one to be maximized. It would seem to me that predictivity is the more important criterion of a classification in terms of Gilmour naturalness, and I would expect that high predictivity should be the goal of a natural classification for general use.

The comparisons of optimality criteria in phenetic and cladistic classifications will continue in an attempt to reach some sort of agreement on which of the classifications is more predictive and stable. But from the point of view of a cladist this comparison misses the fundamental criterion of goodness of a cladistic estimate, which is its closeness to the true cladogeny of the organisms. The exception would seem to be the classificatory criteria of the pattern cladists (Beatty, 1982), whose goals appear to be internal consistency of patterns divorced from the true genealogy of the group of organisms under study. In empirical studies to date with known cladogenies, Sokal (1983b), Fiala (1983), and Baum (1983) have shown that no numerical cladistic method yields more than 80% of common taxonomic subsets with the true cladogram, and that numerous estimates yield only 50% of such subsets. If the numerical cladistic methods are to be used as temporary hypotheses about the true evolution of the group of organisms under study, to be changed and improved as increased knowledge or better techniques reveal more of the cladistic structure, these error rates are acceptable. But if these changing hypotheses are to be used as the bases of classifications, this will result in continuously changing classifications.

At present it is not clear which methods yield more stable and predictive classifications. Rohlf and Sokal (1981) have pointed out that it is unlikely that classifications reflecting the true cladogeny will at the same time be the most predictive ones. Thus, if a method generally recognized as a numerical cladistic approach is in fact shown to be more predictive, it may, in that case, be a poorer reflection of the true cladogeny. It is conceivable that some methods of pattern cladistics may eventually yield such results. Whether such methods are then to be called phenetic or cladistic is probably a question of only academic or polemical interest. Such comparisons between methods will continue to involve much research on determining the best way of comparing classifications, and this topic, which is currently a very active one in numerical taxonomy, is covered by Sackin (see Chapter 2).

Attempts at robust methods for similarity coefficients and cluster analysis may ultimately result in obtaining stable and repeatable phenetic classifications which will not be altered easily by changes in coding and scaling of characters and by addition of subsequent characters or OTUs. But it is probably utopic to hope that complete stability will ever be reached. The question might then well be asked whether for purposes of having a practical, generally useful classification for biologists at large and for the general public, the ultimate in predictivity or stability is really necessary. Some might argue that as long as classifications are strongly Gilmour natural (i.e., predictive), this should be sufficient for general purposes and for phenetic analysis additional refinements will typically not be necessary, whereas for cladistic classifications one would wish to improve them until the true cladogeny has been obtained—if this can ever be done.

Acknowledgements

This is contribution 466 in *Ecology and Evolution* from the State University of New York at Stony Brook. An earlier version of this manuscript benefitted from critical readings by Professors F. J. Rohlf and P. H. A. Sneath. This research was supported by grant DEB (80-03508) from the National Science Foundation, whose continued support is much appreciated.

References

Archie, J. (1980). Definition, criteria, and testing of the predictive value of classifications. Ph. D. dissertation. State Univ. of New York at Stony Brook.

Baum, B. R. (1983). Relationships between transformation series and some numerical cladistic methods at the infraspecific level, when genealogies are known. *In* 'Numerical Taxonomy' (Ed. J. Felsenstein), Proceedings of a NATO Advanced Study Institute, NATO Advanced Study Institute Serial G (Ecological Series), No. 1, pp. 340–345. Springer-Verlag, New York.

Beatty, J. (1982). Classes and cladists. *Systematic Zoology* **31,** 25–34.

Beckner, M. (1959). 'The Biological Way of Thought'. Columbia Univ. Press, New York.

Boyce, A. J. (1964). The value of some methods of numerical taxonomy with reference to hominoid classification. *In* 'Phenetic and Phylogenetic Classification' (Eds. V. H. Heywood and J. McNeill), Systematics Association Publication 6, pp. 47–65. London.

Boyce, A. J. (1965). The methods of quantitative taxonomy with special reference to functional analysis. Ph. D. dissertation. Oxford University.

Cain, A. J., and Harrison, G. A. (1958). An analysis of the taxonomist's judgement of affinity. *Proceedings of the Zoological Society of London* **131,** 85–98.

Cattell, R. B. (1944). A note on correlation clusters and cluster search methods. *Psychometrika* **9,** 169–184.

Cherry, L. M., Case, S. M., and Wilson, A. C. (1978). Frog perspective on the difference between humans and chimpanzees. *Science* **200,** 209–211.

Daget, J., and Hureau, J. C. (1968). Utilisation des statistiques d'ordre en taxonomie numerique. *Bulletin Musée National Histoire Naturelle* **40,** 465–473.

Dunn, G., and Everitt, B. S. (1982). 'An Introduction to Mathematical Taxonomy'. Cambridge Univ. Press, Cambridge.

Eades, D. C. (1965). The inappropriateness of the correlation coefficient as a measure of taxonomic resemblance. *Systematic Zoology* **14,** 98–100.

Eldredge, N., and Cracraft, J. (1980). 'Phylogenetic Patterns and the Evolutionary Process'. Columbia Univ. Press, New York.

Farris, J. S. (1974). Formal definitions of paraphyly and polyphyly. *Systematic Zoology* **23,** 548–554.

Farris, J. S. (1979). The information content of the phylogenetic system. *Systematic Zoology* **28,** 483–520.
Felsenstein, J. (1982). Numerical methods for inferring evolutionary trees. *The Quarterly Review of Biology* **57,** 379–404.
Fiala, K. L. (1983). A simulation model for comparing numerical taxonomic methods. *In* 'Numerical Taxonomy' (Ed. J. Felsenstein), Proceedings of a NATO Advanced Study Institute, NATO Advanced Study Institute Serial G (Ecological Sciences), No. 1. pp. 87–91. Springer-Verlag, New York.
Florek, K., Lukaszewicz, J., Perkal, J., Steinhaus, H., and Zubrzycki, S. (1951a). Sur la liason et la division des points d'un ensemble fini. *Colloquium Mathematicum* **2,** 282–285.
Florek, K., Lukaszewicz, J., Perkal, J., Steinhaus, H., and Zubrzycki, S. (1951b). Taksonomia Wrocłlawska. *Przegl Anthropoliczny* **17,** 193–211.
Gilmour, J. S. L. (1937). A taxonomic problem. *Nature (London)* **139,** 1040–1042.
Gilmour, J. S. L. (1940). Taxonomy and philosophy. *In* 'The New Systematics' (Ed. J. Huxley), pp. 461–474. Oxford Univ. Press (Clarendon), Oxford.
Gilmour, J. S. L. (1951). The development of taxonomic theory since 1851. *Nature (London)* **168,** 400–402.
Gilmour, J. S. L. (1961). Taxonomy. *In* 'Contemporary Botanical Thought' (Eds. A. M. MacLeod and L. S. Cobley), pp. 27–45. Oliver & Boyd, Edinburgh, and Quadrangle Books, Chicago.
Holzinger, K. U., and Harman, H. H. (1941). 'Factor Analysis'. Univ. of Chicago Press, Chicago.
Janowitz, M. F. (1979). A note on phenetic and phylogenetic classifications. *Systematic Zoology* **28,** 197–199.
Jardine, N., and Sibson, R. (1971). 'Mathematical Taxonomy'. Wiley, London.
Michener, C. D., and Sokal, R. R. (1957). A quantitative approach to a problem in classification. *Evolution* **11,** 130–162.
Mickevich, M. F. (1978). Taxonomic congruence. *Systematic Zoology* **27,** 143–158.
Mickevich, M. F. (1980). Taxonomic congruence: Rohlf and Sokal's misunderstanding. *Systematic Zoology* **29,** 162–176.
Minkoff, E. C. (1965). The effects on classification of slight alterations in numerical technique. *Systematic Zoology* **14,** 196–213.
Moss, W. W. (1968). Experiments with various techniques of numerical taxonomy. *Systematic Zoology* **17,** 31–47.
Neel, J. V., Rothhammer, F., and Lingoes, J. C. (1974). The genetic structure of a tribal population, the Yanomama Indians. X. Agreements between representatives of village differences based on different sets of characteristics. *American Journal of Human Genetics* **26,** 281–303.
Rogers, D. J., and Tanimoto, T. T. (1960). A computer program for classifying plants. *Science* **132,** 1115–1118.
Rohlf, F. J., and Sokal, R. R. (1965). Coefficients of correlation and distance in numerical taxonomy. *University of Kansas Science Bulletin* **45,** 3–27.
Rohlf, F. J., and Sokal, R. R. (1980). Comments on taxonomic congruence. *Systematic Zoology* **29,** 97–101.
Rohlf, F. J., and Sokal, R. R. (1981). Comparing numerical taxonomic studies. *Systematic Zoology* **30,** 459–490.
Rohlf, F. J., Colless, D. H., and Hart, G. (1983a). Taxonomic congruence—a reanalysis. *In* 'Numerical Taxonomy' (Ed. J. Felsenstein), Proceedings of a NATO Advanced Study Institute, NATO Advanced Study Institute Serial G (Ecological Series), No. 1, pp. 82–86. Springer-Verlag, New York.

Rohlf, F. J., Colless, D. H., and Hart, G. (1983b). Taxonomic congruence—re-examined. *Systematic Zoology* **32,** 144–158.

Schuh, R. T., and Farris, J. S. (1981). Methods for investigating taxonomic congruence and their application to the Leptopodomorpha. *Systematic Zoology* **30,** 331–351.

Schuh, R. T., and Polhemus, J. T. (1980). Analysis of taxonomic congruence among morphological, ecological, and biogeographic data sets for the Leptopodomorpha (Hemiptera). *Systematic Zoology* **29,** 1–26.

Simpson, G. G. (1961). 'Principles of Animal Taxonomy'. Columbia Univ. Press, New York.

Smith, E. E., and Medin, D. L. (1981). 'Categories and Concepts'. Harvard Univ. Press, Cambridge, Massachusetts.

Sneath, P. H. A. (1957a). Some thoughts on bacterial classification. *Journal of General Microbiology* **17,** 184–200.

Sneath, P. H. A. (1957b). The application of computers to taxonomy. *Journal of General Microbiology* **17,** 201–226.

Sneath, P. H. A. (1961). Recent developments in theoretical and quantitative taxonomy. *Systematic Zoology* **10,** 118–139.

Sneath, P. H. A. (1962). The construction of taxonomic groups. In 'Microbial Classification' (Eds. G. C. Ainsworth and P. H. A. Sneath), pp. 289–332. Cambridge Univ. Press, Cambridge.

Sneath, P. H. A. (1966). A comparison of different clustering methods as applied to randomly-spaced points. *Classification Society Bulletin* **1,** 2–18.

Sneath, P. H. A., and Sokal, R. R. (1962). Numerical taxonomy. *Nature (London)* **193,** 855–860.

Sneath, P. H. A., and Sokal, R. R. (1973). 'Numerical Taxonomy'. Freeman, San Francisco.

Sokal, R. R. (1961). Distance as a measure of taxonomic similarity. *Systematic Zoology* **10,** 70–79.

Sokal, R. R. (1983a). A phylogenetic analysis of the Caminalcules. I. The data base. *Systematic Zoology* **32,** 159–184.

Sokal, R. R. (1983b). A phylogenetic analysis of the Caminalcules. II. Estimating the true cladogram. *Systematic Zoology* **32,** 185–201.

Sokal, R. R. (1983c). A phylogenetic analysis of the Caminalcules. IV. Congruence and character stability. *Systematic Zoology* **32,** 259–275.

Sokal, R. R., and Michener, C. D. (1958). A statistical method for evaluating systematic relationships. *University of Kansas Science Bulletin* **38,** 1409–1438.

Sokal, R. R., and Rohlf, F. J. (1962). The comparison of dendrograms by objective methods. *Taxon* **11,** 33–40.

Sokal, R. R., and Rohlf, F. J. (1970). The intelligent ignoramus, an experiment in numerical taxonomy. *Taxon* **19,** 305–319.

Sokal, R. R., and Rohlf, F. J. (1980). An experiment in taxonomic judgment. *Systematic Botany* **5,** 341–365.

Sokal, R. R., and Sneath, P. H. A. (1963). 'Principles of Numerical Taxonomy'. Freeman, San Francisco.

Tateno, Y., Nei, M., and Tajima, F. (1982). Accuracy of estimated phylogenetic trees from molecular data. I. Distantly related species. *Journal of Molecular Evolution* **18,** 387–404.

Wiley, E. O. (1981). 'Phylogenetics'. Wiley, New York.

Woodger, J. H. (1937). 'The Axiomatic Method in Biology'. Cambridge Univ. Press, Cambridge.

2

Comparisons of Classifications

M. J. SACKIN

Department of Microbiology, University of Leicester, Leicester, UK

Introduction

Why compare classifications? The reason is not much different from that for classifying OTUs in the first place. After all, a classification may itself be considered to be an OTU, and conversely the conception of any object (or OTU) is the result of some sort of classification process, usually mental but sometimes computer assisted.

Although a fairly close conceptual relationship between OTUs and classifications seems central in the motivation for comparing classifications, the two concepts will be distinguished in line with prevailing taxonomic practice. Thus, an OTU will be considered to be an entity usually described in terms of characters and a classification as some sort of grouping or other association between OTUs.

There are many circumstances in which one may wish to compare classifications. Very often one may wish to compare one's own classification with an earlier one done by oneself or by another worker. The differences between the classifications may be due to one or more of the following:

1. Addition (or removal) of characters
2. Addition (or removal) of OTUs, so that the relationships among surviving OTUs may alter
3. Changing the way characters are coded
4. Changing the classification algorithm

These factors are similar to those of Rohlf and Sokal (1980) and of Sokal (see Chapter 1). They need to be defined very broadly to allow, for example, for one or other classification to have been a nonnumerical, classical one. They may be considered as the criteria of stability, stability being one very general aim in classification.

Much has been written about factor 1 above, and it has probably been the most

COMPUTER-ASSISTED BACTERIAL SYSTEMATICS

ISBN 0-12-289665-3

frequently considered criterion of classification stability. In summary, a stable classification usually implies choosing many well-defined, reproducible characters, usually, one hopes, covering most areas of the genome (in biological classifications), ideally choosing the characters at random. Two classifications, based on different subsets of characters chosen in keeping with these considerations, and without any differences in the other three factors, would be expected to be very similar.

Factor 2 refers to the interesting situation whereby the addition or subtraction of OTUs can alter the classificatory relationships among the OTUs common to both studies. All methods of comparing classifications that I know of compare two or more classifications of the same OTUs. If the OTUs are not all common to the classifications under comparison, then it will be necessary to prune away those OTUs that do not appear in all the classifications and just compare the common OTUs. In the case of a dendrogram this is straightforward: the result of pruning OTUs is still a dendrogram, because the (cophenetic) relationships among them are unchanged and are still ultrametric.

Factor 3 has also been much discussed—the question of binary versus multistate characters and of possible transformations to the character states, for example, standardization or ranging.

Factor 4 covers a very wide area, because there are so many methods of classification and indeed different representations of the end product, notably (i) just one partitioning of the OTUs into groups, and (ii) hierarchic clusters in the form of a dendrogram. The term 'dendrogram' is used rather than 'phenogram' because it does not exclude the various forms of cladograms, and this contribution is meant to cover both phenograms and cladograms, that is, dendrograms in general. Sometimes a dendrogram will simply be referred to as a tree.

It would seem reasonable to test stability by altering factors 1 to 4 one at a time so as to be able to ascertain the contribution of each factor in upsetting stability. However, sometimes one may wish to compare a new classification with an earlier one whose method differs in respect of more than one of the four factors, for example, if one of the classifications is a nonnumerical one.

For completeness I think I should mention the case of two or more classifications in which each OTU is a different sample from a population. An example is two classifications of social classes in terms of characteristics of class members, where the two studies have used different actual people as sample members, and I shall do no more than mention the more fundamental question of recognising two OTUs as being 'the same' across two classifications.

For all the above cases one may (a) compare several configurations of OTUs, (b) compare several similarity matrices (this term will be used throughout to include dissimilarity matrices), (c) compare several partitionings of OTUs, or (d) compare several dendrograms.

Most of this contribution will deal with areas (c) and (d) because these are, *par*

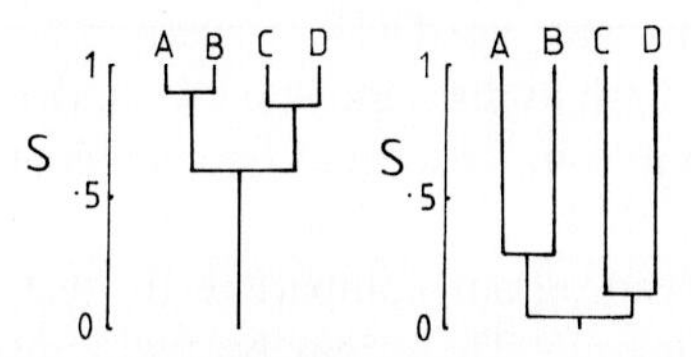

Fig. 1. Two global-order equivalent dendrograms with same similarity scale.

excellence, representations of classifications, although they very often derive from area (a) via area (b) (Rohlf and Sokal, 1981). Moreover, because a dendrogram consists of a nested series of partitions it will be convenient to examine area (c) in the course of describing methods (d).

Also, most of the work to be described is purely descriptive. Only in area (a) has there been any real statistical headway, in the direction of studying OTUs as samples from populations (Gower, 1975; Berge, 1977; Milligan, 1979).

Under area (d) several different types of dendrograms may be compared. First of all one may wish to compare two or more phenograms with the same similarity scale attached (Fig. 1). Typically this would arise for comparisons of classifications which use the same similarity coefficient but not necessarily the same characters or the same clustering method. However, a large degree of variation in the characters may render this type of comparison inappropriate, as, for example, two microbial classifications based on the same set of tests but with earlier times of reading in one classification compared with the other. The results of such a pair of classifications might appear as in Fig. 1, with one phenogram appearing 'squashed' as compared with the other. To prevent the squashing effect masking whatever similarity there may be between the two phenograms, it may be necessary to consider the phenograms without their similarity axes (e.g., see Fig. 2) and compare their shapes or topologies.

Comparisons of just the shapes of dendrograms may themselves be divided into two kinds, according to whether one wishes to distinguish between, for example, the two dendrograms of Fig. 2, in which OTUs A and B join at a higher similarity level than do C and D in the first dendrogram but lower in the second. These two dendrograms are, as elegantly described by Sibson (1972), local-order equivalent (LOE) but not global-order equivalent (GOE). The two dendrograms

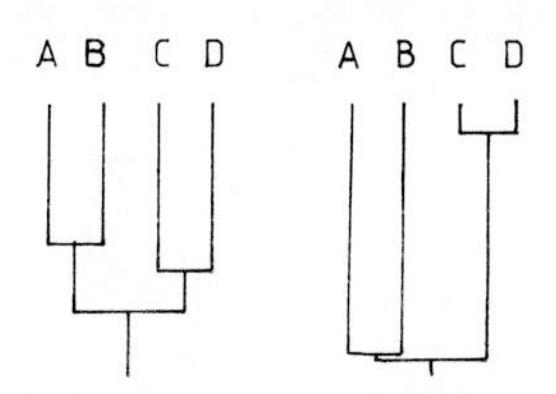

Fig. 2. Two local-order equivalent dendrograms.

of Fig. 1, on the other hand, are GOE. Usually when comparing dendrogram topologies one will not wish to distinguish between dendrograms such as the pair in Fig. 2; that is, one will usually wish to treat as identical dendrograms that are LOE but not GOE.

A further area of dendrogram comparison is that of comparing the many different kinds of cladograms. These may be distinguished by the unit measured along the segments of the cladogram, whether time, estimated mutation rate, minimum mutations, estimated actual mutations, or phenetic similarities between the tips or (terminal) OTUs (Sneath, 1975).

Methods

Comparing Dendrograms: Same Similarity Scale

The methods considered here have been developed and described by Boorman and Olivier (1973) and, in part, by Jardine and Sibson (1971). The two superimposed dendrograms of Fig. 3 may be considered to differ only in the shaded area, and the dissimilarity between the dendrograms may be considered to depend on (i) the degree to which the groupings are different over the similarity range which the shaded area covers, and (ii) the height *a* of the shaded area. The product of these two quantities measures the overall dissimilarity between the two dendrograms, in terms of a *tree metric*. Boorman and Olivier (1973) described four kinds of *partition metrics* (see below) for measuring the grouping differences, and these and others of perhaps less taxonomic interest are also described by Boorman and Arabie (1972) and by Arabie and Boorman (1973).

In general, a tree metric is computed by first finding the partition metric at every similarity level along the dendrograms. If the dendrograms are considered as being superimposed, as in Fig. 3, the groupings, and hence the value of the partition metric, will be seen to be constant within every similarity segment which does not pass any dendrogram joins, for example, the shaded segment in Fig. 3. The value of the tree metric is then the sum over all the segments of the

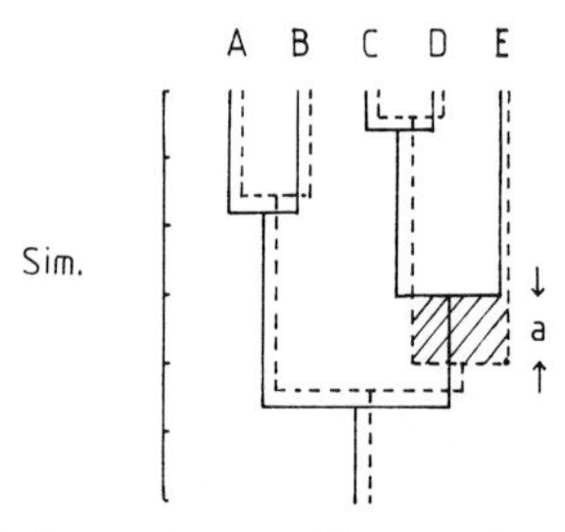

Fig. 3. Two superimposed phenograms to illustrate the tree metrics of Boorman and Olivier (1973).

segment length, multiplied by the value of the partition metric along that segment.

In Fig. 3 the partitionings of the OTUs in the dendrograms along the shaded segment may be represented as {AB, CDE} and {AB, CD, E}. Figure 4 shows the partition metrics for {ABCDEF, G} compared with {EF, ABCDG}. Metric A is the minimum number of OTU moves—or *element moves*—required to get from one partition to the other. In the example of Fig. 4 three moves are required. The minimum number of moves from the first partitioning to the second may be made by transferring OTUs E and F out of the group ABCDEF (two moves because two OTUs are being moved) to form a group of their own, and then (one move) moving G to join the group ABCD.

Metric B differs from A in counting the transfer of a group as one move only. In the example of Fig. 4 the transfer of EF scores one rather than two. In fact, however, only one *set move* is required to get from the first partition to the second, namely the transfer, as a single move, of ABCD from ABCDEF to join with G. Thus B equals 1. In general, A seems more obviously dependent than B on the relative numbers of OTUs that were selected in the various areas of the study, so that, for example, simply replicating an OTU many times can alter A a great deal but would leave B unchanged or only slightly altered.

The partition metric C may be defined as the minimum number of *lattice moves* required to transform one partition into the other. A lattice move is a combining of two groups or a splitting of one group into two. A transfer of a set of OTUs across from one group to another thus takes two lattice moves but only one set move. It can be shown that B can never exceed C for the same two partitions and that the number of lattice moves is minimised by doing all the combining moves followed by all the splitting moves. For a mathematical development see Boorman and Arabie (1972) and Boorman and Olivier (1973). In the example (Fig. 4), two lattice moves are required to transform one of the partitions into the other: (i) combining the two starting groups into one group containing all seven OTUs, and (ii) splitting them into the two groups corresponding to the other partition. Thus C equals 2.

Both B and C seem very useful measures. Intuitively the choice between them

A: Number of OTU moves

B: Number of set moves

C: Number of lattice moves (= number of lumpings + number of splittings)

D: Number of changed pair-bonds

ABCDEF	G	A = 3
versus		B = 1
EF	ABCDG	C = 2
		D = 12

Fig. 4. Partition metrics of Boorman and Olivier (1973).

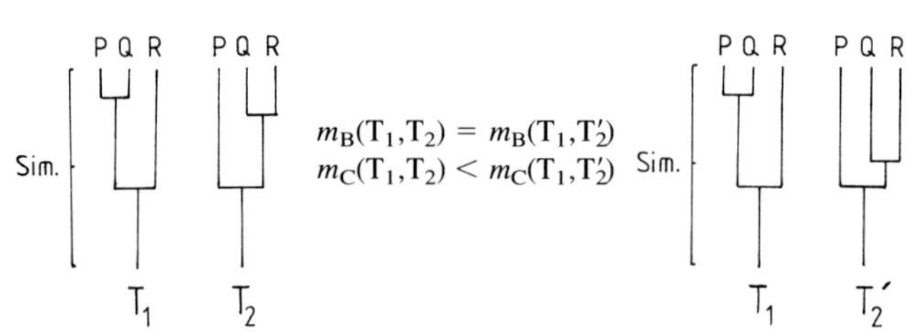

Fig. 5. Tree metrics: m_C, but not m_B, is sensitive to the similarity level at which QR join in T_2 and T_2' when compared with T_1 (see text).

depends on whether one wishes to consider, for example, the two left-hand trees T_1 and T_2 of Fig. 5 to be as dissimilar as the right-hand pair T_1 and T_2', or more dissimilar. The tree metric m_B based on B makes the two pairs equally dissimilar; the tree metric m_C based on C makes T_1 and T_2 more dissimilar than T_1 and T_2' by recognising OTU Q to be more tightly bound to R—and hence further from P—in T_2 than in T_2' (T_1 is the same in both halves of Fig. 5).

The fourth partition metric D is the number of OTU pairs that are in exactly one of the two groupings. In Fig. 4 all eight pairs EF versus ABCD are in the same group in the first grouping but not in the second, and G versus ABCD (four pairs) are in the same group in the second grouping but not the first, whence D equals 12. Computation of D is done in a similar manner to that of C but doing the splittings first, and D is not the total number of lattice moves but the total number of changed pair bonds at each step. D is even more influenced than A by the sizes of the groups involved, and the corresponding tree metric m_D is liable to be dominated by a few large grouping differences near the roots of the two trees at the expense of highlighting differences in fine structure near the tips. This is also apparent from an alternative way of computing m_D without recourse to D at all. It may be shown (Jardine and Sibson, 1971; Boorman and Olivier, 1973) that m_D is the city-block or Manhattan distance between the two cophenetic value matrices, one from each tree, where the elements are considered as characters or variables, that is,

$$m_D = \sum_{i<j} \left| u_{T_1}(i,j) - u_{T_2}(i,j) \right|$$

where $u_{T_1}(i,j)$ is the cophenetic value between OTU i and OTU j in T_1, and similarly for T_2 [the *cophenetic value* between two OTUs is their similarity as implied by a given dendrogram (Sneath and Sokal, 1973)]. Large groups near the root of each tree will correspond to large rectangles of high values of u, and these values, when different between the two trees, may dominate the value of m_D.

One problem, for which there seems no ready remedy, is that a small change in an OTU configuration may lead to a large change in whichever variety of tree metric is used. This is particularly likely if, for example, average-linkage clustering has been performed; it is much less likely with single linkage. It is an aspect

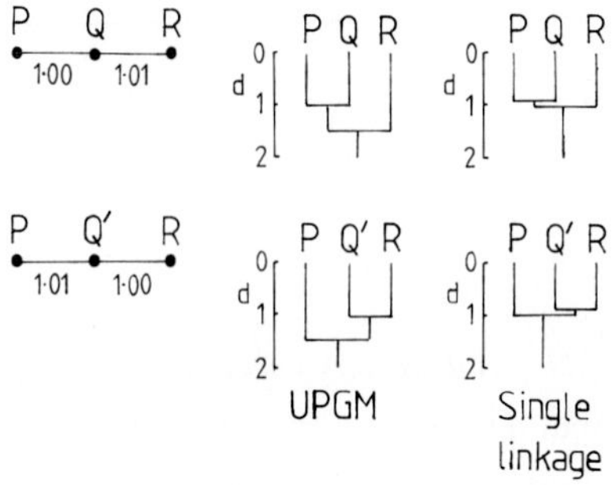

Fig. 6. Instability of average-linkage clustering: simplest case (see text).

of the instability of methods such as average linkage (see Jardine *et al.*, 1967; Jardine and Sibson, 1971). Figure 6 shows the simplest possible bad case of a small perturbation in a configuration radically altering the average-linkage (UPGM) clustering. Here, in the first configuration, in the UPGM phenogram, R loses an early chance to join Q because P gets in first. If this were part of a large configuration, then R may, for example, have an earlier opportunity to join another group than it might have in the second configuration. In single linkage such large differences do not occur. This is precisely the kind of situation that M. J. Sackin, D. Jones, and P. H. A. Sneath (unpublished) have found in a study on coryneform bacteria, and it seemed to point to a possible explanation of the appearance of 'wanderer' strains, a term coined in this context by Dorothy Jones and to which I hope one day to put some statistical substance.

However, I still think that on balance UPGM clustering is much to be recommended, but backed up by single linkage for gaining some insight into which parts of the data do not cluster well, and including the often maligned cophenetic correlation coefficient to gain a bird's-eye view of whether the data were suitable for clustering in the first place. Furthermore, nothing I can offer here removes the necessity for examining both the phenograms and the similarity matrices in great detail for gaining insight into the differences that arise.

Just as the cophenetic correlation coefficient is useful for comparing a phenogram with the similarity matrix from which it is derived, so it may be useful to compare two phenograms by computing the correlation coefficient between all pairs of cophenetic values, notwithstanding the difficulty that the $t(t - 1)/2$ cophenetic values in each phenogram (t OTUs under study) are in varying degrees dependent on each other. Similar considerations apply to the stress measures of Jardine and Sibson (1971), one of which, Δ_1^*, is a scale-free form of m_D.

Comparing Dendrograms: No Similarity Scale

A variety of methods comes under this heading. First of all, Boorman and Olivier (1973) described two ways of forcing a scale onto a pair of trees under com-

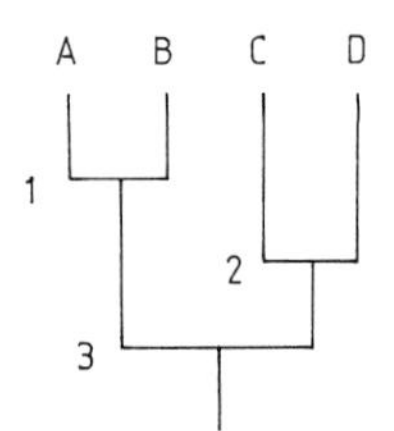

Fig. 7. A ranked tree.

parison and then computing the tree metrics as above. In the first scheme the groups are deemed to join at a 'distance' equal to the rank order of the joins, for the two trees in turn. In this scheme the trees are described as *ranked* trees (see Fig. 7 for a simple example). Trees with an existing similarity scale are termed *valued* trees. If one is not persuaded of a rank ordering of the joins of either or both trees, then one may let the 'distance' at a join be equal to the total number of 'descendant' OTUs emanating from that join (*bare* trees). This might arise when one wishes to treat trees as identical whenever they are LOE. In both these schemes all the tree metrics become available again as well as the cophenetic correlation and the Jardine and Sibson stress measures.

If these strategies seem a little artificial, then one may consider various methods of comparing just the shapes of trees. Some are based on measures of *topological distance* between pairs of tips of a dendrogram (Phipps, 1971; Williams and Clifford, 1971; Bobisud and Bobisud, 1972). The topological distance between two tips is the number of segments passed when travelling from one tip to the other. It is thus one more than the number of nodes (branch points) passed, although some authors use the number of nodes passed, that is, one less than the number of segments traversed. For example, in the dendrogram in Fig. 8 with root R_1, three segments are traversed in passing from A to B. In the dendrogram with root R_2, only two segments are traversed between these two tips. Thus, the difference in the topological distance between these two OTUs is one in tree R_1 compared with tree R_2. The sum of these differences over all pairs of OTUs provides a measure of topological dissimilarity between two trees.

Phipps (1971) compared the elements of topological distance matrices of dendrograms derived from eight different numerical taxonomic methods on the same data. The eight matrices were themselves treated as OTUs (notwithstanding lack of independence of all the elements of each matrix) and subjected to a numerical taxonomic analysis. He also performed a numerical taxonomic analysis on the corresponding eight cophenetic value matrices. The numerical taxonomy on the topological distance matrices gave a higher cophenetic correlation coefficient than did that on the cophenetic value matrices.

A topological dissimilarity measure as just described, though elegant at first sight, has the drawback that two trees intuitively looking very dissimilar may turn out to have very low topological dissimilarity between them. Trees R_1 and

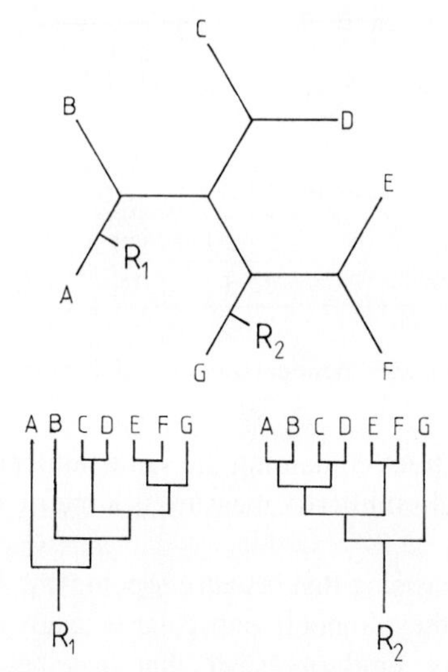

Fig. 8. Two rooted trees deriving from the same unrooted tree by rooting it at R_1 and R_2.

R_2 of Fig. 8 provide an example of this. Though looking very dissimilar, they nevertheless both derive from the same unrooted tree shown at the top of Fig. 8 by rooting this tree in two different places, namely R_1 and R_2 as shown. It may be seen that the topological distance between any pair of OTUs differs by at most one only. In order to allow better for the apparent dissimilarity between, for example, the two trees of Fig. 8, Bobisud and Bobisud (1972) proposed a modification to the topological dissimilarity measure between trees. They added on to this measure the sum over all the OTUs of the difference between the numbers of branches from the OTU to the root. Thus, in tree R_1 OTU A is one branch distant from the root, but in tree R_2 OTU A is four branches from the root. The difference is three, and the Bobisud and Bobisud modification consists of adding all t (t OTUs in all) differences to the topological dissimilarity between the trees.

In fact if the two trees are *dichotomous* (also described as *fully resolved* or *binary*—that is, every branch point has at most two descendant branches arising from it)—then it can be shown (Bobisud and Bobisud, 1972) that two trees with zero (unmodified) topological dissimilarity will always be identical in topology, but for nondichotomous trees this is not necessarily so. Bobisud and Bobisud (1972) give a simple example of two distinct nondichotomous trees with zero topological dissimilarity; they have identical topological distance matrices between the pairs of OTUs (Fig. 9). However, the Bobisud measure of dissimilarity is not zero for these trees. Indeed, the Bobisud measure is a *metric* (Bobisud and

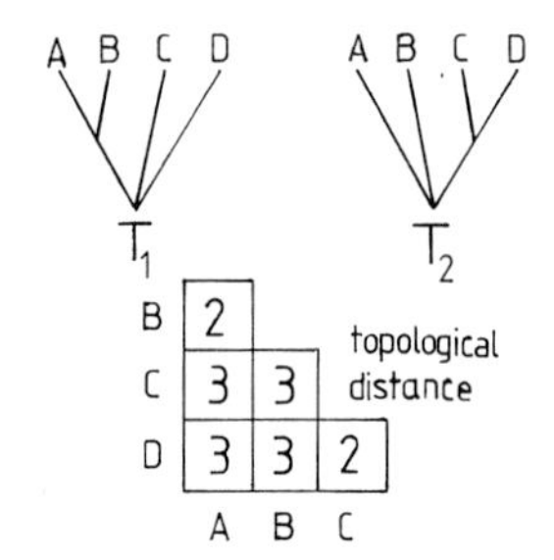

Fig. 9. Two different trees with identical topological distance matrices between the tips.

Bobisud, 1972) over all trees containing the same number of OTUs, whereas the unmodified topological dissimilarity measure is a metric over dichotomous trees only.

It may be worth emphasising that because topological distance methods do not use any similarity scale they do not in particular take any heed of the closeness of branch points along any similarity scale that may be present. Thus, the topological dissimilarity between the first of the three trees shown in Fig. 10 and either of the others would be no greater if in the first tree, for example, A had joined B higher up than where shown in the figure. For any computer-based system of storing and comparing trees, careful programming is needed for distinguishing between borderline cases of the kind shown in Fig. 10. In practice the OTUs A, B, and C may be groups rather than single OTUs. If a similarity scale is available then use of the tree metrics, for example, will ensure that grouping differences affecting very short similarity ranges will have only a very small effect.

One simple and ingenious method of comparing the topologies of trees is due to Farris (1973), and it seems to give considerable insight into the differences between the trees. The method investigates to what extent an individual cluster in one tree has become fragmented in the other tree. There are two ways of computing the fragmentation. An example is shown in Fig. 11, in which tree T_1 is being compared with T_2 according to the two varieties of the method. The figure highlights the fate in tree T_2 of group ABCDE, which appears in T_1. For ease of description one may imagine 'membership of group ABCDE' as a character and

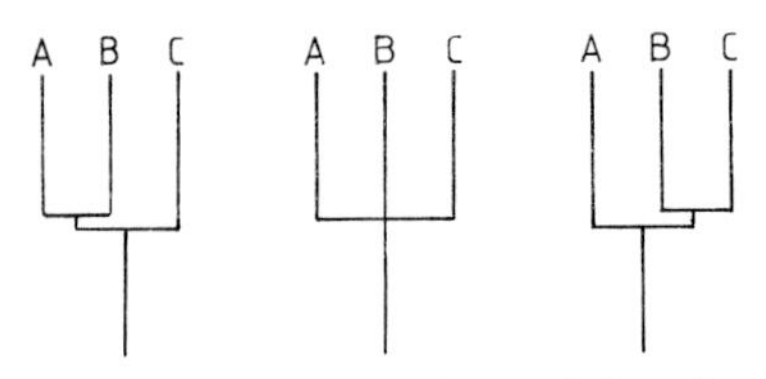

Fig. 10. Topological distance and branch lengths (see text).

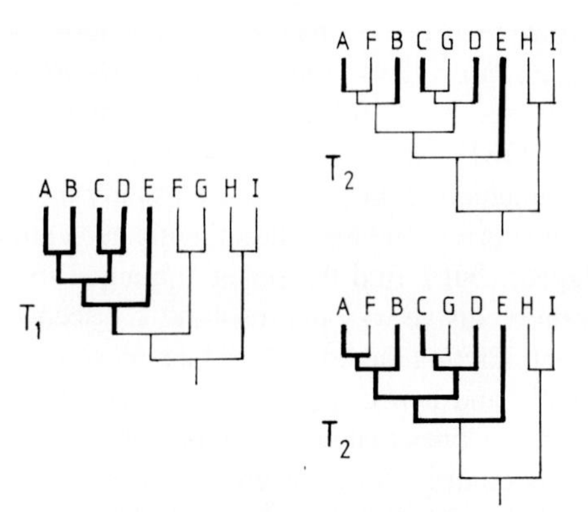

Fig. 11. Farris (1973) distortion measure, illustrating the fate in T_2 of the group ABCDE in T_1: T_2 (above), Camin–Sokal (WISS) model; T_2 (below), Wagner model.

that the tree is an evolutionary tree. The method involves counting the number of times the character has 'evolved' in tree T_2, and the choice between the two varieties of the method depends on whether or not one allows back mutations. Allowing them (Wagner model) seems better for dealing with the odd 'wanderer' (here F or G) deep inside a group. In Fig. 11 membership of group ABCDE is represented by a thick line. The degree of fragmentation of ABCDE in tree T_2 is given by the total number of changes in line thickness in the tree, according to which of the two models is chosen. Similar consideration of all the groups leads to a measure of overall dissimilarity between the two trees. The measure, incidentally, is not symmetrical in that the dissimilarity based on examining the fate in tree T_2 of all the groups in T_1 is not, in general, the same as that when examining the fate in T_1 of the groups in T_2.

Comparing Dendrograms: Consensus Methods

Next we examine a group of methods now under active development which are based on the elegant idea of constructing some sort of tree (a *consensus tree*) containing features common to the two, or, indeed, more than two, trees under comparison and then seeing how much structure there is in the consensus. If the starting trees are very dissimilar in terms of the consensus method used, then the resulting consensus will be a *bush* (Fig. 12).

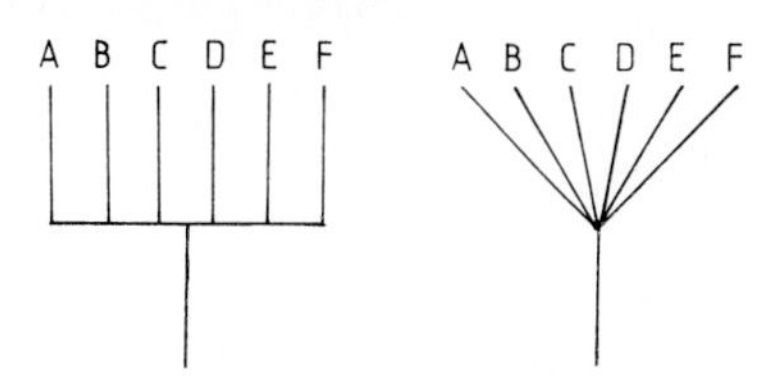

Fig. 12. A bush (two representations).

The paper by Mickevich (1978) has received voluminous discussion in the pages of *Systematic Zoology*. Two consensus methods are used in these articles, the *Adams* consensus tree (CT) (Adams, 1972) and the *majority rule* CT (Margush and McMorris, 1981). Most work has been done on the Adams method in its form which uses unlabelled nodes, because usually in a hierarchic classification the nodes will not correspond to well-authenticated entities. Both methods in themselves are elegant, but I find the results obtained to be rather disturbing, because in some rather elusive way the resulting CT seems to favour aspects of one starting tree, contradicting the other(s). The majority rule CT, at least, seems easier computationally and hence, probably, conceptually too. It is simply the tree whose groups are the ones that appear in an absolute majority of the starting trees (throughout this contribution, the groups on a tree are the sets of tips—OTUs—that emanate from each node or branch point). If just two trees are being compared (see Fig. 13), the majority rule CT will be identical with the *strict* CT (Sokal and Rohlf, 1981; Day, 1983), which is the tree whose groups are those that appear in every one of the starting trees and which corresponds also with the cladogram of replicated components of Nelson (1979).

A difficulty with the majority rule method is that just one badly misplaced OTU can cause major collapse of the majority rule tree, that is, in the direction of a bush. However, there has been much work on loosening the definition of majority rule trees to prevent this kind of collapse while at the same time not reintroducing the difficulties associated with the Adams trees (Rohlf, 1982; McMorris and Neumann, 1983; Neumann, 1983; C. A. Meacham, unpublished work; H.-T. Shao, unpublished thesis; R. Stinebrickner, unpublished manuscript). Most of these methods yield CTs whose groups are intersections of groups of the starting trees; that is, each group in the CT consists of the OTUs in common to groups selected from the starting trees, one group from each tree. The methods differ in their selection of groups from the starting trees. If all combinations of groups, one from each tree, were chosen, then the consensus groupings would not, in general, form a tree. The methods narrow the choice by imposing some sort of common similarity scale on the starting trees; see Neumann (1983) for examples which are similar to the valued, ranked, and bare tree concepts of Boorman and Olivier (1973). In addition, Stinebrickner and Shao both discard from the CT groups whose size is below a preset proportion of the total OTUs in the starting trees from which the CT group derives by intersection.

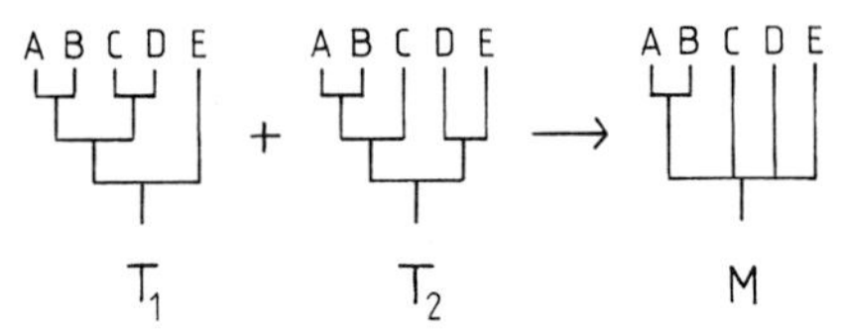

Fig. 13. Majority-rule consensus tree M derived from trees T_1 and T_2.

Various measures, known as *consensus indices,* have been given for assessing the degree of structure or resolution in a consensus tree. A very simple one is the *consensus fork index* of Colless (1980) and is the number of nodes in the consensus tree divided by the maximum possible; it is the same as the *component information* of Nelson (1979). The earliest published index, that of Mickevich (1978), is not fully defined in her paper, but evidently it weights each group by a function of its size, and there are two varieties of her formula, analogous to the two forms of Farris's distortion index (1973). Nelson's *term information* (1979) weights each node by the number of descendant tips emanating from it. Both Mickevich's and Nelson's methods have the side effect of giving a higher consensus index when the consensus tree is very *skew* (Sackin, 1972) than when it is more symmetrical, for the same number of nodes. Thus, for example, two identical symmetrical trees may yield a CT with lower consensus index than two nonidentical asymmetrical trees (see opposing points of view in Colless, 1980, and Mickevich and Farris, 1981). Rohlf (1982) has devised a modification (CI_1) to Mickevich's index that avoids this effect, and he has also devised a measure (CI_2) which is the proportion of all bifurcating trees which contain the groups seen in the consensus tree. This measure was suggested by a probabilistic measure of Nelson (1979).

Day (1983) suggests measuring the complexity of a set of trees by constructing a most parsimonious 'covering tree' whose vertices (nodes and tips) are the starting trees and the corresponding bush. Successive vertices along any path travelling away from the bush will represent trees that contain progressively more groups. In fact, a covering tree is defined so that every group in each starting tree along such a path will also be present in each subsequent tree along the path.

Any suitable measure may be chosen for the branch lengths (or 'weights'), for example, the difference between the consensus fork indices between the pair of trees at the two ends of the branch. Then the most parsimonious covering tree is that with the minimum total branch weights. High complexity will occur when the trees are highly resolved and very different from each other.

Likewise, Day (1983) proposes measuring the fit of a consensus to the trees on which it is based. The method is analogous to that for measuring set complexity except that the CT is used instead of the bush as a vertex in the covering tree.

In both methods the problem of obtaining the most parsimonious covering tree is a computationally intractable problem except when there are only two starting trees or when the number of OTUs in the study is small. The problem of obtaining the most parsimonious tree is an example of an NP-complete problem (Day, 1983), as, incidentally, is the problem of constructing parsimonious (hypothetical) phylogenetic trees.

In a looser sense a consensus concept has been used in epidemiology (Mantel, 1967) to attempt to detect epidemics by collating a pair of similarity matrices on the same OTUs. In Mantel's work the OTUs are occurrences of a disease (e.g., leukaemia), the first similarity matrix is a measure of the spatial distances between the occurrences, and the second matrix represents the times between the occurrences. Where the disease covers a limited period of time this spatial–temporal comparison may be replaced by one in which, for example, a matrix measuring sociological distance is used instead of the time difference matrix. Mantel uses a regression approach for his analysis.

Finally on the subject of consensus methods I shall mention the idea (Milligan, 1979) of combining several configurations of OTUs into one, new 'average' configuration with some axes attached. This is somewhat peripheral to this presentation because it deals with comparing OTUs rather than comparing classifications of OTUs.

Discussion and Conclusions

This presentation does not claim to be exhaustive. Rather, it may be considered as a personal selection of methods of comparing classifications that seem likely to be at least potentially important from the point of view of a practising taxonomist. One practical matter, however, on which I have barely touched, is the question of ease of computation of the methods. Some—in particular some of the partition metrics of Boorman and Olivier (1973)—are distinctly awkward by computer and are possibly easier, though still sometimes laborious, by hand.

It will be interesting to gain experience of the tree metric measures on realistic, large-sized data, and the same applies to the new consensus methods. Further, as R. R. Sokal has stated (see Chapter 1), there has been little work on OTU stability. I know of no detailed study on the effect on stability of the addition or removal of OTUs, be they members of a group, intermediate between two groups, or neither—and in any case more work is required to define these categories.

One particular area of interest for the application of the methods is that in which the classifications for comparison derive from incomplete similarity matrices (Sneath, 1983) in which only the similarities of the OTUs to a restricted set of reference OTUs are known. This developing area is of especial interest in serological and nucleic acid pairing studies in microbiology in which it is usually impractical to obtain complete data on numerous strains.

Finally, it may be useful to sum up with some practical advice for someone who would like to go about comparing two or more classifications. With the state of the art as at present, I would recommend the following:

1. Sit and peer at your phenograms and similarity matrices for a very long time, as ever

2. Find the cophenetic correlation and the Jardine and Sibson (1971) measures between phenograms
3. If the same similarity scale is being used for the trees under comparison, then to try especially m_C, backed up by m_D and m_B, of the Boorman and Olivier (1973) tree metrics
4. The Farris (1973) distortion measures
5. Possibly the Bobisud and Bobisud (1972) metrics (as well as an alternative variety they gave, which I have not detailed here)
6. Some of the new consensus methods

I have included here methods whose implementation is still in their early stages and indeed which are still at an experimental stage but which nevertheless seem very promising.

Acknowledgements

I thank Professor P. H. A. Sneath for advice and Professor R. R. Sokal for drawing my attention to the most recent work on consensus methods in his laboratory and elsewhere, and I thank Drs Meacham, Shao, and Stinebrickner for kindly sending me their unpublished material. Finally I thank Dr D. Jones for periodically reminding me that there are actual bacteria that can be studied by these methods, thereby enhancing my interest in the work.

References

Adams, E. N. III (1972). Consensus techniques and the comparison of taxonomic trees. *Systematic Zoology* **21,** 390–397.

Arabie, P., and Boorman, S. A. (1973). Multidimensional scaling of measures of distance between metrics. *Journal of Mathematical Psychology* **10,** 148–203.

Berge, J. M. F. ten (1977). Orthogonal Procrustes rotation for two or more matrices. *Psychometrika* **42,** 267–276.

Bobisud, H. M., and Bobisud, L. E. (1972). A metric for classifications. *Taxon* **21,** 607–613.

Boorman, S. A., and Arabie, P. (1972). Structural measures and the method of sorting. *In* 'Multidimensional Scaling: Theory and Applications in the Behavioral Sciences' (Eds. R. N. Shepard, A. K. Romney, and S. Nerlove), Vol. 1 (Theory), pp. 225–249. Seminar Press, New York.

Boorman, S. A., and Olivier, D. C. (1973). Metrics on spaces of finite trees. *Journal of Mathematical Psychology* **10,** 26–59.

Colless, D. H. (1980). Congruence between morphometric and allozyme data for *Menidia* species: a reappraisal. *Systematic Zoology* **29,** 288–299.

Day, W. H. E. (1983). The role of complexity in comparing classifications. *Mathematical Biosciences* **66,** 97–114.

Farris, J. S. (1973). On comparing the shapes of taxonomic trees. *Systematic Zoology* **22,** 50–54.

Gower, J. C. (1975). Generalized Procrustes analysis. *Psychometrika* **40,** 33–51.
Jardine, C. J., Jardine, N., and Sibson, R. (1967). The structure and construction of taxonomic hierarchies. *Mathematical Biosciences* **1,** 173–179.
Jardine, N., and Sibson, R. (1971). 'Mathematical Taxonomy'. Wiley (Interscience), Chichester.
Mantel, N. (1967). The detection of clustering and a generalized regression approach. *Cancer Research* **27,** 209–220.
Margush, T., and McMorris, F. R. (1981). Consensus *n*-trees. *Bulletin of Mathematical Biology* **43,** 239–244.
McMorris, F. R., and Neumann, D. A. (1983). Consensus functions defined on trees. *Mathematical and Social Sciences* **4,** 131–136.
Mickevich, M. F. (1978). Taxonomic congruence. *Systematic Zoology* **27,** 143–158.
Mickevich, M. F., and Farris, J. S. (1981). The implications of congruence in *Menidia. Systematic Zoology* **30,** 351–370.
Milligan, G. W. (1979). A note on the use of INDSCAL for the comparison of several classifications. *Systematic Zoology* **28,** 95–99.
Nelson, G. (1979). Cladistic analysis and synthesis: principles and definitions, with a historical note on Adanson's 'Familles des Plantes' (1763–1764). *Systematic Zoology* **28,** 1–21.
Neumann, D. A. (1983). Faithful consensus methods for *n*-trees. *Mathematical Biosciences* **63,** 271–287.
Phipps, J. B. (1971). Dendrogram topology. *Systematic Zoology* **20,** 306–308.
Rohlf, F. J. (1982). Consensus indices for comparing classifications. *Mathematical Biosciences* **59,** 131–144.
Rohlf, F. J., and Sokal, R. R. (1980). Comments on taxonomic congruence. *Systematic Zoology* **29,** 97–101.
Rohlf, F. J., and Sokal, R. R. (1981). Comparing numerical taxonomic studies. *Systematic Zoology* **30,** 459–490.
Sackin, M. J. (1972). 'Good' and 'bad' phenograms. *Systematic Zoology* **21,** 225–226.
Sibson, R. (1972). Order invariant methods for data analysis. *Journal of the Royal Statistical Society B* **34,** 311–349.
Sneath, P. H. A. (1975). Cladistic representation of reticulate evolution. *Systematic Zoology* **24,** 360–368.
Sneath, P. H. A. (1983). Distortions of taxonomic structure from incomplete data on a restricted set of reference strains. *Journal of General Microbiology* **129,** 1045–1073.
Sneath. P. H. A., and Sokal, R. R. (1973). 'Numerical Taxonomy'. Freeman, San Francisco.
Sokal, R. R., and Rohlf, F. J. (1981). Taxonomic congruence in the leptopodomorpha re-examined. *Systematic Zoology* **30,** 309–325.
Williams, W. T., and Clifford, H. T. (1971). On the comparison of two classifications of the same set of elements. *Taxon* **20,** 519–522.

3

Comparison of Transformation and Classification Techniques on Quantitative Data

S. BASCOMB

Department of Medical Microbiology, St. Mary's Hospital Medical School, London, UK

Introduction

Most bacterial taxonomy studies are based on qualitative data, usually expressed as binary or occasionally as ordered or unordered multistate characters. The use of analytical instruments, and the introduction of mechanisation and automation to microbiology, yield basically quantitative data. Although it is possible to convert these quantitative data to binary or multistate characters and use the well-established techniques for clustering of such data (Sokal and Sneath, 1963; Sneath and Sokal, 1973), the conversion inevitably causes a loss of information (Dunn and Everitt, 1982). Methods for classification of quantitative data bases are well documented (Everitt, 1980), but the introduction of quantitative data brings specific problems, generally related to increase in the variability of the data and the origins of this variability.

Quantitative data bases have been used for studying the classification of certain taxa or, more often, for identifying unknown isolates. In the latter case the number of attributes may have been limited to relatively few, often less than is customary in conventional taxonomic studies, and probably insufficient to reveal phenotypic relations between the taxa, but still adequate to provide reasonable identification. Such identification schemes may be preferable to conventional testing because they can be completed in a shorter time, or they may offer other advantages for the investigator who requires identification of a large number of strains routinely. A useful step in such identification studies is the establishment of the grouping which exists in the quantitative data base, with a primary aim of ascertaining that the well-established conventional delineation of the bacterial groups studied is also expressed in the quantitative data base. If this is not the case the system cannot be applied to identification. Justification for applying numerical methods to a data bank consisting of a limited number of characters can be found in Lockhart and Koenig (1965). These workers have shown that

ISBN 0-12-289665-3

exclusion of what they called 'secondary characters' from taxonomic studies yields basically the same clustering as would have been obtained with all characters. This chapter deals with some of the problems inherent in bacterial quantitative data bases and with ways of solving them.

Types of Quantitative Data Bases

Quantitative results used in taxonomic studies can be obtained with instruments measuring a variety of parameters. Here are some examples of such quantitative attributes:

1. Derivatives of organic/fatty acids of the whole organism separated by pyrolysis and gas–liquid chromatography (Jantzen *et al.*, 1974; Drucker, 1976)
2. Organic compounds resulting from breakdown of the whole organism as obtained by high-temperature pyrolysis and mass spectrometry (Meuzelaar *et al.*, 1976)
3. Extent of growth inhibition or promotion caused by antimicrobial agents and other compounds (Friedman and MacLowry, 1973; Sielaff *et al.*, 1976)
4. Relative mobility and quantity of proteins of cell-free extracts separated by electrophoretic techniques (Kersters and De Ley, 1975, 1980)
5. Magnitude of enzymatic activities (Bascomb and Grantham, 1975; Bascomb, 1980; Bascomb and Spencer, 1980)

Initial Data Processing

Sources of Variability of Data

Quantitative data values can be affected by the variability of instrument performance, variability in the performance of reagents, the size of the sample subjected to the analysis, and the nature of the sample. For comparing different bacterial cultures only the last source of variation is of interest. For reliable results it is necessary to remove the other sources of variation or at least to make sure that they are small. A number of techniques are available for this; initial data processing is primarily designed to perform this task. Cluster analysis techniques can then be used on the transformed data to reveal the relationship between OTUs. Table 1 lists the statistical packages that provide cluster analysis procedures.

Table 1. *Statistical packages containing clustering algorithms*

Package	Reference
ARTHUR	Duewer *et al.* (1975)
BMDP	Dixon and Brown (1979)
CLUSTAN	Wishart (1978)
GENSTAT	Alvey *et al.* (1977)
MASLOC	Kaufman and Massart (1981)
TAXPAK	M. J. Sackin (unpublished)

Removal of Variability Due to Instrument and Reagent Performance

The methods used to diminish the effects of instrument and reagent variability depend to some extent on the type of instrument and measurement used. Measuring enzymatic activities, Bascomb and Spencer (1980) included, each day, a number of samples containing different concentrations of every type of reaction product assayed in their continuous-flow system. The measurements on these samples were used to calculate regression coefficients daily for each type of product and to convert absorbance measurements into accepted units of product concentration. Godsey *et al.* (1981), measuring enzyme activities, using a batch-type fluorocolorimeter operated in a kinetic mode, included one machine standard daily to calibrate the instrument, and then used the calibration factor to convert daily measurements into units of product concentration and to calculate enzyme reaction rates. With pyrograms and electrophoretic patterns, constant concentrations of known compounds are usually included on every testing occasion, and their positions used for calibrating the system (Kersters and De Ley, 1980). The calibration factor thus calculated is used for transforming all the data obtained on this occasion.

It is important to realise that every transformation of the data may cause some distortion. Thus, when calibration is done on the basis of a single sample and an error occurs during the measurement of this sample, this error will be introduced into all the results obtained on that occasion. It might be preferable to determine the mean and standard deviation of the calibration sample(s) and restrict acceptability of data, obtained on different occasions, to those where the calibrator falls within predetermined acceptable levels, such as one standard deviation from the mean, assuming the coefficient of variation for calibration samples is not higher than 10%. Healy (1968) briefly described techniques for detection of slow changes in calibration and drifts in the values obtained for assay controls.

Removal of Variability Due to Sample Size

Sample size is the biggest source of extraneous variability in quantitative data, and the removal of its influence has elicited a large number of methods. In studies using instruments for determination of bacterial growth by measurement of light scatter, such as the AutoBac, the sample is standardised before testing by adjusting the opacity of the bacterial suspension to a predetermined range (Barry *et al.*, 1982). A similar method was used by Godsey *et al.* (1981) for measuring enzymatic activities. Such a method will remove the variability, due to sample size, only from bacteria of similar cell size, shape, and physiological state. Thus, suspensions containing similar numbers of rods or spheres will differ in their opacity, as will those containing dividing or resting cells. Moreover, such an approach is possible only if there are sufficient bacteria—usually at least 5–10 colonies.

In other studies, correction for sample size occurs after measurements have been made. With pyrograms and electrophoretic data two normalisation procedures are commonly applied. Individual measurements are normalised with respect to the total information available for each OTU; proportional transformation is expressed as a percentage of total peak area (Jantzen *et al.*, 1982). This method also is not free from distortion. Thus, taking two OTUs with identical large values for character X and a similarly large value for character Y present only in OTU_1, the transformed value for character X of OTU_1 will be considerably smaller than that of X of OTU_2 in spite of their identical unnormalised values. Pyrogram measurements may be normalised with respect to the value of one character chosen because it is present in all samples in large quantities. This transformation is justifiable only if the relative concentration of the chosen character is constant throughout the whole population under study. This is not necessarily so. Moreover, such transformation is possible only if a common character is available.

In enzyme studies it is customary to relate enzyme activity to protein concentration (the specific activity transformation). The problem with this method of standardisation is that estimation of protein at levels below 5 $\mu g\ ml^{-1}$ is not sufficiently accurate, and division by an inaccurate small number will cause a very large error.

Removal of Variability Due to Differences in Magnitude and Units of Different Characters

The logarithmic transformation has been used by Godsey *et al.* (1981) for enzyme data and by Bøe and Gjerde (1980) for fatty acid pyrograms. Such a transformation tends to change the distribution pattern of the results and increase the spread of smaller data values. The assumption behind the use of the logarithmic transformation is that it may make the distribution pattern of results for each test approach a normal distribution pattern, which is a necessary condition for the application of most parametric analytical models. This objective may be achieved when the distribution of the values shows skewness (Healy, 1968), but might not be as effectively achieved with other types of distribution.

Additional needs for transformation arise when measuring overall similarity between OTUs. This is because the different attributes may be measured and expressed in different types of units, and their values may also differ in magnitude. The total similarity of pairs of OTUs may also be greatly affected by the scale used for measurement of each character (Clifford and Stephenson, 1975). Gower (1971) suggested that data could be ranged, the raw measurement X being converted to X' by means of the equation

$$X' = (X - X_{min})/(X_{max} - X_{min})$$

Ware and Hedges (1978) suggested that results obtained through division by the range between X_{max} and X_{min} were likely to be affected by the range of values found in any particular study and may be too dependent on the particular subsample of the population. In the principal component analysis model, test results are standardised to zero mean and units of standard deviation each side of it (Clifford and Stephenson, 1975). Similarly, in the BMDP package, data values are standardised to Z scores, where Z is the difference from the mean divided by the standard deviation for each test,

$$Z = (x - \bar{x})/\mathrm{SD}$$

In GENSTAT, quantitative variates are transformed, as above (Gower, 1971).

Clustering Procedures

Calculation of Resemblance between Pairs of OTUs

After transformation/standardisation of the data, the resemblances of each pair of OTUs are calculated as similarity or distance coefficients. Dunn and Everitt (1982) describe the differences between the options available for calculating the resemblance of OTU pairs on the basis of quantitative characters. Basically, the position of each OTU is visualised in a multi-dimensional space where each character is represented by one axis (dimension), usually orthogonal to existing axes. The distance between OTUs can be calculated by two approaches: summing the differences on each axis (the absolute or city-block distance of Carmichael and Sneath, 1969) or by calculating the Euclidean distance in the multidimensional space, perhaps with ranging or standardising of the characters, as above. However, the relationship between characters is not necessarily orthogonal, particularly if the characters are not independent of each other. The Mahalanobis generalised distance D^2 is reputed to be more appropriate in such circumstances.

Everitt (1980) claimed that Mahalanobis D^2 has the advantage over the Euclidean and city-block measures in that it allows for correlations existing between variables; Dunn and Everitt (1982) also recommended its use in combination with oblique co-ordinates in cases of correlation of character states. However, although all agree that Mahalanobis D^2 generalised distance can be used to calculate the distance between two populations, there is some disagreement as to the acceptability of its use for calculating the distance between two OTUs (Marriot, 1974; Kendall, 1980).

A different measure of the similarity can be obtained by plotting values of each attribute of one OTU against that of the second, and calculating the Pearson product-moment correlation coefficient r (Sneath, 1972; Kersters and De Ley, 1975). Some authors question the validity of using this criterion, particularly if the data include different units (Hand, 1981; Dunn and Everitt, 1982).

Wishart's CLUSTAN package (1966, 1978) contains 12 methods of calculating resemblance between pairs of OTUs based on quantitative characters. Drucker *et al.* (1982), using gas chromotography data, compared seven measures of association of OTU pairs for their ability to identify streptococci and concluded that the Stack coefficient (Stack *et al.*, 1978), which measures the ratio of all character states, was the most efficient. A similar method was suggested by Ware and Hedges (1978). The subjective nature of decision on the choice of similarity coefficient was discussed by Gower (1978).

Relationships between OTUs are usually expressed either as similarities which are on a definite scale range of 0 to 100% (0–1) or else as distances which are unrestricted and can take any positive value (Everitt, 1980).

Clustering Techniques

Once the resemblance of each pair of OTUs has been calculated, the clustering techniques for quantitative data are similar to those applied to qualitative data. They can be divided into hierarchical and non-hierarchical techniques. The well-known hierarchical single-linkage, median, centroid, UPGM, and WPGM procedures are available in most packages. Non-hierarchical techniques include optimisation (Gower, 1974; Barnett *et al.*, 1975) and density search techniques (Carmichael and Sneath, 1969). Basically, such techniques aim to divide the population into K groups, the value of K being determined either by the investigator or reached by iteration of calculation of within- and between-cluster variability. The differences between the various techniques are discussed by Everitt

(1980). The GENSTAT, ARTHUR, BMDP, CLUSTAN, and MASLOC packages all allow non-hierarchic clustering techniques.

The choice of clusters is still subjective, as the cutoff points for species and genera are usually determined arbitrarily. The application of fixed values of similarity to species and genus levels (Colwell and Liston, 1961; Sneath and Sokal, 1962) is more difficult on the distance scales because these have a variable range.

Validation of Clusters

Methods used for checking the validity of the clusters formed include the determination of within- and between-cluster similarity, available in GENSTAT. The BMDP package provides pooled within-cluster covariance and correlation. The MASLOC package contains a validation routine which requires the input of the number of clusters and the similarity level of formation of each cluster; the computer then calculates the position of each OTU on a projected two-dimensional map to permit a subjective assessment of the inter- and intra-cluster relatedness. Comparison of the classification obtained by the use of different computations on the quantitative data with that obtained by other methods, such as conventional testing and identification, can also be used to evaluate the validity of the classification obtained (MacFie *et al.*, 1978; Drucker *et al.*, 1982). Such comparisons should be examined carefully, as it is possible that the studies of OTUs based on completely different sets of characters might reveal genuine but different clusterings.

Sneath (1972) used histograms of the distances of all strains from the centroid of a chosen taxon. A cluster which is reasonably compact will show a peak, while a straggly cluster will give a flattened histogram. The BMDP *K* mean algorithm provides such scattergrams. Darland (1975) used principal component analysis, as applied to qualitative data of 130 isolates of *Escherichia coli* and *Shigella,* to show that the plot of distribution of the first principal component scores produces a bimodal distribution with some overlapping. The method was also used to demonstrate the separation between *Yersinia enterocolitica* and *Y. pseudotuberculosis*. Sneath (1977) provided a method for testing the significance of dichotomous clusters. Drucker *et al.* (1982) used the Andrews plot (Andrews, 1972) to investigate the strains of *Streptococcus milleri;* they suggested that the technique could be used to reveal subgroups as well as atypical strains. A different approach for validation of clusters is to use identification techniques such as discriminant function analysis, available in SPSS (Nie *et al.*, 1975) and BMDP, and then determine the number of OTUs that identify correctly.

Dunn and Everitt (1982) stressed that although algorithms for cluster analysis are available and easy to use, they should be best seen as tools for data exploration. Everitt (1980) applied different methods of clustering to sets of data artificially constructed to represent two clusters, based on two variables, to illustrate some of the problems of clustering. Besides pointing to the fact that a large number of clustering methods are biased towards finding spherical clusters, he emphasised that classification procedures are essentially descriptive techniques and the solution given by such techniques should be used for re-examination of the data. It was therefore interesting to study the effect of choice of algorithm on clustering obtained with a real data set.

Studies on Enzyme Activity Data

The effects of the various processing procedures on the classifications obtained were studied (Bascomb, unpublished data) with activity data of eight enzymes in 154 isolates identified by both conventional and automated methods (Bascomb and Spencer, 1980). The study included strains of *Escherichia, Klebsiella, Proteus,* and *Pseudomonas*. The choice of such a limited data set, which included a small number of tests and taxa, was deliberate to facilitate calculations of variability between and within clusters. There is a practical limitation to the size of the data matrix easily handled by the statistical packages, as computations on quantitative data are much more time-consuming than those on a qualitative data matrix (Gower, 1978).

The computing was performed using the CDC 6400 of Imperial College Computing Centre (ICCC), the CDC 6600 and 7600 computers in the University of London Computing Centre (ULCC), and a PDP/11 computer in St. Mary's Hospital Medical School Computer Unit. Fortran programmes were used in ICCC for the initial data collection and processing; BMDP, CLUSTAN, GENSTAT, and SPSS were used in ULCC. ARTHUR package was used at St. Mary's. Investigation was made into the effect of (a) type of transformation of the data, (b) choice of resemblance coefficient, and (c) choice of clustering method, on agreement between the taxa formed and the classification obtained using conventional data. Division into clusters was performed intuitively, aiming to establish a cutoff point which would maximise the number of strains in each cluster, minimise inclusion of strains of different genera in the same cluster, and produce a small number of clusters.

Test results were transformed using four different types of normalisation (see Tables 2–5), and the transformed values were compared with unmodified raw data. Measures of distance were compared within and between packages. The measures used were Pythagorean and city-block distances of GENSTAT; SUMOFSQ, SUMOFP (P = 1), and CORR of BMDP, as well as correlation, cosine, similarity, size difference, shape difference, nonmetric, and dissimilarity coefficients of CLUSTAN. The BMDP SUMOFSQ gives the Euclidean distance; the SUMOFP (P = 1) gives the city-block distance, while the CORR

Table 2. *Effect of type of data transformation and method of resemblance calculation on within- and between-cluster similarity using GENSTAT*[a]

Cluster[a]	Raw data				Log transformation				Proportional transformation				Specific activity transformation			
Pythagorean distance																
1	96.6				85.8				91.4				97.8			
2	92.7	95.1			76.7	83.6			88.4	94.2			96.6	97.4		
3	87.7	88.5	93.2		75.3	79.3	91.4		85.1	89.0	96.0		92.2	92.2	93.5	
4	89.9	89.0	86.7	98.0	69.9	68.0	64.1	86.1	80.2	85.5	82.2	99.2	95.0	94.0	89.9	97.5
	1	2	3	4	1	2	3	4	1	2	3	4	1	2	3	4
	MW = 95.7		MB = 89.0		MW = 86.7		MB = 72.2		MW = 95.2		MB = 85.1		MW = 96.5		MB = 93.3	
		S = 3.9[b]				S = 4.3				S = 3.0				S = 1.2		
City-block distance																
1	89.6				74.2				81.1				93.3			
2	82.3	86.6			63.0	71.9			75.7	84.6			90.3	91.8		
3	74.4	75.9	84.0		60.2	65.7	81.2		71.8	76.7	88.6		83.3	83.1	85.9	
4	79.0	77.8	74.4	92.2	55.6	54.8	50.9	75.3	68.1	74.6	71.8	95.2	88.1	86.3	80.1	92.7
	1	2	3	4	1	2	3	4	1	2	3	4	1	2	3	4
	MW = 88.2		MB = 77.3		MW = 75.7		MB = 58.4		MW = 87.4		MB = 73.1		MW = 91.0		MB = 85.2	
		S = 7.3[b]				S = 6.2				S = 5.4				S = 3.0		

[a]Cluster designation based on conventional identification of 154 strains. 1, *Escherichia*; 2, *Klebsiella*; 3, *Proteus*; 4, *Pseudomonas*.
[b]MW, Mean within-cluster similarity; MB, mean between-cluster similarity; S, smallest $W - B$ difference.

procedure (usually applied to combine variables into clusters) gives the correlation coefficient r^2.

Clustering was performed using three hierarchic methods (single linkage, centroid, and median) with both BMDP and GENSTAT, as well as the non-hierarchic optimisation techniques, CLASSIFY of GENSTAT and PKM of BMDP.

All classifications were examined for the percentage of strains correctly assigned, definition of correct assignment being the grouping into a cluster containing more than 90% of strains belonging to one genus. A second criterion was the number of 'acceptable' clusters appearing in each genus, the number of 'mixed' clusters, and the total number of clusters (see Table 3). It was assumed that a smaller number of 'mixed' clusters and a smaller total number of clusters represent better agreement with conventional classification.

On the basis of 12 conventional identification tests, the 154 cultures were assigned to four genera: (a) *Escherichia coli* (84), (b) *Klebsiella* (30), all falling into *K. pneumoniae* (*sensu lato*), (c) *Proteus* (25), comprising *P. mirabilis* (13), *P. morganii* (6), *P. rettgeri* (3), and *P. vulgaris* (3), and (d) *Pseudomonas* (15), the majority of which belonged to *P. aeruginosa*.

Identification of the strains using quantitative data for the eight enzyme tests and the discriminant function analysis model of SPSS gave 97.4% agreement with conventional identification (Bascomb, 1983), indicating that the identification based on quantitative characters could be made to agree with that generally expected on the basis of qualitative characters.

The degree of separation between taxa can be indicated by the differences between the within-cluster (W) and between-cluster (B) similarity values. For each conventional taxon (Table 2), W was always higher than B regardless of the type of transformation or method of calculating distance, suggesting that the conventional division into four genera is reflected in the quantitative data base. The values of the mean within-cluster similarity MW, the mean between-cluster similarity MB, and the smallest $W - B$ difference S are also given in Table 2. The values of $W - B$ and of S suggest that the city-block distance measure gives better separation between the clusters. When Pythagorean distance is used the logarithmic transformation gives the highest value of S. With city-block distance the raw data give the highest value of S.

When the same data were used for cluster analysis, the number of clusters formed was more than four, regardless of the type of data-processing procedure, showing up the differences between classification and identification techniques. The latter emphasise the common attributes and attach diminished weights to the variable attributes.

The different packages were compared with regard to the ease of operation and ease of interpretation of output. For people with very little programming experience the BMDP and SPSS packages are probably the easiest to use, GENSTAT and CLUSTAN moderately so, and ARTHUR almost impossible. The last was

```
  VARIABLE
NAME      NO.                                                              (a)
            -------------------------------------------------------------------
X(1)      (  1) 99/99/98/85 83 88 91/73 85 73 81 79 83 82/58/65 64 55
                  /  /  /           /                    /  /
                 /  /  /           /                    /  /
X(3)      (  3)/99/98/85 84 89 92/74 86 73 81 79 83 81/59/66 65 56 54
                 /  /          /                     /  /
                /  /          /                     /. /
X(2)      (  2)/98/85 84 88 91/72 83 71 78 77 81 80/58/64 63 54 52 65
                 /          /                   /  /
                /          /                   /  /.
X(73)     ( 73)/80 80 84 88/67 80 68 77 75 78 77/53/58 56 47 47 62 60
                          /                  /  /
            ----------/                     /  /
X(4)      (  4) 99/99 97/77 76 59 85 84 82 72/63/83 81 82 70 85 87 83
                  /     /                   /  /
                 /     /                   /  /
X(5)      (  5)/99 98/77 75 59 83 81 80 68/64/84 80 81 72 85 86 84 91
                    /                    /  /
            ----/                       /  /
X(6)      (  6) 99/80 80 64 87 84 84 74/65/84 81 80 72 85 86 82 88 79
                  /                   /  /
                 /                   /  /
X(7)      (  7)/78 80 65 84 81 82 72/64/80 77 74 68 82 82 78 85 74 80
                                   /  /
            ---------------------/  /
```

```
    CASE          ORDER OF
NO.  LABEL      AMALGAMATION                                          (b)

  1    1 010  *******-------------------------------------------------
  3    3 010        46.-/     /// /
  2    2 010        79./     /// /
  5    5 010        91.---/// /
  6    6 010        51.///// /
  7    7 010        71.//// /
  4    4 010       120./// /
106  106 020       121.// /
 92   92 020       138./ /
114  114 022       127.-/
 85   85 020       143./
108  108 020       139.-------------------------------------------
 97   97 020       135.-------------------------------------------
  9    9 010       133.-------------------------------------------
 11   11 010        48./// /  /  /
 12   12 010        58.// /  /  /
 10   10 010        81./ /  /  /
 40   40 010       110.-/  /  /
 14   14 010       113./  /  /
123  123 030       108.--/  /
116  116 030        65.-/  /
115  115 030       129./  /
 65   65 010       102.--/
131  131 032       101.-/
 64   64 010       132./
 17   17 010       123.-------------------------------------------
```

Fig. 1. Parts of dendrograms produced by different statistical packages. (a) Correlation and complete linkage of BMDP, dendrogram printed over similarity scale. Tree printed over correlation matrix (scaled 0–100). (b) SUMOFP ($P = 1$) and average linkage of BMDP, OTU group printed under label, no similarity scale. (c, p. 48) City block and single linkage of ARTHUR, OTU identification number and group printed. (d, p. 49) Correlation and single linkage of CLUSTAN. (e, p. 51) City block and centroid of GENSTAT.

```
                          1.00        0.97        0.93        0.90   (c)
                          +-----------+-----------+-----------+-----
137  137  001100   3.00  *********************************
                                                          ****
138  138  001110   3.00  *********************************   *
                                                             *
 85   85  107150   2.00  *******************************     *
                                                        *    *
114  114  000030   2.00  *******************************     *
                                                        *    *
119  119  112130   3.00  *********************              *    *
                                            **          *    *
120  120  112180   3.00  ***********************            *    *
                                              **        *    *
126  126  128320   3.00  ***********************            *    *
                                               *        *    *
125  125  122630   3.00  *********                 *        *    *
                                  ******        *****   *    *
136  136  001090   3.00  *********        **      *    *    *    *
                                          **      *    *    *    *
133  133  001890   3.00  ********************* *    *    *    *
                                           *   * *    *    *    *
122  122  112330   3.00  ****************     ***    **  *    *
                                               *      **  *    *
139  139  001080   3.00  *********************         **  *    *
                                                      **  *    *
117  117  107980   3.00  *********                     **  *    *
                                  *******************  *    *
118  118  108080   3.00  *********                      *  *    *
                                                       *  *    *
```

Fig. 1 (*Continued*)

quite difficult to implement because of some discrepancies in its documentation The BMDP procedures produced both shaded resemblance matrices and dendrograms routinely. The dendrograms could be printed vertically or horizontally on the printout page. The horizontal tree takes less space for a large number of cases. Dendrograms produced using correlation as the measure of similarity were difficult to grasp and interpret (Fig. 1a). The dendrograms produced by the other routines (Fig. 1b) as well as ARTHUR (Fig. 1c), CLUSTAN (Fig. 1d), and GENSTAT (Fig. 1e) were much easier to interpret.

The number of clusters formed, using the quantitative data base, was affected by type of data transformation, method of calculating distance, clustering techniques, and statistical packages used. For complete agreement between qualitative identification results and quantitative classification, the 154 strains should have fallen into four clusters each comprising strains of one taxon only. Table 3 shows the number of 'acceptable' and mixed clusters obtained using the various procedures. The specific activity transformation produced the smallest number of mixed clusters and the smallest number of total clusters and, by this criterion, would be the preferable mode of data transformation. The logarithmically transformed and the non-transformed (raw) data produced only slightly larger num-

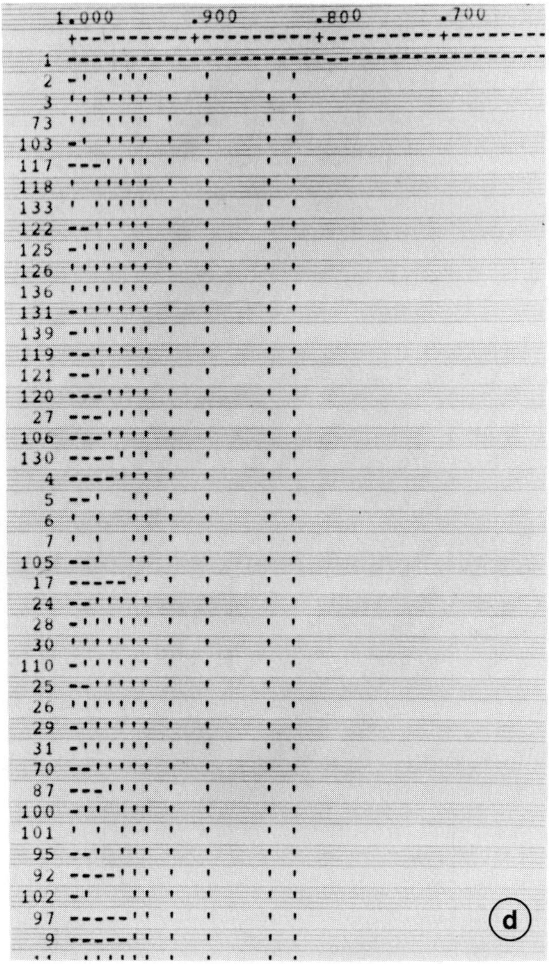

Fig. 1 (*Continued*)

bers of clusters. The number of clusters formed was most affected by the type of data transformation when the distance between OTU pairs was calculated using the Euclidean distance. The percentage of strains appearing in acceptable clusters, using the various combinations, is given in Table 4. Applying this criterion, the specific activity transformation gave poor agreement; best agreement was obtained using the proportional transformation and the BMDP SUMOFSQ distance, 86% of the strains appearing in acceptable clusters. The specific activity transformation yielded the smallest percentage of agreement.

Using the CLUSTAN package, seven different methods for calculating re-

Table 3. *Effect of method of calculating distance, and of statistical package on number of clusters formed, in the study of 154 isolates using single-linkage clustering*

Type of transformation	BMDP								
	SUMOFSQ			SUMOFP ($P = 1$)			CORR		
	Acceptable[a]	Mixed[b]	Total	Acceptable	Mixed	Total	Acceptable	Mixed	Total
Raw	8	20	28	6	6	12	10	5	15
Proportional	5	7	12	4	7	11	9	6	15
Specific activity	3	2	5	8	2	10	10	6	16
Logarithmic	7	4	11	9	5	14	9	7	16

Type of transformation	GENSTAT					
	Pythagorean			City block		
	Acceptable[a]	Mixed[b]	Total	Acceptable	Mixed	Total
Raw	19	2	21	6	1	7
Proportional	19	4	23	16	1	17
Specific activity	4	1	5	13	1	14
Logarithmic	8	5	13	8	2	10
Division by SD	15	2	17	15	1	16

[a]Acceptable cluster: >90% of strains belong to one genus.
[b]Mixed cluster: <90% of strains belong to one genus.

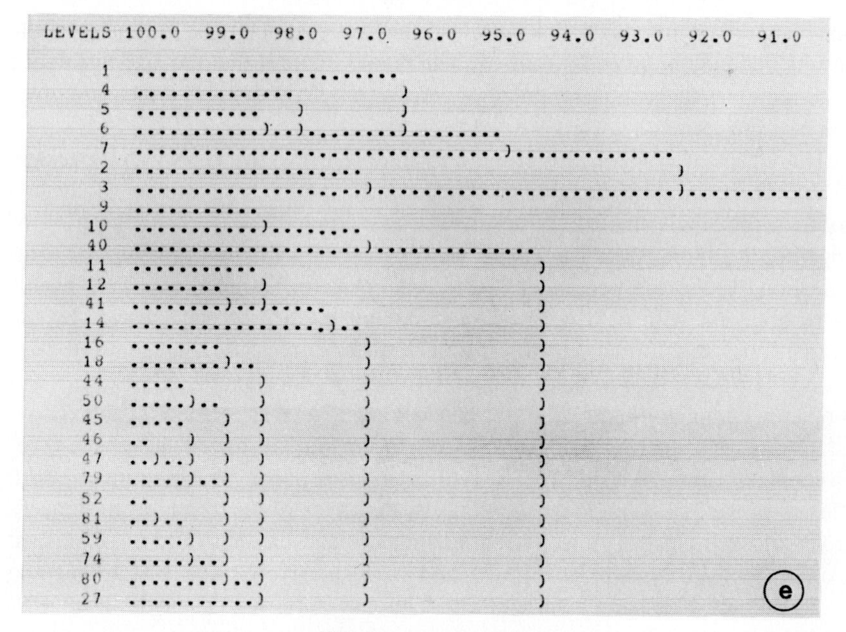

Fig. 1 (*Continued*)

Table 4. *Effect of method of calculating distance and of statistical package on the percentage of strains correctly grouped, using single-linkage clustering*[a]

Type of transformation	BMDP SUMOFSQ	BMDP SUMOFP ($P = 1$)	BMDP CORR
Raw	82	78	74
Proportional	86	83	69
Specific activity	47	75	73
Logarithmic	67	82	72

Type of transformation	GENSTAT Pythagorean	GENSTAT City block
Raw	81	69
Proportional	76	77
Specific activity	34	63
Logarithmic	69	84
Division by SD	74	80

[a]Strains correctly grouped are those assigned to a cluster containing >90% of strains belonging to one genus.

semblance and the single-linkage clustering were applied. A summary of the results obtained is given in Table 5. The results suggest that the size difference criterion gives the smallest number of clusters while the similarity measure gives the best clustering.

The effect of type of clustering was studied using the GENSTAT package (Table 6). Judging by the number of acceptable and the total number of clusters formed, the single-linkage clustering of raw data and the median clustering of logarithmically transformed data showed best agreement with conventional identification. Using the percentage of strains correctly assigned, the best results were obtained with data normalised either by the SD of each test or by the proportional transformation, the city-block distance measure, and centroid clustering. The ARTHUR package using city-block or Mahalanobis D^2 distances and single or complete linkage did not produce such good agreement (Table 7).

Results of clustering using non-hierarchic methods showed the greatest between-package difference. The BMDP provides a listing of the strains of each cluster as well as histograms showing the distances of all strains from the centroids of each cluster. A well-separated cluster and one not fully distinguished from others are shown in Fig. 2a and b, respectively. The BMDP method provided only 42% agreement with conventional clustering (Table 8). The GENSTAT package offers four criteria for producing non-hierarchical clustering. They include sum of squares S, minimal determinant of pooled within-classes dispersion matrix W, maximal total Mahalanobis D^2 distance between classes T, and maximal predictive value P. Maximal predictive value, used for classification of the tribe Klebsielleae and of yeasts (Barnett *et al.*, 1975), can be applied to a qualitative data matrix only. The effect of choice of criterion and K value on agreement with conventional classification is shown in Table 9.

It could be argued that the number of characters used in this study was too small to be manipulated by numerical taxonomy studies. It is also possible that

Table 5. *Effect of method of calculating resemblance, using raw data, on the number of clusters formed and percentage of strains correctly grouped by CLUSTAN package*

Resemblance measure	Type of cluster: Acceptable	Mixed	Total	Strains correctly assigned (%)
Correlation	3	3	6	62
Cosine	5	1	6	67
Similarity	11	5	16	81
Size difference	2	3	5	52
Shape difference	3	3	6	72
Non-metric	18	2	20	60
Dissimilarity	3	3	6	68

Table 6. *Effect of method of hierarchic clustering of GENSTAT, based on similarities obtained using a city-block model, on agreement with conventional identification*

Type of transformation	Number of clusters								
	Method of clustering								
	Single linkage			Centroid			Median		
	Acceptable	Mixed	Total	Acceptable	Mixed	Total	Acceptable	Mixed	Total
Raw	6	1	7	19	1	20	9	2	11
Proportional	16	1	17	15	4	19	23	1	21
Specific activity	13	1	14	8	1	9	13	1	14
Logarithmic	8	2	10	28	0	28	4	3	7
Division by SD	15	1	16	14	1	15	11	1	12

Type of transformation	Percentage of strains correctly clustered		
	Method of clustering		
	Single linkage	Centroid	Median
Raw	69	68	86
Proportional	77	94	73
Specific activity	63	86	61
Logarithmic	84	77	49
Division by SD	80	95	85

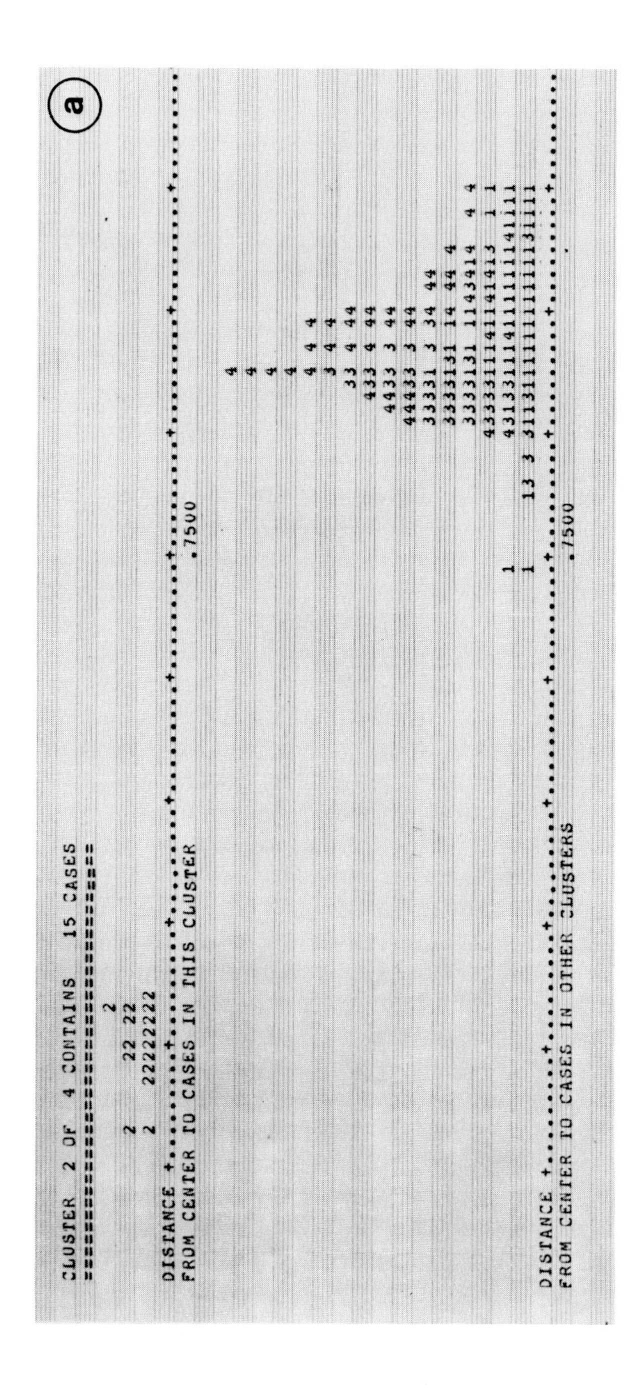
a
CLUSTER 2 OF 4 CONTAINS 15 CASES
DISTANCE
FROM CENTER TO CASES IN THIS CLUSTER
.7500
DISTANCE
FROM CENTER TO CASES IN OTHER CLUSTERS
.7500

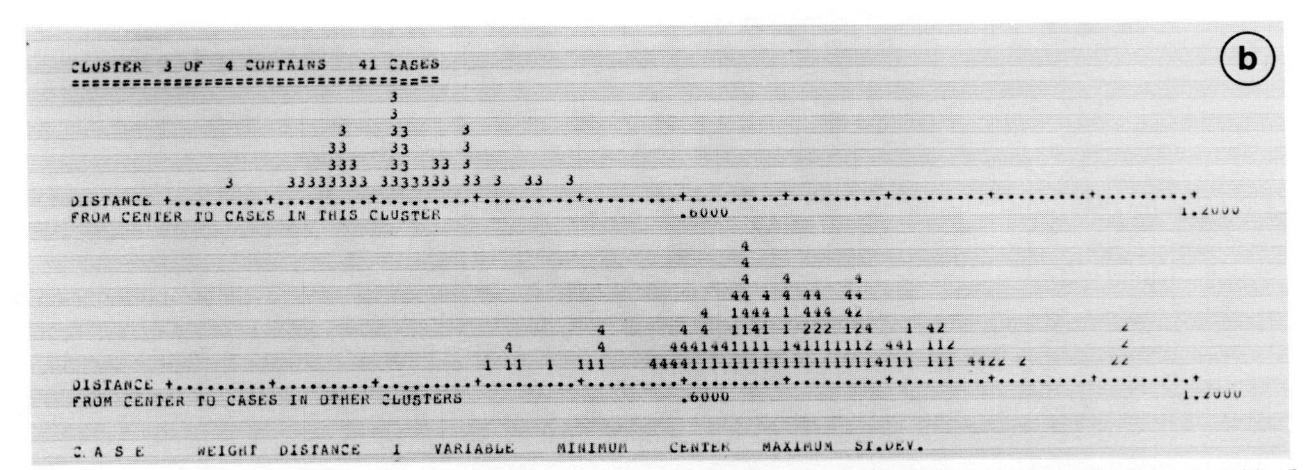

Fig. 2. Output of the *K* mean non-hierarchical clustering routine of BMDP, showing the distance of OTUs from the centroids of clusters. (a) The compact cluster of *Pseudomonas* (No. 2) showing no overlap. (b) A heterogenic cluster (No. 3) containing strains of *E. coli*, *Klebsiella*, and *Proteus* showing considerable overlap with other clusters.

Table 7. *Effect of distance measure and hierarchic clustering method of ARTHUR, using raw data, on percentage of strains correctly grouped*

Clustering	Distance measure	
	Mahalanobis D^{2a}	City block
Single linkage	69.4	75.3
Complete linkage	62.3	74.0

[a]Calculation equivalent to Euclidean distance measure.

the disproportionate number of strains of *E. coli* caused some bias in clustering, particularly with the nonhierarchical methods. However, the distribution of strains amongst the four genera reflects the actual distribution in urine specimens. Moreover, the validity of the four taxa used has been established in numerous previous studies and the purpose of this study was to establish if, and to what extent, the manipulation of the data affected the clustering obtained. Probably the actual effect of using a smaller number of tests would be a general increase in the relative importance of results of each test and a decrease in the level of similarity between clusters.

Furthermore, studies which provide quantitative instrumental data for use in bacterial identification procedures may be limited to a small number of characters. When disagreement between conventional and automated identification occurs, it is important to determine whether such disagreements are due to the inability to obtain differentiation on the basis of the quantitative characters matrix, or because the quantitative matrix contains clusters which differ from those found in the qualitative character matrix. It is therefore necessary to establish which procedures will provide the most reliable information. The present studies suggest that none gives clear-cut results but, for the purpose of classification,

Table 8. *Effect of data transformation on percentage of strains correctly grouped by the BMDP* K *means non-hierarchic clustering algorithm using the Euclidean distance measure (*K *= 4)*

Type of transformation	Correct %
Raw data	9
Proportional	42
Logarithmic	0
Specific activity	7

Table 9. *Effect of chosen number of clusters (K), and type of criterion for optimisation, using raw data and GENSTAT, on percentage of isolates correctly grouped*

Number of clusters	Criterion[a]		
	S	*T*	*W*
4	9	52	13
5	11	52	40
6	11.1	52.6	48
9	16.2	30	74
12	43.5	82.5	75

[a] *S*, Sum of squares; *T*, maximise total Mahalanobis D^2 between classes; *W*, minimum determinant of pooled within-class dispersion matrix.

raw data or SD and logarithmic transformations, the city-block distance measure, and clustering by the centroid method provide the best agreement with qualitative conventional classification. It also appears that the non-hierarchical methods of GENSTAT offer a quick and easy method for preliminary evaluation of the data.

Conclusions

Characterisation of bacteria using instruments yields quantitative data bases which require specific mathematical treatment. A number of statistical packages are available for dealing with quantitative data bases, and it is necessary to apply these techniques experimentally to bacterial data to establish the advantages and shortcomings of each. Particular attention should be given to the initial transformation of data (for the removal of variability due to instrument performance and sample size), as these transformations may introduce new errors into the data base.

The statistical packages contain a variety of methods for calculation of distance between OTU pairs and for formation of clusters. Using an enzyme activity data base it was found that the city-block distance measure was more appropriate than either the Euclidean distance or Pearson's correlation coefficient. Other data bases may show different results.

When used for identification, quantitative data bases appear to agree better with qualitative data bases than when used for classification. This may be related to the smaller number of characters in some quantitative data bases.

The effect of the number of attributes on classifications obtained using quantitative or mixed data bases has not been examined sufficiently. Classifications based on quantitative data are frequently compared with those obtained using qualitative characters which may express quite different parts of the genome. It is impossible to judge, *a priori,* the relative importance of qualitative and quantitative attributes for obtaining 'true' classifications. It would be possible to evaluate objectively the relative merits of qualitative and quantitative characters, only by study of a number of bacterial taxa for all characters and by applying clustering techniques to all characters and to subsamples containing qualitative, quantitative, or a mixture of both types of characters.

Acknowledgements

I thank Mr Graham Dipple for executing the computer studies, Mrs Anne Chibah for preparing the typescript, and the Medical Research Council for financial support.

References

Alvey, N. G., Banfield, D. F., Baxter, R. I., Gower, J. C., Krazanowski, W. J., Lane, P. W., Leech, P. K., Nelder, J. A., Payne, R. W., Phelps, K. M., Rogers, C. E., Ross, G. J. S., Simpson, H. R., Todd, A. D., Wedderburn, R. W. M., and Wilkinson, G. N. (1977). 'GENSTAT. A General Statistical Program'. The Statistics Department, Rothamsted Experimental Station, Harpenden, England.

Andrews, D. F. (1972). Plots of high-dimensional data. *Biometrics* **28,** 125–136.

Barnett, J. A., Bascomb, S., and Gower, J. C. (1975). A maximal predictive classification of Klebsielleae and of the yeasts. *Journal of General Microbiology* **86,** 93–102.

Barry, A. L., Gavan, T. L., Smith, P. B., Matsen, J. M., Morello, J. A., and Sielaff, B. H. (1982). Accuracy and precision of the Autobac System for rapid identification of Gram-negative bacilli: a collaborative evaluation. *Journal of Clinical Microbiology* **15,** 1111–1119.

Bascomb, S. (1980). Identification of bacteria by measurements of enzyme activities and its relevance to the clinical diagnostic laboratory. *In* 'Microbiological Classification and Identification' (Eds. M. Goodfellow and R. G. Board), pp. 359–373. Academic Press, London.

Bascomb, S. (1983). Comparison of mathematical models for identification of bacteria using quantitative characters. *In* 'Les Bacilles à Gram négatif d'intérêt médical et en Santé Publique: Taxonomie—Identification—Applications' (Ed. H. Leclerc), Colloque INSERM, pp. 423–430. INSERM, Paris.

Bascomb, S., and Grantham, C. A. (1975). Application of automated assay of asparaginase and other ammonia-releasing enzymes to the identification of bacteria. *In* 'Some Methods for Microbiological Assay' (Eds. R. G. Board and D. W. Lovelock), pp. 30–54. Academic Press, London.

Bascomb, S., and Spencer, R. C. (1980). Automated methods for identification of bacteria from clinical specimens. *Journal of Clinical Pathology* **33,** 36–46.

Bøe, B., and Gjerde, J. (1980). Fatty acid patterns in the classification of some representatives of the families Enterobacteriaceae and Vibrionaceae. *Journal of General Microbiology* **116,** 41–49.

Carmichael, J. W., and Sneath, P. H. A. (1969). Taxometric maps. *Systematic Zoology* **18,** 402–415.
Clifford, H. T., and Stephenson, W. (1975). 'An Introduction to Numerical Classification'. Academic Press, London.
Colwell, R. R., and Liston, J. (1961). Taxonomic relationships among the pseudomonads. *Journal of Bacteriology* **82,** 1–14.
Darland, G. (1975). Principal component analysis of intraspecific variation in bacteria. *Applied Microbiology* **30,** 282–289.
Dixon, W. J., and Brown, M. B. (Eds.) (1979). 'BMDP-79 Biomedical Computer Programs P-series', 2nd Printing. Univ. of California Press, Berkeley.
Drucker, D. B. (1976). Gas–liquid chromatographic chemotaxonomy. *In* 'Methods in Microbiology' (Ed. J. R. Norris), Vol. 9, pp. 51–125. Academic Press, London.
Drucker, D. B., Hillier, V. F., and Lee, S. M. (1982). Comparison of computer methods for taxonomy of some streptococci using gas chromatographic chemotaxonomic data. *Microbios* **35,** 139–150.
Duewer, D. L., Koskinen, J. R., and Kowalłski, B. R. (1975). 'Documentation for ARTHUR, Version 1-8-75', Chemometrics Society Report No. 2, updated 1981. Infometrix Inc., Seattle, Washington.
Dunn, G., and Everitt, B. S. (1982). 'An Introduction to Mathematical Taxonomy'. Cambridge Univ. Press, Cambridge.
Everitt, B. (1980). 'Cluster Analysis', 2nd Edition. Heinemann, London.
Friedman, F., and MacLowry, J. (1973). Computer identification of bacteria on the basis of their antibiotic susceptibility patterns. *Applied Microbiology* **26,** 314–317.
Godsey, J. H., Matteo, M. R., Shen, D., Tolman, G., and Gohlke, J. R. (1981). Rapid identification of Enterobacteriaceae with microbial enzyme profiles. *Journal of Clinical Microbiology* **13,** 483–490.
Gower, J. C. (1971). A general coefficient of similarity and some of its properties. *Biometrics* **27,** 857–871.
Gower, J. C. (1974). Maximal predictive classification. *Biometrics* **30,** 643–654.
Gower, J. C. (1978). Some remarks on proportional similarity. *Journal of General Microbiology* **107,** 387–389.
Hand, D. J. (1981). 'Discrimination and Classification'. Wiley, London.
Healy, M. J. R. (1968). The disciplining of medical data. *British Medical Bulletin* **24,** 210–214.
Jantzen, E., Bergan, T., and Bøvre, K. (1974). Gas chromatography of bacterial whole cell methanolysates VI. Fatty acid composition of strains within Micrococcaceae. *Acta Pathologica et Microbiologica Scandinavica, Section B: Microbiology* **82,** 785–798.
Jantzen, E., Knudsen, E., and Winsnes, R. (1982). Fatty acid analysis for differentiation of *Bordetella* and *Brucella* species. *Acta Pathologica, Microbiologica et Immunologica Scandinavica, Section B* **90,** 353–359.
Kaufman, L., and Massart, D. L. (1981). 'MASLOC Users' Guide'. Vrije Universiteit, Brussels.
Kendall, M. (1980). 'Multivariate Analysis', 2nd Edition. Charles Griffin, High Wycombe, England.
Kersters, K., and De Ley, J. (1975). Identification and grouping of bacteria by numerical analysis of their electrophoretic protein patterns. *Journal of General Microbiology* **87,** 333–342.
Kersters, K., and De Ley, J. (1980). Classification and identification of bacteria by electrophoresis. *In* 'Microbiological Classification and Identification' (Eds. M. Goodfellow and R. E. Board), pp. 273–297. Academic Press, London.

Lockhart, W. R., and Koenig, K. (1965). Use of secondary data in numerical taxonomy of the genus *Erwinia*. *Journal of Bacteriology* **90,** 1638–1644.

MacFie, H. J. H., Gutteridge, C. S., and Norris, J. R. (1978). Use of canonical variate analysis in differentiation of bacteria by pyrolysis gas–liquid chromatography. *Journal of General Microbiology* **104,** 67–74.

Marriot, F. H. C. (1974). 'The Interpretation of Multiple Observations'. Academic Press, London.

Meuzelaar, H. L. C., Kistemaker, P. G., Eshuis, W., and Engel, H. W. B. (1976). *In* 'Rapid Methods and Automation in Microbiology' (Eds. H. H. Johnston and S. W. B. Newsom), pp. 225–230. Learned Information (Europe), Oxford.

Nie, N. H., Hull, C. H., Jenkins, J. G., Steinbrenner, K., and Brent, D. H. (1975). 'SPSS: Statistical Package for Social Sciences'. McGraw-Hill, New York.

Sielaff, B. H., Johnson, E. A., and Matsen, J. M. (1976). Computer-assisted bacterial identification utilizing antimicrobial susceptibility profiles generated by Autobac 1. *Journal of Clinical Microbiology* **3,** 105–109.

Sneath, P. H. A. (1972). Computer taxonomy. *In* 'Methods in Microbiology' (Eds. J. R. Norris and D. W. Ribbons), Vol. 7A, pp. 29–98. Academic Press, London.

Sneath, P. H. A. (1977). A significance test for clusters in UPGMA phenograms obtained from squared Euclidean distances. *Classification Society Bulletin* **4,** 2–14.

Sneath, P. H. A., and Sokal, R. R. (1962). Numerical taxonomy. *Nature (London)* **193,** 855–860.

Sneath, P. H. A., and Sokal, R. R. (1973). 'Numerical Taxonomy'. Freeman, San Francisco.

Sokal, R. R., and Sneath, P. H. A. (1963). 'Principles of Numerical Taxonomy'. Freeman, London.

Stack, M. V., Donoghue, H. D., and Tyler, J. E. (1978). Discrimination between oral streptococci by pyrolysis gas–liquid chromatography. *Applied and Environmental Microbiology* **35,** 45–50.

Ware, G. C., and Hedges, A. J. (1978). A case for proportional similarity in numerical taxonomy. *Journal of General Microbiology* **104,** 335–336.

Wishart, D. (1966). 'Fortran II Programs for 8 Methods of Cluster Analysis (CLUSTAN 1)', Computer Contribution 39, State Geological Survey. University of Kansas, Lawrence.

Wishart, D. (1978). 'CLUSTAN User Manual', 3rd Edition. Edinburgh University, Edinburgh.

4

Cladistics and the Evolution of Proteins

J. WILLIAMS

Department of Biochemistry, University of Bristol, Bristol, UK

Introduction

In his book on the history of biological ideas Ernst Mayr (1982) pointed out two great revolutions which have occurred in biological systematics. The first occurred between about 1750 and 1850, as naturalists, especially in France, gradually abandoned the old Aristotelian system of classification, in which a major feature had been downward logical division based on the presence or absence of selected single characters which were supposed to represent the essential nature of the organism. Eldredge and Cracraft (1980) refer to this method as an 'A/not-A' dichotomy. Although this method was and is useful in constructing identification keys, it was fundamentally unsatisfactory for classification purposes in that the 'not-A' groups were based only on the absence of features. As belief in the doctrine of essentialism waned, naturalists began to use upward classification, grouping species together according to their degrees of similarity, as expressed in numerous characters. Nevertheless, the use of 'not-A' groups has proved hard to avoid completely, and they were not only prominent in Lamarck's taxonomic work but are still found in present-day biology. In fact there is a continuing debate about whether relationships are revealed by the totality of the phenotype or only by a subset of characters which are valid phylogenetic markers.

The second revolution occurred after 1859, with the general acceptance of Darwin's evolutionary explanation for the hierarchic pattern of groups obtained by taxonomists. In two respects this revolution, too, was incomplete. Thus, D'Arcy Thompson in this century still treated the problem of the relatedness of organisms in a manner similar to that of the pre-evolutionary nature philosophers, using systematic deformations of Cartesian coordinates to transform one structure into another rather than seeking realistic biological theories of change. More important, however, was the fact that evolutionists failed to develop a methodology for studying phylogenetic relationships. Most of such studies were carried out by palaeontologists who generally reached their conclusions by un-

ISBN 0-12-289665-3

analysable intuitive processes, and for groups with poor fossil records the theory of evolution became superfluous to the practice of taxonomy. Thus, after about 1880 a marked decline of interest in phylogenetic problems set in, and this lasted until the middle years of this century.

Since the 1950s there has been a great renewal of interest in systematics associated with two quite different approaches. The first, known variously as numerical taxonomy, phenetic taxonomy, or numerical phenetics, continued the movement away from phylogenetic studies and concentrated instead on the objective classification of species on the basis of their observed similarities and dissimilarities, without any attempt being made to weight these characteristics according to ideas of evolutionary relationships. However, with the later application of phenetic taxonomy to macromolecular sequence data there has emerged a tendency to equate phenetic resemblances and phylogenetic relationships. The other approach, now known generally as cladistics, set out with the object of defining a precise method for determining phylogenetic relationships, and effectively it dates from the publication of Willi Hennig's *Phylogenetic Systematics* in 1966, although, as with phenetic taxonomy, similar ideas had been expressed earlier (Mitchell, 1901). Hennig showed how phylogenetic relationships could be inferred without recourse to palaeontological data by making use only of carefully chosen similarities between species. In this it was diametrically opposed to phenetic taxonomy. There is therefore an interesting symmetry in the fact that later cladistic studies, undertaken with the object of correcting apparent defects in Hennig's methodology, have once more moved away from the phylogenetic goal and approached the original position of phenetic taxonomy (Patterson, 1980, 1982a,b).

These various movements have naturally caused confusion as to the aims and methods of systematists (Charig, 1982), who now themselves present a challenging taxonomic problem. Our aim in the first part of this chapter will be to describe briefly the theory of cladistics. The expression evolutionary cladistics will be used for Hennig's original theory; and transformed cladistics will refer to the later, evolution-free version. This discussion will lead us to support the former theory while recognising the difficulty of applying it to much of the traditional data of biology. The structures of such macromolecules as proteins and nucleic acids, however, offer important advantages in taxonomy, and although they have often been discussed from the standpoint of phenetic taxonomy they have rarely been considered in the context of evolutionary cladistics. Some tentative suggestions relating to this problem will be made later in the chapter.

Classification and Evolution

The relationship between classification and evolution is a central issue in the debate between evolutionary cladists and transformed cladists. The first school holds that classification is a reflection of phylogeny, but for the second school classification represents the primary activity in systematics, phylogeny being unknowable.

Evolutionary Cladistics

Evolutionary cladists and traditional evolutionists agree that the first task is to discover the phylogenetic relationships between species and the second is to convert these relationships into a taxonomy. They disagree, however, over the meaning of the term relationship and hence over the way phylogeny is to be reflected in taxonomy. For Hennig, relationship meant only genealogical relationship or kinship, and he proposed to translate these relationships directly and exactly into taxonomy. Thus, a Hennigian taxonomy is a representation of the sequence of speciation events (cladogenesis) in the history of the group under study. The practical problems raised by cladistic classifications were discussed by Eldredge and Cracraft (1980). For traditional evolutionists, such as Simpson and Mayr, relationship is a vaguer concept and includes not only kinship but also a measure of overall similarity (anagenesis): classification should reflect both these elements in order to convey the greatest amount of information. According to Simpson (1961), the classification should be consistent with the presumed phylogenetic relationships, whereas for Hennig they are one and the same thing. In the case of the salmon, the lungfish, and the cow (Fig. 1) discussed by Gardiner *et al.* (1979), cladogram (a) represents the probable phylogenetic relationship of the three species, but for the purposes of classification the traditional evolutionist would prefer cladogram (c) as being phenetically more informative. This is rejected by the cladist because the group formed by the two fish is defined only by primitive characters; in other words, they are a 'not-A' group. It is clear, especially to the outsider, that this disagreement about the translation of phylogeny into classification is a relatively trivial matter, and one can agree with Maynard Smith's refusal (1982) to get excited about it.

Transformed Cladistics

An important issue, on the other hand, is the belief by transformed cladists and some pheneticists that the object of classification is to reveal the 'natural groups'

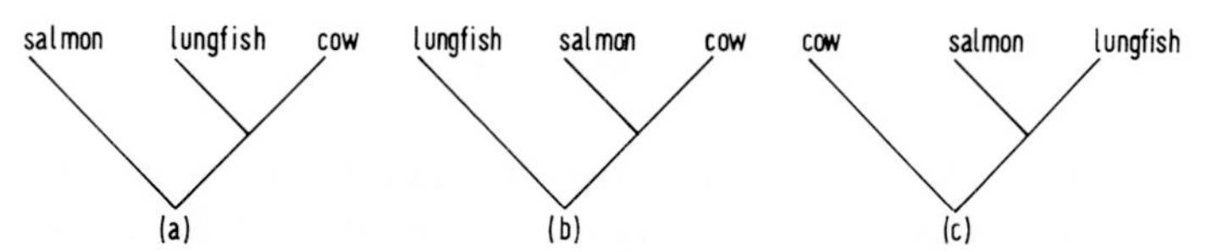

Fig. 1. Three cladograms representing the phylogenetic relationships of the salmon, the lungfish, and the cow.

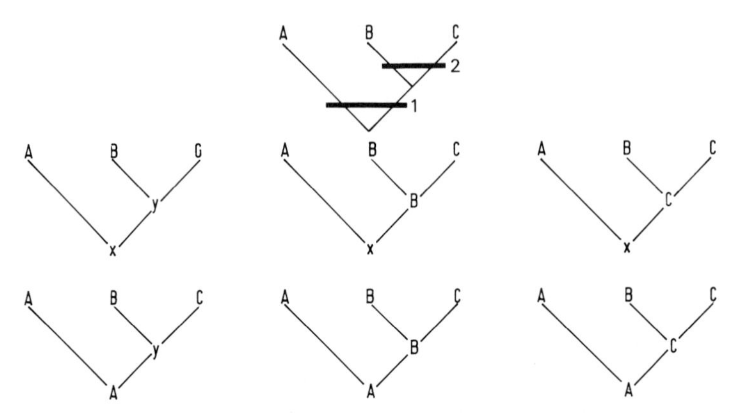

Fig. 2. A diagram showing that a cladogram is not the same thing as a phylogenetic tree. The cladogram at the top is consistent with all six trees shown below. The black bars on the cladogram indicate that species A, B, and C form a group characterized by the common possession of the homologous character 1 and that B and C form a group possessing homologous character 2.

of organisms by unbiased observation in which ideas about evolution have no part to play. The word natural is at least as misleading in this context as 'relationship' was in traditional evolutionary systematics. The cladogram is said by transformed cladists to be a scientific theory, in that it can be tested by studying more characters. If phylogenetic theories are to be entertained at all they must be based on the observed cladogram, but since such theories also require information about extinct ancestors—which is not obtainable from the cladogram—they are inherently untestable, or metaphysical, theories. For example, a given cladogram representing the relationships of three taxa could be explained in evolutionary terms by six different phylogenetic trees, and there is no way of finding which is the correct one or even whether the cladogram is to be explained by evolution at all (Fig. 2). A sharp line is drawn between theories about the pattern of nature and theories about the process by which the pattern was produced. Transformed cladists themselves think that their views represent a return to preevolutionary systematics (Patterson, 1980). Clearly, it is necessary to examine these startling claims in more detail, having first understood the nature of Hennig's original theory.

Evolutionary Cladistics

Hennig envisaged that in a given lineage of organisms a new character would somehow arise and later become modified to produce a transformation series of homologous characters such as a, a', a'', $\cdots$. Any member of the series is said to be *plesiomorphous* (or primitive, or merely earlier) with respect to later members and *apomorphous* (or derived) with respect to earlier members. When speciation occurs the character present in the stem species becomes a homologous character in the daughter species and it is the joint possession of the character by all the descendants of the stem species, which defines a monophyletic group. Such shared derived characters are called synapomorphies. In Fig. 3, (B,C) is a *monophyletic group* defined by the *synapomorphy b′*; species A and B, on the other hand, resemble one another in the joint possession of the plesiomorphous homology *a* (referred to as a *symplesiomorphy*), but they do not make a monophyletic group because species C is excluded. Groups defined only by symplesiomorphies are called *paraphyletic groups*. Species may also come to resemble one another because of having independently acquired the same character: such is the case with species A and C in Fig. 3 (character *c′*), and it is clear that the corresponding characters in the two species are not homologous. A later section will discuss whether convergent similarities can ever be considered to be in fact the same character. For the moment we will simply note that Hennig concurred with all other authors in calling groups such as (A,C) *polyphyletic*.

Thus, depending on the evolutionary history of the characters in question, resemblances between species can lead to three different kinds of groups. Of these, only monophyletic groups portray the phylogenetic relationships of the species and the other two (paraphyletic and polyphyletic) are unnatural or artificial assemblages. The task of phylogenetic analysis is therefore to discover the monophyletic groups and to arrange them hierarchically.

Hennig's treatment of monophyletic and polyphyletic groups is not controversial to evolutionary systematists, but paraphyly perhaps deserves further discussion. The concept was not invented by Hennig, nor is it necessarily linked to the theory of evolution. Richard Owen, for example, who vigorously opposed Darwinism, rejected the invertebrates as a real group because their only defining character would be absence of the notochord. Nevertheless, as Eldredge and Cracraft (1980) have shown, the 'A/not-A' dichotomy is still widespread and is by no means easy to eliminate, since it is often difficult to characterize primitive groups other than by their not possessing certain characters. One should not assume, however, that a 'not-A' group is necessarily paraphyletic, since the

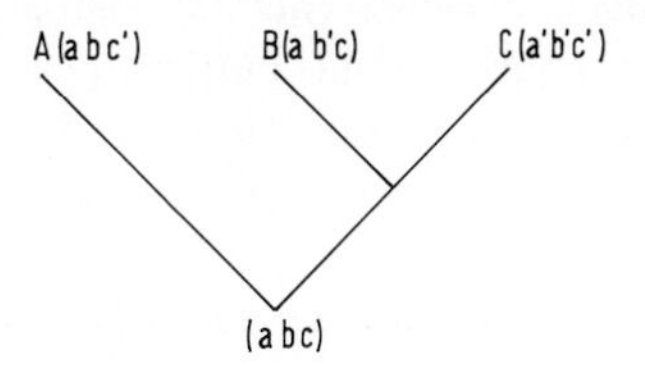

Fig. 3. A diagram showing monophyletic, paraphyletic, and polyphyletic groups. For details see the text.

secondary loss of a character within a group must be either apomorphous or convergent. Charig (1982) points out that in order fully to define a paraphyletic group, two character transformations from plesiomorphous to apomorphous states are needed, one defining an upper boundary and the other a lower boundary. These two transformations could belong to the same character-transformation series or more often to different series. Even with this qualification, paraphyletic groups remain artificial, and symplesiomorphic similarities do not contribute to knowledge of phylogeny.

The Direction of Evolution

Hennig's methodology, if it can be fully applied to any group of species, must infallibly lead to the true phylogenetic relationships between those species, if they are in fact the products of evolutionary branching and divergence. For each branch point it is necessary to have one true synapomorphy; other synapomorphies which arose in the same part of the tree are redundant. In this sense Hennig's system represents a reversal of the historical trend towards the use of many, or ideally of all, characters in systematics. The great problem is that to be able to distinguish synapomorphies from symplesiomorphies one must have information about the *direction* of the *transformation* series. Although Hennig seems to have envisaged transformation series possessing several members, it seems to be more usual in practice—especially for traditional morphological characters—to find only two members. Macromolecular sequence data, on the other hand, typically display multiple character states.

In an impressive example of cladistic analysis, Hennig (1966) showed how a reticulate pattern of brachiopod relationships which had been obtained on the basis of a mixture of apomorphous and plesiomorphous characters became a branching tree-like relationship when only those resemblances judged to be synapomorphies were used. The types of evidence on which decisions about the direction of evolution might be based include the ontogenetic sequence in which characters appear, the frequencies of different character states within and without groups, the geographic distribution of character states, and—in the case of fossils—geological levels. However, none of these approaches is necessarily applicable to a particular type of character or a particular group of organisms. For example, there is no reason to think that the ontogenetic-sequence criterion applies to macromolecular sequence data. Furthermore, it has to be admitted that proposals about ancestral character states can never be finally proven, and if the scientific status of evolutionary cladistics were to depend on absolute proof one would have to agree with its critics that it is unscientific. This criterion, however, is somewhat too sharp, since it would remove a great deal of what is generally regarded as science, without improving our knowledge of the world.

The marked difference between Hennig's biologically orientated approach and that of mathematically minded workers is clearly shown by his treatment of incongruent data. Since there is one real phylogeny, different characters, if correctly interpreted, must all yield cladograms which are consistent with it. It is well known, however, that in practice different characters are often incongruent. Hennig's advice was to recheck the interpretation of the apparent synapomorphies to make sure that the polarity of the transformation series was correct, that an apomorphy had not arisen more than once in the group, and that the characters being compared were actually homologous. If reinterpretation fails to solve the problem, then phylogenetic analysis of the group must be deferred until better data are available. The alternative approach is to find a mathematical solution which minimises the discord, and Felsenstein (1982) discussed various ways of doing this. We shall return to this question later in the chapter.

The Transformation of Cladistics

To cladists such as Patterson (1980, 1982a,b), Hennig's methodology is circular, since a cladogram is constructed using theories about evolution but the only source of such theories is the cladogram itself, as we have already seen. To escape from this tautology it is proposed that evolutionary theories are not, after all, necessary in the construction of cladograms. Thus, the evolutionary concept of homology is not necessary; it is equivalent to synapomorphy and is simply the defining characteristic of 'natural' groups. Symplesiomorphy and its attendant theory about the polarity of transformation is also unnecessary, because symplesiomorphies are simply homologies which refer to a larger group of species than the corresponding synapomorphy. It remains necessary to order homologies so as to avoid the inadvertent creation of unnatural (paraphyletic) groups, but this is said not to imply a direction of transformation but only the distinction between general and particular characters. It was von Baer who first noticed, in the early nineteenth century, that general characters precede particular characters in ontogeny, and this is held to be the best evidence on which homologies may be ordered. Finally, the concept of convergence as an evolutionary process is superfluous, because false synapomorphies of any type can only be detected when they imply incongruent cladistic relations. In place of the discarded evolutionary theories, transformed cladistics requires only the principle of parsimony according to which data are to be grouped in the simplest possible way. In this way transformed cladists claim to discover the natural order of species without the self-justifying effect of having a theory of evolution, or indeed any other kind of theory, built into the process of classification. By extension of this argument, the transformed cladist does not regard a cladogram as being simply and directly interpretable in evolutionary terms, and in order to avoid any unintended evolutionary implications it is sometimes suggested that the cladogram should be depicted as a set diagram (Venn diagram).

Before we attempt to assess the validity of these claims we should look briefly at methods recently developed by numerical taxonomists in which, again by the use of the principle of parsimony, data are grouped in a way which is held to be a direct estimation of phylogenetic relationships.

Quantitative Phyletic Taxonomy

In traditional evolutionary systematics characters are commonly weighted differentially, expressing subjective ideas about their presumed value in indicating phylogenetic relationships. Phenetic taxonomists eschew this practice and accord equal weight to all characters in the interests of objectivity. In aiming at a phenetic classification, all pretensions to express phylogenetic relationships were abandoned, at least in the early phase of phenetic taxonomy. However, under the title *quantitative phyletics* Kluge and Farris (1969) described an approach which combined both objectives. The algorithmic character of the procedure ensured objectivity and by weighting the characters prior to use according to their variability within and between taxa, it was considered that the resulting classification was an estimate of the true phylogeny. Several different types of quantitative phyletics have been developed (Felsenstein, 1982), but all involve in one form or another the principle of parsimony.

Parsimony

Ever since its introduction into evolutionary studies by Edwards and Cavalli-Sforza (1964), the principle of maximum parsimony has been a subject of controversy. In the form in which it was stated by the mediaeval scholastic William of Occam it expresses an attribute which all rational theories possess, and some of those who now support its use in phylogenetic studies argue that it is being used in this sense. However, it seems likely that matters are more complex than this and that parsimony is acting as a theory about evolution.

Felsenstein (1982) divides numerical phyletic methods into two groups: (a) parsimony or minimum evolution methods in which that phylogenetic tree is chosen which entails the smallest number of changes, and (b) compatibility methods in which that phylogeny which is compatible with a majority of characters is chosen. Minimum evolution methods can be further subdivided according to the restrictions they place on the nature of the changes allowed. Thus, in so-called Wagner trees no restrictions are placed on forward or backward changes, in the Camin–Sokal method multiple forward changes are permitted for each character but no backward changes, and in Dollo trees only one forward change is allowed for each character but multiple backward changes are possible.

Several quantitative phyletic methods based on the principle of character com-

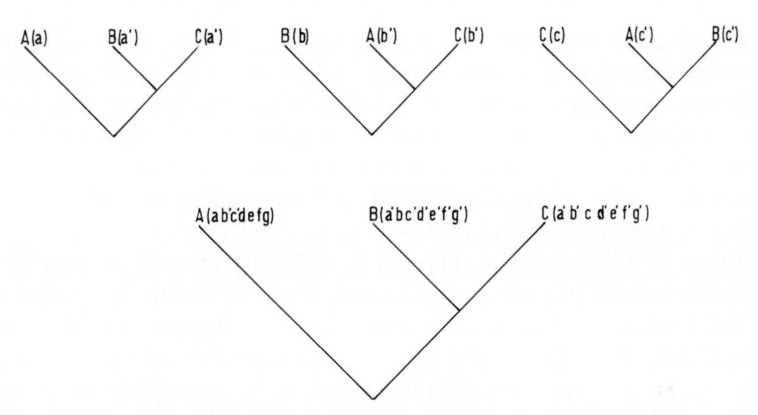

Fig. 4. The three characters *a, b,* and *c* lead to incongruent phylogenetic results. By considering more characters it may be possible to achieve a consensus result.

patibility have been developed. Wiley (1981) equates this principle with that of parsimony. Thus, if there is no evidence on which to resolve character states into plesiomorphous and apomorphous, then any character state which is shared by two of three taxa but not by the third is taken as supporting one of the three possible phylogenetic relationships (Fig. 4). Other characters may imply a different relationship, but since only one can be the true relationship (unless reticulate evolution has occurred), some method of identifying that one is needed. [The extensive occurrence of hybridization in the origin of plant taxa results in a reticulate pattern of evolution. Its relation to cladistic analysis was discussed by Nelson (1983).] This is taken to be the one supported by the most characters. It is described as the most parsimonious solution in the sense that the fewest *ad hoc* hypotheses are needed to explain away invalid shared characters. Larger taxonomic problems can be considered as being conjunctions of three-taxon problems which are to be solved so as to achieve overall parsimony.

Criticisms of Parsimony

It is unlikely that parsimony in its role as a component of rational thought has ever been the target of attack in discussions on phylogeny and statements such as 'Without the parsimony criterion, could we ever make sense of nature? [Patterson, 1980]' appear to mistake the nature of the persistent criticisms. These are directed either at the idea of minimum change as an implied theory of evolutionary change or at the way the parsimony test is applied in particular situations. Kluge and Farris (1969), in rejecting the criticism that 'the use of the parsimony criterion assumes that evolution itself is parsimonious', stated that 'a most parsimonious tree may show a large number of convergent and parallel changes, demonstrating that evolution is not parsimonious'. Nonetheless, they presented a most parsimonious tree of anuran relationships as having some phylogenetic significance, which appears to mean that they were using parsimony as a theory about evolution while at the same time not believing it to be true. Josh Billings' remark, 'The trouble with people is not that they don't know, but that they know so much that ain't so', seems appropriate. In a later section we shall discuss this question in relation to protein evolution.

Criticism of a different sort concerns an ambiguity in the application of parsimony in numerical phyletics. Any character which is incompatible with a given tree, in the sense that it demands one or more extra changes of state, could be said to be a rejection of the theory that the tree is correct. However, it is not clear whether a character which demands two extra changes implies two extra hypotheses (counting rejections) or one extra hypothesis (counting incompatible characters). The process of accepting the least falsified (rejected) theory gives the appearance of scientific validity in Popper's sense but, as Felsenstein (1982) points out, it is not obvious why a theory should be accepted if it has been rejected even once.

An interesting line of reasoning views the construction of phylogenetic trees as a statistical problem of estimating the value of an unknown, in this case the phylogeny, on the basis of resemblances between species. The most parsimonious solution can be seen as a maximum-likelihood estimate of the phylogeny, but this point of view requires that we should have estimates of the probabilities of different reconstructions being correct. This will make possible the assignment of confidence limits to the range of possible solutions. However, the required probability estimates depend on a choice of assumptions about the evolutionary process, as in the other methods of quantitative phyletics discussed.

From this brief survey of quantitative phyletics, two main conclusions may be drawn. Firstly, it is evident that the practical application of the principle of parsimony is less straightforward than might have been expected, even if biology is ignored. Secondly, all the methods proposed for the deduction of phylogenetic relationships include theoretical assumptions about evolution. Ultimately, the possibility of interpreting the past correctly rests on the validity or invalidity of these assumptions. The relationships we deduce will to some extent depend on those theoretical assumptions, and we should therefore consider the transformed cladist's claim to be able to discern the natural order without any theoretical bias.

Theory and Classification

In a much-quoted letter, Darwin wrote, 'About 30 years ago there was much talk that geologists ought only to observe and not theorize. At this rate a man might as well go into a gravel pit and count the pebbles. . . . How odd it is that anyone should not see that all observation must be for or against some view if it is to be of any service'. A familiar intelligence test involves spotting the odd man out in a list of items, and it requires that we classify the items so that all but one fall into the same group. Naturally, there are many ways of doing this, so that what is called intelligence consists here in anticipating which system of classification is required, which involves us in theories about the setters of IQ tests. In general terms, to choose one method of classification out of a number of possible methods implies the belief that that method is more suitable than the others in testing some theory about the entities classified. The method of setting up the classification must therefore be a reflection of the theory.

Does the recognition of the fact that classifications inevitably embody theoretical assumptions imply that phylogenetic studies are forever trapped in a vicious circle, as transformed cladists would have it? Hennig's discussion of this objection to phylogenetic analysis does not attempt to deny that there is an element of circularity but concentrates on the practical improvement in knowledge which results from the 'reciprocal illumination' of the parts of a system in relation to the whole system. By checking, correcting, and rechecking the different types of similarity (morphological, molecular, physiological, etc.) against the overall theory of phylogenetic relationship, an improvement is to be expected in the explanatory power of the theory.

The theoretical basis of transformed cladism does not appear to be as different from that of evolutionists as recent claims suggest. It is assumed that there exists a unique real relationship between species which can be discovered by a suitable interpretation of the pattern of synapomorphies so as to obtain maximal nesting. However, if one rejects the possibility of independent evidence apart from the tree itself, there is no way of distinguishing synapomorphies from other kinds of similarity (Sneath, 1982). Therefore, the expressions synapomorphy and monophyletic group appear to be illogical. In practice transformed cladistics boils down to a phenetic taxonomy based on maximum character compatibility. The suggestion that modern cladists have discarded evolutionary theory and rediscovered preevolutionary systematics (Patterson, 1980) fortunately does not seem to have any valid meaning, since pheneticists and evolutionists alike accept that the groupings of species result from evolution.

The conclusion to which the foregoing discussion has brought us is evident: it is inescapable that phylogenetic inferences are based on what we know (that is, on theories) about the nature of evolution. The subtlety of the different algorithms for making phylogenetic inferences should not obscure the need to get the underlying theories right, as far as this is possible.

Adoption of the view that the construction of classifications is a theory-laden process would have the advantage of clearing away much present confusion. Thus, for transformed cladists a cladogram is a theory but the method of arriving at it is theory-free: it has apparently arisen from brute 'facts'. For the realization that scientific facts do not simply exist but have to be invented we are indebted principally to Fleck (1979). On the other hand, some phenetic taxonomists also appear to hold inconsistent views. Dunn and Everitt (1982) stated that 'the numerical taxonomist does not consider a classification a theory that can be tested', although elsewhere they noted that a major difficulty in taxonomy is which similarity measure to use, since different measures may lead to different final classifications. We have already seen that this implies the existence of alternative sets of theoretical assumptions. Finally, these authors suggested that an important function of a classification is its explanatory power, which invites the comment that if a classification is not a theory it is not obvious how it could explain anything.

Macromolecular Taxonomy

The opinion that proteins and nucleic acids possess important advantages as taxonomic characters over the traditional types of characters was proposed some years ago by many writers and most effectively by Zuckerkandl and Pauling (1965), who began an important debate with Simpson (1964) about whether characters close to the level of action of the genes or distant from this level were the more useful. Simpson's preference for morphological characters was based on the idea that natural selection acts principally at a distance from the genetic level. However, there are difficulties in even defining morphological characters for taxonomy. Of the many attempts to define characters, that of Heywood (1976) is one of the most satisfactory: a character is any attribute abstracted from the total organism for the purpose of study. It makes clear the arbitrary nature of character definition and implies that many characters are likely to be more or less correlated with one another. The causes of correlation could be said to be either genetic or functional. Genes with pleiotropic effects are well known, and the phenomenon was thought by Mayr to be pervasive, for he stated that 'every character of an organism is affected by all genes and every gene affects all characters [Mayr, 1970]', although he admitted this to be a deliberate exaggeration. Functional correlations lead to character complexes in the majority of organ systems. All systems of taxonomy, however, emphasise the importance of uncorrelated characters which should be free, at least potentially, to vary independently of each other. The importance of this in relation to the statistical approach to phylogeny estimation mentioned earlier is particularly clear. Let us note in passing that the common alternative definition of character to that given by Heywood conceives a character to be any property that can vary between organisms (Sneath, 1978). This has the disadvantage of making the characters of an organism dependent on comparisons with other organisms and so of creating many 'negative' characters when widely different organisms are compared. From the evolutionary viewpoint negative characters would appear to be justified only in cases where there has been loss of a previously acquired character. A disturbing feature of both definitions, however, is that no limit is placed on the number of characters possessed by any organism, or more properly any group of organisms. This problem would disappear if characters could be equated with units of genetic information.

If we use the structures of molecules as taxonomic characters, some of these problems are less acute. Certainly protein molecules are not free from functional correlation, but genetic correlation can either be ruled out or its extent defined. The definition of protein characters is, moreover, less arbitrary than that of many morphological characters. This is not quite so obviously the case with the structure of the genes themselves, since at a gross level it would be an oversimplification to regard genes as discrete, non-interacting soloists (Mayr, 1970), and at the level of fine structure the presence of reiterated DNA sequences, regulator genes, and noncoding sequences gives rise to problems. Nevertheless, the general goal of the molecular taxonomist is a taxonomy based on the genome as expressed in the base sequence of DNA.

There was much initial optimism among biochemists that comparative data from proteins and nucleic acids would revolutionize phylogenetic studies. A notable achievement has been the delineation of the archaebacteria on the basis of ribosomal RNA sequences (Fox *et al.*, 1980), but otherwise, traditional approaches continue to dominate. This must be attributed mainly to the great difficulty of obtaining sequence data, and it is clear that progress will be much slower than was at first imagined. Even now it would not be possible to list an inventory of all the proteins, let alone their sequences, of any eukaryote organism. But for some prokaryotes there appears to be no fundamental obstacle in the way of the determination of the complete DNA structure. Nevertheless, in terms of chemical structure, the genome of even the simplest organism presents a formidable task. The chromosome of *Escherichia coli* probably comprises some 2000 genes, of which more than half have been precisely mapped. High-resolution electrophoresis methods indicate at least 1000 polypeptides in this bacterium (O'Farrell, 1975). For human beings, as many as 50,000 structural genes may be present. So far, molecular analyses have sampled only minute parts of this enormous quantity of genetic information, and this thought should obviously temper any general conclusions drawn from the results. Although the goal of molecular taxonomy appears to be utterly impracticable at present, the difficulty is technical and progress can certainly be made. Moreover, it will now be seen that progress in our understanding of the traditional taxonomic characters depends on knowledge of their genetic bases.

The Genetic Basis of Evolutionary Taxonomy

It should not be thought, from the foregoing remarks, that gene sequences represent no more than an additional class of characters which have certain possible advantages over traditional taxonomic characters. On the contrary, knowledge of gene structures is an essential element in phylogenetic analysis, even though this knowledge is very scarce at the moment.

The concept of evolutionary homology implies continuity between a character present in an ancestral species and the same character, either unchanged or more likely modified, in descendant species. Phenotypic characters, however, cannot themselves be continuous, since they must be produced afresh during the development of each organism. Only genes possess physical continuity through successive generations. Julian Huxley suggested in 1942 that problems of homology and convergent evolution required a new, genetically based systematics. On this view, homologous characters are those which are determined by homologous genes. Independent convergent modifications of a character are unlikely to involve the same genes, however closely similar the resulting structures may appear to be. It is in this genetic sense that convergent resemblances may be said to be different characters, while obviously dissimilar but homologous features may be regarded as the same character.

This genetic view of homology could run counter to the more usual phenotypic definition in the case of so-called latent homology (de Beer, 1951). For example, the parallel acquisition of nasal horns in different lines of titanotheres which diverged from a small, unhorned ancestor is an example of parallel evolution at the phenotypic level. At the genetic level, however, all the lines could have inherited the complex of genes responsible for horn development from their unhorned ancestor. The actual growth of the horns, however, required an increase in overall body size. If this interpretation is correct, the horns in the different lines are genetically homologous.

Unfortunately, there is very little known at present about the genetic control of complex phenotypic characters, but it is likely that the production of any character involves the activities of many genes. It does not seem necessary to insist that homologous characters in different organisms are determined by exactly the same complements of genes, since during evolution it is possible for some genes to be lost and others gained. de Beer (1951) noted that stocks of *Drosophila* homozygous for the *eyeless* mutation eventually give rise to some individuals with normal eyes. In these individuals, recombination of the gene complex had effectively suppressed the phenotypic effect of *eyeless,* although the *eyeless* locus itself was still fully active when placed in the wild-type gene complex. From these observations, de Beer concluded that homologous structures are not necessarily determined by homologous genes, and he therefore referred to homology as 'an unsolved problem'. This conclusion appears to be a gross exaggeration, in that a single epistatic gene may well suppress the expression of a large number of normal genes involved in the production of the eye, and other authors (e.g., Futuyma, 1979) have supported the genetic interpretation of homology.

In the great majority of cases, of course, genetic homology cannot be directly known, but it may be inferred from structural similarities between genes, including coding sequences and also the numbers, position, and structure of introns within the genes. Such resemblances will strongly suggest that the genes involved are modified representatives of a common ancestral gene.

The Evolution of Proteins

So far we have considered only the way in which the shared characters of different species can be interpreted so as to reveal phylogenetic affinities. We have seen that Hennig's analysis is the most logical approach available to this problem, and provided it is correctly applied, it must lead to a correct solution. However, the method requires knowledge about the mechanism of evolution of the characters themselves, so that primitive and derived states can be identified and permitted pathways of change between them delineated. This requirement is a stumbling block which effectively prevents application of the method in many cases, such as that of micro-organisms generally. Nevertheless, it is a virtue of Hennig's work that our attention is directed to this important problem, and the final section of this chapter will outline some aspects of the evolution of proteins which appear to be relevant to cladistic analysis.

All our ideas on the evolution of proteins are based on studies of present-day genes and proteins. The principal conclusion from these studies is that new proteins arise during evolution when an ancestral gene duplicates, and the two daughter genes subsequently diverge in structure due to the accumulation of mutations. *Duplication* is also able to bring about increase in molecular size, although in other cases it seems likely that this can result from fusing the products of different genes. The general direction of evolution is therefore likely to be from a few types of small proteins carrying out simple functions to a greater diversity of proteins, many of which are relatively large, performing complex functions. Naturally, this does not exclude possible loss of genes and proteins in particular lines. The great diversity of modern proteins has thus arisen from a few or even just one original protein, but of course it is virtually impossible to prove this belief because of the great differences which now exist between the descendant molecules. Perhaps there are 'missing links', with structures intermediate between those structures we already know, which future research will reveal, but for the moment we have only very speculative theories about the nature of ancestral proteins.

According to the *hypercycle theory* (Eigen and Schuster, 1979), the oldest template-directed polypeptide was probably an RNA polymerase with a β-pleated-sheet structure composed largely of alanine and glycine. Later proteins may have been synthetases for the production of amino acyl tRNA and ribosomal factors for translation. However, there is no structural evidence to support these suggestions. Hall *et al.* (1971) proposed that the ferredoxins are very old proteins because they occur in all types of organisms and could have carried out their electron-transport function in the oxygen-free conditions which are thought to have prevailed during the initial phases of evolution. It has also been pointed out that their limited range of amino acids resembles that found in abiotic synthesis experiments and in meteorites. It is, however, unknown whether ferredoxins or RNA polymerases could be considered to be ancestral to any other types of protein molecules.

Protein Homology

The primary evidence for homology in proteins is the existence of similarity of amino acid sequence in excess of that expected by chance alone. This similarity consists in the possession by two polypeptides of identical amino acids at matching positions, when allowances have been made for inequality of chain length by the use of gaps (Doolittle, 1981). Homology detected in this way will clearly be lost sight of as evolutionary divergence becomes greater, and many authors have endeavoured to extend the range of the method. Thus, it is well known that amino acid replacements in proteins which are definitely homologous tend to involve chemically similar amino acids, and Haber and Koshland (1970) were among the first of many authors to propose weighting factors based on ideas of chemical similarity. It would be difficult to know whether similarity which depended only on such permitted interchanges and without a significant level of identical matches could be attributed to common ancestry rather than convergent modification to support a similar function. Even more problematical is the assumption that similarity of conformation reflects homology (Ptitsyn and Finkelstein, 1980).

Myoglobin/haemoglobin and chymotrypsin/elastase are pairs of homologous proteins whose conformations are much more similar than are their amino acid sequences. Therefore, distantly related proteins might reveal their homology only in conformation and not in their sequences. This has been shown particularly clearly in the case of different types of cytochrome *c* from eukaryotes and prokaryotes (Dickerson, 1980). Although differing markedly in chain length, these cytochromes are structurally 'variations on a common theme', distinguished from one another principally in the peripheral loops, but the alignment of the amino acid sequences was an insoluble problem without a knowledge of the three-dimensional structure. In the case of the lysozymes of birds and of bacteriophage T4, parts of the three-dimensional structures near to the active sites are superimposable but no similarity is detectable in the amino acid sequences, even with the assistance of the structures. Matthews *et al.* (1981) have argued in favour of homology, but it is difficult to exclude the possibility that the two structures are analogues carrying out a common function, and have no common evolutionary ancestor. There is, however, one way in which common ancestry could be made likely: this is to search for and discover some of the postulated intermediate stages. This principle is well known in morphology as the *serial criterion of homology* (Remane, 1952). Our present difficulties may simply reflect lack of data.

According to several studies the most important element in protein conformation is the hydrophobic nucleus composed of α-helices and β-strands (Ptitsyn and Finkelstein, 1980). These aggregates of secondary elements are called *supersecondary structures,* and even in proteins of different functional types only a limited range of such arrangements has been observed. One such supersecondary structure is the so-called Rossmann fold, which consists of two adjacent βαβ units and which occurs in a number of different enzymes such as the NAD-binding domains of various dehydrogenases. Although the amino acid sequences show no similarity, Rossmann *et al.* (1975) have suggested a phylogenetic relationship to explain the structural similarity. Ptitsyn and Finkelstein (1980), on the other hand, have pointed out that the range of available secondary structures may be severely limited by the operation of simple physical principles such as the 'rule' that polar groups which are not hydrogen-bonded in the interior of the protein must be exposed to water. As they remark, the common possession of α-helices or β-pleated sheets would not be an indication of common ancestry. Such similarity is neither homologous nor convergent, but merely accidental.

At the moment it seems likely that detailed similarity of conformation is usually due to evolutionary relationship but that a merely topological similarity in the arrangement of secondary elements is not. However, it is clear that much more has to be learned about the relationship of amino acid sequence and conformation before these evolutionary problems can be resolved.

Theories of Protein Transformation

Problems of protein evolution do not end with the recognition of homologous structures, since there remains the question of defining the pathway of evolutionary change. It will be convenient to distinguish between an essentially mathematical and an essentially chemical approach to this question.

Naive Transformation Theories

Since the mid-1960s the main theory of protein evolution has consisted of a combination of the parsimony dictum with the naive assumption that proteins are nothing but strings of amino acid symbols devoid of structural or functional significance. Genealogical relationships between homologous proteins are represented by uniting existing structures in the smallest possible branching tree-like arrangement, smallness being assessed by the required number of changes in nucleotides or in amino acids. Considerable efforts have been exerted in constructing such trees, and they have had two main uses. Firstly, they have been used in taxonomy, and in the case of well-studied groups like mammals they have given results generally congruent with those obtained from traditional biological data. However, they have not proved decisive in resolving taxonomic disagreements (Romero-Herrera *et al.*, 1978). This present ineffectiveness does not disprove our earlier argument in favour of the use of proteins and genes as taxonomic characters in preference to the traditional characters. It is only relatively recently that such studies have been employed in a conscious effort to solve taxonomic problems and then only with few proteins, chosen because of their abundance and ease of purification (Goodman, 1982). Secondly, such trees have been used as evidence for conflicting theories about the evolution of proteins themselves. The theories that amino acid substitutions take place at a fairly constant rate (Wilson *et al.*, 1977) and that the rate is not constant but declines as the adaptation of a protein approaches optimum (Goodman *et al.*, 1982) have both been upheld by naive transformation trees. It seems clear that such trees are incapable of answering questions about evolutionary rates.

Despite being simplistic, this theory of protein evolution may be reasonably adequate when there are only a few changes, and these have small effects on protein conformation and function. Its main advantage may be its ability to estimate the minimum number of changes needed to transform one structure into another, but it will tend to underestimate multiple changes occurring at a site, as well as convergent changes.

An alternative approach is the theory of *stochastic evolution,* which attempts to represent the probabilistic nature of mutation and fixation. Unlike the previous theory, this method does not attempt to describe the pathway of evolution in detail; indeed this is held to be impossible, and Holmquist *et al.* (1982) commented sharply that the only ancestral states correctly inferred by the parsimony method are the invariant ones. The main use of this theory appears to be the estimation of the magnitude of the constraint imposed by natural selection on random events.

Chemical Theories of Protein Evolution

To devise theories of protein evolution which are more realistic than those referred to above, we need to take account of the nature of protein molecules. In general terms, we could characterize proteins as being complex macromolecules which carry a multitude of functional chemical groups. The internal and external interactions of these groups with one another and with water molecules and small ions cause the protein to take up a three-dimensional form which has marginal stability under normal physiological conditions. In many cases, but not in all, this conformation is nonrigid and transitions between alternative stable forms are an essential feature of the biological function of the protein. This view argues against the general use of individual amino acid sites as independent characters, although some externally directed side groups may approach this condition if they are not involved in inter-protein interactions or in the binding of ligands.

These generalisations can help us to arrive at tentative theories about the pathways of protein evolution.

The concept of *molecular stability* appears likely to be of central importance, instability being one of the commonest effects of deleterious gene mutations. By comparing the amino acid sequences of dehydrogenases and ferredoxins in mesophilic organisms with those found in thermophiles, Argos *et al.* (1980) concluded that increased thermal stability is acquired from the additive effects of many small changes which increase the hydrophobicity of internal residues and the helix-forming propensities of residues in helices and improve the packing properties of internal residues. It would be of interest to know whether these conclusions are confirmed by studies on the proteins of cryophilic organisms. In these intracellular proteins changes in stability could presumably be effected by merely changing lysine into arginine and glycine or serine into alanine.

A rather different situation appears to exist in the case of the extracellular *serine proteases* and *blood-clotting factors* (Hartley, 1974, 1979). Here we have a family of homologous enzymes with a remarkably conserved three-dimensional conformation despite many amino acid changes. The core residues, whose interactions determine the overall structure, are not strictly conserved as might have been expected. Rather, there have been multiple compensated changes so that, for example, an important hydrophobic contact in the core of one enzyme becomes replaced by a hydrogen bond in another. A detailed comparison of the packing arrangements in α-chymotrypsin and elastase is given by Sawyer *et al.* (1978). Hartley has remarked that the evolution of these proteins cannot be understood in terms of step-by-step amino acid replacements in the core regions, because the intermediate stages would probably be unable to pack in a stable way. Only when all the changes had taken place would stability be restored. Hartley resolved this dilemma by proposing that following a gene duplication one of the genes would be freed from the impact of natural selection: it would become 'silent' and thus be granted breathing space in which to acquire multiple compensated changes. Subsequent correction of the mutation which had silenced the gene would allow the multiply changed protein to be produced. Although this theory is ingenious and plausible there appears to be no evidence for it, and it may not be necessary, for it will be here suggested that there exists a mechanism whereby 'destabilising' amino acid changes can become acceptable in the evolution of extra-cellular proteins.

Disulphide Bridges in Protein Evolution

Disulphide bridges (cystine residues) are common structural elements in extracellular proteins but are only occasionally found in intracellular proteins. In a comprehensive review Thornton (1981) pointed out that in families of homologous proteins the cystines are much more strictly conserved than almost any other amino acid and that, in general, variations in cystines are only encountered when the general level of sequence similarity is low. In contrast, free thiol groups in proteins (cysteine residues) are more common in intracellular than in extracellular proteins, and—with the exception of ligand-binding cysteines—they are less well conserved than amino acid residues in general. We would expect that the disulphide bridges in extracellular globular proteins play some particularly important role, so that once formed they cannot easily be dispensed with. An important observation is that in those cases where there is variation of cystine residues the presence or absence of the bridge is not associated with any marked change in three-dimensional structure: for example, the two 'extra' disulphides of trypsin can be fitted into the structure of elastase with no distortion of the latter.

From experiments involving the reductive cleavage of disulphides in proteins, it appears that the principal function of the cross-links is to stabilise the folded structure, although they play no part in determining that structure. It is difficult to say how large this effect is. Thornton (1981) quoted an expression for the entropy reduction which results from introducing a cross-link into a disordered chain, but, in the absence of information on the entropy of the folded native structures, the entropy increase on denaturation cannot be estimated. Furthermore, the increased stability of the cross-linked structure as compared with the non-cross-linked structure is a question of the free-energy changes on denaturation and not simply a matter of entropy levels. Despite this uncertainty, the evolutionary conservation of disulphides is probably largely a result of their stabilising role. De Haën *et al.* (1975) noted that both gain and loss of disulphide bridges will be rare events, because both require two simultaneous events if the disadvantage of a free thiol group is to be avoided. The net effect will be that loss of a bridge occurs less often than a gain.

Thus, we have in disulphide bridges an indicator of the direction of evolutionary change. If a protein lacks a bridge which is present in homologous proteins, we shall regard that as representing the primitive state rather than secondary loss. The difficulty of forming a bridge leads to the conclusion that a shared disulphide is likely to be a synapomorphy (De Haën *et al.*, 1975), but the possibility that an apparently common bridge has in fact been independently acquired in different lineages must also be considered. Strict identity of the locations of a bridge in the three-dimensional structures of proteins will be convincing evidence for homology. Matching which is only approximate will suggest the possibility of convergence. All other things being equal, we may even use the criterion of parsimony to aid the distinction between homology and convergence.

Origin of Disulphide Bridges

We must now ask why new bridges appear. Since the formation of a bridge is not associated with a change in the three-dimensional structure, we may suppose that it was made selectively advantageous by changes which threatened the stability of that structure. Thus, when the disulphide-bridged structure is compared with a homologous but unbridged structure we will expect that, in addition to the bridge itself, there will be 'destabilising' changes present in the first structure. But it would probably be wrong to suppose that all these destabilising changes occurred prior to the appearance of the bridge, since the new bridge will confer excess stability which permits further potentially destabilising changes to take place. (I am grateful to Dr Hilary Muirhead for pointing this out to me.) We thus distinguish two types of destabilising events: those which preceded the bridge and were solely responsible for its appearance, and those which followed its arrival. The second type clearly play no part in favouring the formation of the bridge, but they have an important part in discouraging its loss.

It may be difficult to recognise the destabilising changes, as the following example from cytochrome *c* shows. Bullfrog cytochrome *c* is unusual in possessing a disulphide bridge (residues 20–102), the cysteine residues replacing valine and threonine, respectively (Dayhoff, 1978). Horse cytochrome *c* has no disulphides, but the bridge can be placed in its structure without deformation. The nature of the postulated destabilising changes is not clear: most likely is the tyrosine which replaces histidine-33 in bullfrog cytochrome, since this residue is very close to the position of the bridge. On the other hand, in bonito cytochrome, histidine-33 is replaced by tryptophan and this is not associated with a disulphide bridge. Unfortunately, nothing appears to be known about the effect of cleaving the disulphide on the properties of the protein. It is of interest that bakers' yeast cytochrome *c* has cysteine at position 102 and histidine at position 20. Further study of this protein would be valuable, since it represents the allegedly disadvantageous thiol intermediate in bridge formation and loss. It would be unwise to deny totally the possibility of loss of a disulphide. We can, for instance, envisage a disulphide acting as a temporary scaffold until the destabilising changes have corrected themselves and then being able to disappear without harmful consequences.

Clearly several unresolved difficulties remain, but the theory that disulphides are gained in evolution and only rarely lost could be useful in cladistic analyses of proteins. The theory makes predictions about the patterns of disulphides to be expected in proteins and about the probable deleterious effects of disulphide cleavage on the biological usefulness of proteins. It is therefore, a testable theory. We shall now briefly examine the application of this idea to the evolution of two families of proteins.

Evolution of Transferrins

Since this subject has been discussed in more detail elsewhere (Williams, 1982), we shall confine attention here to the disulphide bridge patterns. Unfortunately,

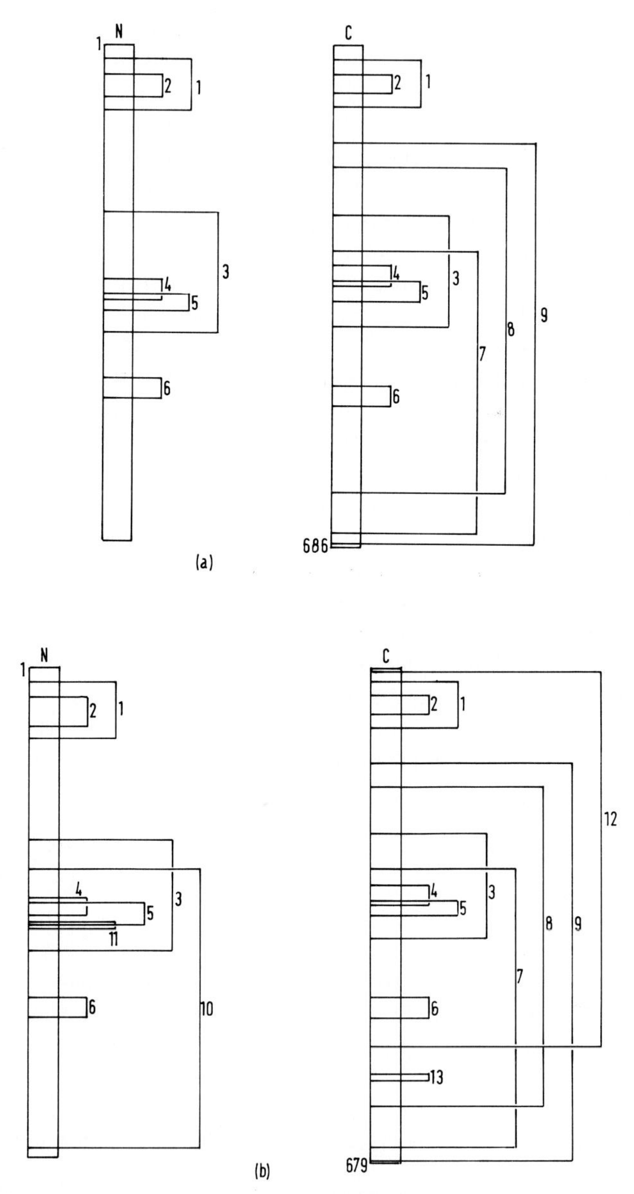

Fig. 5. Schemes showing the disulphide bridge patterns of transferrins. The bridges are numbered. (a) The N-domain and the C-domain of hen ovotransferrin. (b) The N-domain and the C-domain of human transferrin; the pattern for human lactoferrin is the same, except for the absence of disulphides 10, 11, and 12 and the presence of an unmatched half-cystine within the loop of bridge 6 in the N-domain (referred to in Fig. 6 as bridge 14).

there is only little information on the primary structure of transferrins, and our first picture may require drastic revision later. All the currently known transferrins are iron-binding proteins with a duplicate structure, in that the polypeptide chain folds up so that the N-terminal and C-terminal halves form independent globular domains which are homologous to each other. Sequence data are only available for human transferrin, human lactoferrin, and hen ovotransferrin, but each of the domains has its own unique disulphide pattern (Fig. 5). The rule that disulphides are apomorphous gives the phylogenetic relationships shown in Fig. 6. We tentatively suggest that the acquisition of disulphide number 10 in the N-terminal domain of human transferrin was independent of that of disulphide number 7 in the C-terminal domain. They occur in similar positions but their precise locations are not the same, although there will not be certainty about this until X-ray crystallographic studies have been completed.

Although there has been clear conservation of residues which are thought to be responsible for metal binding, the amino acid sequences of human and hen transferrins differ in about 50% of their residues. For myoglobin and cytochrome *c* there are about 30 and 10% differences, respectively. The amino acid sequence of transferrin is therefore highly variable, although the function of the protein is thought to be essentially the same in all vertebrates. Thus, transferrin appears to have accepted many amino acid changes, and by acquiring extra disulphide bridges it has been enabled to withstand the destabilising effects of these. In a

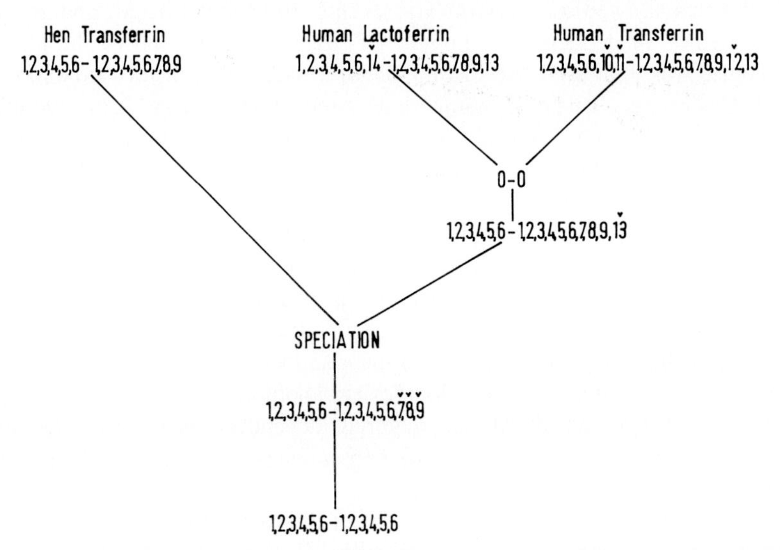

Fig. 6. A scheme of the evolution of transferrin, based on disulphide patterns. Following gene duplication of an ancestral molecule to give two domains, divergence occurred and new bridges appeared (marked by arrowheads). The gene duplication which preceded the divergence of lactoferrin and serum transferrin is indicated by 0-0.

molecule with a higher proportion of essential residues the disulphide pattern would have remained constant.

Evolution of Serine Proteases

The serine proteases are a widely distributed family of homologous enzymes. In vertebrates they are of two main types: the digestive enzymes trypsin, chymotrypsin, and elastase, secreted by the pancreas, and the blood-clotting enzymes factor Xa, plasmin, and thrombin, which are secreted by the liver. In silkworms there is the trypsin-like cocoonase, and homologous proteases also occur in bacteria. They do not appear to have been found in fungi or plants.

In structure the digestive serine proteases consist of a chain of 220 to 250 residues folded up in a characteristic conformation. In the blood-clotting enzymes the trypsin-like section is fused to a variable number of 'kringle' domains, which are involved in binding the enzyme to phospholipid membranes, and to a calcium-binding segment. The blood-clotting enzymes are therefore much larger and contain 450 to 800 residues.

Figure 7 shows the pattern of disulphide bridges in the 'protease' part of these enzymes. De Haën *et al.* (1975) noted that 'Cys events' (the formation or loss of a disulphide) are relatively rare in protein evolution and proposed a method of deducing phylogenetic relationships from such events. Pairwise comparisons of the enzymes allowed a matrix of 'Cys events' to be constructed, and an ingenious method of deducing the number of events in each limb of the possible topologies was then used. No distinction was made between forward and backward 'Cys events'. Using the numbering system shown in Fig. 7, the evolutionary scheme deduced by De Haën *et al.* (1975) is as follows: the ancestral protein was of high molecular weight (>60,000) and possessed disulphides numbers 1, 2, 3, and 7. These characteristics were retained in the hepatic branch, and some of the derivatives (e.g., plasmin) became even larger by duplication of the 'kringle' domains. In the pancreatic branch, on the other hand, there was a reduction in size but disulphide number 4 was gained. Further differentiation was marked by the gain of disulphides numbers 5 and 6 in trypsin and the independent loss of disulphide number 7 in trypsin and elastase. Alternative schemes, based on the assumption that disulphides are apomorphic, are shown in Fig. 8. One or two convergent events are required, depending on whether plasmin is associated with chymotrypsin to satisfy parsimony or with thrombin to acknowledge the presence of 'kringle' domains.

Hartley (1974) argued that comparisons of the packing of internal residues in elastase and chymotrypsin suggest the impossibility of effecting single amino acid changes without first relaxing selection pressure. On the other hand, the formation of disulphide number 7 may well have conferred sufficient excess

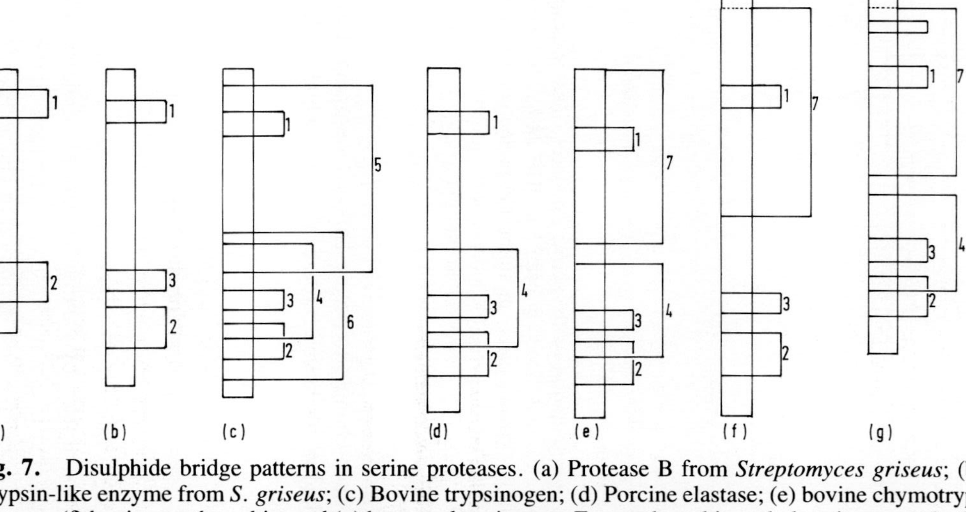

Fig. 7. Disulphide bridge patterns in serine proteases. (a) Protease B from *Streptomyces griseus*; (b) Trypsin-like enzyme from *S. griseus*; (c) Bovine trypsinogen; (d) Porcine elastase; (e) bovine chymotrypsinogen; (f) bovine prothrombin; and (g) human plasminogen. For prothrombin and plasminogen only the C-terminal segment, corresponding to the 'trypsin-like' part of the enzyme, is represented.

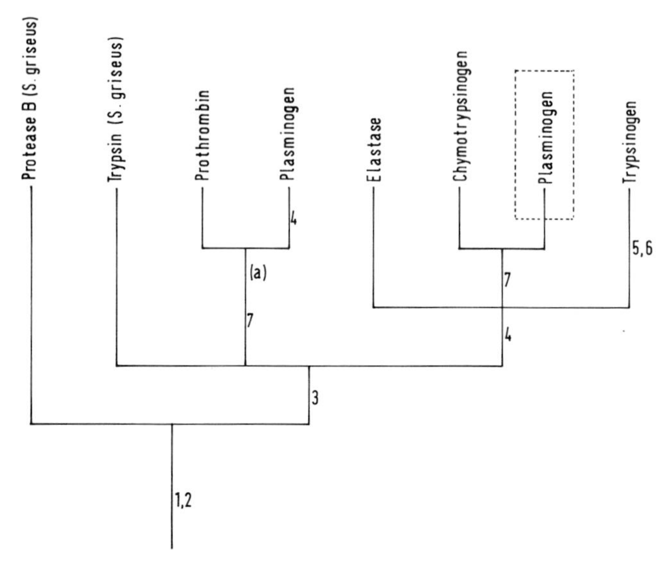

Fig. 8. Cladistic representation of the serine proteases, incorporating the assumption that disulphides, once formed, have not been lost. The dashed box around plasminogen shows a 'parsimonious' arrangement, grouping the proteolytic segment of the enzyme with the pancreatic enzymes to avoid the independent formation of disulphide number 4. Letter (a) marks the fusion of phospholipid-binding kringle domains to the proteolytic segment of the hepatic blood-clotting factors.

stability to allow such changes to occur. This is a long-range bridge, and Thornton (1981) remarked on the large decrease of entropy which such bridges cause in denatured polymers.

Conclusions

There is no infallible method by which evolutionary relationships can be discovered. The approach used by Hennig appears to be the most useful which is available at the moment in that it focuses our attention on the most fundamental problems. These concern the best definition of characters for evolutionary studies and the construction of theories which describe the transformations of these characters under the operations of mutation and natural selection. These are difficult tasks, but in the study of the biological distributions and structures of proteins there is the possibility of progress, even though absolute certainty is unattainable.

Acknowledgements

I am very grateful to the following colleagues, who have given me the benefit of their knowledge in discussions and correspondence: Drs M. Levine, H. Muirhead, C. Patterson, and P. J. Shaw, Professor P. H. A. Sneath, Dr J. M. Thornton, and Dr J. Warwicker.

References

Argos, P., Rossman, M. G., Grau, U. M., Zuber, H., Frank, G., and Tratschin, J. D. (1980). Thermal stability and protein structure. *In* 'The Evolution of Protein Structure and Function' (Eds. D. S. Sigman and M. A. B. Brazier), pp. 159–169. Academic Press, New York.

Charig, A. J. (1982). Systematics in biology: a fundamental comparison of some major schools of thought. *In* 'Problems of Phylogenetic Reconstruction' (Eds. K. A. Joysey and A. E. Friday) pp. 363–440. Academic Press, New York.

Dayhoff, M. O. (1978). 'Atlas of Protein Sequence and Structure' Vol. 5, Suppl. 3. National Biomedical Research Foundation, Washington, D.C.

de Beer, G. R. (1951). 'Embryos and Ancestors'. Oxford Univ. Press, London.

De Haën, C., Neurath, H., and Teller, D. C. (1975). The phylogeny of trypsin-related serine proteases and their zymogens. *Journal of Molecular Biology* **92,** 225–259.

Dickerson, R. E. (1980). The cytochromes *c:* an exercise in scientific serendipity. *In* 'The Evolution of Protein Structure and Function' (Eds. D. S. Sigman and M. A. B. Brazier) pp. 173–202. Academic Press, New York.

Doolittle, R. F. (1981). Similar amino acid sequences: chance or common ancestry? *Science* **214,** 149–159.

Dunn, G., and Everitt, B. S. (1982). 'An Introduction to Mathematical Taxonomy' Cambridge Univ. Press, Cambridge.

Edwards, A. W. F., and Cavalli-Sforza, L. L. (1964). Reconstruction of evolutionary trees. *In* 'Phenetic and Phylogenetic Classification' (Eds. V. H. Heywood and J. McNeill), Systematics Association Publication No. 6, pp. 67–76.

Eigen, M., and Schuster, P. (1979). 'The Hypercycle'. Springer, Berlin.

Eldredge, N., and Cracraft, J. (1980). 'Phylogenetic Patterns and the Evolutionary Process'. Columbia Univ. Press, New York.

Felsenstein, J. (1982). Numerical methods for inferring evolutionary trees. *The Quarterly Review of Biology* **57,** 379–404.

Fleck, L. (1979). 'Genesis and Development of a Scientific Fact'. Univ. of Chicago Press, Chicago.

Fox, G. E., Stackebrandt, E., Hespell, R. B., Gibson, J., Maniloff, J., Dyer, T., Wolfe, R. S., Balch, W., Tanner, R., Magrum, L. J., Zablen, L. B., Blakemore, R., Gupta, R., Leuhrsen, K. R., Bonen, L., Lewis, B. J., Chen, K. N., and Woese, C. R. (1980). The phylogeny of prokaryotes. *Science* **209,** 457–463.

Futuyma, D. J. (1979). 'Evolutionary Biology'. Sinauer Associates, Inc., Sunderland, Massachusetts.

Gardiner, B. G., Janvier, P., Patterson, C., Forey, P. L., Greenwood, P. H., Miles, R. S., and Jefferies, R. P. S. (1979). The salmon, the lungfish and the cow: a reply. *Nature (London)* **277,** 175–176.

Goodman, M. (1982). 'Macromolecular Sequences in Systematics and Evolutionary Biology'. Plenum, New York.
Goodman, M., Romero-Herrera, A. E., Dene, H., Czelusniak, J., and Tashian, R. E. (1982). Amino acid sequence evidence on the phylogeny of primates and other eutherians. *In* 'Macromolecular Sequences in Systematic and Evolutionary Biology' (Ed. M. Goodman) pp. 115–191. Plenum, New York.
Haber, J. E., and Koshland, D. E. (1970). An evaluation of the relatedness of proteins based on comparison of amino acid sequences. *Journal of Molecular Biology* **50,** 617–639.
Hall, D. O., Cammack, R., and Rao, K. K. (1971). Role for ferredoxins in the origin of life and biological evolution. *Nature (London)* **233,** 136–138.
Hartley, B. S. (1974). The evolution of enzymes. *In* 'Evolution in the Microbial World' (Eds. M. J. Carlile and J. J. Skehel) pp. 151–182. Cambridge Univ. Press, Cambridge.
Hartley, B. S. (1979). Evolution of enzyme structure. *Proceedings of the Royal Society of London, Series B* **205,** 443–452.
Hennig, W. (1966). 'Phylogenetic Systematics'. Univ. of Illinois Press, Urbana.
Heywood, V. H. (1976). 'Plant Taxonomy', 2nd Edition, The Institute of Biology's Studies in Biology, No. 5. Arnold, London.
Holmquist, R., Pearl, D., and Jukes, T. H. (1982). Nonuniform molecular divergence. *In* 'Macromolecular Sequences in Systematic and Evolutionary Biology' (Ed. M. Goodman) pp. 281–315. Plenum, New York.
Huxley, J. S. (1942). 'Evolution. The Modern Synthesis'. Allen & Unwin, London.
Kluge, A. G., and Farris, J. S. (1969). Quantitative phyletics and the evolution of anurans. *Systematic Zoology* **18,** 1–32.
Matthews, B. W., Remington, S. J., Grutter, M. G., and Anderson, W. F. (1981). Relation between hen egg-white lysozyme and bacteriophage T4 lysozyme: evolutionary implications. *Journal of Molecular Biology* **147,** 545–558.
Maynard Smith, J. (1982). 'Evolution Now. A Century after Darwin'. MacMillan, London.
Mayr, E. (1970). 'Populations, Species and Evolution'. Harvard Univ. Press, Cambridge, Massachusetts.
Mayr, E. (1982). 'The Growth of Biological Thought'. Harvard Univ. Press, Cambridge, Massachusetts.
Mitchell, P. C. (1901). On the intestinal tract of birds, with remarks on the valuation and nomenclature of zoological characters. *Transactions of the Linnean Society* (*Zoological Series 2*) **8,** 173–275.
Nelson, G. (1983). Reticulation in cladograms. *Advances in Cladistics* **2,** 105–111.
O'Farrell, P. H. (1975). High resolution two-dimensional electrophoresis of proteins. *Journal of Biological Chemistry* **250,** 4007–4021.
Patterson, C. (1980). Cladistics. *Biologist* **27,** 234–240.
Patterson, C. (1982a). Cladistics and classification. *New Scientist* 29th April, 303–306.
Patterson, C. (1982b). Morphological characters and homology. *In* 'Problems of Phylogenetic Reconstruction' (Eds. K. A. Joysey, and A. E. Friday), pp. 21–74. Academic Press, London.
Ptitsyn, O. B., and Finkelstein, A. V. (1980). Similarities of protein topologies: evolutionary divergence, functional convergence or principles of folding? *Quarterly Reviews of Biophysics* **13,** 339–386.
Remane, A. (1952). 'Die Grundlagen des natürlichen Systems, der vergleichende Anatomie und der Phylogenetik'. Geest und Portig, Leipzig.

Romero-Herrera, A. E., Lehmann, H., Joysey, K. A., and Friday, A. E. (1978). On the evolution of myoglobin. *Philosophical Transactions of the Royal Society of London, Series B* **283,** 61–163.

Rossman, M. G., Liljas, A., Branden, C.-I., and Banazak, L. J. (1975). Evolutionary and structural relationships among dehydrogenases. *In* 'The Enzymes', 3rd Edition, (Ed. P. D. Boyer), Vol. 11, pp. 61–102. Academic Press, New York.

Sawyer, L., Shotton, D. M., Campbell, J. W., Wendell, P. L., Muirhead, H., Watson, H. C., Diamond, R., and Ladner, R. C. (1978). The atomic structure of crystalline porcine pancreatic elastase at 2.5 Å: comparisons with the structure of α-chymotrypsin. *Journal of Molecular Biology* **118,** 137–208.

Simpson, G. G. (1961). 'Principles of Animal Taxonomy'. Columbia Univ. Press, New York.

Simpson, G. G. (1964). Organisms and molecules in evolution. *Science* **146,** 1535–1538.

Sneath, P. H. A. (1978). Classification of microorganisms. *In* 'Essays in Microbiology' (Eds. J. R. Norris and M. H. Richmond). Wiley, Chichester.

Sneath, P. H. A. (1982). Review of *Systematics and Biogeography,* by G. Nelson and N. Platnick. *Systematic Zoology* **31,** 208–217.

Thornton, J. M. (1981). Disulphide bridges in globular proteins. *Journal of Molecular Biology* **151,** 261–287.

Wiley, E. O. (1981). 'Phylogenetics: The Theory and Practice of Phylogenetic Systematics'. Wiley, New York.

Williams, J. (1982). The evolution of transferrin. *Trends in Biochemical Sciences* **7,** 394–397.

Wilson, A. C., Carlson, S. S., and White, T. J. (1977). Biochemical evolution. *Annual Reviews of Biochemistry* **46,** 573–640.

Zuckerkandl, E., and Pauling, L. (1965). Evolutionary divergence and convergence in proteins. *In* 'Evolving Genes and Proteins'. (Eds. V. Bryson and H. J. Vogel), pp. 97–166. Academic Press, New York.

5

Computer-assisted Analysis of Data from Co-operative Studies on Mycobacteria

L. G. WAYNE

Tuberculosis Research Laboratory, Veterans Administration Medical Center, Long Beach, California, USA, and Department of Microbiology and Immunology, California College of Medicine, University of California, Irvine, California, USA

Introduction

When I first became involved in classification of mycobacteria about 30 years ago, most efforts were directed towards developing virulence tests, which were intended to distinguish 'tubercle bacilli' from all other mycobacteria. These efforts were soon confounded by reports of cases of human disease that were indistinguishable from tuberculosis, but associated with mycobacteria that appeared to be quite different from classical tubercle bacilli in culture. The seminal paper of Timpe and Runyon in 1954, in which the so-called atypical mycobacteria were divided into four groups (referred to as Runyon's groups) on the basis of growth rate and pigment pattern, made it clear that no single feature could be used to determine whether a given mycobacterial isolate was the etiologic agent of the disease of the patient from whom it was recovered.

Tuberculosis is an indolent disease, mycobacteria are indolent organisms, and, in self-defence, mycobacteriologists also tend to become indolent, or at least very patient. However, the problems raised by the recognition of the clinically significant atypical mycobacteria jarred us out of our indolence and propelled us into the use of numerical taxonomic methods rather early after their introduction (Sneath, 1957).

Initial attempts to establish quantitative histogram profiles of classical tubercle bacilli and of the members of Runyon's groups with the few available tests were promising, but this approach became unwieldy as data accumulated. Ruth Gordon had already concluded that a monothetic approach was no longer appropriate to bacterial classification, and had effected a major consolidation of the rapidly growing species of the genus *Mycobacterium,* that is, those that require only 2–5 days for visible growth from dilute inocula (Gordon and Smith, 1953, 1955; Gordon and Mihm, 1959). She used a large series of tests that were based on

ISBN 0-12-289665-3

techniques in common use for other bacterial genera. In 1962, Bojalil and colleagues published (Bojalil *et al.*, 1962) the first computer-assisted numerical taxonomic (NT) analysis of the genus *Mycobacterium*. They included not only the rapid growers, which had been studied mainly by soil microbiologists, but also the slow growers which were of interest to clinical microbiologists. Bojalil and colleagues used the same types of tests as had Gordon, and their NT analysis confirmed Gordon's intuitive conclusions about the classification of rapid growers. On first glance they appeared to have been successful with slow growers as well. However, Bojalil *et al.* (1962) had used the Jaccard coefficient, in which negative matches were excluded. Since the slow growers gave almost uniformly negative responses to most of the tests employed, the actual similarity scores used to classify the slow growers were based on very few features and, in fact, reflected a simple reiteration of the Runyon groups.

By this time a number of investigators had recognized that new tests would have to be developed for classifying slow growers. As each pursued his favourite approach, divergent classifications began to evolve. The need for coordinated efforts became apparent (Wayne, 1964), and the time was ripe for cooperative studies.

Co-operative Studies

Development and Organisation

In the earlier stages of my own NT research (Wayne, 1967), we had very limited access to computers and to sophisticated programming; although all strain pair comparisons were done by computer, strain sequencing or clustering was done by a manual ordering of data cards and trial-and-error repetitive generation of triangular matrices. In some regards I feel that manual clustering may still be a useful supplement to computer-assisted clustering.

As primitive as they were, our initial experiences, which included some success in the recognition and description of several species within each of Runyon's Groups II and III, were sufficient to convince a number of investigators that a pooling of data derived from their new tests might be fruitful.

The International Working Group on Mycobacterial Taxonomy (IWGMT) initiated its first co-operative studies in 1967. The full set of guiding policies have been described elsewhere (Wayne, 1981), and here only some key features will be mentioned in the presentation of our experiences. The participants included both basic scientists and individuals whose primary responsibility was to provide diagnostic clinical services. Therefore, a permissive approach was adopted towards accession of data. Replicate sets of 50 to 100 strains selected for each study were sent to each of the participants, all of whom had agreed to examine them according to the technical procedures that they preferred and to forward their data to the coordinator of that study. No attempt was made to dictate which tests should be run or how they were to be carried out. The basic scientists had their own ideas about each test, and the people in the service laboratories had set routines that could not be changed easily. In retrospect the permissive approach appears to have been correct, since it ensured maximum participation, with relatively few dropouts over the course of 15 years. There has of course been some turnover of participants, but that has been the result largely of retirement and job relocation rather than loss of interest.

Initial Co-operative Studies (Restricted Set)

The first eight studies involved a total of 49 laboratories in 19 countries, with an average of approximately 75 strains per study. Four of these involved slowly growing mycobacteria, including the *M. tuberculosis* complex and Runyon's Groups I–III (Wayne *et al.*, 1971; Meissner *et al.*, 1974; Wayne *et al.*, 1978; Kleeberg *et al.*, in preparation); two studies dealt with rapid growers (Kubica *et al.*, 1972; Saito *et al.*, 1977), and two were directed towards clarifying relationships between members of the genus *Mycobacterium* and a number of so-called allied genera (Goodfellow *et al.*, 1974; Goodfellow *et al.*, in preparation). I shall limit the remainder of my comments to studies of the slow growers, mentioning only that experiences with the rapid growers were similar, but that attempts to use the same panels of tests for members of different genera have led to poor resolution into stable groups, and that molecular and chemotaxonomic approaches appear to offer the best promise at the inter-generic level (Goodfellow *et al.*, personal communication).

In recent years, Dr. Micah Krichevsky has generously provided us access to highly sophisticated programming and computer equipment, and we have been able to consolidate data from the various cooperative studies and generate pooled NT matrices based on modal data. In deciding whether data from different laboratories, employing different techniques and even different criteria for scoring a given test positive, can be combined to create modal scores, it is critical that a computer specialist and a laboratory investigator with expertise in the area under investigation, work very closely together. The composite matrices derived from the results of the four initial studies on slow growers (Fig. 1) illustrates some of the problems encountered in mycobacterial taxonomy.

To begin with, considering the matrix in Fig. 1, in which symbols represent matching (M) scores of 60% M or greater, it is evident that the slow growers resolve into two superclusters, corresponding to the '*M. tuberculosis* complex' and 'all others'. When the cutoff is raised to 80% M the TB complex resolves into subclusters that appear to correspond to the hierarchic levels of named species in the other group. However, information from DNA homology studies

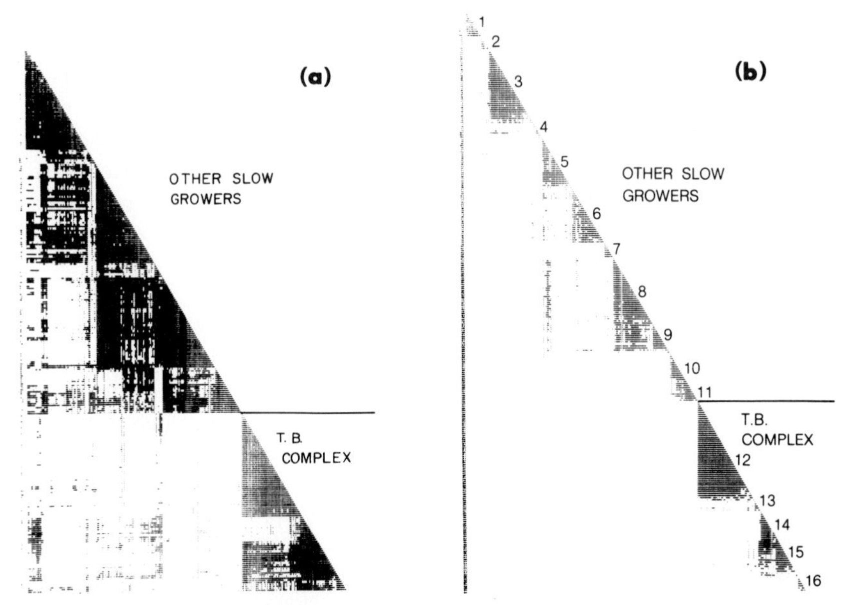

Fig. 1. Numerical taxonomy matrix of members of the tuberculosis complex and of other species of slowly growing mycobacteria. Symbols in (a) represent inter-strain matching scores of 60% or greater. Symbols in (b) represent inter-strain matching scores of 80% or greater. Clusters correspond to (1) *Mycobacterium gastri,* (2) *M. marinum;* (3) *M. kansasii;* (4) *M. flavescens;* (5) *M. scrofulaceum;* (6) *M. gordonae;* (7) *M. xenopi;* (8) *M. intracellulare;* (9) *M. avium;* (10) *M. terrae;* (11) *M. nonchromogenicum;* (12) *M. tuberculosis;* (13) *M. microti;* (14) *M. bovis;* (15) *M. africanum,* and (16) bacillus Calmette–Guérin (BCG). From Wayne (1982), reproduced with permission of the *American Review of Respiratory Disease.*

(Baess and Weis-Bentzon, 1978; Baess, 1979), immunodiffusion analysis (Stanford and Grange, 1974), and immunologic distance measurements among mycobacterial catalases (Wayne and Diaz, 1979, 1982) all provide strong reasons for concluding that the TB complex represents a single species, with *M. tuberculosis, M. bovis, M. microti,* and the ill-defined *M. africanum* representing biovars or subspecies at most. The appearance of superclusters at 60% M and of subclusters of the TB complex at 80% M both reflect the traditional preoccupation of mycobacteriologists with *M. tuberculosis* and *M. bovis* (Wayne, 1964). *Mycobacterium tuberculosis* is far and away the most important and prevalent pathogen among the cultivable mycobacteria; consequently major emphasis was long placed on development of tests that could distinguish this organism from all others. This introduced a bias into the tests that were selected by participants in the cooperative studies, and contributed to the appearance of the superclusters. Furthermore, when *M. tuberculosis* and *M. bovis* were considered to be the only important human pulmonary pathogens, emphasis was placed on developing tests which would distinguish between the two, because the methods for control and eradication of disease caused by these two pathogens were different. This, in turn, contributed a bias in test selection that amplified the differences between the subclusters in the TB complex, so that these subclusters appeared to be equivalent to species among the other slow growers. Most basic scientists who work with mycobacteria now agree that *M. tuberculosis* and *M. bovis,* along with the intermediate taxa mentioned earlier, belong in one genetic species. However, they have shown restraint in formally proposing reduction to a single species, largely because of the confusion it would cause in clinical circles. I have found that my comments on the unity of this complex at meetings exert a chemotactic effect on veterinarians in the audience, who converge on me in horror or rage at the thought that *M. bovis* might simply disappear into the senior species *M. tuberculosis. Mycobacterium bovis* will not disappear, for the reasons outlined above, although nomenclatural adjustments concerned with species status may occur.

Bacteriologists who work with pathogens are constantly faced with a conflict between practical nomenclatural stability and scientific purity. The problem arises again but in a somewhat different form as we consider the two species *M. avium* and *M. intracellulare,* clusters 8 and 9 in the matrix shown in Fig. 1b. These two species do not resolve at 80% M, but do so at 85% M (Meissner *et al.*, 1974). However, of all the features examined, only eight yielded modal frequencies greater than 50% for one of these two species and less than 50% for the other. In fact, only three features—aryl sulfatase, nitrite reduction, and growth on propanol—exhibited a *difference* in modal frequencies of 70% or greater between the two species (Meissner *et al.*, 1974). This was the lowest discriminatory power we considered to be of any use for identification of strains.

Certain types of tests are usually excluded from our NT analyses. In the case of exclusionary tests like serotyping, where usually only 1 of 20 or more possible character states will be found in a given strain, the result would be greatly diluted in impact. It is useful also to reserve some kinds of tests to see how an independent data set correlates with the results of phenetic analysis (Jones and Sneath, 1970).

Good agreement is seen between agglutinating serovar and thin-layer chromatography patterns of surface lipids (Meissner *et al.*, 1974; Jenkins, *et al.*, 1971). Most strains of *M. avium* fall into avian serovars 1–3, and most strains of *M. intracellulare* fall into one of approximately 25 other serovars (McClatchy, 1981). The agreement between thin-layer chromatography and seroagglutination is explained by the observation of Brennan *et al.* (1978) that low molecular weight polar peptidoglycolipid surface antigens are responsible for specificity of the agglutination reaction. These techniques provide a means for making very fine subdivisions within the *M. avium–M. intracellulare* complex, but do not help resolve the question of whether *M. avium* and *M. intracellulare* are separate species. Using immunodiffusion techniques, Stanford and Grange (1974) could find no significant differences in antigenic composition between cell extracts of *M. avium* and *M. intracellulare,* and consequently concluded that they represent a single species. Our own studies on immunologic distances between the T-type catalases from *M. avium* and *M. intracellulare* show little structural divergence, thus supporting the unity of the two taxa (Wayne and Diaz, 1979). Baess (1979), on the other hand, has presented evidence for two DNA homology groups within the complex, thus supporting continued recognition of the two species perhaps with reassignment of some of the serotypes.

The ambiguity of the data led some of the participants in the IWGMT Group III Cooperative Study (Meissner *et al.*, 1974) to disagree with the recommendation to reduce *M. intracellulare* to synonymy with *M. avium.* This highlights the importance of another of the IWGMT policies. On entering a study, all participants agree that once their data have been entered, they may not be withdrawn. In a project conducted in a single laboratory, an investigator may choose to defer publication when the findings are in some way dissatisfying. However, when a co-operative study has been completed, if the conclusions based on the whole data set are in conflict with the opinions of one of the participants, the whole study would be disrupted if that participant's data were withdrawn. Therefore, the members have the option of withdrawing their names from authorship of reports, but not of withdrawing their data. A protection for participants whose data may be in conflict with corresponding data from other laboratories is the policy of not identifying specific data in terms of the individual contributor. This protection is needed to compensate for the loss of the power to defer publication that can usually be exercised by an individual investigator.

In the case of the IWGMT study which dealt with *M. avium* and *M. intracellulare,* a sizable minority of members objected to the recommended synonymy. This led to a decision, which I understand became the occasion for some hilarity in the British scientific press, to incorporate in the published paper a minority statement that took issue with the conclusion in the body of the paper.

The studies on which this pooled data matrix (Meissner *et al.*, 1974) was based were all of a restricted type: almost all of the strains examined corresponded to recognized species for which many representatives were available, and once a set of strains had been analyzed, the study was terminated. These studies provided a very broad data base for description and circumscription of the more commonly encountered species. We are now conducting open-ended studies, which do not terminate but simply expand. They are designed to accommodate, by continuous accession and distribution, strains of uncommonly encountered and/or poorly defined named species as well as strains that are not recognizably members of any named species. Although this open-ended study is permissive, allowing members to choose their own test precedures, we did ask that each participant advise us from the start as to which tests were to be employed, and to avoid changing procedures.

Current Co-operative Studies (Open-ended)

In order to try to relate the results of these investigations to those of the earlier ones, strains from the earlier studies were selected as markers for recognized taxa and the available data on these, as well as from strains that had not linked into clusters, were transferred to the open-ended study file. This was not successful because differences in the tests selected for use in the two series of studies caused marked distortions in the linkage levels (Wayne *et al.*, 1981). Therefore, the actual strains of interest from the earlier studies were recovered, recoded, and distributed as unknowns along with new accessions for the open-ended study. We do not yet have data on the reexamined marker strains, but the open-ended study has provided useful information on several taxa that were not represented in the first studies (Wayne *et al.*, 1981, 1983).

The primary motivation for initiating the open-ended study was the difficulty that medical microbiologists had in identifying occasional isolates that deviated from the expected results in one of the key properties useful for distinguishing *M. scrofulaceum* from the *M. avium–M. intracellulare* complex. Most strains of atypical mycobacteria that fail to hydrolyze Tween 80 within 10 days fell into one of these species. They were distinguished from one another by the scotochromogenicity, high catalase activity, and urease reactions of *M. scrofulaceum.* When a strain was either positive or negative for all three of these features, an identification could be made with confidence. When a mixed pattern was observed, a decision became difficult and Hawkins proposed (1977) the designation '*M. avium–intracellulare–scrofulaceum* (MAIS) intermediate' as a temporary expedient until the taxonomy of such organisms could be worked out. The earliest accessions into the open-ended cooperative study of slow growers included many of these forms.

Of 49 cultures that were entered into the open-ended study up to the time of preparation of the second report (Wayne *et al.*, 1983), 11 fell into a cluster that could be identified as *M. simiae* (Fig. 2), a species that had been previously poorly defined. This cluster embraced a tight subcluster, consisting mainly of strains that agglutinated with *M. avium* serovar 18 antiserum, and a looser subcluster, members of which yielded erratic seroagglutination results, but included the '*simiae* 1' serovar. It is of some interest that all strains of serovar 18 exhibited negative niacin tests and all others were positive. This helps to explain why serovar 18 was originally designated as belonging in the '*M. avium* complex', since a positive niacin reaction was one of the key features on which the original definition of *M. simiae* was based (Karasseva *et al.*, 1965). We now see that niacin-negative forms, possibly representing a subspecies, do exist, and

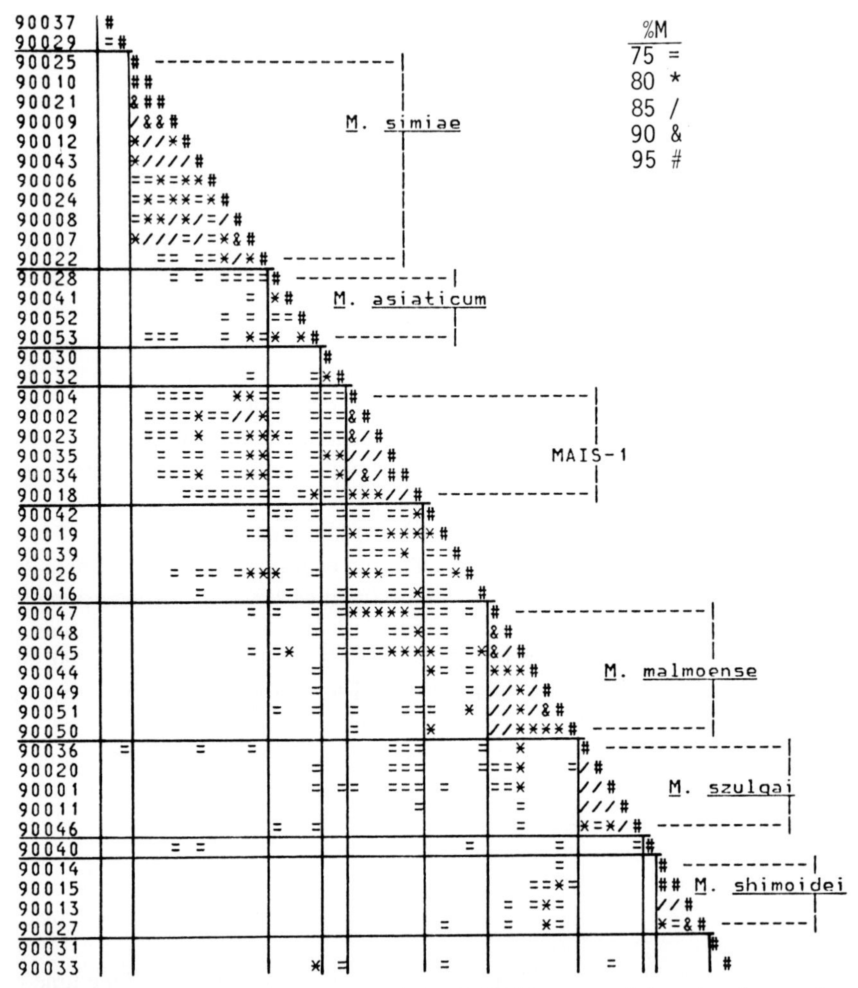

Fig. 2. Numerical taxonomy matrix of the IWGMT open-ended study of unusual or infrequently encountered strains of slowly growing mycobacteria. From Wayne *et al.* (1983), reproduced with permission of the *International Journal of Systematic Bacteriology*.

they need not be confused with the '*M. avium* complex', nor need they be considered 'MAIS intermediates'.

A very loose cluster, which linked best to the marker strain for *M. asiaticum,* is seen on the NT matrix (Fig. 2), but the internal match is too low to clarify the definition of this species.

Six strains formed a tight cluster that linked to the *M. intracellulare* marker strain at 82.4% M, and these, which are labelled 'MAIS-1', may be considered as strains of *M. intracellulare* which deviate from the norm for the species.

The remaining three clusters recognized in this study would not be considered MAIS intermediates because they are made up of strains which hydrolyze Tween 80. The *M. malmoense* cluster is very homogeneous and is composed of disease-causing strains from a restricted region of Europe. This cluster will probably become less compact as strains from other geographic regions enter the study (R. C. Good, personal communication). The cluster embracing the human pathogen *M. szulgai* is quite compact, and this study provides features, in addition to thin-layer chromatography (Marks *et al.*, 1972), on which to base a practical differentiation from *M. flavescens,* a nonpathogen, which it resembles closely. Of the four cultures in the *M. shimoidei* cluster, three were isolated at various times from the same patient in Japan, but the fourth strain was from a patient in Australia. All members of this cluster differ in many respects from those of any other cluster, suggesting the validity of the species, but development of a predictively useful description must await the isolation and examination of more strains of *M. shimoidei.*

Reproducibility of Results

Thus far the discussion has been limited to what we refer to as horizontal NT studies, that is, those based on examinations of large numbers of strains in a wide variety of tests, whether of the restricted-set type, or the more recent open-ended studies. Because of the permissive policy on test selection, frequency data on well over 100 features became available for most clusters, and some insight was gained into the inter-laboratory consistency of results of some of the more popular tests. It was only after completion of the first three of the horizontal studies dealing with slow growers that projects were undertaken to confirm reproducibility of the more useful differential tests that had emerged.

By determining modal feature results (i.e., results obtained by the majority of the participants) for each strain in each test that was performed in three or more laboratories, we were able to determine which laboratories yielded results that were most frequently in agreement with the majority. These laboratories were invited to submit detailed protocols for the test procedures. All laboratories were then invited to participate in studies of small numbers of strains, with each investigator agreeing to perform the selected tests exactly as described in the protocols. Of 40 tests examined in the two reproducibility studies completed to date, 11 tests yielded inter-laboratory agreement in excess of 90% (Wayne *et al.*, 1974, 1976). In the remaining tests, certain laboratories often exceeded 90% agreement with the majority findings. The bimodal distributions of agreement scores suggested either deviation from the protocols or the use of unsatisfactory reagents in the low-scoring laboratories. Our continuing co-operative studies remain permissive in that no one is required to follow the protocols devised for the reproducibility studies. However, the availability of these protocols makes it possible to evaluate modified or simplified diagnostic procedures in terms of their agreement with standard methods.

In addition to the formal studies on test reproducibility, we do on occasion analyze the results of the permissive studies in such a way as to be able to advise each participant in confidence on how his or her performance compares to that in other laboratories in key tests. In several instances, laboratories that had performed significantly less well than most others were then able to identify the sources of their difficulties and upgrade the reproducibility of their results in subsequent studies. This is a bonus which helps ensure continued enthusiastic participation in the long-range programme.

All strains used in all of the IWGMT studies are distributed simply with code numbers. When a study is completed, the strains are identified by these code numbers in reports and publications, and no further data on the strains so identified are added to the NT data files. If further study of a strain is needed, it is recoded for blind redistribution.

One feature of our agreement with participants is that anyone who has contributed data to any of our studies can request any or all raw data from the files of any of the completed studies to be used for any supplementary analysis that may come to mind. At that point the only secrecy that continues to be maintained is the identity of the laboratories providing the individual sets of data. The contributing laboratories are identified only by code in these files.

Development of Identification Matrices

Krichevsky and I have used data from these files to develop probability matrices based on modal feature frequencies in the most powerful tests, for the identification of strains at different hierarchical levels.

Our first matrix, which we designated a taxonomic probability matrix, was intended to identify strains at the species level. Features were selected which demonstrated at least a 90% difference in frequency between a chosen pair of clusters in the pooled NT matrix, and at least two such features were sought for each cluster pair. A total of 34 features were used and probability products (Dybowski and Franklin, 1968) calculated for each of the 14 well-defined taxa in the original IWGMT pooled NT diagram (Fig. 1b). The likelihood scores for a strain against each of the possible taxa in the matrix are based only on tests actually performed, and normalized to yield an identification (ID) score for each taxon. An ID score of .99 or greater to a given taxon means that the likelihood of the feature pattern of the strain being found in that taxon is at least 99-fold greater than in all other possible taxa combined. The ID score is thus a measure of discrimination. However, the possibility remains that a strain may fit one taxon far better than all other taxa in the matrix, and still not belong to that taxon, simply because taxa exist that are not described in the matrix. Therefore, we introduced a new measure of fit, termed the *R* score. The *R* score is a ratio of the best possible likelihood score for the most likely taxon, to the observed likelihood of the strain in question with the tests employed. We established that an *R* score greater than 100 was highly associated with an 'erroneous' diagnosis in terms of actual NT cluster location of a strain (Wayne *et al.*, 1980).

The first matrix, based on the earlier closed-set IWGMT studies, could not account for a number of species that were added to our studies later, but strains of these species were at least recognized as unidentifiable rather than being misdiagnosed. It also required a rather large set of tests to discriminate between even the 14 taxa that were included in the matrix. We have since developed matrices for clinical diagnostic purposes, which require far fewer tests and can accommodate most of the additional species described in the later open-ended studies. This was achieved by eliminating the discrimination between a pair of species when such a discrimination was achieved at the expense of extra tests and was not needed in the clinical decision-making process.

Excluding epidemiologic questions, there is no need for a clinician to know whether an isolate is *M. avium* or *M. intracellulare,* since management of a patient is the same for either. For epidemiologic purposes, agglutination serotyping is far superior to any discrimination by biochemical tests. Therefore, these two species were combined into the '*M. avium* complex' in the clinical matrix, with feature frequencies based on the pooled data from both clusters. *Mycobacterium terrae, M. triviale,* and *M. nonchromogenicum* are rarely or never the cause of human pulmonary disease: they are phenetically so similar to one another that they are easily treated in the matrix as members of the '*M. terrae* complex'. The *M. tuberculosis* supercluster was reduced to two taxa in the matrix, *M. tuberculosis* and *M. bovis*: with this arrangement, *M. microti* might be identified as *M. tuberculosis* and *M. africanum* as *M. bovis*. In fact, *M. microti* and *M. africanum,* as well as some strains clustering by NT with *M. tuberculosis* and *M. bovis,* did not identify as either, but their two highest ID scores were to *M. tuberculosis* and *M. bovis,* with the sums of these two ID scores exceeding .99. In this special case, a number of strains could safely be called 'atypical' members of the TB complex; the majority of strains of *M. tuberculosis* and *M. bovis* could be identified to the correct species.

Table 1 summarizes the categories of results and the distribution of 343 strains according to these categories. Employing a 23-feature matrix, incorporating 14 species or complexes, 91% of the strains were either correctly identified (82.5%) or correctly recognized as not belonging to any of the taxa in the matrix (8.5%). Only 2.9% of the strains were identified according to both ID and *R* scores as belonging to taxa that did not agree with their NT clustering behaviour, and these were treated as errors. The balance of 6.1% either completely failed to be

Table 1. *Categories of results with 343 strains of mycobacteria examined in a 23-feature diagnostic probability matrix*[a]

I. Identified (ID $\geq$ .99; $R < 100$) (special case, 'TB complex', ID-TB + ID-*bovis* $\geq$ .99, regardless of R)	
Correct: agrees with NT	82.5%
Error: disagrees with NT	2.9%
II. Unclassified ($R > 100$, regardless of ID)	
Correct: strain outside of any NT cluster	7.0%
Failure: strain falls in a cluster	2.9%
III. Not identified (ID < .99)	
Correct: strain outside of any NT cluster	1.5%
Missing data: additional tests suggested	2.0%
Failure: strain falls in a cluster	1.2%

[a]ID, Identification score; NT, numerical taxonomic study.

identified, or the program indicated tests that were missing and might, if included, bring the ID score above the threshold. We have also run a trial on a shortened diagnostic matrix that requires only 13 tests. This matrix yielded a comparable proportion of errors, but only 83.5% correct diagnoses. A choice between the 13- and 23-feature diagnostic matrices in a medical laboratory would depend on overall workload, availability of referral laboratories, and the distribution of the different taxa in the geographic region served by that laboratory.

Conclusions

An ongoing 15-year series of international co-operative studies on mycobacterial systematics has yielded

1. A broad baseline description of at least 15 species of slow growers and 9 of rapid growers, as well as a segregation of the '*rhodochrous*' taxon, which has since been placed in the genus *Rhodococcus;* a new study started under the direction of Dr. Jenkins to clarify the classification of members of the rapidly growing *M. fortuitum–M. chelonae* complex, which cause serious infections after surgery or other trauma
2. Definition of some problem areas where in-depth studies are needed at the molecular level to resolve the hierarchic status of members of the *M. avium* and of the *M. tuberculosis* complexes
3. Information on reproducibility of techniques for some of the most powerful tests used for mycobacterial differentiation and identification
4. Diagnostic strategies to translate our efforts from the realm of joyful inquiry for its own sake, into practical information for the management of human disease

References

Baess, I. (1979). Deoxyribonucleic acid relatedness among species of slowly-growing mycobacteria. *Acta Pathologica et Microbiologica Scandinavica* **87,** 221–226.

Baess, I., and Weis-Bentzon, M. (1978). Deoxyribonucleic acid hybridization between different species of mycobacteria. *Acta Pathologica et Microbiologica Scandinavica* **86,** 71–76.

Bojalil, L. F., Cerbón, J., and Trujillo, A. (1962). Adansonian classification of mycobacteria. *Journal of General Microbiology* **28,** 333–346.

Brennan, P. J., Souhrada, M., Ullom, B., McClatchy, J. K., and Goren, M. B. (1978). Identification of atypical mycobacteria by thin-layer chromatography of their surface antigens. *Journal of Clinical Microbiology* **8,** 374–379.

Dybowski, W., and D. A. Franklin. (1968). Conditional probability and the identification of bacteria: a pilot study. *Journal of General Microbiology* **54,** 215–229.

Goodfellow, M., Lind, A., Mordarska, H., Pattyn, S., and Tsukamura, M. (1974). A co-operative numerical analysis of cultures considered to belong to the '*rhodochrous*' taxon. *Journal of General Microbiology* **85,** 291–302.

Gordon, R. E., and Mihm, J. M. (1959). A comparison of four species of mycobacteria. *Journal of General Microbiology* **21,** 736–748.

Gordon, R. E., and Smith, M. M. (1953). Rapidly growing, acid fast bacteria. I. Species descriptions of *Mycobacterium phlei* Lehmann and Neumann and *Mycobacterium smegmatis* (Trevisan) Lehmann and Neumann. *Journal of Bacteriology* **66,** 41–48.

Gordon, R. E., and Smith, M. M. (1955). Rapidly growing acid fast bacteria. II. Species description of *Mycobacterium fortuitum* Cruz. *Journal of Bacteriology* **69,** 502–507.

Hawkins, J. (1977). Scotochromogenic mycobacteria which appear intermediate between *Mycobacterium avium/intracellulare* and *M. scrofulaceum*. *American Review of Respiratory Disease* **116,** 963–964.

Jenkins, P. A., Marks, J., and Schaefer, W. B. (1971). Lipid chromatography and seroagglutination in the classification of rapidly growing mycobacteria. *American Review of Respiratory Disease* **103,** 179–187.

Jones, D., and Sneath, P. H. A. (1970). Genetic transfer and bacterial taxonomy. *Bacteriological Reviews* **34,** 40–81.

Karasseva, V., Weiszfeiler, J., and Krasznay, E. (1965). Occurrence of atypical mycobacteria in *Macacus rhesus*. *Acta Microbiologica Academiae Scientiarum Hungaricae* **12,** 275–282.

Kubica, G. P., Baess, I., Gordon, R. E., Jenkins, P. A., Kwapinski, J. B. G., McDurmont, C., Pattyn, S. R., Saito, H., Silcox, V., Stanford, J. L., Takeya, K., and Tsukamura, M. (1972). A co-operative numerical analysis of rapidly growing mycobacteria. *Journal of General Microbiology* **73,** 55–70.

Marks, J., Jenkins, P. A., and Tsukamura, M. (1972). *Mycobacterium szulgai*—a new pathogen. *Tubercle* **53,** 210–214.

McClatchy, J. K. (1981). The seroagglutination test in the study of nontuberculous mycobacteria. *Review of Infectious Disease* **3,** 867–870.

Meissner, G., Schröder, K. H., Amadio, G. E., Anz, W., Chaparas, S., Engel, H. W. B., Jenkins, P. A., Käppler, W., Kleeberg, H. H., Kubala, E., Kubin, M., Lauterbach, D., Lind, A., Magnusson, M., Mikova, Z., Pattyn, S. R., Schaefer, W. B., Stanford, J. L., Tsukamura, M., Wayne, L. G., Willers, I., and Wolinsky, E. (1974). A co-operative numerical analysis of nonscoto- and nonphotochromogenic mycobacteria. *Journal of General Microbiology* **83,** 207–235.

Saito, H., Gordon, R. E., Juhlin, I., Käppler, W., Kwapinski, J. B. G., McDurmont, C., Pattyn, S. R., Runyon, E. H., Stanford, J. L., Tárnok, I., Tasaka, H., Tsukamura,

M., and Weiszfeiler, J. (1977). Cooperative numerical analysis of rapidly growing mycobacteria. The second report. *International Journal of Systematic Bacteriology* **27,** 75–85.

Sneath, P. H. A. (1957). The application of computers to taxonomy. *Journal of General Microbiology* **17,** 201–226.

Stanford, J. L., and Grange, J. M. (1974). The meaning and structure of species as applied to mycobacteria. *Tubercle* **55,** 143–152.

Timpe, A., and Runyon, E. H. (1954). The relationship of ''atypical'' acid-fast bacteria to human disease: a preliminary report. *Journal of Laboratory and Clinical Medicine* **44,** 202–209.

Wayne, L. G. (1964). The mycobacterial mystique: deterrent to taxonomy. *American Review of Respiratory Disease* **90,** 255–257.

Wayne, L. G. (1967). Selection of characters for an Adansonian analysis of mycobacterial taxonomy. *Journal of Bacteriology* **93,** 1382–1391.

Wayne, L. G. (1981). Numerical taxonomy and cooperative studies: roles and limits. *Review of Infectious Disease* **3,** 822–827.

Wayne, L. G. (1982). Microbiology of the tubercle bacilli. *American Review of Respiratory Disease* **125** (Suppl.), 31–41.

Wayne, L. G., and Diaz, G. A. (1979). Reciprocal immunological distances of catalases derived from strains of *Mycobacterium avium, Mycobacterium tuberculosis* and closely related species. *International Journal of Systematic Bacteriology* **29,** 19–24.

Wayne, L. G., and Diaz, G. A. (1982). Serological, taxonomic, and kinetic studies of the T and M classes of mycobacterial catalase. *International Journal of Systematic Bacteriology* **32,** 296–304.

Wayne, L. G., Dietz, T. M., Gernez-Rieux, C., Jenkins, P. A., Käppler, W., Kubica, G. P., Kwapinski, J. B. G., Meissner, G., Pattyn, S. R., Runyon, E. H., Schröder, K. H., Silcox, V. A., Tacquet, A., Tsukamura, M., and Wolinsky, E. (1971). A cooperative numerical analysis of scotochromogenic slowly growing mycobacteria. *Journal of General Microbiology* **66,** 255–271.

Wayne, L. G., Engbaek, H. C., Engel, H. W. B., Froman, S., Gross, W., Hawkins, J., Käppler, W., Karlson, A. G., Kleeberg, H. H., Krasnow, I., Kubica, G. P., McDurmont, C., Nel, E. E., Pattyn, S. R., Schröder, K. H., Showalter, S., Tárnok, I., Tsukamura, M., Vergmann, B., and Wolinsky, E. (1974). Highly reproducible techniques for use in systematic bacteriology in the genus *Mycobacterium:* tests for pigment, urease, resistance to sodium chloride, hydrolysis of Tween 80 and galactosidase. *International Journal of Systematic Bacteriology* **24,** 412–419.

Wayne, L. G., Engel, H. W. B., Grassi, C., Gross, W., Hawkins, J., Jenkins, P. A., Käppler, W., Kleeberg, H. H., Krasnow, I., Nel, E. E., Pattyn, S. R., Richards, P. A., Showalter, S., Slosarek, M., Szabo, I., Tárnok, I., Tsukamura, M., Vergmann, B., and Wolinsky, E. (1976). Highly reproducible techniques for use in systematic bacteriology in the genus *Mycobacterium:* tests for niacin and catalase and for resistance to isoniazid, thiophene 2-carboxylic acid hydrazide, hydroxylamine and *p*-nitrobenzoate. *International Journal of Systematic Bacteriology* **26,** 311–318.

Wayne, L. G., Andrade, L., Froman, S., Käppler, W., Kubala, E., Meissner, G., and Tsukamura, M. (1978). A cooperative numerical analysis of *Mycobacterium gastri, Mycobacterium kansasii,* and *Mycobacterium marinum. Journal of General Microbiology* **109,** 319–327.

Wayne, L. G., Krichevsky, E. J., Love, L. L., Johnson, R., and Krichevsky, M. I. (1980). Taxonomic probability matrix for use with slowly growing mycobacteria. *International Journal of Systematic Bacteriology* **30,** 528–538.

Wayne, L. G., Good, R. C., Krichevsky, M. I., Beam, R. E., Blacklock, Z., Chaparas, S. D., Dawson, D., Froman, S., Gross, W., Hawkins, J., Jenkins, P. A., Juhlin, I., Käppler, W., Kleeberg, H. H., Krasnow, I., Lefford, M. J., Mankiewicz, E., McDurmont, C., Meissner, G., Morgan, P., Nel, E. E., Pattyn, S. R., Portaels, F., Richards, P. A., Rüsch, S., Schröder, K. H., Silcox, V. A., Szabo, I., Tsukamura, M., and Vergmann, B. (1981). First report of the cooperative, open-ended study of slowly growing mycobacteria by the International Working Group on Mycobacterial Taxonomy. *International Journal of Systematic Bacteriology* **31,** 1–20.

Wayne, L. G., Good, R. C., Krichevsky, M. I., Beam, R. E., Blacklock, Z., David, H. L., Dawson, D., Gross, W., Hawkins, J., Jenkins, P. A., Juhlin, I., Käppler, W., Kleeberg, H. H., Krasnow, I., Lefford, M. J., Mankiewicz, E., McDurmont, C., Nel, E. E., Portaels, F., Richards, P. A., Rusch, S., Schröder, K. H., Silcox, V. A., Szabo, I., Tsukamura, M., Vanden Breen, L., and Vergmann, B. (1983). Second report of the cooperative open-ended study of slowly growing mycobacteria by the International Working Group on Mycobacterial Taxonomy. *International Journal of Systematic Bacteriology* **33,** 265–274.

6

The Contribution of Numerical Taxonomy to the Systematics of Gram-negative Bacteria

M. T. MACDONELL AND R. R. COLWELL

Department of Microbiology, The University of Maryland, College Park, Maryland, USA

Introduction

At the turn of the century, quantitative methods were applied to the taxonomy of plants and animals (Heinke, 1898), but progress was relatively slow until the advent of the electronic computer. In 1957, publication of papers on numerical taxonomy by Sneath (1957) and Michener and Sokal (1957) catalyzed a renewed interest in quantitative analysis of taxonomic data. These papers are now considered landmarks in this field. By our estimate, more than 3000 papers have been published since on microbiological applications alone. Between 1898 and 1957, taxonomic studies did make use of numerical models, but none were as successful in generating research as the seminal publications of Sneath (1957) and Michener and Sokal (1957), who addressed a number of significant problems. These included a challenge of *a priori* assumptions associated with weighting of characters (Sneath and Sokal, 1973) and treatment of the vexing problems of insufficient observations, either from employment of too few tests or an unacceptably small sample of organisms, as well as the need for clustering algorithms.

Numerical taxonomy, based on phenetic similarity and adhering to empirical neo-Adansonian criteria (Sneath and Sokal, 1973), in which (i) as many test characters as possible are observed for a large number of samples, (ii) every character is considered to be of equal importance, and (iii) overall similarity is considered to be a function of the sum of individual similarities, has passed the rigors of nearly three decades of scrutiny and intense debate. More recently, technological advances, such as the refinement of molecular genetic techniques, have opened new vistas in microbial taxonomy. The result is that new levels of data gathering, from which taxonomic information may be obtained, can now be achieved. It is gratifying that the molecular genetic evidence, in general, has provided a strong underpinning of numerical taxonomy constructs.

COMPUTER-ASSISTED
BACTERIAL SYSTEMATICS

ISBN 0-12-289665-3

Numerical Taxonomic Methods

Variations on a Theme

The advent of the data-processing era, in which widespread employment of computers for the purpose of manipulating large volumes of information is commonplace, has allowed successful applications of numerical taxonomy and an astounding growth in the area of microbial systematics. With the facilities now available, new methods in numerical taxonomy have evolved. Modifications, in general, have focussed on similarity coefficients and cluster analyses. In fact, a large number of defined similarity coefficients have been published in the intervening years. An interesting finding of one study, undertaken to evaluate the efficacy of 36 similarity coefficients (Austin and Colwell, 1977), showed that, for analysis of microbial taxonomic data, roughly half did not improve resolution. That is, despite an enormous variety in the algorithms, similar results were generally obtained. We hasten to add, however, that some were more sensitive than others in the resolution of clusters. Coefficients most frequently employed in microbiology are those of Gower (1971), Cattell (pattern coefficient; Cattell, 1949), and Jaccard (1908), as well as the simple-matching coefficient (Sokal and Michener, 1958). The majority of studies in which results obtained using the Jaccard coefficient are compared wtih those using the simple matching coefficient show the Jaccard coefficient to be more discriminating, that is, yielding discrete clusters at lower levels of similarity than the simple-matching coefficient (Azad and Kado, 1980; Gray and Stewart, 1980; Green and Bousfield, 1982; Ralston-Barrett *et al.*, 1976).

The most significant advances since the earliest applications of numerical taxonomy to microbiology have been in cluster analysis. An excellent discussion of the critical aspects of clustering has been published (Sneath and Sokal, 1973). In practice, two methods of clustering emerged and were very widely used until recently: (i) *single linkage*, by which strains are clustered at the highest level of relatedness, and (ii) *complete linkage*. From these has evolved *unweighted pair-group arithmetic average clustering* (UPGMA), which generates clusters at a level of average similarity between strains, offering unbiased clustering of related groups. This clustering method has emerged as the preferred one for generating clusters in taxonomic studies of micro-organisms.

Variations of the Theme: Identification

Alternative approaches have been developed which are aimed at resolving taxonomic relationships among a set of bacterial strains and/or identifying taxa using matrix algorithms to estimate the probability that given isolates are representative. One approach (Beers and Lockhart, 1962), essentially Adansonian in principle, involves the use of a 'distance' coefficient to partition dissimilar (rather than group similar) isolates, using the criterion of mutual phenetic 'distance'. This approach employs a parameter unusual to most similarity-clustering techniques, in that estimates are made of the proportion of a group in possession of any given property. Comparison of results of partitioning, using distance coefficients, with those grouping isolates on the basis of overall similarity indicates that the two approaches yield similar results (Beers *et al.*, 1962). Alternatively, monothetic approaches to identification of isolates rely on the comparison of results of a defined set of tests with those of reference strains, or those corresponding to the definition of a given species. Although such computer-assisted identification methods serve a useful function in clinical diagnoses, they are based on *a priori* assumptions, rendering them less useful for taxonomy. 'Key characters' lists available from numerical taxonomy studies are given in Tables 1–5.

Other methods, closely related to the above, used to identify strains, include computer-generated *identification matrices* and *numerical profiles*. The first is analogous to computer-assisted identification, except that one compares test results of an unknown isolate with extensive collections of test results from defined reference strains, thus circumventing the necessity to access a computer in order to arrive at a tentative identification. Otherwise, it is identical to computer-assisted identification and has only limited value for taxonomic studies. This approach is used predominantly in clinically oriented, rapid-identification test methods.

The second method, the numerical profile (Griffiths and Lovitt, 1980), represents an extreme in numerical taxonomy, since the number of test characters employed is quite small, usually ranging from 9 to 21 (but always a multiple of 3). Although it is not immediately apparent, by severely restricting the number of test characters, a weight of +1 is assigned to a small subset of all possible characteristics, and a weight of 0 is assigned to the remainder. Thus, the usefulness of this method as a taxonomic tool is reduced. Nevertheless, numerical profiles are helpful in cases where bacteria associated with specific ecological niches need to be partitioned into physiological types or other such groupings.

Comparison between Numerical Taxonomy and Taxonomies Based on Traditional Criteria

Of the hundreds of numerical taxonomies of micro-organisms which have been published, the majority have correlated to some degree with 'classically derived' taxonomic schemata. Those showing poor correlation more frequently have involved certain groups of the less familiar Gram-negative bacteria, mainly the nonfermentative rods and taxa associated with the aquatic environment. As a

Table 1. *Key characteristics of selected species of the families enterobacteriaceae and pseudomonadaceae*[a,b]

	Escherichia coli	*Enterobacter cloacae*	*Enterobacter aerogenes*	*Klebsiella pneumoniae*	*Klebsiella aerogenes*	*Shigella sonnei*	*Shigella flexneri*	*Shigella boydii*	*Shigella dysenteriae*	*Serratia marcescens*	*Yersinia enterocolitica*	*Yersinia fredricksenii*	*Yersinia intermedia*	*Yersinia kristensenii*	*Yersinia pseudotuberculosis*	*Erwinia herbicola*	*Hafnia alvei*	*Proteus mirabilis*	*Proteus vulgaris*	*Proteus rettgeri*	*Providencia stuartii*	*Providencia alcalifaciens*	*Morganella morganii*	*Pseudomonas fluorescens*	*Zymomonas* spp.
ADH	−	+	+	−	−	−	−	−	−	−	−	−	−	−	−	−	−	−	−	−	v	v	−		
LDC	+	−	d	+	+	−	−	−	−	+	−	−	−	−	−	d	+	−	−	−	−	−	−		d
ODC	d	+	+	−	−	d	−	−	−	+	+	+	+	+	+	d	+	+	−	d	−	−	+		−
H_2S	−	−	−	−	−	−	−	−	−	−	−	−	−	−	−	−	−	+	d	−	−	−	+	d	−
Indole	+	−	−	−	−	+	+	−	d	−	d	d	d	d	d	−	−	−	+	+	+	+	+	d	−
Voges–Proskauser		d	d	d	+	−	−	−	−	+	+	+	+	+	+	+	d	+	+	−	−	−	−	−	
Glucose (gas)		+	+	+	+	−	−	−	−	−	d	d	d	d		−	+	d	d	−	−	d	+	−	+
Gelatinase		d	−	−	−	−	−	−	−	+	−	−	−	−	−	d	d	+	d	−	−	−	−	+	−
Acid																									
Adonitol	−	+	+	+	+	−	−	−	−	+	d	−	−	−			−	−	−	+	−	d	−		
Arabinose	+	d	+	+	+	+	−	−	−	−	+	+	+	+		+	+	+	d	d	−	−	d	−	
Glucose	+	+	+	+	+	+	+	+	+	+	+	+	+	+	+	+	+	+	+	+	+	+	+		
Inositol	−	d	+	+	+	−	−	−	−	+	d	d	d	d		d	−	−	−	+	+	−	−	d	
Mannitol	+	+	+			+	+	d	−		+	+	+	+	+	+	+	−	−	+	−	−	−	+	
Mannose	+					+	+	+	+		+	+	+	+	+			−	−	+	+	+	+		
Sucrose	d	d	−	+	+	+	d	d	−	+	+	+	+	−	+	+	+	d	+	d	d	d	−	d	
Xylose	+	+	+	+	+	−	−	d	d	−	d	d	−	d			+	+	+	d	−	d	d		
Lactose	+	+	+	+	+	+	−	−	−	d	d	+	+	−	d	+	d	−	−	−	−	−	−	−	
Maltose	+	+	+	+	+	+	d	−	d		+	+	+	+	+		+	+	+	+	+	−	−		
Cellobiose	−	+	+	+	+	−	−	−	−	d	+	+	+	+	+		−	−	−	−	−	−	−		

Melibiose	+	+	+	+	+	+	d	−	−	d	−	−	+	−	−		−	+	−	+	−	d	−		
Sorbitol	+	+	+	+	+	−	d	+	d	+	+	+	+	+		+	−	d	−	d	−	−	−	−	
Raffinose	d	d	d	+	+	d	d	−	−	−	−	−	+	−	−	+	−	−	−	−	−	−	−	d	
Rhamnose	+	+	+	+	+	+	−	−	d	d	−	+	+	−	−		+	d	−	+	−	−	−		
Arbutin						+	−	−	−		d	+	+	−	d			−	d	d	−	−	−		
Salicin	d	+	d	+	+	−	−	−	−	−	d	+	+	−	+		d	d	d	d	−	−	−		
Trehalose	+	+	+	+	+	+	+	+	+	+	+	+	+	+	+		+	+	d	−	+	d	d	d	
Glycerol	+	+	+	+	+	d	d	+	+	+	+	+	+	+	+	+	+	d		+	+	+	−	+	
Starch	+	−	−	−	−	d	−	−	−	−							−	d		+	+		+	−	
Malonate											−	−	−	−	−										
Aesculin hydrolysis	d	+	+	d	+	−	−	−	−	−	d	+	+	−	d	+	d	−	d	d	−	−	−		
Lecithinase	−	−	−	−	−	−	−	−	−	−	−	−	+	−	+	−	−	+	+	−		−	d	d	
Nitrate reduction	+	+	+							+	+	+	+	+	+	+	+						+	d	−
Citrate	+	+	+	+	+	−	−	−	−	+	−	−	−	−	−	+	d	−	−	+		+	−		
Oxidase	−	−	−	−	−	−	−	−	−	−	−	−	−	−	−	−	−			+			+	+	−
Urease	−	d	d	+	+	−	−	−	−	+	+	+	+	+	+	d	−	+	+	+	−	−	+	d	d
ONPG	+	d	+			+	−	d		+	+	+	+	+		+	−	−	−	−	−	d	−	+	
O/F	f	f	f	f	f	f	f	f	f	f	f	f	f	f	f	f	f	f	f	f	f	f	f		
Growth																									
5% (w/v) NaCl	+	+	+	+	+	+	+	+	+	+						+	+	d		−	−		−	+	
4°C		d	d			−	−	−	d	−						d	+	d	+	+	d	d	+	d	
42°C	+	d	d	−	d	+	d	+	+	d						−	−	−		−	−	d	−	d	
Swarming (0.01% peptone)	−	−																+	+	−	−	−	d		
TDA	−	−	−	−	−	−	−	−	−	−						−	−	+	+	+	+	d	−	−	
Catalase	+	+	+			+	+	d	d	+	+	+	+	+	+	+	+	+	+	+	+	+	+	+	+
Melanin from tyrosine											−	−	−	−	−										
Motility						−	−	−	−									d	+	+	d	d	+		d
Sensitivity: penicillin																		d	−	−	−	d	d		
Pellicle formation						−	−	−	−									+	d	d	+	−	d		

[a]Taken from Austin *et al.*, 1981; Broom and Sneath, 1982; De Ley and Swings, 1976; Dodd and Jones, 1982; Gavini *et al.*, 1981; Grimont *et al.*, 1977; Goodfellow *et al.*, 1976; Harvey and Pickett, 1980; Hussong *et al.*, 1981; Johnson *et al.*, 1975; Kaneko and Hashimoto, 1982; Kapperud *et al.*, 1981; McKell and Jones, 1976.
[b]+, 90% or more strains positive; −, 90% or more strains negative; d, 11–89% of strains positive; v, strain instability (not equivalent to d); f, fermentative; [ ], insufficient data.

Table 2. *Key differentiating characteristics of* Cytophaga *and* Flavobacterium[a,b]

	Cytophaga	*Flavobacterium*
Gelatinase	+	−
Acid		
Glucose	+	−
Mannitol	+	−
Mannose	+	−
Sucrose	+	−
Xylose	+	−
Fructose	+	−
Lactose	+	−
Nitrate reduction	d	−
Elongated forms	+	d
Spreading on *Cytophaga* agar	+	−
Growth		
30°C	d	−
25°C	+	+
45°C × 45 min	+	−
37°C × 48 hr	+	−
Inorganic salts and glucose	+	−
Monomorphic	+	−
Pleomorphic	−	+
Orange/yellow pigmentation	−	d
Sensitivity: 0/129	+	−
Swarming (0.01% peptone)	+	−
No swarming	d	+

[a]Taken from Agbo and Moss, 1979; Christensen, 1980; Colwell, 1969; Floodgate and Hayes, 1963; Hayes, 1977; Holmes *et al.*, 1981; McMeekin *et al.*, 1972; Thurner and Busse, 1978.
[b]For key to symbols, see footnote to Table 1.

result, these have received a great deal more attention from numerical taxonomists. A benefit deriving from this situation is that numerical taxonomic methods have become more widely known among soil and water microbiologists and, therefore, employed by them in their studies.

The extent to which taxonomic schemata based on numerical taxonomic analyses correspond to those of traditional conventionally determined taxonomies, often is a function of whether or not a polythetic taxonomy (numerical taxonomy) has been constructed, compared to a monothetic (classical or 'alpha') taxonomy. It should not be surprising, therefore, that large divergences become evident (see Table 6). In microbiology, virtually all of the conventional taxonomic methods are based on methods designed for identifying bacteria of medical significance (Table 5). Diagnostic tables, which are of a clinical heritage, are available for identification of many bacterial species, but the unfortunate feature

Table 3. *Key differentiating characteristics of* Agrobacterium, Rhizobium, *and* Chromobacterium[a,b]

	Agrobacterium	*Chromobacterium*	*Rhizobium*
Acid			
Sucrose	+	d	+
Lactose	+	−	d
Cellobiose	+	d	d
Melibiose	+	−	d
Raffinose	−	−	d
Gas from glucose	−	−	−
Lecithinase	−	+	−
Nitrate reduction	+	+	d
Oxidase	+	d	+
Urease (48 hr)	−	−	−
O/F	f	d	d
Flagella			
Polar	−	+	d
Peritrichous	+	+	d
Melanin from tyrosine	+	+	−
Methylene blue reduction	+	d	−
H_2S (from cysteine)	+	+	+
Growth			
pH 4.0	d	−	d
42°C	−	d	−
Sensitivity:			
0/129	−	−	d
Penicillin	−	d	d
Aesculin hydrolysis	+	d	d
Motility	+	+	+
Starch hydrolysis	−	−	−
Mucilagenous colonies	+	−	+
Voges–Proskauer	−	−	−
Clearing: Tween 80	−	+	−
Litmus milk reduction	+	d	+
Gelatinase	+	d	−
Citrate	−	d	−
Catalase	+	+	+
ADH	−	d	−
LDC	−	−	−
ODC	−	−	−

[a]Taken from Kersters *et al.*, 1973; Moffett and Colwell 1968; Moss *et al.*, 1978; White, 1972.
[b]For key to symbols, see footnote to Table 1.

common to diagnostic keys is the tendency for the user to assume that for each bacterial species there exists a unique set of minimum characters, the presence of which is sufficient for allocation of an unidentified isolate to a given taxon. As a result, nomenclatural problems have been created, an example of which was the

Table 4. *Key differentiating characteristics for selected species of* Serratia *and* Citrobacter[a,b]

	Serratia marcescens	*Serratia liquifaciens*	*Serratia plymuthica*	*Serratia marinorubrum*	*Citrobacter freundii*	*Citrobacter diversus*	*Citrobacter amalonitica*
ADH	−	d	−	−			
LDC	+	+	−	d	−	−	−
ODC	+	+	−	−	−	+	+
H_2S (from thiosulphate)					+	−	−
Indole	−	−	−	−	−	+	+
Voges–Proskauer	+	+	+	+	−	−	−
Gelatinase	+	+	+	+	−		
Urease	−	−	−	−			
Gas (glucose)	−	+	+	−	+	−	
Acid							
Adonitol					−	−	+
Arabinose					+		
Glucose	+	+	+	+	+	+	+
Inositol					−		
Raffinose					−	−	−
Aesculin hydrolysis	+	+					
Nitrate reduction	+	+	+	+			
Oxidase	−	−	−	−	−	−	−
ONPG	+	+	+	+			
O/F	f	f	f	f	f	f	f
Growth							
5% (w/v) NaCl	+	+	d	+	−		
4°C	−	+	+	−			
42°C	d	−	−	−	−		
Motility	+	+	+	+			
Clearing: Tween 80	+	+	+	+			

[a]Taken from Austin *et al.*, 1981; Grimont and DuLong de Rosnay, 1972; Grimont *et al.*, 1977; Sakazaki *et al.*, 1976.
[b]For key to symbols, see footnote to Table 1.

separation of *Vibrio* and *Beneckea,* which was not based on overall phenetic similarity but on 'key characteristics', in this case, production of a chitinase and apparent lack of lateral flagella (West and Colwell, 1984). In aquatic microbiology, other such examples can be cited. Austin *et al.* (1981) found that there was no correlation between the results of numerical taxonomy-derived clusters of

Table 5. *Key differentiating characteristics for some medically significant Gram-negative bacteria*[a,b]

	Brucella spp.	*Bordetella bronchiseptica*	*Bordetella pertussis*	*Bordetella parapertussis*	*Acinetobacter lwoffi*	*Acinetobacter anitratus*	*Moraxella* spp.	*Alcaligenes faecalis*	*Francisella* spp.
Acid									
Glucose	+	−	−	−	−	−	−		
Starch	−	−	−	−	d	d	+	−	−
Nitrate reduction	d	+	−	−	−	−	d	+	−
Oxidase	+	+	+	−	−	−	+		
Urease	+	+	−	+	d	d	d	d	−
O/F								f	
Melanin from tyrosine	−	−	−	+	−	−	d	−	−
Anaerobic growth with nitrate	+	−	−	−	−	−	−		
Motility	+	−	−	−	−	−	−	+	−
Clearing: Tween 80	d	+	−	+	+	+	d		
Sensitivity:									
Penicillin	+	−	+	−	d	−	+		
0/129	−	−	+	−	d	d	d	−	−
Sulphonamide	−	d	−	+	+	+	+	+	−
Pellicle formation	+	+	−	+	−	+	d	+	+
Diffuse brown pigment	−	−	−	+	−	−	d	−	−
Growth									
5% (w/v) bile salts	d	+	−	+	+	+	d	+	
10% (w/v) bile salts	d	+	−	+	+	+	d	+	−
0.1% (w/v) phenol agar	d	+	−	−	+	+	d	+	d
0.32% (w/v) tellurite	−	−	−	−	−	−	−	+	−
10°C		d	−	−	d	+	d	d	

[a]Taken from: Austin *et al.*, 1981; Johnson and Sneath, 1973; Mallory *et al.*, 1977; Pagel and Seyfried, 1976; Pichinoty *et al.*, 1978; Thornley, 1967.
[b]For key to symbols, see footnote to Table 1.

lactose-fermenting Gram-negative rods cultured from positive MPN tubes and results obtained applying traditional taxonomic methods, such as IMViC, diagnostic keys, and rapid-identification strips. Indeed there are reports in which results of numerical taxonomy and those of traditional methods are in direct conflict (Grimont and DuLong de Rosnay, 1972; McKell and Jones, 1976; Sakazaki *et al.*, 1976; Thurner and Busse, 1978).

Table 6. *A chronology of taxon assignments attributable to the results of numerical taxonomic studies*

Traditionally derived classification	Revised classification based on NT results	References
Rhizobium rubi, Agrobacterium rubi	*Rhizobium rubi*	Moffett and Colwell (1968)
Rhizobium meliloti	*Rhizobium leguminosarum*	
Agrobacterium tumefaciens, Agrobacterium radiobacter, Agrobacterium rhizogenes	*Rhizobium radiobacter*	
Vibrio cuneatis (ATCC 6972)	*Pseudomonas fluorescens*	Colwell *et al.* (1968)
Vibrio 'el tor'	*Vibrio cholerae* biotype el tor	Colwell (1970)
Spirillum minitulum, Spirillum halophilum, Spirillum linum, Spirillum atlanticum	*Spirillum volutans*	Carney *et al.* (1975)
'Kauffman Group I Salmonellae	*Salmonella enteritidis*	Johnson *et al.* (1975)
Proteus morganii	*Morganella morganii*	McKell and Jones (1976)
Zymomonas congolensis	*Zymomonas mobilis*	De Ley and Swings (1976)
Pseudomonas putrifaciens, Pseudomonas rubescens	*Alteromonas putrifaciens*	Lee *et al.* (1977)
Vibrio cholerae biotype *proteus*	*Vibrio metschnikovii*	Lee *et al.* (1978)
'*Chromobacterium* sp.'	*Chromobacterium fluviatile*	Moss *et al.* (1978)
Pseudomonas thomasii	*Pseudomonas pickettii*	King *et al.* (1979)
Haemophilus vaginalis	*Gardnerella vaginalis*	Greenwood and Pickett (1980)
Flavobacterium heparinum	*Cytophaga heparina*	Christensen (1980)
'Group F *Vibrio*'	*Vibrio fluvialis*	Lee *et al.* (1981)
Vibrio succinogenes	*Wollinella succinogenes*	Tanner *et al.* (1981)
Yersinia enterocolitica	*Yersinia enterocolitica* *Yersinia kristensenii* *Yersinia frederiksenii* *Yersinia intermedia*	Kaneko and Hashimoto (1982)

New Levels of Information

Since the beginning of the 'computer revolution' in microbial taxonomy in the early 1960s, enormous advances in molecular biology have occurred. Nucleic acid methods, including DNA–DNA and DNA–RNA hybridization, and nucleic acid sequencing, have attained prominence in microbial taxonomy. These methods permit access to greater detail in molecular structure, from which information of taxonomic importance can be gleaned. The first of this new generation of technology to emerge was the determination of bacterial DNA base composition (Lee *et al.*, 1956; Belozersky and Spirin, 1960), which provided a direct, although crude, assay of the bacterial genome (see Table 7). Results of comparisons of early 'base ratio' determinations demonstrated clearly the significance of G + C molar ratios in taxonomy and now provide a powerful and routine method for discriminating between phenetically similar but genetically unrelated strains. Compilations of bacterial base compositions provided early support for numerically derived, as contrasted with classical, taxonomic schemata (Colwell and Mandel, 1964; Thornley, 1967). Within a decade, DNA–DNA hybridization methods extended the sensitivity with which the bacterial genome could be probed and provided the means by which primary structures of two distinct genomes could be directly compared, thus permitting the extraction of information of taxonomic value, depending on the degree of DNA homology shared by the two strains being compared.

Nucleic acid methods have attained additional sophistication, with determination of the linear sequence of nucleotide bases in genetic material now possible (Sanger *et al.*, 1977; Donis-Keller, 1979; Peattie, 1979; Maxam and Gilbert, 1980). The large quantities of information thus obtained can be stored on computers and retrieved to compare sequence data for given species. The data can be added to as new sequences are determined. It is encouraging that the correlation established between results of numerical taxonomy and those of DNA–DNA or DNA–RNA hybridizations has been, with very few exceptions, remarkably strong (Colwell, 1970; Johnson and Ault, 1978; Azad and Kado, 1980; Champion *et al.*, 1980). In cases where poor correlation between results of numerical taxonomic analysis and those of nucleic acid homology determinations have occurred, significant departure from established Adansonian procedures is evi-

Table 7. *Levels of molecular genetic information useful in taxonomy*

Level	Direct	Indirect
1	DNA sequences DNA–DNA hybridization	DNA base composition
2	RNA sequences rRNA–DNA hybridization	mRNA–DNA hybridization
3	Serology	Membrane protein electrophoresis
	(some biochemical assays)	
4		Uptake assays Sole carbon source assays Most biochemical assays

dent. For example, studies which employed insufficient numbers of test characters (Johnson *et al.*, 1968), or those where there was preselection of a sample population of high phenotypic similarity (Harvey and Pickett, 1980), have yielded little or no correlation.

Other developments important to taxonomy which have occurred over the years since 1960 include pyrolysis spectrometry (Gutteridge *et al.*, Chapter 14; Gutteridge and Puckey, 1982), comparison of membrane lipid and protein profiles, and more sensitive enzymatic assays. Each provides a means of increasing the sensitivity of numerical methods to reflect natural relationships.

Standardization of Methods

The simplest and most effective way in which accidental bias in numerical taxonomic studies can be controlled is by use of standardized tests and carefully detailed methods. Although it is impossible to eliminate spurious bias and/or weighting in defined protocols, taxonomic test methods can nevertheless be standardized for all studies in which defined protocols are employed, and such methods provide 'core characteristics' (Colwell and Wiebe, 1970). A battery of core characteristics can be augmented to include test characters and methods appropriate for a broader spectrum of Gram-negative bacteria, for example, chemolithotrophs (Hutchinson *et al.*, 1966, 1969), soil bacteria (Moffett and Colwell, 1968; White, 1972; Kersters *et al.*, 1973) and nutritionally fastidious bacteria (Johnson and Sneath, 1973). Such core characteristics provide a common language through which numerical taxonomic studies can be interlocked (see Hill, 1975). From such interlocking, compatible numerical taxonomic studies employing results of tests on thousands of strains can be merged, thereby providing a basis for placing into perspective taxonomic relationships amongst widely divergent groups of bacteria.

Standardization of data requires that the investigator be able to discriminate between those data which provide significant taxonomic information and those which do not. Johnson and Sneath (1973) observed that for *Moraxella* species, production of acid from carbohydrates was linked to production of acid from glucose. Furthermore, a nonspecific aldose dehydrogenase in *Moraxella* species catalyzes the oxidation of nearly a dozen carbohydrates (Baumann *et al.* 1968). Johnson and Sneath (1973) concluded that, for *Moraxella* strains, carbohydrate acidification tests could not represent independent characters. The subsequent deletion of carbohydrate acidification data resulted in the elimination of a significant interaction, the effect of which was an increase in the number of negative matches. Comparison of cluster saltation obtained before and after elimination of interacting characters provides dramatic evidence of potential error that can arise from the use of nonindependent tests (see Johnson and Sneath, 1973).

Internal controls are critical to all taxonomic analyses and can be established in several ways. For example, bacterial strains designated as 'marker' or reference strains, or related strains serving as reference clusters, can be included in a set of bacteria under study. Thus, point calibration for multiple data sets and information on relative performance of a given experimental design are provided. An analysis by Sneath and Johnson (1972), of the influence of errors in microbiological tests, as well as the influence of the inclusion of incorrectly identified reference strains, provides a useful means of estimating probable error. It also provides a means of controlling error and supplies information on practical limits within which probable error can be tolerated in interlocked studies.

Numerical Taxonomy of Gram-negative Bacteria

Genera for Which Numerical Taxonomic Data Are Available

A survey of the major journals indicates that results of numerical taxonomic studies are now available for at least 50 genera of Gram-negative bacteria. The results indicate that as of 1970 approximately 14 genera had been studied, whereas by 1980 the number had increased to 45 and has increased to more than 50 since 1980 (see Table 8). A number of the genera were subjected to major taxonomic revision as a result of these studies. For example, data obtained from numerical taxonomic studies were instrumental in the definition of the genus *Alteromonas* (Lee *et al.*, 1977; Gray and Stewart, 1980; Gillespie, 1981), abolition of the genus *Beneckea* (Baumann *et al.*, 1980) and reorganization of the genus *Vibrio* (Colwell, 1974; Carney *et al.*, 1975; Lee *et al.*, 1981), reorganization and redefinition of the genus *Pseudomonas* (Colwell *et al.*, 1965; Austin *et al.*, 1978; Sneath *et al.*, 1981; Molin and Ternstrom, 1982), resolution of the genera *Agrobacterium* and *Rhizobium* (Moffett and Colwell, 1968; White, 1972; Kersters *et al.*, 1973), reorganization of the genus *Proteus* into the genera *Proteus, Providencia,* and *Morganella* (Johnson *et al.*, 1975; McKell and Jones, 1976), resolution of the genera *Cytophaga, Flexibacter,* and *Flavobacterium* (Floodgate and Hayes, 1963; McMeekin *et al.*, 1972; Hayes, 1977), identification of species belonging to the genus *Serratia* (Grimont and DuLong de Rosnay, 1972; Grimont *et al.*, 1977), validation and redefinition of the genus *Erwinia* (Goodfellow *et al.*, 1976; Azad and Kado, 1980), resolution of the species of the genus *Yersinia* (Harvey and Pickett, 1980; Kapperud *et al.*, 1981; Kaneko and Hashimoto, 1982), and reorganization of the genus *Bacteroides* (Barnes and Goldberg, 1968; Johnson and Ault, 1978). Significantly, several genera, such as *Gardnerella* (Greenwood and Pickett, 1980), *Morganella,* and *Providencia* (Johnson *et al.*, 1975; McKell and Jones, 1976), arose from numerical taxonomic studies.

Table 8. *Genera of Gram-negative bacteria subjected to numerical taxonomy analysis: a chronology of studies by genus*

1961–1971	1971–1981	1981–1983
Acinetobacter		
	Actinobacillus	
Aerobacter		
Aeromonas		
Agrobacterium		
	Alcaligenes	
	Alteromonas	
Bacteroides		
	Bordetella	
	Brucella	
		Campylobacter
	Chromobacterium	
	Citrobacter	
Cytophaga		
	Edwardsiella	
		Eikenella
	Enterobacter	
	Erwinia	
Escherichia		
Flavobacterium		
Flexibacter		
	Francisella	
	Gardnerella	
	Haemophilus	
		Halomonas
	Klebsiella	
	Levinea	
	Listeria	
	Moraxella	
	Morganella	
	Neisseria	
	Paracoccus	
	Pasteurella	
	Photobacterium	
	Plesiomonas	
	Proteus	
	Providencia	
Pseudomonas		
Rhizobium		
	Salmonella	
	Saprospira	
	Serratia	
	Shigella	
	Spirillum	
Thiobacillus		
Vibrio		
		Wollinella
Xanthomonas		
		Yersinia
	Zymomonas	

Taxa for Which Numerical Taxonomic Studies Have Most Frequently Been Undertaken

In the early 1970s there was a tendency for numerical taxonomists to focus on easily cultured, mesophilic, nonfastidious, non-spore-forming bacteria. Not unexpectedly, the genus *Pseudomonas* attracted the greatest attention, being the focus of more than two dozen studies (Fig. 1), the results of which have radically altered the profile of the genus. Based on results of numerical taxonomy, the genus has undergone transformation from a collection of loosely related and

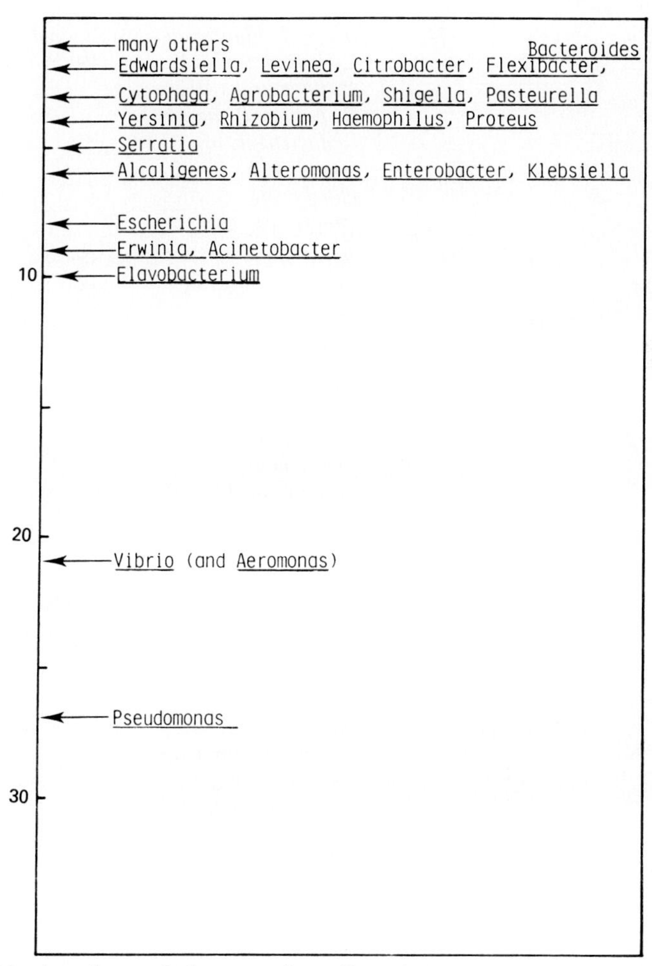

Fig. 1. Number of numerical taxonomic studies per genus on each of 24 most frequently studied bacterial genera since 1971.

Table 9. *Commonality of clusters identified and classified in a variety of NT studies of the genus* Pseudomonas

Target group	Number of strains	Number of tests	Major clusters	References
Marine strains	31	98	*Alteromonas* and *Pseudomonas* spp.	Gray and Stewart (1980)
Strains associated with fish	134	200	*Alteromonas putrifaciens, P. fragi*	Gillespie (1981)
Psychrophilic strains	218	174	*P. fragi, P. fluorescens* I, *P. fluorescens* II, *P. fluorescens* III, *P. putida, P. aureofaciens–chlororaphis, A. putrifaciens, Aeromonas* spp.	Molin and Ternstrom (1982)
'*P. alcaligenes*-like'	18	142	*P. alcaligenes, P. pseudoalcaligenes*	Ralston-Barrett *et al.* (1976)
'Marine strains'	38	165	*Alteromonas* and *Pseudomonas* spp.	Lee *et al.* (1977)
Genus *Pseudomonas*	401	155	'Fluorescent pseudomonad spp.', *P. cepacia–pseudomallei, P. acidovorans*-like, *P. solanacearum, P. mallei*	Sneath *et al.* (1981)
Genus *Pseudomonas* strains	80	134	*P. elongata*-like, *P. fragi*-like, 'fluorescent pseudomonad spp.', *A. denitrificans*	Colwell and Liston (1961)
Phytopathogenic strains	143	200	*P. syringae*-like, 'fluorescent pseudomonad spp.' *P. stutzerii*-like, *P. cepacia*-like	Sands *et al.* (1970)
Genus *Pseudomonas*	118	150	*P. aeruginosa, P. syringae*-like, *P. fluorescens, P. fragi–P. denitrificans, Flavobacterium–Xanthomonas, A. putrifaciens*	Colwell *et al.* (1965, 1969)

largely uncharacterized bacteria, many of which shared only the features of nonfermentative metabolism, into several genera, whose component species represent, for the most part, well-defined clusters (Table 9).

The group of bacteria which has been the second most frequently studied by numerical taxonomic methods are the vibrios, for which results of more than 18 major analyses have been published. For *Vibrio* taxonomy, the results have had even greater impact than those of the genus *Pseudomonas*. Since 1970, the genus has undergone extensive revision related, to a great extent, to numerical taxonomic studies. In 1974 the genus comprised 5 species compared with 29 species now validly defined (see Table 10). A number of *Vibrio* species arose

Table 10. *Chronology of the speciation of the genus* Vibrio

1974[a]	1983[b]	Past synonyms
V. anguillarum	*V. anguillarum*	
V. cholerae	*V. cholerae*	
V. costicola	*V. costicola*	
V. fischeri	*V. fischeri*	*Photobacterium fischeri*
V. parahaemolyticus	*V. parahaemolyticus*	
	V. albensis	*V. cholerae* biotype *albensis*
	V. alginolyticus	*V. parahaemolyticus* biotype 2
	V. campbelli	*Beneckea campbelli*
	V. damsela	
	V. diazotrophicus	
	V. fluvialis	Group F *Vibrio*
	V. gazogenes	*Beneckea gazogenes*
	V. harveyi	*Lucibacterium harveyi*
	V. logei	*Photobacterium logei*
	V. metschnikovii	*V. cholerae* biotype *proteus*
	V. mimicus	'Sucrose-negative *V. cholerae*'
	V. natriegens	*Beneckea natriegens*
	V. nereis	*Beneckea nereida*
	V. nigripulchritudo	*Beneckea nigrapulchrituda*
	V. pelagius	*Beneckea pelagia*
	V. proteolyticus	*Aeromonas hydrophila* subsp. *proteolytica*
	V. splendidus	*Beneckea splendida*
	V. succinogenes	
	V. vulnificus	*Beneckea vulnifica*
	V. orientalis	
	V. aestuarianus	
	V. tubiashi	
	V. furnissii	
	V. carchariae[c]	

[a]Buchanan and Gibbons (1974).
[b]Skerman *et al.* (1980), and periodic validation lists of new names and new combinations previously published in the *International Journal of Systematic Bacteriology*.
[c]Grimes *et al.* (1984).

from the abolition of the genera *Beneckea* and *Lucibacterium*. However, several new species have resulted from the resolution into new species, by cluster analysis, of previously misidentified strains. A review of the taxonomy of the genus *Vibrio* has been published (West and Colwell, 1984).

Taxonomic Profiles of Gram-negative Bacteria

A means by which the impact of numerical taxonomy on the systematics of Gram-negative bacteria can be evaluated is the comparison of taxonomic schemes as they have evolved since the late 1950s. Not all changes in the taxonomy of the Gram-negative bacteria can be attributed directly to the influence of numerical taxonomy, but numerical taxonomy has had a significant influence.

Anomalies have been reported. For example, DNA–DNA reassociation experiments with *Escherichia coli* and *Shigella flexneri* yielded relative binding ratios (RBR) of 80 to 90% (Brenner *et al.*, 1969, 1978). Results of numerical taxonomic studies, in which *E. coli* and *S. flexneri* were included, showed these species to be related at ~75% S (Dodd and Jones, 1982) to ~87% S (Johnson *et al.*, 1975), suggesting that both species belong to the same genus. For the sake of expediency for clinical microbiologists they have been retained in separate genera. This has not been the case for several other members of the family Enterobacteriaceae, such as *Serratia* (Grimont and DuLong de Rosnay, 1972; Grimont *et al.*, 1977) and *Proteus* (Johnson *et al.*, 1975; McKell and Jones, 1976), for which significant revisions were recently proposed (see Table 11).

Correlation between Genetic and Phenetic Data

While it is improbable that closely related (i.e., genetically similar) strains will produce dissimilar phenetic patterns, there is a much higher probability that distantly related strains may produce similar phenetic patterns. This follows, since dissimilar gene products (enzymes) may catalyze similar reactions. The degeneracy of the genetic code, where a single base substitution (mutation) has about 66.7% chance of being expressed on translation, can have an important effect. If expressed, a mutation can give rise to an alteration in phenotype only in the event that the function of a catalytic site of an enzyme or functional group, in the case of a structural protein, is impaired. However, even single base substitutions can produce inordinately large perturbations in DNA hybridization reactions, since a linear relationship does not exist between percentage of sequence homology and RBR. For example, 95% homology between two pieces of DNA does not mean that 95% of their base sequences are equivalent (Wallace *et al.*, 1979).

The effects of base substitutions (point mutations) on hybridization reactions are dependent on (i) physical location of the substitution, (ii) the relative prox-

Table 11. *Historical perspective for the establishment of the genera* Proteus *and* Serratia

1974[a]	1983[b]	Comments
Proteus		
P. rettgeri	*P. rettgeri*	
P. vulgaris	*P. vulgaris*	
P. mirabilis	*P. mirabilis*	
	P. myxofaciens	New species
P. inconstans	*Providencia inconstans*	Redefined genus
	Providencia alcalifaciens	Redefined genus
	Providencia stuartii	Redefined genus
	Providencia rustigianii	New species
P. morganii	*Morganella morganii*	Redefined genus
Serratia		
S. marcescens	*S. marcescens*	
	S. fonticola	New species
	S. liquifaciens	Redefined species
	S. marinorubra	New species
	S. odorifera	New species
	S. rubidea	Redefined species
	S. plymuthica	New species
	S. proteamaculans	New species

[a]Buchanan and Gibbons (1974).
[b]Skerman *et al.* (1980), and periodic validation lists of new names and new combinations previously published in the *International Journal of Systematic Bacteriology*.

imity or density of substitutions, and (iii) the total number of substitutions. Therefore, there will be a tendency for DNA reassociation results to be somewhat more sensitive to sequence differences than phenetic studies. In fact, this is what is observed when percentage phenetic similarity (%S) is compared with RBR, that is, 'percentage homology' (%H), for a large number of isolates.

Correlations between results of numerical taxonomic analyses and those of DNA base composition determinations, DNA–DNA, and DNA–RNA hybridization studies have been, in the main, excellent. Studies for which a high degree of correlation between nucleic acid and numerical taxonomic results is evident include those accomplished with the genera *Pseudomonas* (Lee *et al.*, 1977; Gray and Stewart, 1980). *Haemophilus* (Thornley, 1967; Greenwood and Pickett, 1980), *Vibrio* and *Aeromonas* (Colwell and Mandel, 1964; Colwell, 1970; Tanner *et al.*, 1981), *Edwardsiella, Enterobacter, Escherichia, Klebsiella, Morganella, Proteus, Providencia, Salmonella,* and *Serratia* (Johnson *et al.*, 1975; Dodd and Jones, 1982), *Erwinia* (Azad and Kado, 1980), *Bacteroides* (Johnson and Ault, 1978), *Alcaligenes* (Pichinoty *et al.*, 1978), and *Cytophaga* and *Flavobacterium* (Hayes, 1977).

For those few examples in which poor correlation has been demonstrated

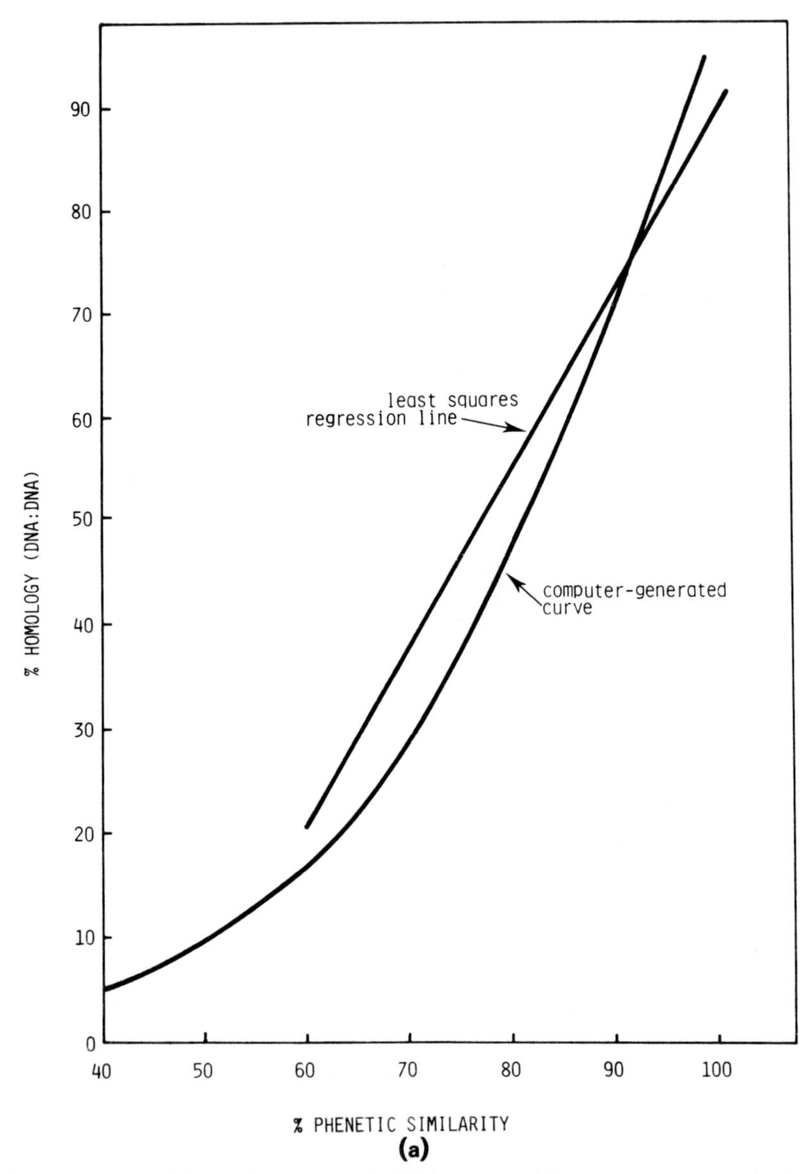

Fig. 2. (a) Linear and curvilinear least squares regression lines generated for 82 observations from published data (r = .74) comparing DNA–DNA homology with phenetic similarity (r = .82) as obtained by numerical taxonomy. (b) Partial decomposition of (a) into least squares regression plots (%S versus %H) to show correlations for each of four genera. ——, Specific genus; – – –, least squares line.

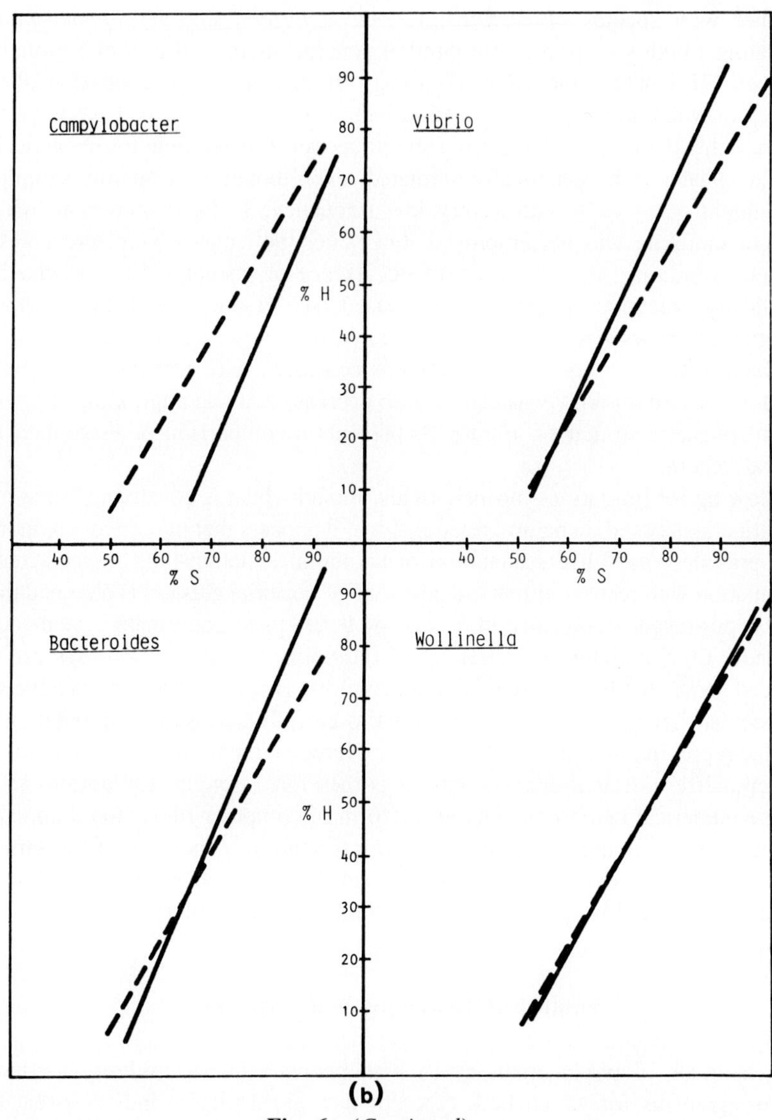

Fig. 6 (*Continued*)

between results of DNA base composition and/or DNA–DNA hybridization and those of numerical taxonomy, reasons for the lack of correlation are apparent in the experimental design. The results of Johnson *et al.* (1968) showed little correspondence between base composition and numerically defined clusters. However, their method of analysis resulted in the clustering of *Vibrio* species together with species of *Bacillus, Corynebacterium,* and *Hyphomicrobium,* indicating a highly suspect partitioning of genera. Significant lack of correlation between NT clusters and DNA homology groups also has occurred in those studies in which the sample population had been preselected, that is, using the criterion that the results fit a given phenetic model. It is possible for phenetically similar isolates to be genetically unrelated, even though in a random sampling this should occur only with a very low frequency. In those studies in which random sampling was not employed, but rather the isolates were screened for phenetic similarity, the frequency of occurrence of phenotypically similar but genetically unrelated isolates was increased (see Harvey and Pickett, 1980).

Figure 2a shows a regression line generated from observations in the literature in which percentage homology (%H) was compared with percentage similarity (%S). Genetic homology was observed to decrease at a rate approximately twice that of phenetic similarity. Figure 2b presents a comparison of these data for several genera.

Allowing for limitations intrinsic to any model which is constructed to assign classifications based on natural relationships, it appears that numerical taxonomy does provide a useful representation of taxonomic relationships. Indeed, more information with respect to relative positions of bacterial clusters is obtained than can be portrayed graphically in a two- or three-space coordinate system. The difficulty of visualizing the topology of multidimensional relationships can be reduced somewhat by the use of Euclidean plots. However, these plots have not yet been sufficiently used for their merits to be fully assessed. Manipulation of graphic representation of results can never overcome the basic human inability to conceptualize *n*-dimensional models. It is therefore strongly desirable to accumulate numerical taxonomic data in the form of computer files, stored on cards or disks, and allowing for the addition of information, expansion of the sample population, and reevaluation of the data set from a fresh perspective. The *n*-dimensional model is, after all, in the realm of the computer.

Numerical Taxonomy and Phylogeny

A comparison of phylogenetic (%P) with phenetic (%S) similarity for those micro-organisms for which both types of data are available indicates that the overall relationship between the two approaches is strong, particularly in the case of *Escherichia* and *Yersinia,* and among *Pseudomonas, Acinetobacter, Alcaligenes,* and *Pasteurella.* This finding suggests that analysis by numerical taxonomy may provide a useful prediction of phylogeny (Fig. 3), which is most encouraging even though there are only a small number of genera for which phylogenetic representation of rRNA relatedness has been demonstrated as yet.

Conserved regions in ribosomal (r) RNA are unequalled as predictors of phylogeny, because of evolutionary conservation in several regions of 5 and 16 S

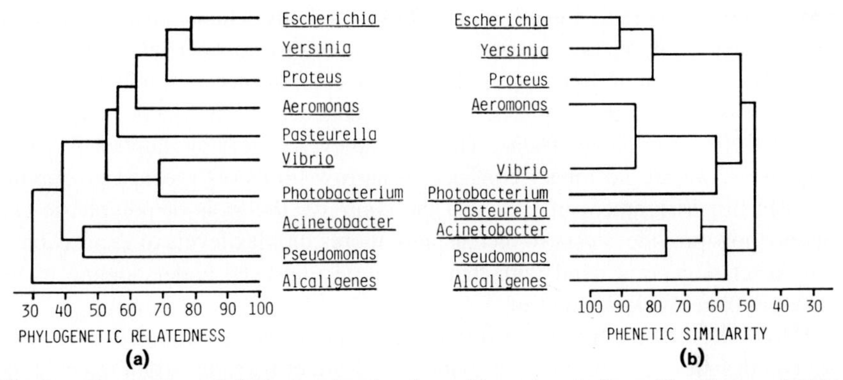

Fig. 3. Comparison of phylogenetic relatedness (S_{AB}, a), as indicated by ribosomal RNA studies, with phenetic similarity (%S, b), as indicated by numerical taxonomy.

rRNA. Early work was done with catalogues of oligomers prepared from 16 S rRNA digests, but, as sequencing methods for RNA improved to the point where the primary structures of rRNA molecules could be determined unambiguously, interest in oligomer catalogues declines sharply. Oligomer populations resulting from 16 S digests are not necessarily unique, and *n*-mers, not bases, are catalogued, 16 S oligomer populations are unable to provide as sensitive an assay of phylogeny as sequences. Whereas oligomer catalogues from 5 S rRNA digests are statistically useless, the 120 bases of 5 S rRNA can be sequenced readily, although 5 S rRNA contains less than one-tenth the number of nucleotide bases contained in the 16 S rRNA molecule. Thus, the 5 S rRNA sequence perhaps may be somewhat less interesting than that of 16 S rRNA, but unequivocal determination of the 1600 bases present in 16 S rRNA molecules is not a trivial problem. Compilations of 5 S rRNA sequences indicate that sequence differences in less highly conserved regions may be expected even at the species level and may therefore allow drawing of inferences about phylogeny for individual genera. At present, the total published library of bacterial 5 S rRNA sequences is small, less than 100 in number, but is steadily increasing.

An excellent preliminary analysis of *bacterial phylogeny,* based largely on 16 S rRNA catalogues, has been published (Fox *et al.*, 1980). Comparisons between results published by Fox *et al.* (1980) and those of numerical taxonomic studies yield the dendrograms shown in Fig. 3.

Conclusions

Methods employed in bacterial systematics have evolved to a level of sophistication sufficient for us now to make a transition from speculation as to which of all possible combinations of genes in bacteria are successful (i.e., viable and reproducing) to a prediction of such. Those methods which have allowed for development of polyphasic taxonomy in microbiology now permit a compilation of profiles of genetic expression. Individual components of a polyphasic taxonomy amount to what might be termed instantaneous 'snapshots', each taken from a different reference frame. Thus, for an ever-increasing number of microorganisms, we are no longer confined to narrow limits of 'greatest probability profiles' for assignment of taxa. On the contrary, the tools of polyphasic taxonomy now provide access to increasingly more complex levels of expression of the bacterial genome, and with that has come improved understanding of the range in phenotypic expression.

DNA–DNA and DNA–RNA hybridization methods, developed over the past several decades, now routinely provide a measure of bacterial strain relatedness. Gene maps prepared for several species of bacteria are now available, and data on gene sequences appear in the literature at such an accelerating rate that soon substantial libraries of genetic information will be available and subject to numerical taxonomic analysis. Access to DNA sequences of entire genomes is likely in the near future, and therefore, direct comparisons between biovars, species, and genera will be possible. Direct access to genome sequences will elevate bacterial systematics substantially beyond our present limited ability to glean chronology in microbial evolution from rRNA sequences.

The Future

Comparative sequencing of informational macromolecules, such as 5 S rRNA, are essentially vectorial; that is, a more 'primitive' sequence becomes a more 'modern' sequence by passing through intermediate ones. Base sequences can be manipulated mathematically by using the tools of vector analysis in order to provide an estimation of the 'direction' in which an RNA sequence, a gene, or a species is evolving.

Advances achieved by the application of data processing to bacterial systematics have been impressive. The dawn of a technology sufficiently sophisticated to provide, for the first time, predictive capabilities is new. Just as bacterial systematics took a quantum leap forward in the late 1950s with the incorporation of computer technology, it is now at the threshold of another quantum leap, through the integration of molecular genetics, systematics, and computer sciences. We should all find ahead of us an exciting era in the field of microbial systematics.

References

Agbo, J. A. C., and Moss, M. O. (1979). The isolation and characterization of agarolytic bacteria from a lowland river. *Journal of General Microbiology* **115,** 355–368.

Austin, B., and Colwell, R. R. (1977). Evaluation of some coefficients for use in numerical taxonomy of microorganisms. *International Journal of Systematic Bacteriology* **27,** 204–210.

Austin, B., Goodfellow, M., and Dickinson, C. H. (1978). Numerical taxonomy of phylloplane bacteria isolated from *Lolium perenne*. *Journal of General Microbiology* **104,** 139–155.

Austin, B., Hussong, D., Weiner, R. M., and Colwell, R. R. (1981). Numerical taxonomy analysis of bacteria isolated from the completed "most probable numbers" test for coliform bacilli. *Journal of Applied Bacteriology* **51,** 101–112.

Azad, H. R., and Kado, C. I. (1980). Numerical and DNA:DNA reassociation analysis of *Erwinia rubrifaciens* and other members of the Enterobacteriaceae. *Journal of General Microbiology* **120,** 117–129.

Barnes, E. M., and Goldberg, H. S. (1968). The relationships of bacteria within the family Bacteroidaceae as shown by numerical taxonomy. *Journal of General Microbiology* **51,** 313–324.

Baumann, P., Doudoroff, M., and Stanier, R. Y. (1968). A study of the *Moraxella* group: II. Oxidative-negative species (genus *Acinetobacter*). *Journal of Bacteriology* **95,** 1520–1541.

Baumann, P., Baumann, L., Bang, S. S., and Woolkalis, M. J. (1980). Reevaluation of the taxonomy of *Vibrio, Beneckea,* and *Photobacterium:* abolition of the genus *Beneckea*. *Current Microbiology* **4,** 127–132.

Beers, R. J., and Lockhart, W. R. (1962). Experimental methods in computer taxonomy. *Journal of General Microbiology* **28,** 633–640.

Beers, R. J., Fisher, J., Megraw, S., and Lockhart, W. R. (1962). A comparison of methods for computer taxonomy. *Journal of General Microbiology* **28,** 641–652.

Belozersky, A. N., and Spirin, A. S. (1960). Chemistry of the nucleic acids of microorganisms. *In* 'The Nucleic Acids' (Eds. E. Chargaff and J. N. Davidson), Vol. 3, pp. 147–185. Academic Press, New York.

Brenner, D. J., Fanning, G. R., Johnson, K. E., Citarella, R. V., and Falkow, S. (1969). Polynucleotide sequence relationships among members of Enterobacteriaceae. *Journal of Bacteriology* **98,** 637–650.

Brenner, D. J., Farmer, J. J., III, Fanning, G. R., Steigerwalt, A. G., Klyken, P., Wathen, H. G., Hickman, S. W., and Ewing, W. H. (1978). Deoxyribonucleic acid relatedness of *Proteus* and *Providencia* species. *International Journal of Systematic Bacteriology* **28,** 269–282.

Broom, A. K., and Sneath, P. H. A. (1981). Numerical taxonomy of *Haemophilus*. *Journal of General Microbiology* **126,** 123–149.

Buchanan, R. E., and Gibbons, N. E. (1974). 'Bergey's Manual of Determinative Bacteriology', 8th Edition. Williams & Wilkins, Baltimore, Maryland.

Carney, J. F., Wan, L., Lovelace, T. E., and Colwell, R. R. (1975). Numerical taxonomy study of *Vibrio* and *Spirillum* spp. *International Journal of Systematic Bacteriology* **25,** 38–46.

Cattell, R. B. (1949). r_p and other coefficients of pattern similarity. *Psychometrika* **14,** 279–298.

Champion, A. B., Barrett, E. L., Palleroni, N. J., Soderberg, K. L., Kunisawa, R., Contopoulou, R., Wilson, A. C., and Doudoroff, M. (1980). Evolution in *Pseudomonas fluorescens*. *Journal of General Microbiology* **120,** 485–511.

Christensen, P. (1980). Description and taxonomic status of *Cytophaga heparina* (Payza and Korn) comb. nov. (basionym: *Flavobacterium heparinum* Payza and Korn 1956). *International Journal of Systematic Bacteriology* **30,** 473–475.

Colwell, R. R. (1969). Numerical taxonomy of the flexibacteria. *Journal of General Microbiology* **58,** 207–215.

Colwell, R. R. (1970). Polyphasic taxonomy of the genus *Vibrio:* numerical taxonomy of *Vibrio cholerae, Vibrio parahaemolyticus,* and related *Vibrio* species. *Journal of Bacteriology* **104,** 410–433.
Colwell, R. R. (1974). Occurrence and biology of *Vibrio parahaemolyticus. Microbiology 1974.* American Society for Microbiology, Washington, D.C.
Colwell, R. R., and Liston, J. (1961). Taxonomic relationships among the pseudomonads. *Journal of Bacteriology* **82,** 1–14.
Colwell, R. R., and Mandel, M. (1964). Adansonian analysis and deoxyribonucleic acid base composition of some Gram-negative bacteria. *Journal of Bacteriology* **87,** 1412–1422.
Colwell, R. R., and Wiebe, W. J. (1970). Core characteristics for use in classifying aerobic, heterotrophic bacteria in numerical taxonomy. *Bulletin of the University of Georgia Academy of Science* **28,** 165–185.
Colwell, R. R., Citarella, R. V., and Ryman, I. (1965). Deoxyribonucleic acid base composition and Adansonian analysis of heterotrophic, aerobic pseudomonads. *Journal of Bacteriology* **90,** 1148–1149.
Colwell, R. R., Moffett, M. L., and Sutton, M. D. (1968). Computer analysis of relationships among phytopathogenic bacteria. *Phytopathology* **58,** 1207–1215.
Colwell, R. R., Citarella, R. V., Ryman, I., and Chapman, G. B. (1969). Properties of *Pseudomonas iodinum. Canadian Journal of Microbiology* **15,** 851–857.
De Ley, J., and Swings, J. (1976). Phenotypic description, numerical analysis, and proposal for an improved taxonomy and nomenclature of the genus *Zymomonas* Kluyver and Van Neil 1936. *International Journal of Systematic Bacteriology* **26,** 146–157.
Dodd, C. E. R., and Jones, D. (1982). A numerical taxonomic study of the genus *Shigella. Journal of General Microbiology* **128,** 1933–1957.
Donis-Keller, H. (1979). Specific enzymatic cleavage of RNA. *Nucleic Acids Research* **7,** 179–192.
Floodgate, G. D., and Hayes, P. R. (1963). The Adansonian taxonomy of some yellow pigmented marine bacteria. *Journal of General Microbiology* **30,** 237–244.
Fox, G. E., Stackebrandt, E., Hespell, R. B., Gibson, J., Maniloff, J., Dyer, T. A., Wolfe, R. S., Balch, W. E., Tanner, R. S., Magrum, L. J., Zablen, L. B., Blakemore, R., Gupta, R., Bonen, L., Lewis, B. J., Stahl, D. A., Luehrsen, K. R., Chen, K. N., and Woese, C. R. (1980). The phylogeny of the procaryotes. *Science* **209,** 457–463.
Gavini, F., Izard, D., Trinel, P. A., LeFebvre, B., and LeClerc, H. (1981). Etude taxonomique d'enterobacteries appartenant ou apparentees a l'especie *Escherichia coli. Canadian Journal of Microbiology* **27,** 98–106.
Gillespie, N. C. (1981). A numerical taxonomic study of *Pseudomonas*-like bacteria isolated from fish in southeastern Queensland and their association with spoilage. *Journal of Applied Bacteriology* **50,** 29–44.
Goodfellow, M., Austin, B., and Dickinson, C. H. (1976). Numerical taxonomy of some yellow-pigmented bacteria isolated from plants. *Journal of General Microbiology* **97,** 219–233.
Gower, J. C. (1971). A general coefficient of similarity and some of its properties. *Biometrics* **27,** 857–874.
Gray, P. A., and Stewart, D. J. (1980). Numerical taxonomy of some marine pseudomonads and alteromonads. *Journal of Applied Bacteriology* **49,** 375–383.
Green, P. N., and Bousfield, I. J. (1982). A taxonomic study of some Gram-negative facultatively methylotrophic bacteria. *Journal of General Microbiology* **128,** 623–638.
Greenwood, J. R., and Pickett, M. J. (1980). Transfer of *Haemophilus vaginalis* Gardner

and Dukes to a new genus, *Gardnerella: G. vaginalis* (Gardner and Dukes) comb. nov. *International Journal of Systematic Bacteriology* **30,** 170–178.
Griffiths, A. J., and Lovitt, R. (1980). Use of numerical profiles for studying bacterial diversity. *Microbial Ecology* **6,** 35–43.
Grimes, D. J., Stemmler, J., Hada, H., May, E. B., Maneval, D., Hetrick, F. M., Jones, R. T., Stoskopf, M., and Colwell, R. R. (1984). *Vibrio* species associated with mortality of sharks held in captivity. *Microbial Ecology* **10,** 271–282.
Grimont, P. A. D., and DuLong de Rosnay, H. L. C. (1972). Numerical study of 60 strains of *Serratia. Journal of General Microbiology* **72,** 259–268.
Grimont, P. A. D., Grimont, F., DuLong de Rosnay, H. L. C., and Sneath, P. H. A. (1977). Taxonomy of the genus *Serratia. Journal of General Microbiology* **98,** 36–66.
Gutteridge, C. S., and Puckey, D. J. (1982). Discrimination of some Gram-negative bacteria by direct probe mass spectrometry. *Journal of General Microbiology* **128,** 721–730.
Harvey, S., and Pickett, M. J. (1980). Comparison of Adansonian analysis and deoxyribonucleic acid hybridization results in the taxonomy of *Yersinia enterocolitica. International Journal of Systematic Bacteriology* **30,** 86–102.
Hayes, P. R. (1977). A taxonomic study of flavobacteria and related Gram-negative yellow pigmented rods. *Journal of Applied Bacteriology* **43,** 345–367.
Heinke, F. (1898). Naturgeschichte des Herings. I. Die Lokalformen und die Wanderungen des Herings in den europaischen Meeren. *Abhandlung Deutsch Seefischerei-Vereins* **2,** 1–223.
Hill, L. R. (1975). Interlocking numerical taxonomies. *International Journal of Systematic Bacteriology* **25,** 245–251.
Holmes, B., Owen, R. J., and Weaver, R. E. (1981). *Flavobacterium multivorum,* a new species isolated from human clinical specimens and previously known as group IIk, biotype 2. *International Journal of Systematic Bacteriology* **31,** 21–34.
Hussong, D., Damaré, J. M., Weiner, R. M., and Colwell, R. R. (1981). Bacteria associated with false-positive most-probable-number coliform test results for shellfish and estuaries. *Applied and Environmental Microbiology* **41,** 35–45.
Hutchinson, M., Johnstone, K. I., and White, D. (1966). Taxonomy of the acidophilic thiobacilli. *Journal of General Microbiology* **44,** 373–381.
Hutchinson, M., Johnstone, K. I., and White, D. (1969). Taxonomy of the genus *Thiobacillus:* the outcome of numerical taxonomy applied to the group as a whole. *Journal of General Microbiology* **57,** 397–410.
Jaccard, P. (1908). Nouvelles recherches sur la distribution florale. *Bulletin de la Societe Vaudoise des Sciences Naturelles* **44,** 223–270.
Johnson, J. L., and Ault, D. A. (1978). Taxonomy of the *Bacteroides* II. Correlation of phenotypic characteristics with deoxyribonucleic acid homology groupings for *Bacteroides fragilis* and other saccharolytic *Bacteroides* species. *International Journal of Systematic Bacteriology* **28,** 257–268.
Johnson, R., and Sneath, P. H. A. (1973). Taxonomy of *Bordetella* and related organisms of the families Achromobacteriaceae, Brucellaceae, and Neisseriaceae. *International Journal of Systematic Bacteriology* **23,** 381–404.
Johnson, R., Colwell, R. R., Sakazaki, R., and Tamura, K. (1975). Numerical taxonomy study of the Enterobacteriaceae. *International Journal of Systematic Bacteriology* **25,** 12–37.
Johnson, R. M., Katarski, M. E., and Weisrock, W. P. (1968). Correlation of taxonomic criteria for a collection of marine bacteria. *Applied Microbiology* **16,** 708–713.
Kaneko, K., and Hashimoto, N. (1982). Five biovars of *Yersinia enterocolitica*

delineated by numerical taxonomy. *International Journal of Systematic Bacteriology* **32,** 275–287.

Kapperud, G., Bergan, T., and Lassen, J. (1981). Numerical taxonomy of *Yersinia enterocolitica* and *Yersinia enterocolitica*–like bacteria. *International Journal of Systematic Bacteriology* **31,** 401–419.

Kersters, K., De Ley, J., Sneath, P. H. A., and Sackin, M. (1973). Numerical taxonomic analysis of *Agrobacterium. Journal of General Microbiology* **78,** 227–239.

King, A., Holmes, B., Phillips, I., and LaPage, S. P. (1979). A taxonomic study of clinical isolates of *Pseudomonas pickettii, "P. thomasii"* and "group IVd" bacteria. *Journal of General Microbiology* **114,** 137–147.

Lee, K. Y., Wahl, R., and Barbu, E. (1956). Contenu en bases puriques et pyrimidiques des DNA des bacteries. *Annales de l'Institut Pasteur* **91,** 212–224.

Lee, J. V., Gibson, D. M., and Shewan, J. M. (1977). A numerical taxonomic study of some *Pseudomonas*-like marine bacteria. *Journal of General Microbiology* **98,** 439–451.

Lee, J. V., Donovan, T. J., and Furniss, A. L. (1978). Characterization, taxonomy, and emended description of *Vibrio metschnikovii. International Journal of Systematic Bacteriology* **28,** 99–111.

Lee, J. V., Shread, P., Furniss, A. L., and Bryant, T. N. (1981). Taxonomy and description of *Vibrio fluvialis* sp. nov. (synonym group F vibrios, group EF6). *Journal of Applied Bacteriology* **50,** 73–94.

Mallory, L. M., Austin, B., and Colwell, R. R. (1977). Numerical taxonomy and ecology of oligotrophic bacteria isolated from the estuarine environment. *Canadian Journal of Microbiology* **23,** 733–750.

Maxam, A. M., and Gilbert, W. (1980). Sequencing end-labeled DNA with base-specific chemical cleavages. *Methods in Enzymology* **65,** 499–560.

McKell, J., and Jones, D. (1976). A numerical taxonomy study of *Proteus–Providence* bacteria. *Journal of Applied Bacteriology* **41,** 143–161.

McMeekin, T. A., Steward, D. B., and Murray, J. G. (1972). The Adansonian taxonomy and the deoxyribonucleic acid base composition of some Gram-negative, yellow pigmented rods. *Journal of Applied Bacteriology* **35,** 129–137.

Michener, C. D., and Sokal, R. R. (1957). A quantitative approach to a problem in classification. *Evolution* **11,** 130–162.

Moffett, M. L., and Colwell, R. R. (1968). Adansonian analysis of the Rhizobiaceae. *Journal of General Microbiology* **51,** 245–266.

Molin, G., and Ternstrom, A. (1982). Numerical taxonomy of psychrophilic pseudomonads. *Journal of General Microbiology* **128,** 1249–1264.

Moss, M. O., Ryall, C., and Logan, N. A. (1978). The classification and characterization of chromobacteria from a lowland river. *Journal of General Microbiology* **105,** 11–21.

Pagel, J. E., and Seyfried, P. L. (1976). Numerical taxonomy of aquatic *Acinetobacter* isolates. *Journal of General Microbiology* **95,** 220–232.

Peattie, B. A. (1979). Direct chemical methods for sequencing RNA. *Proceedings of the National Academy of Science of the United States of America* **76,** 1760–1764.

Pichinoty, F., Veron, M., Dandel, M., Durand, M., Job, C., and Garcia, J.-L. (1978). Etude physiologique et taxonomique du genre *Alcaligenes: A. denitrificans, A. odorans* et *A. faecalis. Canadian Journal of Microbiology* **24,** 743–753.

Ralston-Barrett, E., Palleroni, N. J., and Doudoroff, M. (1976). Phenotypic characterization and deoxyribonucleic acid homologies of the '*Pseudomonas alcaligenes*' group. *International Journal of Systematic Bacteriology* **26,** 421–426.

Sakazaki, R., Tamura, K., Johnson, R., and Colwell, R. R. (1976). Taxonomy of some recently described species in the family Enterobacteriaceae. *International Journal of Systematic Bacteriology* **26,** 158–178.

Sands, D. C., Schroth, M. N., and Hildebrand, D. C. (1970). Taxonomy of phytopathogenic pseudomonads. *Journal of Bacteriology* **101,** 9–23.

Sanger, F., Nicklen, S., and Coulson, A. R. (1977). DNA sequencing with chain terminating inhibitors. *Proceedings of the National Academy of Science of the United States of America* **74,** 5463–5467.

Skerman, V. B. D., McGowan, V., and Sneath, P. H. A. (1980). Approved lists of bacterial names. *International Journal of Systematic Bacteriology* **30,** 225–420.

Sneath, P. H. A. (1957). The application of computers to taxonomy. *Journal of General Microbiology* **17,** 201–226.

Sneath, P. H. A., and Johnson, R. (1972). The influence on numerical taxonomic similarities of errors in microbiological tests. *Journal of General Microbiology* **72,** 377–392.

Sneath, P. H. A., and Sokal, R. R. (1973). 'Numerical Taxonomy'. Freeman, San Francisco.

Sneath, P. H. A., Stevens, M., and Sackin, M. J. (1981). Numerical taxonomy of *Pseudomonas* based on published records of substrate utilization. *Antonie van Leeuwenhoek* **47,** 423–448.

Sokal, R. R., and Michener, C. D. (1958). A statistical method for evaluating systematic relationships. *University of Kansas Science Bulletin* **38,** 1409–1438.

Tanner, A. C. R., Badger, S., Lai, C.-H., Listgarten, M. A., Visconti, R. A., and Socransky, S. S. (1981). *Wolinella* gen. nov., *Wolinella succinogenes* (*Vibrio succinogenes* Wolin *et al.*) comb. nov., and description of *Bacteroides gracilis* sp. nov., *Wolinella recta* sp. nov., *Campylobacter concisus* sp. nov., and *Eikenella corrodens* from humans with periodontal disease. *International Journal of Systematic Bacteriology* **31,** 432–445.

Thornley, M. J. (1967). A taxonomic study of *Acinetobacter* and related genera. *Journal of General Microbiology* **49,** 211–257.

Thurner, K., and Busse, M. (1978). Numerisch taxonomische untersuchungen en Enterobakterien aus Oberflaechenwasser. *Zentralblatt für Bakteriologie, Mikrobiologie und Hygiene, Abteilung 1, Originale, Reihe B* **167,** 262–271.

Wallace, R. B., Shaffer, J., Murphy, R. F., Bonner, J., Hirose, T., and Itakura, K. (1979). Hybridization of synthetic oligodeoxyribo-nucleotides to ϕX 174: the effect of single base pair mismatch. *Nucleic Acids Research* **6,** 3543–3557.

West, P. A., and Colwell, R. R. (1984). Identification of Vibrionaceae—an overview. *In* 'Vibrios in the Environment' (Ed. R. R. Colwell), pp. 285–363. Wiley (Interscience), New York.

White, L. O. (1972). The taxonomy of the crown-gall organism *Agrobacterium tumefaciens* and its relationship to rhizobia and other agrobacteria. *Journal of General Microbiology* **72,** 565–574.

Note Added in Proof

Since compilation of this chapter, the named species of the Vibrionaceae have been analyzed by 5 S rRNA comparative sequence analysis, and a phylogeny proposed (MacDonnell and Colwell, 1985). Furthermore, the partitioning of the *Aeromonas* species into the family Aeromonadaceae also has been proposed. It will be interesting to observe the development of a phylogeny of bacteria, in general, as the new data unfold.

7

Numerical Taxonomy of Lactic Acid Bacteria and Some Related Taxa

F. G. PRIEST

Department of Brewing and Biological Sciences, Heriot–Watt University, Edinburgh, UK

E. A. BARBOUR

Technical Centre, Scottish and Newcastle Breweries PLC, Edinburgh, UK

Introduction

Numerical taxonomy has proved to be a popular and successful approach to bacterial classification and identification. Despite the acknowledged value and usefulness of numerical techniques, however, some groups of bacteria have been largely ignored. *Lactobacillus* is a typical example, and yet these common bacteria are economically important and widely distributed in various dairy products, fermented beverages, and ensilages, and are found in association with humans, animals, and plants (London, 1976). This article is concerned with the evaluation of numerical phenetic studies that have focussed on, or included, *Lactobacillus* and related strains.

Relationships between *Lactobacillus* and Some Other Gram-positive Taxa

Lactobacilli are nonsporulating, Gram-positive rods of varying length. They are strictly fermentative, producing either a mixture of lactic acid, CO_2, acetic acid, and/or ethanol (heterofermentation), or almost entirely lactic acid (homofermentation) from glucose. They are usually catalase negative, aciduric, and have complex nutritional requirements. Current classifications of *Lactobacillus* are still based on the pioneering work of Orla Jensen (1919), who divided the genus into three groups on the basis of growth temperature and end product fermentation. Heterofermentative strains were placed in '*Betabacterium*', and homofer-

ISBN 0-12-289665-3

mentative strains were split into two taxa. Those that grew at 45°C but not 15°C constituted '*Thermobacterium*', and those with the opposite temperature relationships were placed in '*Streptobacterium*'. Subsequently the heterofermentative strains were divided into two: those which are fermentatively active, and a group of species which are inert to most carbohydrates, are acidophilic, and tolerate organic acids and ethanol at high concentrations (Sharpe, 1979).

Orla Jensen (1919) recognised three other genera of lactic acid bacteria: *Streptococcus*, '*Betacoccus*', and '*Tetracoccus*'. Of these, only *Streptococcus* remains valid for the homofermentative, facultatively anaerobic cocci. '*Betacoccus*' is now *Leuconostoc* and includes heterofermentative cocci occurring in pairs or short chains which appear to be physiologically similar to the heterofermentative lactobacilli. '*Tetracoccus*' has been renamed *Pediococcus* and contains those homofermentative cocci which occur in pairs and tetrads through division in two planes. The lactic acid bacteria share considerable physiological and biochemical similarity and have long been recognised as a 'natural' group. The greatest diversity is revealed in the base composition of the DNA of *Lactobacillus* species (32–53% G + C), which form a heterogeneous taxon in need of some taxonomic revision. *Streptococcus* (34–46% G + C), *Leuconostoc* (37–46% G + C), and *Pediococcus* (34–44% G + C) show a narrower range of C + G content, and each conforms to the guideline of less than 10–12% variation within a genus (Bradley, 1980; Priest, 1981).

It is well known that numerical phenetic analyses are usually of little value in determining relationships above the genus level. Indeed, suprageneric relationships are best determined from analyses of macromolecules that are commonly distributed amongst micro-organisms. DNA is an obvious choice, but bacterial chromosomes have diverged to such an extent that base sequences from bacteria belonging to different genera have little in common and relatedness cannot readily be detected in this way. Ribosomal (r) RNA however, appears to have been highly conserved throughout evolution, and molecules even from the most distantly related bacteria have sequences in common. Comparison of 16 S rRNA molecules therefore allows a comprehensive phylogenetic classification of micro-organisms to be prepared (Stackebrandt and Woese, 1981). Since the molecule is too large to be sequenced directly, it is first hydrolysed enzymically; the oligonucleotide products are then separated by electrophoresis and sequenced. The sequences (catalogues) are usually compared using a similarity coefficient and data clustered using the unweighted pair-group method to provide a dendrogram (Fox *et al.*, 1977).

Analysis of lactobacilli and other low G + C content (<55 mol %) Gram-positive bacteria using the 16 S rRNA cataloguing technique revealed three rRNA families. The early phenetic studies were validated when *Lactobacillus* strains were recovered in the same suprafamily as *Leuconostoc* and *Pediococcus*. This aggregate group also contained *Kurthia zopfii* and was related on the one hand to the streptococci, which formed a homogeneous rRNA taxon, and on the other to the *Bacillus* group. The latter also encompassed *Peptococcus, Planococcus, Sporolactobacillus, Sporosarcina, Staphylococcus,* and *Thermoactinomyces* (Stackebrandt and Woese, 1981). The influence of numerical taxonomy on the classification of the three suprafamilies is considered below.

The *Bacillus* Group

Bacillus systematics has been reviewed (Berkeley and Goodfellow, 1981), but some of the more important developments are considered here. The genus *Bacillus* is heterogeneous as indicated by the wide range of DNA base composition (34–68% G + C) amongst representative species (Priest, 1981). Since bacterial genera seldom display more than 10 to 12% variation in G + C content, it is clear that *Bacillus* species should be assigned to more than one genus. Two extensive numerical phenetic analyses of bacilli provide a framework for such a classification (Logan and Berkeley, 1981; Priest *et al.*, 1981), but the results of these studies need to be confirmed and extended by chemical and nucleic acid studies. It is, however, encouraging that the two numerical phenetic surveys, which employed completely different data sets, gave congruent results.

Logan and Berkeley (1981) performed 139 tests on 600 cultures and analysed the data using the Gower coefficient S_G and average-linkage clustering. Six aggregate clusters were recovered: Group I comprised *B. cereus, B. thuringiensis,* and related organisms; Group II contained strains of *B. alvei, B. brevis, B. firmus,* and *B. lentus;* Group III included *B. pasteurii* and *B. sphaericus;* Group IV consisted of *B. megaterium, B. subtilis,* and related organisms; Group V contained *B. circulans, B. macerans,* and *B. polymyxa;* and Group VI comprised the thermophilic species *B. coagulans* and *B. stearothermophilus.*

The numerical analysis of Priest *et al.* (1981) was based on data published by Gordon *et al.* (1973); consequently only 35 characters were included. Computation by the simple-matching coefficient S_{sm} and unweighted pair-group method with averages (UPGMA) clustering revealed six aggregate clusters that were largely in accord with those of Logan and Berkeley. Some differences were apparent: for example, *B. alvei* clustered with *B. polymyxa* and *B. macerans* rather than with *B. firmus,* and *B. coagulans* clustered with the *B. polymyxa* group rather than with *B. stearothermophilus.* These data emphasise the need to confirm numerical phenetic classifications with chemotaxonomic and genetic data, particularly when considering relationships between taxa of probable generic rank. However, the large measure of agreement found between the two studies indicates that numerical taxonomy is providing a generally sound classification for aerobic endospore-forming bacilli.

Other numerical taxonomic studies of bacilli have concentrated on particular groups. Bonde (1975, 1981) and Boeyé and Aerts (1976) found that *Bacillus* isolates from marine sediments were variants of established species such as *B. cereus, B. megaterium,* and *B. pumilus.* Bonde (1981) also recovered a number of unidentified clusters that had properties in common with species that are common in the marine environment, such as '*B. epiphytus*' and '*B. cirroflagellosus*'. Psychrotrophic bacilli have also been examined using numerical techniques. Gyllenberg and Laine (1970) classified 41 strains into five clusters using 60 characters and the S_{sm} coefficient with both single-linkage and average-linkage algorithms. The results showed that, unlike the bacilli from marine habitats, the psychrotrophic strains represented taxa which were distinct from established mesophilic species. Similarly, Sheard and Priest (1981) examined 14 reference mesophilic strains and 75 psychrophilic bacilli isolated from various foods for 107 characters and analysed the data using the S_{sm} coefficient and average-linkage (UPGMA) clustering. The results are shown in abbreviated form in Fig. 1. It can be seen that most of the food isolates clustered with the psychrotrophic taxa: *B. insolitus,* '*B. psychrophilus*', and '*B. psychrosaccharolyticus*'. Some new psychrotrophic taxa were indicated, but only in the case of one strain of *B. macerans* was there an indication of a cold-tolerant variant of a mesophilic species.

These examples of numerical analyses of bacilli show the insight into the taxonomic structure of a large heterogeneous taxon that can be provided by comprehensive studies. However, they also indicate the value of the smaller study of isolates from a particular ecological niche and how this might be used to classify the organisms present and relate them to established taxa.

Ribosomal RNA cataloguing has suggested close relationships between some taxa that had hitherto been considered so distinct as to be placed in different families. The genus *Thermoactinomyces* is a good example. The thermophilic bacteria in this genus characteristically produce an extensive substrate mycelium that carries aerial hyphae, and they have consequently been classified with the actinomycetes (Küster, 1974). The discovery that thermoactinomycetes produced endospores which resemble those of the Baciliaceae (Cross *et al.*, 1968; Cross, 1981) questioned this relationship. A numerical phenetic study of monosporic actinomycetes (McCarthy and Cross, 1983) included strains of *Thermoactinomyces, Actinomadura, Micropolyspora, Saccharomonospora,* and *Thermomonospora.* The thermoactinomycetes were recovered as a discrete cluster showing low relatedness to the true actinomycete genera. This result is consisitent with the proposal (Goodfellow and Cross, 1984) that the genus *Thermoactinomyces* be transferred to the family Bacillaceae on the basis of a wealth of data including that from 16 S rRNA cataloguing studies. The relatively low G + C content of thermoactinomycetes (52–55%; Craveri *et al.*, 1965, 1973) supports the relationship with *Bacillus* species particularly with the thermophilic taxa that have G + C contents in the range 47–65% (Sharp *et al.*, 1980)

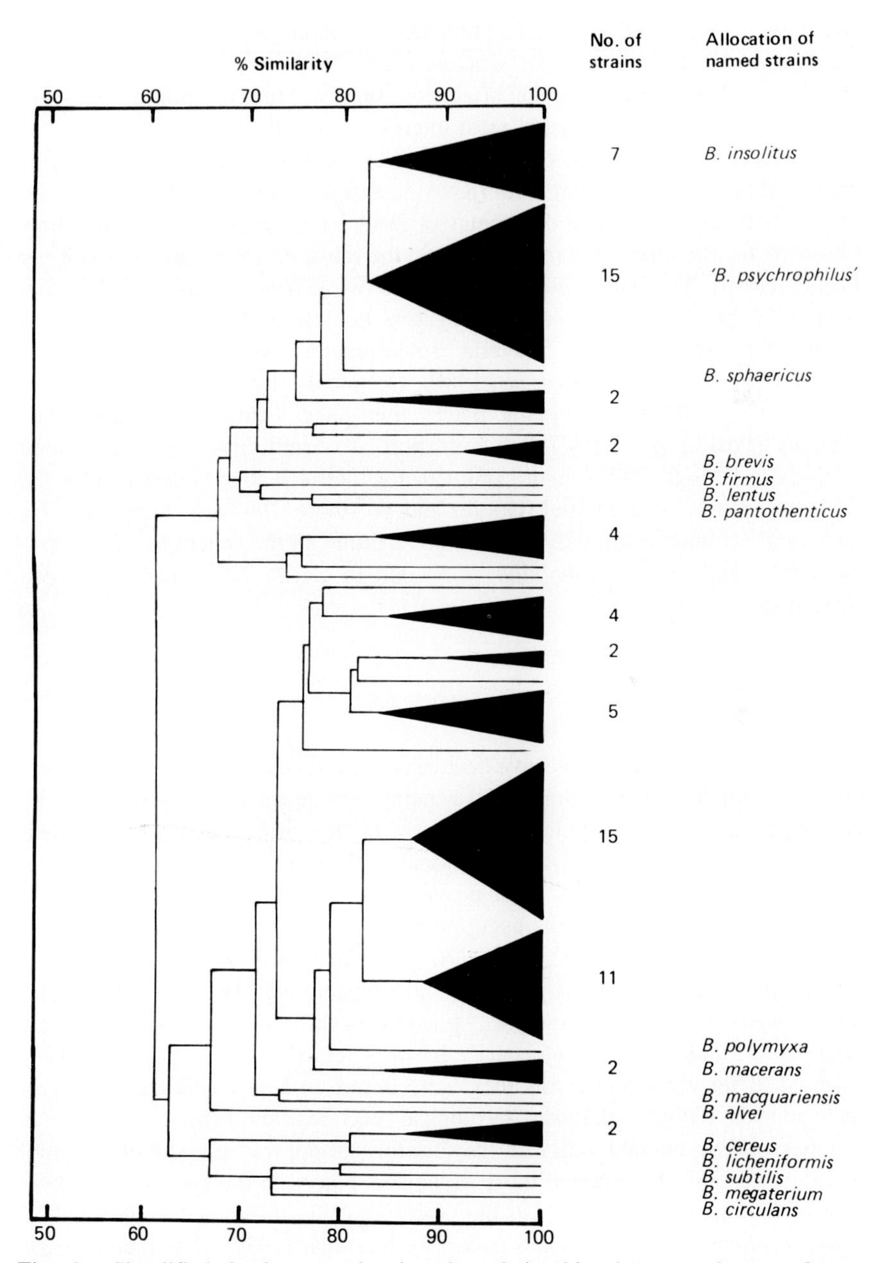

Fig. 1. Simplified dendrogram showing the relationships between clusters of psychrotrophic *Bacillus* isolates and reference strains based on the S_{sm} coefficient and average-linkage (UPGMA) cluster analysis.

and helps to underline their separation from actinomycetes which have much higher G + C contents (>63%; Craveri *et al.*, 1973; Bradley and Mordarski, 1976). It is likely that a numerical taxonomic survey of thermoactinomycetes and thermophilic bacilli would yield most interesting results.

Similarly, the relationships between *Bacillus* and *Lactobacillus* have yet to be examined in detail. It is clear from rRNA cataloguing studies that these taxa are phylogenetically related and their range of DNA base composition is compatible (32–65% for *Bacillus,* with most strains in the range 32–55%, and 33–52% for *Lactobacillus*). The borderline between these two genera became even less distinct with the isolation of catalase-negative homofermentative (producing D-lactate) endospore-forming bacteria subsequently assigned to the genus *Sporolactobacillus* (Skerman *et al.*, 1980). Despite its physiological similarity to *Lactobacillus,* rRNA cataloguing places this genus firmly within the genus *Bacillus* (Fox *et al.*, 1977). Several sporolactobacilli have been described (Uchida and Mogi, 1973), and a protocol for their selective isolation from the environment has been devised (Doores and Westhoff, 1983). A comprehensive numerical taxonomic study of representative strains of the genera *Bacillus, Lactobacillus,* and *Sporolactobacillus* is needed to clarify relationships between these taxa.

Streptococcus

The systematics of this genus have been reviewed in detail (Jones, 1978). However, the application of numerical phenetic techniques to the genus will be considered in order to highlight two points: (1) the broad correlation between numerical classification and the traditional division of the genus into several physiological groups, and (2) the provision of useful identification schemes based on probabilistic matrices using phenetic data.

In her review, Jones (1978) divided the genus *Streptococcus* into seven groups based partly on the original physiological divisions of Sherman (1937). The groups were designated 'pyogenic', 'pneumococcal', 'oral', 'faecal', 'lactic', 'anaerobic', and 'other streptococci'. It was stressed that the groupings were established for convenience, as they were based on a few characters such as pathogenesis, oxygen tolerance, serological reactions, and habitat.

Jones (1978) pointed out that the 'faecal' group was particularly heterogeneous, because it contained those organisms placed in the enterococcal division of Sherman (1937) and also the physiologically different *S. bovis* and *S. equinus* allocated to the viridans group by Sherman (1937). The enterococcal species '*S. avium*', *S. faecalis, S. faecium,* and the newly designated *S. gallinarum* have been recovered as an independent aggregate cluster in several numerical phenetic studies (Raj and Colwell, 1966; Seyfried, 1968; Davis *et al.*,

1969; Bridge and Sneath, 1983a,b). The distinctness of the enterococcal group is supported by DNA–RNA hybridization, which is negligible between *S. faecalis* and viridans streptococci (Weissman *et al.*, 1966). Kalina (1970) suggested that the enterococci merited genus rank. The relationships of *S. bovis* and *S. equinus* to the enterococci have also been clarified by numerical taxonomy. In a study of 122 serological group D streptococci, *S. bovis* and *S. equinus* clustered separately and distinct from *S. faecalis* (Jones *et al.*, 1972), and in a comprehensive numerical analysis of the genus, these organisms clustered with *S. salivarius* and the paraviridans group (Bridge and Sneath, 1983b), supporting Sherman's original proposal. Moreover, *S. bovis* shares very little DNA homology with the enterococci (Farrow *et al.*, 1983).

The complex and confusing state of the taxonomy of the oral streptococci has been reviewed (Hardie and Marsh, 1978). '*Streptococcus milleri*', '*S. mitior*', *S. mutans, S. salivarius,* and *S. sanguis* are commonly found in the mouth and constitute the oral (viridans) group. The few numerical studies that have been carried out on these bacteria (Carlsson, 1968; Coleman, 1968; Drucker and Melville, 1971) generally support the division of the oral streptococci into these taxa. Bridge and Sneath (1983b) recovered them as a large complex cluster divided into the paraviridans (containing *S. bovis* and *S. equinus*) and viridans groups with intermediate clusters of *S. lactis* and *S. thermophilus*. These divisions were also apparent in a principal coordinates analysis of the centroids of individual clusters (Bridge and Sneath, 1983b). Thus, the concept of the viridans group and the species within it has been largely supported by numerical phenetic studies.

The lactis group includes *S. cremoris* and *S. lactis,* which are often considered to be synonymous. Certainly, the properties of the fructose diphosphate aldolases are very similar (London and Kline, 1973), and there is considerable sequence homology between their genomes (Garvie *et al.*, 1981). Although Seyfried (1968) found substantial differences between *S. lactis* and *S. cremoris* strains subjected to numerical analysis, the studies of Bridge and Sneath (1983b) and Feltham (1979) showed them to be phenetically very similar and attributed minor differences to the slow growth rate of *S. cremoris*.

Finally, the pyogenic streptococci have been recovered as a distinct aggregate cluster in several numerical taxonomic studies (Coleman, 1968; Seyfried, 1968; Feltham, 1979; Bridge and Sneath, 1983b). Few *S. pneumoniae* strains have been studied. Bridge and Sneath (1983b) included only one strain.

Identification of streptococci has traditionally relied on serological methods, but many strains are untypable using current schemes (Hardy *et al.*, 1978). Consequently, Feltham and Sneath (1982) have prepared a probabilistic identification matrix for streptococci using data largely derived from a numerical taxonomic study (Feltham, 1979). This matrix includes 32 *Streptococcus* taxa and *Aerococcus viridans,* and provides percentage probabilities for 60 test results.

The matrix has been used to identify successfully 146 of 160 (93.6%) reference strains and 68 of 80 (85%) field strains. Of the 22 unidentified strains, most were of the viridans group, indicating that these bacteria are inadequately classified. Sneath devised several statistical analyses for the assessment of the quality of probabilistic identification matrices of which CHARSEP (Sneath 1979a) can be used to determine those characters with high discriminatory power. Of the 60 tests, only 25 provided good discrimination and could be considered useful for the identification of streptococci. OVERMAT (Sneath, 1980) can be used to calculate the overlap between taxa, and within the *Streptococcus* matrix there was considerable overlap (1–10%) between several clusters (Feltham and Sneath, 1982). It is interesting to compare this to a similar matrix for micrococci and staphylococci in which, of 60 characters, 29 had good discriminatory power and the greatest overlap between two taxa (*M. luteus* 3 and *M. luteus* 4) was 3%. No other overlap exceeded 1.6% (Feltham and Sneath, 1982). Although these better parameters may be partly due to the use of more strains and better tests, the high overlap figures for the *Streptococcus* taxa suggest that these bacteria may be relatively phenetically homogeneous, with the divisions between clusters not being so pronounced as in the Micrococcaceae and other taxa. This point will also be considered in respect to the genus *Lactobacillus.*

Lactobacillus Group

This group contains the genera *Lactobacillus, Leuconostoc,* and *Pediococcus.*

Pediococcus

This genus contains six species commonly associated with lactic fermentations of vegetables, grain mashes, and yeast fermentations. Despite some nomenclatural problems (Garvie, 1974a), the taxonomy of the genus is now well established (Back and Stackebrandt, 1978). A recent comprehensive study of 830 strains, based on physiological and DNA pairing data, supported the integrity of *P. acidilactici, P. damnosus, P. dextrinicus, P. halophilus, P. parvulus,* and *P. pentosaceus,* but also highlighted a new species, '*P. inopinatus*'. All of the species were genotypically distinct; the highest inter-species reassociation at a nonrestrictive incubation temperature was 30–36% between DNA from *P. damnosus* and *P. parvulus* strains. The '*P. inopinatus*' DNA showed ~40% reassociation with both *P. damnosus* and *P. parvulus* DNA (Back, 1978), but such a degree of reassociation is sufficiently low to warrant separate species status (Bradley, 1980).

Pediococci are important spoilage agents of beers. Traditionally, *P. damnosus* was considered to be the most serious spoilage agent, but *P. pentosaceus* can also be a problem. It is therefore important that rapid and reliable identification schemes are available to distinguish these and related bacteria. With this in mind, Lawrence and Priest (1981) examined 96 Gram-positive cocci from beer and the brewery environment. Thirty-six of the test strains were pediococci; the remainder were micrococci and staphylococci. All strains were examined for 139 biochemical, morphological and physiological features based on the API 50L system. Additional tests included characteristics relevant to the study such as the ability of strains to grow at low pH and in the presence of hops, and their ability to grow in, and consequently spoil, ale- and lager-type beers. Data were coded as two-state characters and analysed by the simple-matching (S_{sm}), Jaccard (S_J), and pattern similarity (S_p) coefficients with unweighted average-linkage (UPGMA) clustering. The S_p coefficient removed variation due to differences in vigor between strains (Sneath, 1968), a factor that is considered particularly relevant when comparing bacteria that are as metabolically diverse as pediococci and micrococci. Moreover, since many of the isolates had been recently isolated from beers, the S_p coefficient should allow for negative results that might arise from poor growth on laboratory media. In fact, all three computations gave virtually the same results. Of particular importance, the distribution of strains to clusters was unaffected by the type of computation. Such robustness is indicative of a sound and stable classification.

The pediococci were recovered in five clusters at 80% similarity with three unidentified strains (Fig. 2). Most of the brewery isolates formed a homogeneous phenon with the type strain of *P. damnosus*. The reference strains of *P. dextrinicus, P. halophilus,* and *P. parvulus* each formed single-member clusters substantiating their species status and suggesting that these organisms are not common in breweries. The cluster labelled *P. pentosaceus* contained a variety of named strains. In addition to authentic *P. pentosaceus* strains and two brewery isolates, a strain received as *P. damnosus* NCDO 1833, the type strain of *P. acidilactici* NCDO 1859, and *P. acidilactici* NCIB 6990 were recovered in this cluster. The assignment of NCDO 1859 to *P. pentosaceus* is in accord with DNA reassociation data which showed 90–100% homology between this strain and authentic strains of *P. pentosaceus* and only 20% homology with other strains labelled *P. acidilactici* (Back and Stackebrandt, 1978). Thus, the type of strain of *P. acidilactici* would appear to be a strain of *P. pentosaceus*. Such congruence between DNA reassociation and numerical taxonomy emphasises the validity of both approaches for the classification of pediococci.

With regard to the value of these studies for the brewer, only *P. damnosus* strains were able to grow in beer and thus cause spoilage, although other Gram-positive cocci, in particular *P. pentosaceus* and *Micrococcus kristinae* strains, were sufficiently tolerant of the low pH and hop constituents in beer to survive for long periods without multiplying. From the numerical data, an identification table was constructed to separate the seven taxa of Gram-positive cocci com-

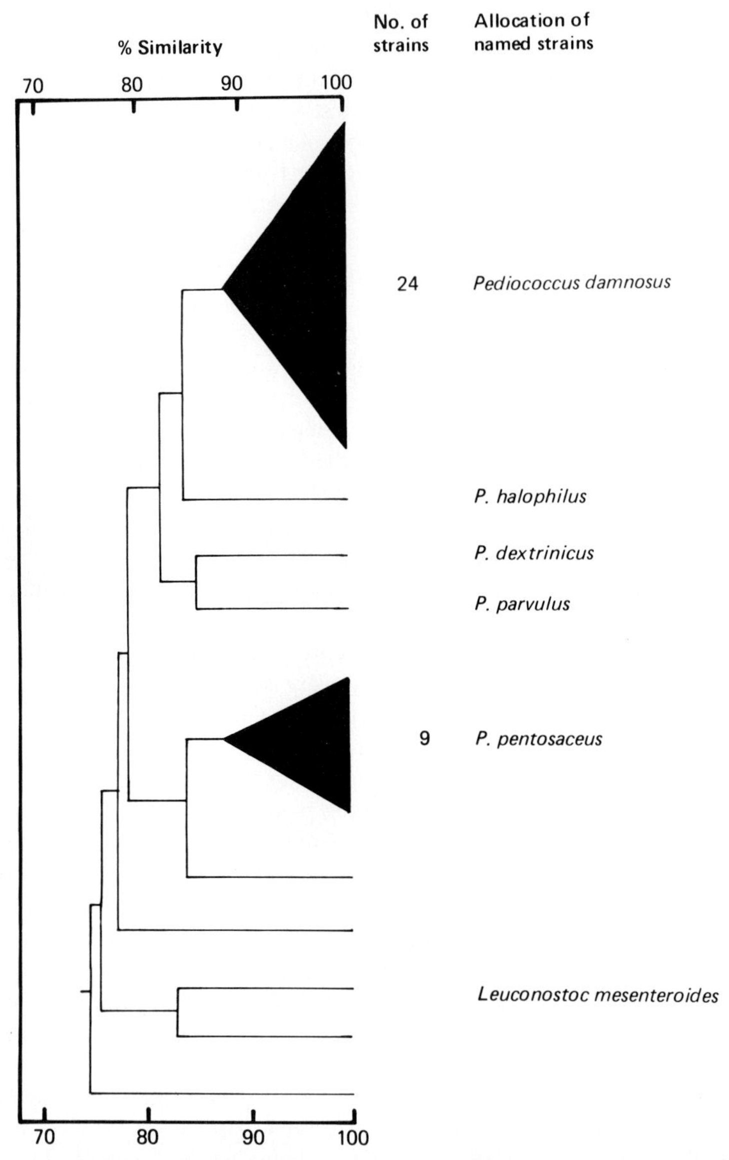

Fig. 2. Simplified dendrogram showing the relationships between Gram-positive cocci isolated from beer and brewery plant and reference strains of *Pediococcus* based on the S_{sm} coefficient and average-linkage (UPGMA) cluster analysis.

monly encountered in breweries (Lawrence and Priest, 1981). This was subsequently converted into a probabilistic indentification matrix for use with a desktop computer. This computer-assisted identification scheme has since been tested on a range of newly isolated brewery cocci and found to be effective for the identification and thus the evaluation of the spoilage potential of these bacteria.

Leuconostoc

At one time heterofermentative cocci were classified with streptococci, largely because of morphological similarity. As more was learned about leuconostocs their similarity with the heterofermentative lactobacilli, in particular *L. confusus* and *L. viridescens,* was recognised (Sharpe *et al.*, 1972; Garvie, 1976). However, morphological considerations predominated and in the eighth edition of *Bergey's Manual of Determinative Bacteriology* (Garvie, 1974b), *Leuconostoc* was classified with *Streptococcus* in the family Streptococcaceae. Some numerical phenetic surveys have included strains of *Leuconostoc* (Bridge and Sneath, 1983b; Shaw and Harding, 1984), but there are no published studies which concentrate on these organisms. Indeed, the current classification of the genus has been largely derived from nucleic acid analyses and serological studies of enzymes, with traditional phenotypic characterization based on a small sample of features.

At present, six *Leuconostoc* species are cited on the *Approved Lists of Bacterial Names* (Skerman *et al.*, 1980) and of these the organism associated with wine, *Leuconostoc oenos,* is unique in being acidophilic and ethanol tolerant; the other five species show phenetic similarity to the heterofermentative lactobacilli. Extensive DNA reassociation studies have indicated six 'genospecies'; several of these cannot be equated with the species that appear on the *Approved Lists of Bacterial Names* (Garvie 1976; Hontebeyrie and Gasser, 1977). Three of the DNA homology groups, however, corresponded with the established species *Leuconostoc lactis, L. oenos,* and *L. paramesenteroides. Leuconostoc mesenteroides* strains comprised three DNA homology groups: one included strains labelled as *L. cremoris* and *L. dextranicum;* the other two probably represent new species. These DNA homology groups correlate perfectly with taxa based on immunologic studies of the D-lactate dehydrogenase and glucose-6-phosphate dehydrogenase (Hontebeyrie and Gasser, 1975; Gasser and Hontebeyrie, 1977). There is little evidence of overlap between leuconostocs and heterofermentative lactobacilli. Phenetic similarities have been observed between leuconostocs and *L. confusus* and *L. viridescens* (Sharpe *et al.*, 1972), but they have not been fully supported by DNA reassociation. DNA–rRNA hybridization studies failed to demonstrate any significant homology between *Leuconostoc* and the two *Lactobacillus* species (Garvie, 1981).

A small numerical taxonomic study of leuconostocs and some heterofermentative lactobacilli from whisky distilleries has been completed (Pleasants and Priest, unpublished). The majority of strains involved were cocci or short rods isolated from grain mashes and fermentations in a Scotch whisky distillery by plating samples onto yeast, glucose, citrate agar, and acid tomato agar (Garvie, 1967), supplemented with cycloheximide (20 μg ml^{-1}) to suppress yeast growth. After incubation at 35°C for 2 days, heterofermentative cocci and short rods were retained as presumptive *Leuconostoc* strains. These strains, reference strains of *Leuconostoc* and *Lactobacillus,* and some heterofermentative lactobacilli isolated from whisky distilleries, were examined for 104 features encompassing a selection of biochemical, morphological, and physiological tests. Data were analysed with S_{sm} and S_J coefficients and complete-linkage clustering. Similar numerical classifications were obtained in each analysis; an abbreviated dendrogram from the S_{sm} computation is shown in Fig. 3. The 73 strains were recovered in two aggregate clusters representing *Leuconostoc* and *Lactobacillus,* a result that confirms the status of *Leuconostoc* as a separate genus. The only misplaced strain was *L. fermentum* NCDO 1750, which was recovered in the *Leuconostoc* aggregate cluster. This heterofermentative *Lactobacillus* has a high G + C content (53%) compared with leuconostocs (39–42%; Garvie, 1974b) and probably represents a misplaced strain that would cluster with the lactobacilli if more strains were studied. The previously reported lack of affinity (Garvie, 1981) between *L. confusus, L. viridescens,* and the *Leuconostoc* strains was therefore confirmed.

Within the *Leuconostoc* phenon, several clusters were apparent at the 79% similarity level. Cluster 1 contained most of the distillery isolates and two strains from frozen peas received as *L. mesenteroides*. Strains in this cluster typically produced dextran from sucrose. The type of *L. dextranicium* and *L. mesenteroides* clustered at 75% similarity to give an aggregate phenon. Since each of the type strains shared considerable DNA homology with one of the DNA homology groups of *L. mesenteroides* (Hontebeyrie and Gasser, 1977), it would seem probable that the aggregate cluster represents *L. mesenteroides.* However, the enlarged cluster also includes the type strain of *L. paramesenteroides,* an organism that shares limited (~20%) DNA homology with *L. mesenteroides.* This association is not particularly surprising, since *L. mesenteroides* and *L. paramesenteroides* are phenotypically very similar. Indeed, for many years *L. paramesenteroides* was considered to be a non-dextran-forming variant of *L. mesenteroides* (Sharpe, 1979). Cluster 2 (Fig. 3) is apparently homogeneous but has surprisingly few cluster-specific features. Selected strains are currently being examined in DNA base composition and homology studies, to determine and evaluate the status of cluster 2, which might represent one of the DNA homology groups identified within *L. mesenteroides* by Hontebeyrie and Gasser (1977).

It can be concluded that numerical taxonomy can be used to confirm and

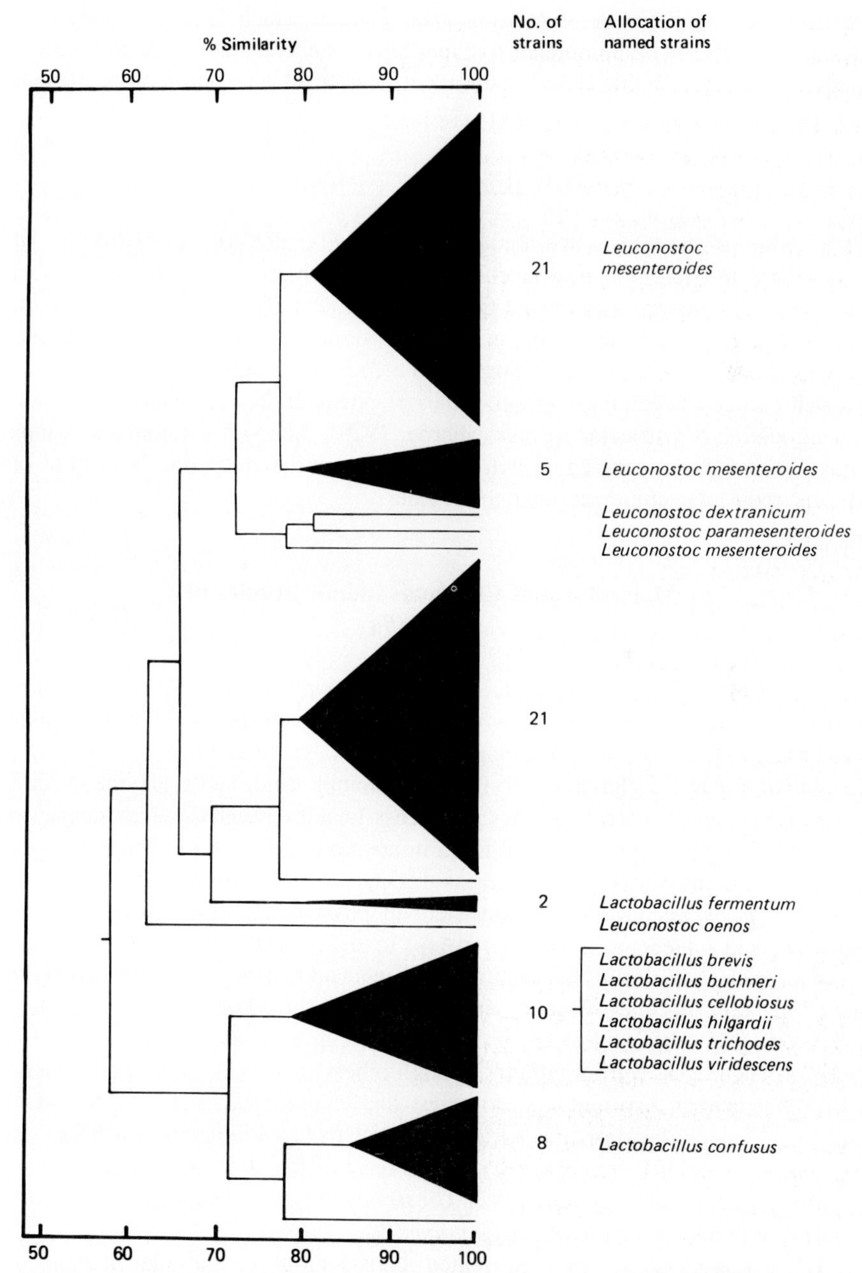

Fig. 3. Simplified dendrogram showing the relationships between presumptive *Leuconostoc* strains isolated from Scotch whisky distilleries, reference strains of *Leuconostoc* and of *Lactobacillus*, based on the S_{sm} coefficient and complete-linkage cluster analysis.

extend the classification of *Leuconostoc,* a taxon which is largely based on molecular data. More comprehensive numerical phenetic studies are now necessary to provide suitable data for the identification of *Leuconostoc* and related taxa.

Lactobacillus

The value of numerical taxonomy in the classification and identification of *Lactobacillus* strains can now be considered within the context of the systematics of lactic acid bacteria and related Gram-positive taxa. The current classification of *Lactobacillus* is largely influenced by the studies of Orla-Jensen (1919) and involves the division of the genus into the four subgroups or subgenera mentioned earlier. Within each group, tables of phenetic data are available for the identification of particular species (Sharpe, 1979). Although the four subgroups have been widely accepted, it will be useful to review them briefly in light of recent molecular and chemotaxonomic data.

Molecular and Chemotaxonomic Studies of *Lactobacillus*

Considerable use has been made of cell wall and nucleic acid analyses and enzyme patterns in the taxonomy of *Lactobacillus* (Williams, 1975; Williams and Shah, 1979). The wall peptidoglycans of lactobacilli all have a tetrapeptide, L-alanine (LAla), D-glutamic acid (DGlu), a diaminoacid, and D-alanine (DAla), attached to muramic acid moieties, and cross-linked through the diaminoacid in position 3 of one peptide and the DAla in position 4 of another. Three types of diaminoacid and variation in the cross-linking occurs. Most species of the subgenus '*Thermobacterium*' are consistent in possessing L-lysine (LLys) as the diaminoacid linked via D-aspartic (DAsp) to DAla (Kandler, 1970). However, two anaerobic species in this group, *L. ruminis* and *L. vitulinus,* do not conform to this pattern as they contain diaminopimelic acid (DAP) in place of LLys (Sharpe *et al.,* 1973). Several species in '*Streptobacterium*' also possess the LLys, DAsp, DAla configuration, although others such as *L. plantarum* contain DAP but neither Asp nor Lys. Amongst the betabacteria most have the basic LLys, DAsp, DAla form, but variation occurs in *L. cellobiosus* which has ornithine in place of Lys, and *L. viridescens* have different cross-linkage arrangements (Schleifer and Kandler, 1972). Clearly there is no correlation between cell wall types and the subgenera of *Lactobacillus.*

DNA reassociation has contributed considerably to the clarification of *Lactobacillus* taxonomy at the species level but is of little value for the establishment of higher rank taxa. It is likely that rRNA cataloguing or rRNA–DNA hybridization will be most useful in this respect, but to date no studies have been published. The DNA base composition values available for the genus *Lactobacillus* indicate considerable heterogeneity within the subgenera. Thermobacteria can be divided into two groups, the low (34–36%) G + C species, *L. acidophilus, L. helveticus, L. jensenii,* and *L. salivarius,* and the high (about 50%) G + C species, *L. bulgaricus, L. delbrueckii,* and *L. leichmanii* (Rogosa, 1974). Similarly, '*Streptobacterium*' comprises '*L. yamanashiensis*' (Carr *et al.*, 1977) and *L. xylosus* with low base compositions (33–35%) and several species with a high G + C content (44–47%; Rogosa, 1974). Finally, '*Betabacterium*' contains several species with low (42–45%) and high (~53%) G + C contents (Rogosa, 1974). Enzyme patterns have been used extensively for *Lactobacillus* classification (Williams and Shah, 1979), but they are most useful at the species level and do not provide suitable information for subdivision of the genus into groups of species.

In summary it would seem that, although the traditional subdivision of the genus *Lactobacillus* has its uses for identification, the DNA base composition and cell wall analyses indicate that it is not soundly based. In view of the value of numerical taxonomy for the delineation of aggregate clusters/subgenera in *Bacillus* (Logan and Berkeley, 1981; Priest *et al.*, 1981) and *Streptococcus* (Bridge and Sneath, 1983b), it might also be expected to be useful for *Lactobacillus*.

Numerical Taxonomic Studies of *Lactobacillus*

Numerical taxonomy has played a minor role in the classification of *Lactobacillus*. Hauser and Smith (1964) examined 59 lactobacilli isolated from cheese and nine reference strains for 109 features. The data were analysed using the S_J coefficient and the single-linkage algorithm, and five clusters were recognised. Many of the cheese isolates identified with the reference cultures of *L. brevis, L. casei,* and *L. plantarum*. The clusters formed had low S values, 53% in the case of the *L. brevis* cluster, indicating considerable heterogeneity.

Other numerical taxonomic studies have dealt with strains from wine. Barre (1969) examined 65 wine strains and four reference strains for 73 characters and analysed the data using the S_{sm} coefficient and the average-linkage cluster analysis technique. The resultant dendrogram showed two aggregate clusters at 65% similarity; one contained three clusters of streptobacteria (including *L. casei* and *L. plantarum*) and the other, four clusters of betabacteria (*L. buchneri, L. fermentum,* and two unidentified phena). Most of the wine isolates clustered with *L. buchneri*. Phenetic evidence for the three subgenera of *Lactobacillus* was also provided in a numerical analysis of 30 lactobacilli by Seyfried (1968). Using 70 characters, the S_{sm} coefficient, and the average-linkage cluster analysis technique, the 30 strains were recovered in three aggregate clusters, at 75 to 80% similarity, which conformed exactly to the three subgenera of *Lactobacillus*.

Numerical taxonomic studies of other Gram-positive taxa have often included some *Lactobacillus* reference strains. In a comprehensive examination of *Listeria* and related bacteria, Wilkinson and Jones (1977) recovered 25 *Lactobacillus* strains in four phena. Two streptobacteria (*L. casei* and *L. plantarum*) were joined by *L. salivarius* ('*Thermobacterium*') in one cluster. *Lactobacillus mali* formed a distinct homogeneous phenon. The remaining thermobacteria (*L. acidophilus, L. bulgaricus, L. delbrueckii, L. helveticus, 'L. jugurt'*, and *L. lactis*) were barely distinguished from the heterofermentative strains of *L. brevis, L. buchneri, L. cellobiosus*, and *L. fermentum*. This study was therefore largely in accord with traditional views on the taxonomy of the lactobacilli, as was the numerical analysis of Piot *et al.* (1980), who included several lactobacilli in a study of *Gardnerella vaginalis*.

Finally, two numerical phenetic studies of lactobacilli from meats should be mentioned. Laban *et al.* (1978) examined 190 strains, mainly isolated from French sausages, using the API 50L system and clustered the resultant data by single-linkage analysis. Six clusters were recovered but a large number of intermediate strains made it difficult to highlight tests that allowed the clusters to be distinguished. Nevertheless, the lactobacilli from sausage differed considerably from those of dairy origin. In a numerical taxonomic study of lactic acid bacteria from vacuum-packed meats, Shaw and Harding (1984) examined 100 isolates of lactic acid bacteria and 23 reference strains for 79 characters, and analysed the data with the S_{sm} coefficient and the average-linkage algorithm. The meat isolates largely comprised two clusters at 78% similarity, both entirely composed of 'streptobacteria'. Strains of one phenon had a low G + C content (33.2–36.9%) and could not be identified. The other contained 57 strains which had a higher G + C content (40.7–43.7%) and were provisionally identified with '*L. bavaricus*' or *L. sake*. Several interesting points emerged from this study. First, although the two major clusters were clearly separated in the dendrogram, there were few features that could be used to distinguish them reliably. Second, if the clusters were defined at the same similarity level as that at which the reference stains of different *Lactobacillus* species formed distinct clusters, then each of the two clusters of meat isolates became an aggregate cluster encompassing several individual clusters. Indeed, Shaw and Harding pointed out that subclusters were evident in the two clusters but concluded that supporting chemotaxonomic evidence would be required to justify any further division. Finally, the reference *Lactobacillus* strains did not cluster entirely according to the three suggested subgenera of *Lactobacillus*. Thus, although this study resolved many of the problems associated with the classification and identification of lactic acid bacteria from vacuum-packed meat, it also raised some interesting points with regard to the taxonomy of the lactobacilli which will be explored a little further below.

Numerical Taxonomic Studies of Lactobacilli from Scotch Whisky Distilleries

Scotch whisky is manufactured from malted barley or, in the case of grain whisky for use in blending, a mixture of malted barley and some starchy materials such as unmalted cereals or corn (maize). Although there is considerable variation in the manufacture of Scotch whisky between different distilleries, the process essentially involves grinding the malt and 'mashing' it in hot water at 60 to 85°C for ~90 min. During this period, enzymes from the malt hydrolyse polymeric material (mainly starch and proteins) from the malt and adjuncts. The hot liquid is collected, cooled, fermented with *Saccharomyces cerevisiae,* distilled, and matured in oak casks for at least 3 years (Simpson, 1968). Since the extract from the mash is not boiled as it is in a beer brewery, the fermentation has a mixed microbial flora derived from those bacteria on the malt and grain that can survive the mashing procedure and grow in the fermentation. In practice this is largely lactobacilli because, although spores will be transferred into the fermentation, the pH is too low for germination to occur. Generally lactobacilli are not detrimental to the fermentation, and their metabolism may have a positive contribution to the development of the flavour of the final product. However, under certain conditions lactobacilli may grow prolifically, reaching about 5×10^9 ml^{-1} at the end of the fermentation. Under these circumstances, yeast growth is inhibited and the final ethanol yield will be reduced (Dolan, 1976).

There have been some attempts to identify the offending lactobacilli using traditional schemes (MacKenzie and Kenny, 1965; Bryan-Jones, 1975). These have been largely unsuccessful, but although the bacteria were not identified, it was concluded that they probably represented variants of recognised species such as *Lactobacillus brevis, L. casei, L. debrueckii, L. fermentum,* and *L. plantarum.* In view of this unsatisfactory situation, an analysis of 146 distillery isolates and 32 reference strains of *Lactobacillus* was undertaken with a view to classifying the test strains, using numerical techniques and providing an identification scheme (Barbour and Priest, 1983).

The bacteria were isolated from mashes and fermentations from distilleries throughout Scotland, and examined for 169 features representing a range of biochemical, morphological, and physiological criteria. Analysis of test error (Sneath and Johnson, 1972) indicated a probability of test error of 5.6%, so the most error-prone characters (17) and those characters which were invariant for all bacteria (45) were excluded from the computations. This reduced test error to 3% for the remaining 107 characters. Data were analysed initially using the S_{sm} coefficient and average-linkage (UPGMA) clustering, but the resultant dendrogram showed considerable chaining and few distinct clusters. The data were therefore computed using a more powerful clustering algorithm, complete linkage, and more distinct phena were recovered (Fig. 4). However, of the 27 phena

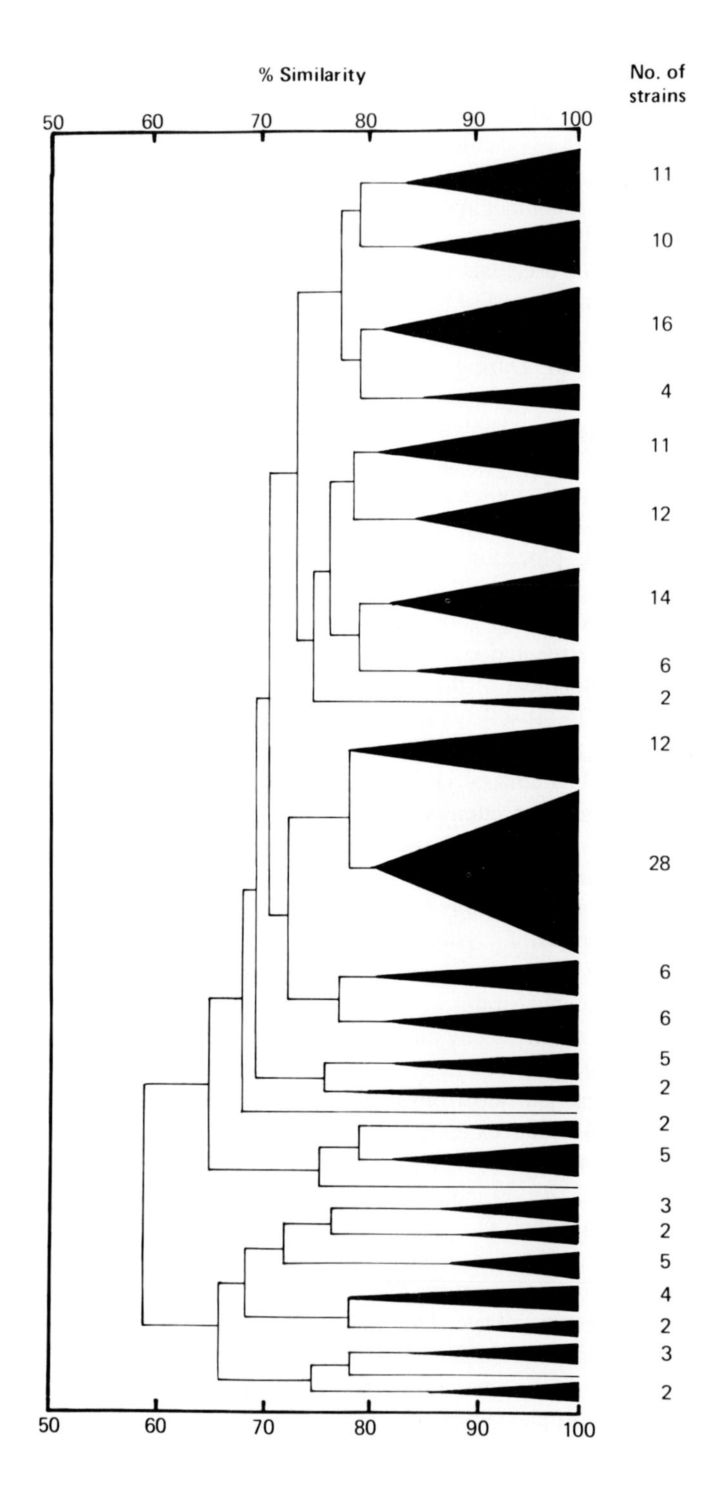

% Similarity
No. of strains
50
60
70
80
90
100
11
10
16
4
11
12
14
6
2
12
28
6
6
5
2
2
5
3
2
5
4
2
3
2
50
60
70
80
90
100

defined at 82% similarity, few could be identified because the reference strains failed to cluster with the distillery isolates but were recovered in eight clusters. Since, with the exception of *L. casei* var. *rhamnosus* and *L. fermentum*, the clusters contained more than one reference strain, their identity was uncertain. However, the results indicated that the distillery isolates were not variants of established species but probably represented undescribed taxa.

The distribution of the reference strains to clusters did not correlate well with the 'subgenera' of *Lactobacillus*. 'Thermobacteria' were distributed between four clusters, as were various 'betabacteria'. Streptobacteria were recovered in three clusters (Fig. 4). Hierarchic clustering algorithms give poor representation of relationships between major clusters, although they reproduce distances between close neighbours faithfully. Sneath and Sokal (1973) recommend that both cluster analysis and an *ordination* technique should be used for numerical classification, with the latter providing an understanding of the taxonomic structure in greater detail. The data for the reference strains were therefore analysed using detrended correspondence analysis (Hill, 1979). A plot of the first two axes is shown in Fig. 5. The groups formed were different from those obtained using complete-linkage cluster analysis but again did not correlate with the traditional division of the genus into subgenera.

A third problem to emerge from this study was also apparent in the numerical analysis of lactobacilli isolated from meats (Shaw and Harding, 1984). This concerned the percentage similarity at which to define clusters of approximate species rank. If phena are delineated at 82% similarity as in Fig. 4, the results suggest that the lactobacilli are overclassified, since 23 species names could be reduced to 8. If, on the other hand, clusters are defined at a sufficiently high similarity level that most reference strains are separated, the results indicate a multitude of species-rank taxa in the distillery environment. There is no simple solution to this problem, but DNA base composition and reassociation and chemotaxonomic analysis would appear to be essential.

When the data were computed using the S_p and S_J coefficients with average-linkage and complete-linkage cluster analysis, the distribution of strains to clusters varied considerably. This instability indicates that the assignment of strains to clusters was largely random (it should be remembered that clustering al-

---

Fig. 4. Simplified dendrogram showing the relationships between *Lactobacillus* isolates from Scotch whisky distilleries and reference strains of *Lactobacillus* based on the S_{sm} coefficient and complete-linkage cluster analysis. Allocation of reverence strains to clusters is as follows: cluster 1, *L. bulgaricus, L. confusus;* cluster 5, *L. casei* subsp. *casei, L. casei* subsp. *pseudoplantarum, L. cellobiosus,* '*L. jugurt*'; cluster 6, *L. casei* subsp. *tolerans, L. collinoides, L. coryneformis, L. coryneformis* subsp. *torquens, L. curvatus, L. fructosus, L. plantarum, L. trichodes,* '*L. yamanashiensis*'; cluster 8, *L. fermentum;* cluster 9, *L. casei* subsp, *alatosus, L. delbrueckii,, L. helveticus, L. lactis, L. leichmannii;* cluster 18, *L. fructivorans, L. mali.*

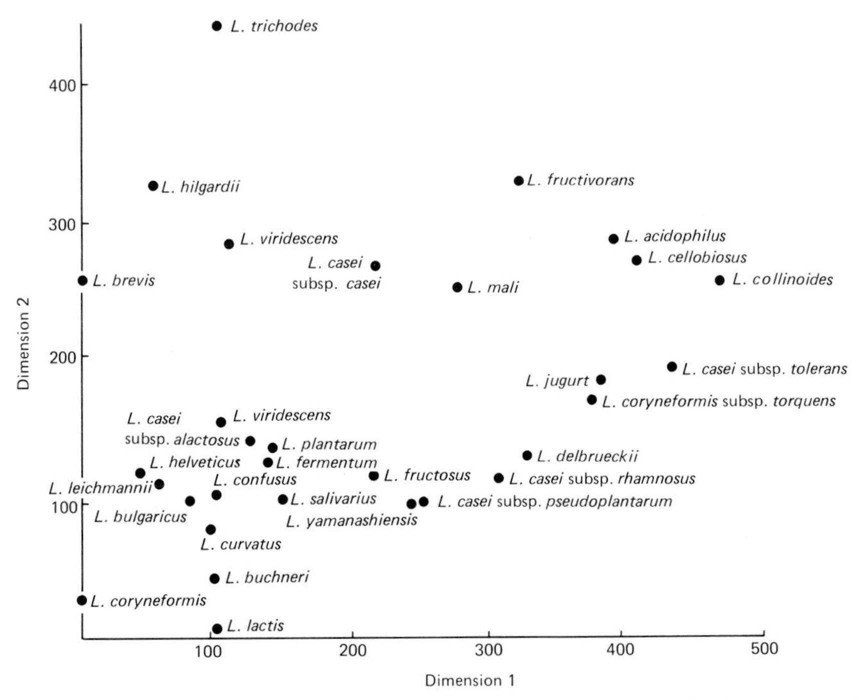

Fig. 5. Ordination plot generated by detrended correspondence analysis of some named strains of *Lactobacillus*.

gorithms will generate clusters from random data) and corroborates the original S_{sm} average-linkage computation which showed few distinct clusters. It would seem that, on the basis of these data, this collection of strains is phenetically homogeneous and that numerical taxonomy is not providing an accurate or useful classification. To explore this possibility further, CHARSEP (Sneath, 1979a) was used to indicate those tests that best separated the taxa. Of the 107 characters, only 27 had a VSP (Sneath, 1979a) index greater than 25% and could be considered to have good discriminatory potential. None were excellent (VSP >70%). Of the 27 characters, only 15 had a CSP (Sneath, 1979a) index greater than 50%, the recommended value (Sneath, 1979a). This compares most unfavourably with data for Micrococcaceae; of 60 characters used in an identification matrix for *Micrococcus* and *Staphylococcus* (Feltham and Sneath, 1982), 29 had good discrimination (VSP >20%), and nine of these were excellent (VSP >70%). Similarly, for streptococci (Feltham and Sneath, 1982), of 60 characters, 25 had a VSP index >20%, and 6 of these were excellent (VSP >70%). These figures show that the data base for lactobacilli has few characters diagnostic for the clusters generated by the S_{sm}–complete-linkage computation, and provides further evidence for the phenetic homogeneity of the distillery isolates and reference strains of *Lactobacillus*.

The 15 characters which had a VSP index >25% and a CSP index >50% were used to construct a preliminary identification matrix that was tested with OVERMAT (Sneath, 1980) to detect overlap between clusters. Many pairs of clusters showed overlap of more than 8%, the point at which clusters begin to merge, and several pairs overlapped by 30 to 50%. These figures are very high compared to the maximum overlap of 3% shown by two clusters in the identification matrix for *Micrococcus* and *Staphylococcus*, and 9.8% in the *Streptococcus* matrix (Feltham and Sneath, 1982). These figures for overlap are heavily influenced by the number of characters used to define the taxa: the fewer the characters the greater the overlap, and in the *Lactobacillus* matrix only 15 characters were used. By increasing this number the overlap between clusters could be reduced, the problem being that of the 107 characters; most had very low diagnostic value. Together, these various findings indicate that there is little phenetic diversity amongst this collection of lactobacilli from distilleries and the reference strains, and that the clusters that have been generated by numerical analysis are essentially random and have little taxonomic value.

Concluding Remarks

There are two common approaches to the evaluation of a numerical taxonomic study. The numerical classification can be examined for stability with regard to different computational algorithms and various statistical analyses can be used; in particular, the dendrogram can be compared to the similarity matrix using the cophenetic correlation coefficient (Rohlf, 1970), cluster overlap can be determined (Sneath, 1979b), and tests for the validity of partitioning have been suggested (Dubes and Jain, 1979). Alternatively, the classification can be compared to those derived from chemical, genetic, or serological criteria. In this review of the taxonomy of *Lactobacillus* and related taxa, both approaches have been used; in particular, numerical classifications have been compared with those derived from DNA base composition and reassociation data where available. But evaluation of numerical phenetic classifications in the context of DNA analyses assumes that a constant proportion of the genome is expressed by microorganisms belonging to different taxa, and that a given amount of genetic diversity will be reflected in a parallel amount of phenetic diversity. The studies reported in this article suggest that this assumption may be invalid.

The genus *Bacillus* is genetically and physiologically diverse, and, in general, numerical studies have been supported by DNA reassociation data. For example, strains that are phenotypically similar such as *B. cereus*, '*B. cereus* var. *mycoides*', and *B. thuringiensis* cluster together at high similarity levels in numerical phenetic analyses (Logan and Berkeley, 1981; Priest *et al.*, 1981) and have essentially homologous chromosomes (Somerville and Jones, 1972), while strains such as '*B. amyloliquefaciens*', *B. licheniformis*, *B. pumilus*, and *B. subtilis* that cluster into species-rank taxa in numerical phenetic studies have relatively few DNA sequences in common (O'Donnell *et al.*, 1980). However, the limited numerical studies of *Leuconostoc* and *Pediococcus* and the more extensive studies of *Streptococcus* show a trend towards phenetic homogeneity, although genetic diversity, as revealed by DNA base composition and reassociation, is similar to that in *Bacillus*. This phenetic homogeneity is revealed by high similarity levels at which species-rank clusters are defined and, in the case of *Streptococcus*, considerable overlap between taxa in a probabilistic identification matrix for these bacteria prepared by Feltham and Sneath (1982). These authors attribute this overlap to an inadequate data base and the small sizes of some taxa. However, it could be argued that the 'inadequate' data base is a reflection of the phenetic similarity of the taxa. Nevertheless, numerical classification is effective for these genera in that the clusters which are created can be substantiated by DNA analyses and chemotaxonomy.

Phenetic homogeneity seems to be most apparent in *Lactobacillus*, posing some difficult problems of which perhaps the most important is the inability of numerical analysis to create phenetically sound clusters. This is reported in this article in connection with lactobacilli isolated from Scotch whisky distilleries and was evident to a lesser extent in a study of *Lactobacillus* isolates from vacuum-packed meats (Shaw and Harding, 1984). It was not apparent in some earlier studies, particularly those involving isolates from Cheddar cheese (Hauser and Smith, 1964) and wine (Barre, 1969) and some studies of named strains (Seyfried, 1968; Wilkinson and Jones, 1977). The reasons for this are not clear but may relate to the range of organisms studied; in general, the smaller the study the more successful it appears to have been.

The apparent phenetic homogeneity of the distillery isolates and reference strains of *Lactobacillus* is not a result of a limited data base, since the same (but slightly reduced) battery of tests was used for the *Leuconostoc* computation (Fig. 3), in which it is evident that reference strains and distillery isolates of *Leuconostoc* formed clusters at ~80% similarity, whereas the reference strains of *Lactobacillus* cluster at >90% similarity. It must therefore be a reflection of the bacteria. The chaining effect observed in the S_{sm}, UPGMA dendrogram of the distillery lactobacilli, the variable allocation of strains to clusters when different computations were employed, the lack of diagnostic features using CHAR-SEP, and the high overlap between clusters in the preliminary identification matrix, all point towards there being very little phenetic variation amongst the strains studied, although—within the reference strains at least—the range of DNA base composition indicates considerable genetic diversity. This indicates that genetic variation within *Lactobacillus* cannot be adequately tested as biochemical, morphological, and physiological features in the traditional fashion and used for numerical analysis as it can in other taxa. Either there is very limited expression of the genome due to adaptation to a limited and specialised habitat, or convergent evolution has provided similar apparent phenotypes despite diverse genotypes. Whether this is peculiar to those lactobacilli from particular habitats such as distilleries and meat products, or is representative of the genus as a whole, will only be known when more comprehensive numerical studies of *Lactobacillus* have been completed.

References

Back, W. (1978). Zur taxonomie der Gattung *Pediococcus. Brauwissenschaft* **31,** 237–250, 312–320, 336–343.

Back, W., and Stackebrandt, E. (1978). DNS/DNS-Homologiestudien innerhalb der Gattung *Pediococcus. Archives of Microbiology* **118,** 79–85.

Barbour, E. A., and Priest, F. G. (1983). Numerical classification of lactobacilli from Scotch whisky distilleries. *In* 'Current Developments in Malting, Brewing and Distilling' (Eds. F. G. Priest and I. Campbell), pp. 289–291. Institute of Brewing, London.

Barre, P. (1969). Taxonomie numérique de lactobacilles isolés du vin. *Archives für Mikrobiologie* **68,** 74–86.

Berkeley, R. C. W., and Goodfellow, M. (eds.) (1981). 'The Aerobic Endospore-forming Bacteria: Classification and Identification'. Academic Press, London.

Boeyé, A., and Aerts, M. (1976). Numerical taxonomy of *Bacillus* isolates from North Sea sediments. *International Journal of Systematic Bacteriology* **26,** 427–441.

Bonde, G. J. (1975). The genus *Bacillus.* An experiment with cluster analysis. *Danish Medical Bulletin* **22,** 41–61.

Bonde, G. J. (1981). *Bacillus* from marine habitats: allocation to phena established by numerical techniques. *In* 'The Aerobic Endospore-forming Bacteria: Classification and Identification' (Eds. R. C. W. Berkeley and M. Goodfellow), pp. 181–215. Academic Press, London.

Bradley, S. G. (1980). DNA reassociation and base composition. *In* 'Microbiological Classification and Identification' (Eds. M. Goodfellow and R. G. Board), pp. 11–26, Academic Press, London.

Bradley, S. G., and Mordarski, M. (1976). Association of polydeoxyribonucleotides of deoxyribonucleic acids from nocardioform bacteria. *In* 'The Biology of the Nocardiae' (Eds. M. Goodfellow, G. H. Brownell, and J. A. Serrano), pp. 310–336. Academic Press, London.

Bridge, P. D., and Sneath, P. H. A. (1983a). *Streptococcus gallinarum* sp. nov. and *Streptococcus oralis* sp. nov. *International Journal of Systematic Bacteriology* **32,** 410–415.

Bridge, P. D., and Sneath, P. H. A. (1983b). Numerical taxonomy of *Streptococcus. Journal of General Microbiology* **129,** 565–597.

Bryan-Jones, G. (1975). Lactic acid bacteria in distillery fermentations. *In* 'Lactic Acid Bacteria in Beverages and Food' (Eds. J. G. Carr, C. V. Cutting, and G. C. Whiting), pp. 165–176. Academic Press, London.

Carlsson, J. (1968). A numerical taxonomic study of human oral streptococci. *Odontologisk Revy* **18,** 55–74.

Carr. J. G., Davies, P. A., Dellaglio, F., and Vescovo, M. (1977). The relationship between *Lactobacillus mali* from cider and *Lactobacillus yamanashiensis* from wine. *Journal of Applied Bacteriology* **42,** 219–228.

Coleman, G. (1968). The application of computers to the classification of streptococci. *Journal of General Microbiology* **50,** 149–158.

Craveri, R., Hill, L. R., Manachini, P. L., and Silvestri, L. G. (1965). Deoxyribonucleic acid base composition among thermophilic actinomycetes: the occurrence of two strains with low GC content. *Journal of General Microbiology* **41,** 335–339.

Craveri, R., Manachini, P. L., Aragozzini, F., and Merendi, C. (1973). Amino acid composition of the proteins from mesophilic, thermofacultative and thermophilic actinomycetes. *Journal of General Microbiology* **74,** 201–204.

Cross, T. (1981). The monosporic actinomycetes. *In* 'The Prokaryotes' (Eds. M. P. Starr, H. Stolp, H. G. Trüper, A. Balows, and H. G. Schlegel), pp. 2091–2102. Springer-Verlag, New York.

Cross, T., Walker, P. D., and Gould, G. W. (1968). Thermophilic actinomycetes producing resistant endospores. *Nature (London)* **220,** 352–354.

Davis, G. H. G., Fomin, L., Wilson, E., and Newton, K. G. (1969). Numerical taxonomy of *Listeria,* streptococci and possibly related bacteria. *Journal of General Microbiology* **57,** 333–348.

Dolan, T. C. S. (1976). Some aspects of the impact of brewing science on Scotch malt whisky production. *Journal of the Institute of Brewing* **82,** 177–181.

Doores, S., and Westhoff, D. C. (1983). Selective method for the isolation of *Sporolactobacillus* from food and environmental sources. *Journal of Applied Bacteriology* **54,** 373–380.

Drucker, D. B., and Melville, T. H. (1971). The classification of some oral streptococci of human or rat origin. *Archives of Oral Biology* **16,** 845–853.

Dubes, R., and Jain, A. K. (1979). Validity studies in clustering methodologies. *Pattern Recognition Jounal* **11,** 235–254.

Farrow, J. A. E., Jones, D., Phillips, B. A., and Collins, M. D. (1983). Taxonomic studies on some group D streptococci. *Journal of General Microbiology* **129,** 1423–1432.

Feltham, R. K. A. (1979). A taxonomic study of the genus *Streptococcus. In* 'Pathogenic Streptococci' (Ed. M. T. Parker), pp. 247–248. Redbooks Ltd., Chertsey, England.

Feltham, R. K. A., and Sneath, P. H. A. (1982). Construction of matrices for computer-assisted identification of aerobic Gram-positive cocci. *Journal of General Microbiology* **128,** 713–720.

Fox, G. E., Pechman, K. R., and Woese, C. R. (1977). Comparative cataloging of 16S ribosomal ribonucleic acid: molecular approach to prokaryotic systematics. *International Journal of Systematic Bacteriology* **27,** 44–57.

Garvie, E. I. (1967). *Leuconostoc oenos* sp. nov. *Journal of General Microbiology* **48,** 431–438.

Garvie, E. I. (1974a). Nomenclatural problems of the pediococci. Request for an opinion. *International Journal of Systematic Bacteriology* **24,** 301–306.

Garvie, E. I. (1974b). Genus *Leuconostoc* Van Tiegham. *In* 'Bergey's Manual of Determinative Bacteriology' (Eds. R. E. Buchanan and N. E. Gibbons), 8th Edition, p. 510. Williams & Wilkins, Baltimore.

Garvie, E. I. (1976). Hybridization between the deoxyribonucleic acids of some strains of heterofermentative lactic acid bacteria. *International Journal of Systematic Bacteriology* **26,** 116–122.

Garvie, E. I. (1981). Subdivisions within the genus *Leuconostoc* as shown by RNA/DNA hybridization. *Journal of General Microbiology* **127,** 209–212.

Garvie, E. I., Farrow, J. A. E., and Phillips, B. A. (1981). A taxonomic study of some strains of streptococci which grow at 10°C but not at 45°C including *Streptococcus lactis* and *Streptococcus cremoris. Zentralblatt für Bakteriologie Parasitenkunde, Infektionskrankheiten und Hygiene, Abteilung 1: Originale, Reihe C* **2,** 151–165.

Gasser, F., and Hontebeyrie, M. (1977). Immunological relationships of glucose-6-phosphate dehydrogenase of *Leuconostoc mesenteroides* NCDO 768 (= ATCC 12291). *International Journal of Systematic Bacteriology* **27,** 6–8.

Goodfellow, M., and Cross, T. (1984). Classification. *In* 'The Biology of the Actinomycetes' (Eds. M. Goodfellow, M. Mordarski, and S. T. Williams), pp. 7–164. Academic Press, London.

Gordon, R. E., Haynes, W. C., and Pang, C. H.-N. (1973). 'The genus *Bacillus*'. United States Department of Agriculture, Washington, D.C.

Gyllenberg, H. G., and Laine, J. J. (1970). Numerical approach to the taxonomy of psychrotrophic bacilli. *Annales Medicinae Experimentalis et Biologiae* **49,** 62–66.

Hardie, J. M., and Marsh, P. D. (1978). Streptococci and the human oral flora. *In* 'Streptococci' (Eds. F. A. Skinner and L. B. Quesnel), pp. 157–206. Academic Press, London.

Hardy, M. A., Dalton, H. P., and Allison, M. J. (1978). Laboratory identification and epidemiology of streptococcal hospital isolates. *Journal of Clinical Microbiology* **8,** 534–544.

Hauser, M. M., and Smith, R. E. (1964). The characterization of lactobacilli from Cheddar cheese. II. A numerical analysis of the data by means of an electronic computer. *Canadian Journal of Microbiology* **10,** 757–762.

Hill, M. O. (1979). DECORANA, A FORTRAN program for detrended correspondence analysis and reciprocal averaging. Cornell Univ. Press, Ithaca, New York.

Hontebeyrie, M., and Gasser, F. (1975). Comparative immunological relationships of two distinct sets of isofunctional dehygrogenases in the genus *Leuconostoc. International Journal of Systematic Bacteriology* **25,** 1–6.

Hontebeyrie, M., and Gasser, F. (1977). Deoxyribonucleic acid homologies in the genus *Leuconostoc. International Journal of Systematic Bacteriology* **27,** 9–14.

Jones, D. (1978). Composition and differentiation of the genus *Streptococcus. In* 'Streptococci' (Eds. F. A. Skinner and L. B. Quesnel), pp. 1–49. Academic Press, London.

Jones, D., Sackin, M. J., and Sneath, P. H. A. (1972). A numerical taxonomic study of the streptococci of serological group D. *Journal of General Microbiology* **72,** 439–450.

Kalina, A. P. (1970). The taxonomy and nomenclature of enterococci. *International Journal of Systematic Bacteriology* **20,** 185–189.

Kandler, O. (1970). Amino acid sequence of the murein and taxonomy of the genera *Lactobacillus, Bifidobacterium, Leuconostoc* and *Pediococcus. International Journal of Systematic Bacteriology* **20,** 491–507.

Küster, E. (1974). *Thermoactinomyces. In* 'Bergey's Manual of Determinative Bacteriology', (Eds. R. E. Buchanan and N. E. Gibbons), 8th Edition, pp. 855–856. Williams & Wilkens, Baltimore, Maryland.

Laban, P., Favre, C., Ramet, F., and Larpent, J. P. (1978). Lactobacilli isolated from French saucisson (taxonomic study). *Zentralblatt für Bakteriologie Parasitenkunde, Infektionskrankheiten und Hygiene, Abteilung 1: Originale, Reihe B* **166,** 105–111.

Lawrence, D. R., and Priest, F. G. (1981). Identification of brewery cocci. *In* 'Proceedings of the European Brewery Convention', Copenhagen, pp. 217–227. IRL Press Ltd., London.
Logan, N. A., and Berkeley, R. C. W. (1981). Classification and identification of members of the genus *Bacillus* using API tests. *In* 'The Aerobic Endospore-forming Bacteria: Classification and Identification' (Eds. R. C. W. Berkeley and M. Goodfellow), pp. 105–140. Academic Press, London.
London, J. (1976). The ecology and taxonomic status of the lactobacilli. *Annual Review of Microbiology* **30,** 279–301.
London, J., and Kline, K. (1973). Aldolases of lactic acid bacteria: a case history in the use of an enxyme as an evolutionary marker. *Bacteriological Reviews* **37,** 453–478.
McCarthy, A. J., and Cross, T. (1983). A taxonomic study of *Thermomonospora* and other monosporic actinomycetes. *Journal of General Microbiology* **130,** 5–25.
Mackenzie, K. G., and Kenny, M. C. (1965). Non-volatile organic acid and pH changes during the fermentation of distiller's wort. *Journal of the Institute of Brewing* **71,** 160–165.
O'Donnell, A. G., Norris, J. R., Berkeley, R. C. W., Claus, D., Kaneko, T., Logan, N. A., and Nozaki, R. (1980). Characterization of *Bacillus subtilis Bacillus pumilus, Bacillus lichenformis* and *Bacillus amyloliquefaciens* by pyrolysis gas–liquid chromatography and by deoxyribonucleic acid (DNA)–DNA hybridization, biochemical tests and API systems. *International Journal of Systematic Bacteriology* **30,** 448–459.
Orla Jensen, S. (1919). 'Lactic Acid Bacteria'. Andre Fred Host and Son, Copenhagen.
Piot, P., Van Dyck, E., Goodfellow, M., and Falkow, S. (1980). A taxonomic study of *Gardnerella vaginalis* (*Haemophilus vaginalis*) Gardner and Dukes 1955. *Journal of General Microbiology* **119,** 373–376.
Priest, F. G., (1981). DNA homology in the genus *Bacillus*. *In* 'The Aerobic Endospore-forming Bacteria: Classification and Identification' (Eds. R. C. W. Berkeley and M. Goodfellow), pp. 33–57. Academic Press, London.
Priest, F. G., Goodfellow M., and Todd, C. (1981). The genus *Bacillus:* a numerical analysis. *In* 'The Aerobic Endospore-forming Bacteria: Classification and Identification' (Eds. R. C. W. Berkeley and M. Goodfellow), pp. 91–103. Academic Press, London.
Raj, H., and Colwell, R. R. (1966). Taxonomy of enterococci by computer analysis. *Canadian Journal of Microbiology* **12** 353–362.
Rogosa, M. (1974). Genus *Lactobacillus* Beijerinck. *In* 'Bergey's Manual of Determinative Bacteriology' (Eds. R. E. Buchanan and N. R. Gibbons), 8th edition, pp. 576–593. Williams & Wilkins, Baltimore, Maryland.
Rohlf, F. J. (1970). Adaptive hierarchical clustering schemes. *Systematic Zoology* **19,** 58–82.
Schleifer, K. H., and Kandler, O. (1972). Peptidoglycan types of bacterial cell walls and their taxonomic implications. *Bacteriological Reviews* **36,** 407–477.
Seyfried, P. L. (1968). An approach to the classification of lactobacilli using computer-aided numerical analysis. *Canadian Journal of Microbiology* **14,** 313–318.
Sharpe, M. E. (1979). Identification of the lactic acid bacteria. *In* 'Identification Methods for Microbiologists' (Eds. F. A. Skinner and D. W. Lovelock), pp. 233–259. Academic Press, London.
Sharpe, M. E., Garvie, E. I., and Tilbury, R. H. (1972). Some slime-forming heterofermentative species of the genus *Lactobacillus*. *Applied Microbiology* **23,** 389–397.
Sharpe, M. E., Latham, M. J., Garvie, E. I., Zirngibl, J., and Kandler, O. (1973). Two

new species of *Lactobacillus* isolated from the bovine rumen. *Lactobacillus ruminis* sp. nov. and *Lactobacillus vitulinus* sp. nov. *Journal of General Microbiology* **77,** 37–49.
Sharp, R. J., Bown, K. J., and Atkinson, A. (1980). Phenotypic and genotypic characterisation of some thermophilic species of *Bacillus*. *Journal of General Microbiology* **117,** 201–210.
Shaw, B. G., and Harding, C. D. (1984). A numerical study of lactic acid bacteria from vacuum packed beef, pork, lamb and bacon. *Journal of Applied Bacteriology* **56,** 25–40.
Sheard, M. A., and Priest, F. G. (1981). Numerical classification of some psychrotrophic bacilli isolated from frozen foods. *Journal of Applied Bacteriology* **51,** xxii–xxiii.
Sherman, J. M. (1937). The streptococci. *Bacteriological Reviews* **1,** 3–97.
Simpson, A. C. (1968). The manufacture of Scotch whisky. *Process Biochemistry* **3,** 9–12.
Skerman, V. B. D., McGowan, V., and Sneath, P. H. A. (1980). Approved lists of bacterial names. *International Journal of Systematic Bacteriology* **30,** 225–420.
Sneath, P. H. A. (1968). Vigour and pattern in taxonomy. *Journal of General Microbiology* **54,** 1–11.
Sneath, P. H. A. (1979a). BASIC program for character separation indices for an identification matrix of percent positive characters. *Computers and Geosciences* **5,** 349–357.
Sneath, P. H. A. (1979b). BASIC program for a significance test for two clusters in Euclidean space as measured by their overlap. *Computers and Geosciences* **5,** 143–155.
Sneath, P. H. A. (1980). BASIC program for determining overlaps between groups in an identification matrix of percent positive characters. *Computers and Geosciences* **6,** 262–278.
Sneath, P. H. A., and Johnson, R. (1972). The influence of test errors on numerical taxonomic similarities in microbiological tests. *Journal of General Microbiology* **72,** 377–392.
Sneath, P. H. A., and Sokal, R. R. (1973). 'Numerical Taxonomy. The Principles and Practice of Numerical Classification' Freeman, San Francisco.
Somerville, H. J., and Jones, M. L. (1972). DNA competition experiments within the *Bacillus cereus* group of bacilli. *Journal of General Microbiology* **73,** 257–265.
Stackebrandt, E., and Woese, C. R. (1981). The evolution of prokaryotes. *Symposium of the Society for General Microbiology* **32,** 1–31.
Uchida, K., and Mogi, K. (1973). Cellular fatty acid spectra of *Sporolactobacillus* and some other *Bacillus–Lactobacillus* intermediates as a guide to their taxonomy. *Journal of General and Applied Microbiology* **19,** 129–140.
Weissman, S. M., Reich, P. R., Somerson, N. L., and Cole, R. M. (1966). Genetic differentiation by nucleic acid homology. IV. Relationships among Lancefield groups and serotypes of streptococci. *Journal of Bacteriology* **92,** 1372–1377.
Wilkinson, R. J., and Jones, D. (1977). A numerical taxonomic survey of *Listeria* and related lactic acid bacteria. *Journal of General Microbiology* **98,** 399–421.
Williams, R. A. D. (1975). A review of biochemical techniques in the classification of the lactobacilli. *In* 'Lactic Acid Bacteria in Beverages and Food' (Eds. J. G. Carr, C. V. Cutting, and G. C. Whiting), pp. 351–367. Academic Press, London.
Williams, R. A. D., and Shah, H. N. (1979). Enzyme patterns in bacterial classification and identification. *In* 'Microbiological Classification and Identification' (Eds. M. Goodfellow and R. G. Board), pp. 299–315. Academic Press, London.

8

Delineation and Description of Microbial Populations Using Numerical Methods

M. GOODFELLOW

Department of Microbiology, The Medical School, Newcastle upon Tyne, UK

C. H. DICKINSON

Department of Plant Biology, The University, Newcastle upon Tyne, UK

Introduction

It is usually taken for granted that ecological studies of higher animals and plants involve accurate identification which is based on sound classification. Similar assumptions cannot be made about most ecological studies of micro-organisms. Indeed, a cursory examination of the recent literature on microbial ecology might lead some to question whether there is any liaison between those involved in contemporary microbial systematics and ecology.

Much of the frustration that besets microbial ecologists arises from the difficulty, if not the sheer impossibility, of identifying microbes *in situ*. This continuing problem imposes many constraints on the development of microbial ecology, but it also acts as a spur to studies of the processses mediated by micro-organisms. This branch of ecology, which has progressed from percolation and enrichment techniques to the use of fermenters and the construction of simulation models, usually ignores questions about the identity of individual microbial taxa involved in the processes under examination. However, despite this apparent deficiency it is clear that process-oriented studies have yielded valuable data about such phenomena as litter decomposition, the nitrogen cycle, and energy flow through ecosystems.

Measurements of input and output can give a superficial understanding of the activities of microbes, but a complete analysis of any ecosystem will only be obtained when complementary syn- and autecological studies are directed towards elucidating the roles played by specific microbes, acting either singly or within communities. Information derived from such studies is not only of aca-

ISBN 0-12-289665-3

demic interest but is important for diagnostic purposes, when, for instance, habitats are affected by extreme environmental conditions or pollutions. It can also be important to describe the microbial populations in natural habitats to determine changes due to diurnal and seasonal factors, higher animal or plant activity, or agronomic practices. In other spheres it is often important to discover the identity of organisms responsible for biodeterioration, food spoilage, or the production of odours and taints in water supplies. Thus, few microbial ecologists would take exception to the view stated by Cross *et al.* (1976) that workers interested in (particular) organisms have to provide information on their numbers, distribution, activities, growth, survival, and dissemination before their role in microbial processes within ecosystems can be elucidated.

Acceptance of this argument raises the perennial question of how sufficient information can be obtained for the identification of an acceptable sample of the microbial taxa that occur in such heterogeneous habitats as forests, rivers, and oceans. Even if it is assumed that the current classification of the organisms is sound, accurate identification invariably involves cultural testing and microscopic examination of numerous isolates. The situation is especially acute with bacteria, yeasts, and many unicellular algae, for—unlike filamentous fungi, most multicellular algae, and many protozoa—they can rarely be identified from their morphological features alone. In addition, the range of micro-organisms in many habitats is so diverse that generating meaningful data on particular taxa usually necessitates the identification of large numbers of isolates. This is time-consuming and is not without its difficulties. The microbial ecologist is therefore confronted by a classic dilemma which has still to be resolved, if for no other reason than that skilled labour and resources are invariably limited. As Williams *et al.* (1969) noted, the emphasis in microbial ecology is either on autecological studies of a tiny but well-defined component of the microbial community or on synecological studies of the whole, albeit taxonomically ill-defined assemblage of organisms.

Early Attempts to Group Micro-organisms

It is now clearly appreciated that most habitats defined on macroecological grounds are markedly heterogeneous when they are considered on size and time scales appropriate to micro-organisms. Such habitats consist of arrays of microhabitats, each of which may be occupied by an assemblage of diverse microorganisms (White, 1983). It is not surprising that the combination of large populations and wide taxonomic diversity led to problems when attempts were made to describe the microbial populations in soils, lakes, and seas. Nevertheless, in the early part of this century much time and effort was devoted to general microbiological surveys designed to determine the contribution of microorganisms to phenomena such as soil fertility and biodegradation.

Assessments of microbial numbers have usually involved some variation of the dilution plate technique, which though simple to apply is often extraordinarily difficult to interpret (Jensen, 1967; Kaper *et al.*, 1978). Even greater difficulties were encountered with the qualitative aspects of these surveys (Lochhead and Taylor, 1938). In particular many early attempts to identify bacteria from natural habitats floundered as isolates were unresponsive to the biochemical tests that had been used to such good effect in the classification and identification of medically important bacteria. These difficulties forced many bacteriologists to abandon attempts to apply the Linnaean binomial system, and instead they created artificial groupings based on limited numbers of characters, such as cell morphology and staining reaction (Conn, 1917; Taylor and Lochhead, 1937), morphology and gelatin digestion (Conn, 1948), pigmentation and glucose metabolism (Stout, 1960), and nutritional requirements (Lochhead and Chase, 1943).

The most widely used of these 'alternative' classifications were those based on nutrition (Lochhead and Chase, 1943; Taylor, 1951; Rouatt, 1967; Jensen, 1971). The *raison d'être* of these schemes was twofold. The nutritional tests, which were simple to perform and which were meant to reflect conditions prevailing in the soil, enabled bacteriologists to determine the growth requirements of each isolate. This knowledge then permitted speculation about the likely distribution and function of the organisms in the ecosystem. However, subsequent elevation of these ecophysiological groups to the status of taxonomic pigeonholes was difficult to sustain, as it invariably transpired that each grouping accommodated a miscellany of Linnaean taxa. This latter fact meant that significant patterns of variation in species composition often went unnoticed.

Comparable ecological groupings have been devised for soil fungi (Burges, 1939; Garrett, 1956), but they have supplemented rather than replaced classical taxonomic studies. This, perhaps, explains why artificial classifications of fungi have thrived, to such an extent that the terms sugar fungus or root-invading fungus convey a wealth of meaning about the biology and ecology of appropriate taxa. It is unfortunate (in this respect) that comparable progress has not been made in ecological studies of other groups of micro-organisms such as the microalgae and protozoa.

At present there are no ecological or nutritional groupings of microbes which can serve as acceptable alternatives to conventional classification. It may be argued that the accurate identification of even a small proportion of the microbes in heterogeneous habitats is beyond the scope of individual scientists, but the fact remains that in microbiology, as in botany and zoology, the precision of the Linnaean system is usually needed if different workers are to correlate their findings effectively. Although individual workers are unlikely to have sufficient experience to identify isolates to the rank of species in more than a few genera, it should be possible for trained microbial ecologists to recognise at least the common genera or higher taxonomic categories without too much difficulty.

Soil
(330 species)

Actinomadura *Mycobacterium*
Actinoplanes *Nitrobacter*
Arthrobacter *Nitrosomonas*
Azotobacter *Nocardia*
Bacillus *Nocardioides*
Bdellovibrio *Nocardiopsis*
Beijerinckia *Promicromonospora*
Cellulomonas *Pseudomonas*
Clostridium *Pseudonocardia*
Cytophaga *Rhodococcus*
Janthinobacterium *Streptomyces*
Micrococcus *Streptosporangium*
Micromonospora *Thiobacillus*

Freshwater
(200 species)

Aeromonas *Metallogenium*
Aquaspirillum *Naumanniella*
Beggiatoa *Ochrobium*
Blastobacter *Pelodictyon*
Caulobacter *Planctomyces*
Chlorobium *Rhodopseudomonas*
Ectothiorhodospira *Siderocapsa*
Enterobacter *Sphaerotilus*
Gallionella *Spirillum*
Leptothrix *Thiopedia*

Seawater
(102 species)

Cristispira *Nitrosococcus*
Desulfovibrio *Oceanospirillum*
Flexithrix *Photobacterium*
Leucothrix *Planococcus*
Lucibacterium *Thiothrix*
Nitrococcus *Vibrio*

Fig. 1. Diagrammatic illustration of the relative density of the bacterial floras in soil, fresh water, and the seas, together with lists of the common genera therein. From Buchanan and Gibson (1974).

Some idea of the scale of the problem facing those trying to identify bacteria from aquatic and terrestrial habitats can be gleaned from Fig. 1. Species lists, however, not only reflect the isolation procedures used but must also be interpreted with care as they may include alien as well as indigenous taxa.

In addition to the foregoing, it must also be remembered that the ecology of a species, as exhibited by its distribution, its survival mechanisms, and the role it plays in the ecosystem, is a product of many factors including its developmental and morphological characteristics and its biochemical and physiological mechanisms. Hence it is imperative that microbial classifications are sufficiently exact for closely related taxa, which may occupy different ecological niches, to be distinguished. However, until recently the identification of bacteria using dichotomous keys and diagnostic tables, as found in successive editions of *Bergey's Manual of Determinative Bacteriology,* was frequently so ineffective that isolates resembling one another in a few 'key' characters but which differed markedly in other properties fell into the same genus. This problem became so acute that several genera, such as *Cytophaga, Erwinia, Flavobacterium,* and *Pseudomonas,* became 'sinks' for heterogeneous collections of Gram-negative bacteria.

Divisive–monothetic classifications have frequently been used in microbial ecology, for instance in the classification of bacteria from the rhizosphere (Vágnerová *et al.*, 1960; Rouatt and Katznelson, 1961), the phylloplane (Stout, 1960), oxidation ponds (Pretorius, 1964), the soil (Sperber and Rovira, 1959; Brisbane and Rovira, 1961), the rumen (Hungate, 1966), frozen vegetables (Splittstoesser *et al.*, 1967), and cannery environments (Bean and Everton, 1969). Again, however, the groupings obtained were often heterogeneous. Mathematical procedures to effect diverse *monothetic classifications* have been developed (Gyllenberg, 1963, 1964; Gyllenberg *et al.* 1963; Rypka *et al.* 1967; Niemalä *et al.* 1968) but were similarly flawed.

Problems such as those outlined raise a question mark concerning the ability of microbial ecologists to attain the level of taxonomic precision regarded as the norm in higher plant and animal ecology. Clearly new approaches were required to determine the kinds, characteristics, discriminating features, and roles of the indigenous microflora in natural habitats.

Enter, Numerical Taxonomy

The introduction of miniaturized and multipoint inoculation techniques and the availability of computers for handling the resulting coded data opened up the possibility of using numerical methods for determining the composition of microbial populations in natural habitats. Both taxonomic and ecological investigations can generate a great deal of data, but to enable conclusions to be drawn some form of condensation, highlighting significant relationships, is necessary. In numerical taxonomy the starting point is usually an $n \times t$ matrix, where n represents the number of characters collected for each strain and t is the number of strains. In what can be termed conventional numerical taxonomy, similarities between strains are determined by calculating similarity coefficients, the so-called Q analyses, and the strains are then sorted into hierarchic, essentially nonoverlapping, taxa using clustering algorithms (Sneath, 1978; Jones and Sackin, 1980). The results of these analyses are usually presented as dendrograms or shaded diagrams (see Priest and Barbour, Chapter 7).

Primary data may also be arranged into correlation or R matrices, based on $n \times n$ characters, and analyzed using non-hierarchical ordination methods that represent data in reduced number of dimensions. Ordinations are sometimes derived from Q matrices (see Alderson, Chapter 9). The most widely used ordination methods in microbiology are factor and principal components analyses. The latter represent a branch of multivariate statistics concerned with the internal relationships of a set of variables. The application of non-hierarchical methods to bacterial classification is considered elsewhere (see Alderson, Chapter 9). The power of multivariate analysis, especially factorial analysis, in unravelling the complexities of heterogeneous microbial populations was highlighted in a number of reviews that also considered the underlying statistics (Gyllenberg, 1964; Sundman and Gyllenberg, 1967; Rosswall and Kvillner, 1978). The mathematical basis of factor analysis was described by Harman (1960). The results of ordinations are usually displayed as two-dimensional taxonomic maps or as three-dimensional taxometric models (see Alderson, Chapter 9).

In conventional numerical phenetic surveys, large numbers of randomly selected isolates and appropriate marker strains are examined for many properties chosen to effect a classification (Goodfellow *et al.*, 1976a; D. Jones, 1978). In contrast, factor and principal components analyses are not used to classify or identify individual isolates, but rather to describe or characterise whole microbial populations in terms of the tests applied and the environmental parameters measured. These multivariate techniques are designed to express a number of characters, the observed variables, in terms of a small number of successively derived hypothetical variables, termed the axes, the principal components, or factors, thereby allowing the maximum variance in the original data set to be ascertained. This procedure is well suited to the analysis of complex populations such as those found in natural habitats, since the direction of variation in the observed combinations of characters is given by means of the axes or factors. The latter can be used to interpret either taxonomical or ecological relationships depending on the nature of the data and the purpose at hand. Factors that have been described in ecological terms include those for carbohydrate metabolism (Sundman, 1970), food deterioration (Gyllenberg and Ecklund, 1967), and humification (Sundman,

1970). Sundman (1970) also suggested that the oxidase–nitrate reductase factors might be an index for soil fertility, as it was found to be heavily loaded for a fertile grassland soil but not for a forest humus soil.

Factor and Principal Components Analysis in Determining the Structure and Function of Microbial Populations

Description of Microbial Populations from Natural Habitats

The conversion of primary taxonomic data into descriptive terms, such as binomial names, usually involves a much greater loss of information than occurs following factorization of data from correlation coefficient matrices. In addition, in studies of the dynamics of microbial ecosystems it may be more important to know about the physiological potential of the whole population, or the relationships of specific components of the microbial community to particular environmental factors, than to be able to identify individual microbes. Indeed, descriptive numerical approaches which use all of the test data can be expected to yield more ecologically relevant information than is usually obtained from conventional procedures.

Factor and principal components analyses can be seen as a logical development of the nutritional and functional groupings of soil bacteria (Lochhead and Chase, 1943; Taylor, 1951). The tests used for characterising bacterial populations have to be functional and more varied than those used in earlier studies, and in practice they are similar to those employed in conventional numerical phenetic investigations. For example, in surveys of bacteria in temperate soil, tests for growth at 55°C or in the presence of teepol are clearly of less relevance than the ability of isolates to degrade complex organic molecules, such as various phenolic compounds that are of little taxonomic value (according to current practice) but of considerable ecological importance. As in conventional numerical taxonomy, tests should be of good quality, although the number of characters is usually small in comparison with the number of isolates as the numerical figures that represent the correlation between tests are proportional to the number of isolates. Prior to factorization, truistic factors and characters with low communalities should be removed from correlation coefficient matrices (Sundman and Gyllenberg, 1967). In an investigation on soil bacteria, Debette *et al.* (1975) were able to reduce the number of tests employed from 51 to 21 without significant loss of information.

The two multivariate methods under consideration have been used to describe bacterial populations from a number of habitats, especially soil (Table 1).

Table 1. *Numerical description of microbial populations using factor and principal components analysis*

Source of strains	Selective factors	Data matrix		References
		Strains	Tests	
Clay, mull, and sandy soils under grass, rape, and wheat	Yeast extract tryptone, soil extract agar, plus growth stimulators, incubation at 25 to 28°C for 10 to 12 days	465	32	Gyllenberg (1964); Gyllenberg and Rauramaa (1966)
Milk and other food	Procedures selective for pseudomonads	215	24	Gyllenberg (1965a)
Milk	Procedures selective for coliforms	278	21	
Soil	Procedures for the isolation of heterotrophic, non-spore-forming bacteria	176	35	Gyllenberg (1967)
Milk, meat, and eggs	Procedures selective for psychrophilic pseudomonads	481	61	Gyllenberg and Eklund (1967)
Meat stored in air and under carbon dioxide	Nutrient gelatin potassium dihydrogen phosphate agar, 28°C, 3–5 days	203	24	Pohja and Gyllenberg (1967)
Soils rich in organic matter	Soil suspensions, enrichment culture at 20°C, soil extract agar	765	32	Sundman and Carlberg (1967); Sundman (1968)
Rhizosphere and nonrhizosphere soil	3-week-old flax–soil extract agar at 28°C for 12 days	400	98	Skyring and Quadling (1969)

Acid forest humus, grassland, and arable soil	Soil extract agar at 21°C for 14 days	467	42	Sundman (1970)
Anaerobic digester	Prereduced digester supernatant agar plus growth stimulators in carbon dioxide, hydrogen atmosphere at 30°C for 8 days	92	45	Toerien (1970a)
Soil	Bunt and Rovira's mineral medium, 28°C, 8 days	76	47	Soumare *et al.* (1973)
Fresh water	River water polluted with cadmium	77	11	De Leval *et al.* (1976)
Wild and farmed diseased fish	Procedures for the relative isolation of diseased fish	169	28	Håstein and Smith (1977)
Podsol soil under spruce	Soil extract, yeast extract, tryptone salts agar, plus antifungal antibiotics, 22–24°C, 10 days	684	28	Niemelä and Sundman (1977)
Freshwater, marine, and highly saline habitats	Modified Starkey's medium supplemented with sodium chloride, hydrogen atmosphere, 30 and 50°C	92	50	Skyring *et al.* (1977)
Rhizosphere and nonrhizosphere soils	Wheat–peptone yeast extract agar plus actidione, 28°C, 14 days	600	79	Deavin *et al.* (1981)
Fresh water	Glucose nitrogen minimal medium plus casamino acids, 20°C, 7 days	1300	82	Bell *et al.* (1982)
Fresh water	Casein peptone starch agar, 20°C	1250	173	Holder-Franklin and Wuest (1983)

Gyllenberg (1964) plotted isolates sampled to represent a given population as points in multidimensional space where each of the two-state tests employed to characterise the isolates constituted a single dimension. Principal components analysis was then applied to reduce the number of dimensions so that the population was represented as points in a three-dimensional Euclidian factor space. In a development of this approach Sundman (1968) found that some factor estimates were constant features of the soil population of bacteria, and in a further study (Sundman, 1970) she described four populations of bacteria from soil using discriminating factors which accounted for 56% of the total factor variance. The similarity or difference between populations was expressed as the squared distance between the population means in the selected discriminating six-factor space. Although these early studies were restricted to relatively small numbers of aerobic, heterotrophic bacteria from soil, they did show that factor analysis provided a means of comparing bacterial populations without the need to cluster individual isolates.

Quadling (1967) introduced a two-staged principal components method for stepwise condensation of the information obtained with sets of dichotomous tests. Skyring and Quadling (1969) used the method to compare populations of bacteria, isolated from rhizosphere and nonrhizosphere soils, in terms of the clusters observed in the various populations. They found that isolates which predominated in the rhizosphere were not represented amongst the test strains from nonrhizosphere soil. Deavin *et al.* (1981) also distinguished between populations of bacteria from rhizosphere and nonrhizosphere soil.

Several studies have been designed to establish the influence of soil management techniques on the functional characteristics of the bacterial flora. Gyllenberg and Rauramaa (1966) found a relatively homogeneous bacterial population in a field undergoing crop rotation as compared with that in a soil repeatedly sown with cereal and treated with herbicides. This finding suggested that continuous cereal cultivation which was facilitated by the use of selective weedkillers did not reduce the complexity of the population structure of bacteria. Sundman and Carlberg (1967) devised a formula for determining the radius of a cloud of isolates and found that the radii of bacterial populations showed a tendency to decrease with reduced fertility of the soil. In a subsequent study, Sundman (1970) was unable to find any difference in the physiological diversity of bacterial populations in two cultivated arable soils and an uncultivated soil.

The effects of pesticides on populations of soil bacteria have been the subject of several investigations. Gyllenberg and Rauramaa (1966) discovered a significant trend towards more homogeneous populations in soils amended with either the herbicide MCPA or straw. Skyring and Quadling (1969) did not find any differences between populations from untreated soils and those treated with Vapam (sodium *N*-methyldithiocarbamate), and Torstensson and Rosswall (1977) did not observe any marked changes in the functional characteristics of bacterial populations in soils treated over a period of years with the herbicides MCPA and 2,4-D. Similarly, the effect of other stress factors including clear-cutting (Niemelä and Sundman, 1977), eutrophication (Persson and Rosswall, 1978), and manure decomposition (Rosswall, 1976) have been studied. Factor analysis has also been used to relate variations in bacterial populations to toposequence variations in soil properties (Hagedorn and Holt, 1975b).

Principal components analysis has also been used to determine the characteristics of bacterial populations in substrates other than soil and to detect novel species (Moss *et al.*, 1978) and infra-specific variation amongst bacteria (Darland, 1975). Thus, Håstein and Smith (1977) revealed two major groups of some epidemiological importance among strains of *Vibrio anguillarum* from diseased wild and farmed fish, whereas Kapperud *et al.* (1981) concluded that *Yersinia enterocolitica sensu stricto, Y. frederiksenii,* and *Y. intermedia* formed a phenotypic continuum. Pohja and Gyllenberg (1967) highlighted three distinct differences in the activities and ecological behaviour of organisms predominating in meat kept under carbon dioxide or in air during the early stages of storage. Similarly, Toerien (1970a) established several relationships between the characteristics of bacterial isolates, recorded as *operational ecological units,* from anaerobic digesters. In addition to detecting ecological relationships, the two multivariate analysis techniques yield data that can be interpreted within a taxonomic context (see Alderson, Chapter 9). Gyllenberg and Eklund (1967) recognised factors typical of *Pseudomonas fluorescens* and *P. fragi* in an analysis of psychrophilic pseudomonads from meat and dairy products, and generalised '*Pseudomonas*' and '*Bacillus*' factors have also been recognised by Pohja and Gyllenberg (1967) and Sundman (1970).

Physico-Chemical Factors Influencing Microbial Populations

Multivariate analysis based on quantitative parameters has been used to show the effect of environmental factors on the structure and function of microbial populations in both freshwater and marine habitats. Erkenbrecher and Stevenson (1977) used correlation coefficients and regression analyses to establish the influence of tidal flux on the microbial community in salt marsh creeks. They found that the ebb and flood tides had different microbial characteristics, with the population in the former resembling that of the sediment.

Väätänen (1980) applied factor analysis to data previously subjected to regression analysis in an ambitious attempt to unscramble the influence of different environmental factors on microbial communities in the Tvärminne archipelago in southern Finland. Seventy-one samples taken from the sampling station over a 2-year period were examined for 10 environmental and 16 microbiological parameters and the correlations between the parameters factorized. Eight factors considered in detail accounted for between 60 and 98% of the variance of the microbiological parameters. The bacterial populations were especially sensitive to water temperature, to phytoplankton blooms, and to freshwater outflows that carried bacterial endospores, faecal bacteria, fluorescent pseudomonads, and yeasts into the brackish waters of the sampling station. It was observed that environmental processes affecting the micro-organisms were generally easier to identify from factor analysis than by regression analysis.

Factor analysis has been used to measure the effect of environmental parameters on spatial and temporal changes in bacterial populations in freshwater habitats exposed to considerable seasonal fluctuation. In a series of ecological surveys Holder-Franklin and colleagues (Holder-Franklin *et al.*, 1978; Holder-Franklin, 1981; Bell *et al.*, 1982; Holder-Franklin and Wuest, 1983) observed marked diurnal and seasonal changes in populations of Gram-negative bacteria in river water. The fluctuations mainly involved changes in the relative numbers of different, but possibly overlapping, biovars. The ability of fractions of what may amount to a continuum of bacterial variation to grow and multiply in response to subtle environmental pressures may be of considerable ecological importance. Clearly novel strategies are needed to detect new patterns of bacterial variation, as it seems most unlikely that all bacteria fall into neat, well-separated clusters (see Sneath, Chapter 16).

Bell *et al.* (1982) also compared heterotrophic bacterial populations, as characterised by total counts and species numbers (using conventional numerical taxonomy), by *diversity indices,* and by factor analysis. The latter proved to be the most effective way of revealing correlations between bacterial characteristics and environmental parameters. In addition, the bacterial communities of two contrasting rivers were profiled by four factors, namely, fermentative metabolism, inorganic nitrogen metabolism, oxidative-fluorescence metabolism, and lack of starch hydrolysis. In one river the oxidative-fluorescence factor correlated positively with rainfall, which suggested that oxidative bacteria were wash-in components from surrounding land. In the second river the oxidative–fermentative factor correlated negatively with the amount of sunshine which was received. There was also evidence of an intricate algal–bacterial interaction in these freshwater habitats. Similarly, Jooste (1979) considered that anaerobic and coliform bacteria were major food sources for Copepoda and Cladocera, respectively. The occurrence of the former was associated with higher water temperatures and that of the latter with dissolved oxygen.

Jones (1977) demonstrated the effect of stratification and deoxygenation on lacustrine populations; the latter were also influenced by seasonal changes in nutrient concentrations in the epilimnion and the development of anoxic conditions in the hypolimnion. Most of the variation in the bacterial populations was explained by 5 of the 18 regressor variables, and the factors considered likely to provide additional information included measures of predation and changes in the lake retention time. In a further study, Jones (1978) examined the effect of certain physical and chemical variables on the distribution of planktonic bacteria in two stratified eutrophic lakes. Once again a significant amount of the bacterial variation was accounted for and, in particular, iron bacteria were found to grow best in the oxycline where there was sufficient oxygen for aerobic growth and a plentiful supply of reduced iron.

The effect of eutrophication on bacterial populations in Swedish lakes was examined by Rosswall and Persson (1982) using factor analysis. A clear difference was found between the bacterial populations in the eutrophic and oligotrophic lakes in samples taken in the spring, a result confirmed by discriminant analysis and by the significant differences in the total numbers of bacteria in samples from the two types of lakes. The spring populations showed more diversity than the autumnal ones, although in each case the bacterial community was heterogeneous.

Numerical Taxonomic Studies

The introduction of Adansonian principles and computer methods to bacterial systematics was seen to provide a strategy for overcoming many of the perennial problems associated with the classification and identification of bacteria from natural habitats (Brisbane and Rovira, 1961; Pfister and Burkholder, 1965). Strains randomly chosen from isolation plates could be examined for many equally weighted properties and assigned to numerically defined taxa on the basis of overall similarity. Defined clusters might represent novel taxa or be identified by marker or reference cultures of established taxa included in the analysis. Once a numerical classification had been constructed, cluster-specific characters could be weighted for identification. Simple, but accurate, dichotomous keys and diagnostic tables could then be used in the identification of further isolates by workers with little detailed knowledge of the taxa in question. Such identification schemes also allow the examination of sufficient replicate samples to permit statistical treatment of the data. It was thus hoped that they would provide a practical way of surveying the bacterial flora of diverse habitats and of monitoring changes in populations due to seasonal factors, pathogen damage, and agricultural practices. Detailed ecological and taxonomic studies could then be carried out on isolates of particular interest, such as representatives of the most common or novel taxa, or on taxa showing restricted distribution patterns.

To a limited extent some of these objectives have been fulfilled. It is evident that when microbial communities contain strains with independent and correlated characters, then the standard statistics applied in numerical taxonomy can be used to detect and define representative samples of the constituent taxa. It has also been shown that when there is sufficient consistency within, and differences between, the defined taxa, then characters can be extracted from the data base and weighted for identification. It is also possible to define the 'average' organism within each cluster (Silvestri *et al.*, 1962; Gyllenberg, 1963; Liston *et al.*, 1963) and to use such 'median' organisms for the identification of unknown isolates (Ercolani, 1978).

Many numerical taxonomic studies have been published concerning bacteria from natural habitats, but most can be described as either broad or restricted ecological surveys. Although no sharp division can be drawn between the two types, the broad surveys are designed to determine the predominant bacterial populations in ecologically complex communities, whereas restricted studies have been largely confined to a consideration of closely circumscribed habitats or particular groups of bacteria. No attempt will be made to review the various studies exhaustively; studies will be selected to illustrate points of particular taxonomic or ecological importance.

Broad Ecological Surveys

Construction and Evaluation of Numerical Classifications

Most studies of the bacterial composition of ecologically complex communities have centred on aerobic, heterotrophic, mesophilic organisms. Oligotrophic bacteria, unlike their copiotrophic counterparts, have received little attention (Mallory *et al.*, 1977; Witzel *et al.*, 1982b), although many of the former can convert to saprophytic growth on rich media (Kuznetsov *et al.*, 1979). It is well known that isolation media and incubation regimes exert a profound influence on the numbers and kinds of bacteria isolated (Vickers *et al.*, 1984; Williams *et al.*, 1984), and it seems likely that the regular use of a small number of relatively nonselective media and standardisation on a restricted range of growth conditions accounts for the fact that relatively few of the taxa found in natural habitats are normally included in broad ecological investigations. In particular, organisms that require prolonged incubation or the use of selective isolation procedures have been neglected. Further, the tendency to carry out tests at temperatures well above those experienced by bacteria *in situ* may mean that certain taxa have been excluded from these studies (Kaneko *et al.*, 1979; see Sneath, Chapter 16).

The nature of the numerical taxonomic study influences the number of strains that can be studied, as do the computer facilities available (Sneath, 1978; Jones and Sackin, 1980). The number of strains included in ecological surveys should be as high as can be competently handled, but to account for strain variation and sampling error at least 60, and preferably many more, should be examined. In order to have enough information to be discriminatory, at least 50 and preferably several hundred characters are desirable, but with higher numbers any gain in information has to be offset against the effort involved and the efficiency achieved in amassing the data. However, it has not always been appreciated that abundance of characters is not an end in itself; they should also be of high quality. Unit characters should represent a judicious spread of taxonomic criteria and should include ecologically significant properties. In practice between 37 and 1789 strains have been studied in broad numerical phenetic surveys with the number of tests employed ranging from 39 to 300 (Tables 2 and 3).

Once data matrices have been prepared, similarity coefficients and clustering algorithms can be employed to assign the test strains to clusters defined on the basis of overall similarity (Sneath, 1978; Jones and Sackin, 1980). Most workers have used either the simple-matching coefficient S_{sm}, which counts both positive and negative similarities between strains and expresses them as a fraction of the characters studied, or the Jaccard coefficient S_J, which is the same as the S_{sm} coefficient except that negative matches are ignored; clustering has usually been achieved using either single- or average-linkage algorithms (Tables 2 and 3). The former defines the similarity between two clusters as the similarity of the two most similar strains, one in each cluster; the latter takes the average of all of the similarities across the two groups. A variant of the average-linkage algorithm, the unweighted pair-group method with averages (UPGMA), is often preferred (Austin *et al.*, 1977a,b,c, 1978; Mallory *et al.*, 1977; Ventosa *et al.*, 1982; Witzel *et al.*, 1982b).

The results of clustering have usually been presented either as dendrograms or as shaded diagrams, although those derived from single-linkage clustering have occasionally been presented as *minimum spanning trees* (Byrom, 1971; Lowe and Gray, 1972; Hissett and Gray, 1973; Delarras *et al.*, 1979). The latter, unlike dendrograms, show the pairs of organisms responsible for the fusion of clusters. The choice between the S_J and S_{sm} coefficients is somewhat arbitrary, although essentially similar classifications have been obtained where both coefficients have been employed (Lowe and Gray, 1972; Hissett and Gray, 1973; Austin *et al.*, 1977a,b,c, 1979a; Mallory *et al.*, 1977; Austin, 1982). However, in ecological surveys of relatively unreactive, but diverse, strains the S_J coefficient has the advantage that it avoids unrelated strains being assigned to common clusters based on negative correlation (Goodfellow, 1969; Ercolani, 1978). A number of workers have presented their findings in linkage maps derived from distance matrices (Seiler and Busse, 1977; Seiler *et al.*, 1980; Seiler and Hennlich, 1983).

The pattern difference or dissimilarity coefficient D_p (Sneath, 1968a) seeks to measure the differences between pairs of strains by excluding the component of the various differences that is due merely to differences in vigor. Distortions of phenetic similarities due to vigor have been demonstrated (Priest *et al.*, 1973; Goodfellow *et al.*, 1976b; Ercolani, 1983; Hookey, 1983). The D_p coefficient calculates the product of the number of tests for which one strain gives a positive

Table 2. *Numerical phenetic surveys of bacteria from terrestrial habitats*[a]

Selective factors	Data matrix		Statistics		Strains assigned to major clusters (%)	Identity of major taxa	References
	Strains	Tests	Resemblance coefficient	Cluster technique			
Bacteria associated with rhizosphere and soil							
Clover and rye grass rhizosphere soils; yeast extract, soil extract agar (YESEA); 26°C, 14 days	64	39	S_{sm}	SL	0	*Agrobacterium, Arthrobacter, Pseudomonas, Rhizobium*	Brisbane and Rovira (1961)
Soil–trypticase soy yeast extract agar, 25°C, 10 days	177	84	S_{sm}	SL	47	*Arthrobacter*	Hagedorn and Holt (1975a)
Clover and wheat rhizosphere soils; YESEA, 27°C, 14 days	383	37	S_{sm}	SL	46	*Agrobacterium, Arthrobacter, Mycoplana, Pseudomonas*	Rovira and Brisbane (1967)
Pine forest, A and C horizons; peptone yeast extract agar (PYEA) plus antifungal antibiotics, 25°C, 14 days	527	106	S_J, S_{sm}	SL	65	*Arthrobacter, Bacillus, Flavobacterium, Micrococcus, Pseudomonas, Streptomyces*	Goodfellow (1969)

Pine forest, A and C horizons; PYEA plus antifungal antibiotics, 25°C, 14 days; Gram-positive coccobacilli	209	179	S_J, S_{sm}	SL	90	*Arthrobacter*, *Micrococcus*, *Staphylococcus*	Lowe and Gray (1972)
Mixed deciduous forest, litter and mineral soil; PYEA plus antifungal antibiotics, 25°C, 14 days	400	131	S_J, S_{sm}	SL	65	*Arthrobacter*, *Bacillus*, *Pseudomonas*	Hissett and Gray (1973)
Soil; procedures selective for nonfermentative, Gram-negative rods	165	51	S_J	AL	70	*Acinetobacter*, *Pseudomonas*	Debette *et al.* (1975)
Soil; diagnostic sensitivity test (DST) agar plus tetracyclines and antifungal antibiotics, 25°C, 21 days	73	61	S_{sm}	AL	74	*Nocardia*	Orchard *et al.* (1977)
Sand dune grass rhizosphere soil; PYEA plus actidione, 25°C, 21 days	254	109	S_J, S_{sm}	SL	64	*Bacillus*, *Pseudomonas*	Schofield and Whalley (1978)

(*continued*)

Table 2 (*Continued*)

Selective factors	Data matrix		Statistics		Strains assigned to major clusters (%)	Identity of major taxa	References
	Strains	Tests	Resemblance coefficient	Cluster technique			
Soil DST agar plus tetracyclines and antifungal antibiotics, 25°C, 21 days	197	137	S_p, S_{sm}	AL, SL	78	*Nocardia*	Orchard and Goodfellow (1980)
Soil from Tienshan Mountain, China	268	38	S_{sm}	SL	85	*Bacillus*, *Pseudomonas*	Dasi *et al.* (1981)
Soil, water, and biodeteriorating rubber, 55°C for 6 min; DST agar plus methacyline and antifungal antibiotics, 30°C, 21 days	396	107	S_J, S_p, S_{sm}	AL, SL	79	*Nocardia*	Hookey (1983)
Bacteria associated with plants							
Legume root nodules	121	100	S_{sm}	SL	100	*Agrobacterium*, *Arthrobacter*, *Pseudomonas*, *Rhizobium*	Graham (1964)
Pine leaves and litter; PYEA plus antifungal antibiotics, 25°C, 14 days	113	77	S_{sm}	SL	76	*Bacillus*, *Micrococcus*, *Pseudomonas*	Goodfellow *et al.* (1976a)
Grass seeds and leaves; glucose yeast	124	158	S_J, S_p, S_{sm}	AL	94	Enterobacteriaceae including *Erwinia*	Goodfellow *et al.* (1976b)

extract agar (GYEA), 25°C, 14 days							
Rye grass leaves; GYEA, 25°C, 7 days	688	135	S_{sm}	AL	43	*Listeria, Pseudomonas, Staphylococcus, Xanthomonas*	Austin *et al.* (1978)
Olive leaves; yeast tryptone glucose extract agar (YTGEA), 28°C, 5 days	1789	210	S_J	SL	97	*Bacillus, Erwinia, Micrococcus, Pseudomonas, Xanthomonas*	Ercolani (1978)
Healthy and diseased cabbage cotyledons; GYEA, 25°C, 9 days	720	98	S_{sm}	AL	57	*Bacillus, Curtobacterium, Enterobacter, Pseudomonas*	Al-Hadithi (1979)
Olive leaves; YTGEA, 28°C, 5 days	1050	60	S_p, S_{sm}	AL, SL	92	*Pseudomonas syringae* pathovar *savastanoi*	Ercolani (1983)
Bacteria associated with waste material							
Activated sludge produced from whey	262	54	DC	SL	57	*Arthrobacter, Brevibacterium, Corynebacterium, Microbacterium, Mycobacterium*	Seiler and Busse (1977)
Activated sludge	383	135	S_{sm}	LM, SL	85	*Arthrobacter, Cellulomonas, Curtobacterium, Corynebacterium, Microbacterium, Mycobacterium, Nocardia, Rhodococcus*	Seiler *et al.* (1980)

(*continued*)

Table 2 *(Continued)*

Selective factors	Data matrix		Statistics		Strains assigned to major clusters (%)	Identity of major taxa	References
	Strains	Tests	Resemblance coefficient	Cluster technique			
Pig manure; Eugor agar, 30°C, 7 days, pleomorphic strains	131	52	EC	AL, LM	81	*Brevibacterium, Corynebacterium, Curtobacterium, Rhodococcus*	Seiler and Hennlich, (1983)
Bacteria associated with food							
Frozen vegetables; tryptone glucose extract agar, 32°C, 2 days	111	114	S_J	SL	75	Corynebacteriaceae	Splittstoesser *et al.* (1967)
Chlorinated cannery cooling waters	132	60	S_J	SL	73	*Brevibacterium, Flavobacterium, Xanthomonas*	Bean and Everton (1969)
Food canneries; procedure for selective isolation of flavobacteria	250	96	S_{sm}	SL	41	*Flavobacterium*	Byrom (1971)
Meat products; nutrient agar, 22°C, 3 days	59	78	S_J	SL	58	*Cytophaga, Flavobacterium*	McMeekin *et al.* (1972)
Meat and dairy products	149	61	ED	SL	80	*Micrococcus, Staphylococcus*	Delarras *et al.* (1979)
Processed beef at 20°C for 14 days; agar plates: mesophiles, 32°C for 2 days; psychrophiles, 7°C for 10 days	706	108	DĆ	AL	92	*Bacillus, Lactobacillus, Microbacterium, Micrococcus, Staphylococcus, Streptococcus*	Lee *et al.* (1982)

[a] Abbreviations: DC, distance coefficient; DĆ, Dice coefficient; ED, Euclidean distance; LM, linkage maps; S_J, Jaccard coefficient; S_{sm}, simple-matching coefficient; S_p, pattern coefficient; AL, average-linkage algorithm; SL, single-linkage algorithm.

result and the second shows a negative one, and the number for the reciprocal situation. However, where the many positive responses of a vigorous organism coincide with the few of the less vigorous one, this product is small and gives rise to a misleading pattern difference. This matching effect can be detected by comparing the relative numbers of positive test results given by each strain and confirmed by reference to similarity values calculated from the S_J and S_{sm} coefficients (Goodfellow *et al.*, 1978). A variant of the D_p coefficient which excludes negative matches is available (Sackin, 1981). Other distance coefficients that have been employed are shown in Tables 2 and 3.

Hierarchical clustering techniques impose a hierarchical structure upon data which may not always be warranted. Although the results of many numerical taxonomic studies indicate that there may be sharp limits to species variation (see MacDonell and Colwell, Chapter 6), there has been a tendency to overlook the fact that many isolates from natural habitats do not form tight clusters (Tables 2 and 3). In such cases the pattern of variation may take the form of a few dense clusters in a sparse scattering of single isolates (Sneath, 1968b); there is also evidence that some taxa may form elongated clusters consisting of overlapping subclusters (see Sneath, Chapter 16). It is important that more attention be given to the implications of such variation patterns, and studies based on hierarchical methods should include a test for the hierarchicalness of the data. The most widely used measure of hierarchicalness is the cophenetic correlation coefficient (Jones and Sackin, 1980). It has been shown that UPGMA clustering gives the highest cophenetic correlation of all the standard clustering methods (Farris, 1969).

Numerical classifications have to be interpreted carefully as similarity values between strains can be distorted by factors which not only include test and sampling error but also test reproducibility, test error, and the statistics used (Sneath and Johnson, 1972; Austin and Colwell, 1977; Sneath, 1978). It is now recommended that duplicate strains, amounting to ~5% of the total, should be included to allow an internal check on test error and test reproducibility. Several workers have already taken the precaution of determining the robustness of established clusters using different similarity coefficients and clustering algorithms (Orchard and Goodfellow, 1980; Ercolani, 1983; Hookey, 1983; Kaper *et al.*, 1983), and such good practice should be encouraged. Ideally, the quality of test data should also be assessed by determining cluster overlap (Sneath, 1977, 1979a) and mean intra- and inter-cluster similarities (Goodfellow, 1969; Hookey, 1983), and by comparing the results of numerical classifications with those based on chemical, genetic, and serological techniques (Jones and Sackin, 1980; MacDonell and Colwell, Chapter 6). In many broad numerical taxonomic surveys such *criteria of goodness* have been neglected in favour of examining additional strains, Such omissions need to be made good in future investigations.

Table 3. *Numerical phenetic surveys of bacteria from aquatic habitats*[a]

Selective factors	Data matrix		Statistics		Strains assigned to major clusters (%)	Identity of major taxa	References
	Strains	Tests	Coefficient	Cluster technique			
Bacteria associated with freshwater habitats							
Chlorella culture; brain–heart infusion agar, 37°C, 2 days	37	140	S_{sm}	AL	57	*Acinetobacter, Bacillus, Flavobacterium, Pseudomonas*	Litchfield *et al.* (1969)
Pond water and pond-reared shrimps; standard methods agar, 28°C, 2 days	88	163	S_{sm}	SL	28	Corynebacteriaceae	Vanderzant *et al.* (1972)
Eutrophic lake; Taylor's formulation agar, 20°C, 10 days	30	51	S_J	AL	77	*Aeromonas/Vibrio, Cytophaga, Flavobacterium*	Lighthart (1974)
Eutrophic lake and river; nutrient agar and seven other media, 27°C, 7 days	311	92/105	S_{sm}	SL	75	Aerobic, Gram-negative heterotrophs	Witzel *et al.* (1981)
Eutrophic lake; nutrient agar, 25°C, 7 days	124	105	S_{sm}	SL	82	Aerobic, Gram-negative heterotrophs	Witzel *et al.* (1982a)
Eutrophic lake; enrichment agar media for oligocarbophilic bacteria, 26°C, 14–28 days	58	130	S_{sm}	SL	86	*Caulobacter, Hyphomicrobium, Pseudomonas, Vibrio*	Witzel *et al.* (1982b)

Fish farm; tryptone soy agar, 15°C, 7 and 14 days	722	124	S_J, S_{sm}	AL	55	*Acinetobacter, Aeromonas, Alcaligenes, Enterobacter, Escherichia, Hafnia, Pseudomonas, Serratia, Vibrio, Yersinia*	Austin (1982)
Bacteria associated with estuarine nabitats							
Water and sediment; liquid enrichment, oil medium, 15°C, 21 days	132	48	S_J, S_{sm}	AL	54	Enterobacteriaceae, *Micrococcus, Nocardia, Pseudomonas, Sphaerotilus*	Austin *et al.* (1977a)
Water and sediment; liquid enrichment, oil medium, 15°C, 21 days	462	48	S_J, S_{sm}	AL	30	*Bacillus, Erwinia, Klebsiella, Lactobacillus, Leucothrix, Micrococcus, Moraxella, Nocardia, Pseudomonas, Streptomyces, Vibrio, Xanthomonas*	Austin *et al.* (1977b)
Water and sediment; glucose tryptone yeast extract agar (GTYEA) plus heavy metals, 15°C, 8 days	324	112	S_J, S_{sm}	AL	84	*Bacillus, Erwinia, Mycobacterium, Pseudomonas*	Austin *et al.* (1977c)
Water and sediment; beach water agar–GTYEA, 15°C, 28 days	164	119	S_J, S_{sm}	AL	60	*Hyphomonas, Listeria, Planococcus*	Mallory *et al.* (1977)

(*continued*)

Table 3 (*Continued*)

Selective factors	Data matrix		Statistics		Strains assigned to major clusters (%)	Identity of major taxa	References
	Strains	Tests	Coefficient	Cluster technique			
Water, sediment, and wood pilings; marine agar, 25°C, 7 days	338	107	S_J, S_{sm}	SL	61	*Bacillus*, *Hyphomicrobium*, *Hyphomonas*, *Pseudomonas*, *Vibrio*	Austin *et al.* (1979a)
Water; chitin, fish protein, marine, and thiosulphate citrate bile salt agars, 25°C, 14 days	160	115	S_J, S_{sm}	SL	41	*Acinetobacter—Moraxella*, *Caulobacter*, *Pseudomonas*, *Vibrio*	Austin *et al.* (1979b)
Estuarine water and sediment; procedures selective for aeromonads and vibrios	227	129	S_J, S_{sm}	AL, SL	57	*Aeromonas*, *Vibrio*	Kaper *et al.* (1983)
Marine studies—bacteria associated with fish and other animals							
Procedures selective for yellow-pigmented marine bacteria	62	104	S_J	SL	55	*Cytophaga*, *Flavobacterium*	Floodgate and Hayes (1963)
Marine animals; marine agar, enrichment cultures, basal medium agar, 18–22°C, 6 days	145	49	S_{sm}	CL	100	*Beneckea*, *Photobacterium*	Baumann *et al.* (1971)

Gut and light organs of macrourid fish; seawater nutrient agar, 20°C, 48 days	191	73	S_J	SL	76	*Moraxella, Photobacterium, Pseudomonas, Vibrio*	Singleton and Skerman (1973)
Skin of fresh cod, haddock, and plaice	167	90	S_J	AL	76	*Brevibacterium, Curtobacterium, Corynebacterium*	Bousfield (1978)
Fish farm; seawater and thiosulphate citrate bile salt sucrose agars, 15°C, 14 days	661	110	S_J, S_{sm}	AL	42	*Acinetobacter, Photobacterium, Vibrio*	Austin (1982)
Marine studies—benthic organisms							
Deep-sea sediments; marine agar, psychrophilic medium, 25°C, 14–21 days	38	116	S_{sm}	SL	55	*Aeromonas, Pseudomonas*	Quigley and Colwell (1968)
Seawater and sediments; nutrient agar	704	44	S_{sm}	SL	75	*'Achromobacter', Flavobacterium*	Bölter (1977)
Seawater and sediment, Beaufort Sea; marine agar, 4°C for 21 days, 20°C for 14 days	564	300	S_J	SL	56	*Flavobacterium, Microcyclus, Vibrio*	Kaneko *et al.* (1979)
Seawater and sediment, Gulf of Alaska, marine agar, 5°C for 21 days, 20°C for 14 days	1257	300	S_J	SL	26	*Beneckea—Vibrio, Flavobacterium, Microcyclus, Moraxella—Acinetobacter*	Hauxhurst *et al.* (1980)

(*continued*)

Table 3 (*Continued*)

Selective factors	Data matrix		Statistics		Strains assigned to major clusters (%)	Identity of major taxa	References
	Strains	Tests	Coefficient	Cluster technique			
Marine studies—plankton							
Sediment, Canadian Arctic; procedures selective for cold-tolerant Gram-negative bacteria	46	120	S_J	SL	90	*Cytophaga*	Quadling and Colwell (1964)
Seawater, Antarctic Sea and North Atlantic Ocean; seawater agar-enrichment culture, 5°C, 14 days	151	124	S_J, S_{sm}	SL	35	*'Achromobacter'*, *Flavobacterium*, *Pseudomonas*	Pfister and Burkholder (1965)
Seawater; procedures for the selective isolation of vibrios	208	50	S_J	SL	74	*Vibrio* spp.	Johnson *et al*. (1968)
Seawater, Ligurian Sea; peptone yeast extract agar, 18°C	229	51	CC	AL	88	*Achromobacter*, *Bacillus*, *Corynebacterium*, *Flavobacterium*, *Micrococcus*, *Pseudomonas*, *Vibrio*	Melchiorri-Santolini (1968)

Seawater; marine agar, 25°C, 6 days	80	73	S_J, S_{sm}	SL	61	*Micrococcus, Staphylococcus*	Ezura and Sakai (1970)
Seawater; marine agar, enrichment methods, basal medium agar, 18–22°C, 6 days	218	145	S_{sm}	SL	97	*Alcaligenes, Alteromonas, Pseudomonas*	Baumann *et al.* (1972)
Solar saltern; halophile medium, 23°C, 7 days	116	46	S_J, S_{sm}	SL	80	*Halobacterium, Halococcus*	Colwell *et al.* (1979)
Solar saltern, halophile medium, 35°C, 10 days	108	85	S_J	AL	80	*Alcaligenes, Alteromonas, Chromobacterium, Flavobacterium, Vibrio*	Ventosa *et al.* (1982)
Seawater and fresh water	135	215	S_{sm}	AG	69	*Achromobacter—Alcaligenes, Bacillus, Pseudomonas, Vibrio*	Delabré *et al.* (1973)
Seawater, marine animals, and fresh water; procedures selective for spirillas and vibrios	107	175	S_J	SL	94	*Spirillum, Vibrio*	Carney *et al.* (1975)

[a] Abbreviations: CC, correlation coefficient, S_J, Jaccard coefficient; S_{sm}, simple-matching coefficient; AG, aggregation coefficients; AL, average-linkage algorithm; CL, complete-linkage algorithm; SL, single-linkage algorithm.

Even greater problems of interpretation arise in studies which include strains with markedly different physiological requirements. For instance, it may not be prudent to compare acidophilic and neutrophilic bacteria, as it is not known whether the differences observed between them are real or merely artefacts resulting from common tests performed under two contrasting pH conditions (Williams *et al.*, 1983a). Similar problems may be encountered in interpreting numerical studies which encompass aerobic and anaerobic bacteria (Melville, 1965) or mesophilic and thermophilic organisms (Goodfellow and Pirouz, 1982). Such problems have to be overcome, and more thought needs to be given to doing so.

Detection and Identification of Microbial Populations

If natural selection results in the formation of groups of bacteria adapted to specific microhabitats, then clusters of bacteria defined using numerical taxonomic techniques can be expected to be found in particular habitats. Thus, in a soil with several distinct horizons, the occurrence of specific bacteria throughout the profile could indicate a microhabitat common to all horizons whereas restriction of bacterial taxa to one horizon could suggest that this contains a unique microenvironment. In practice, numerical phenetic surveys have been successfully used to characterise the bacterial populations in different soil types (Rovira and Brisbane, 1967; Goodfellow, 1969; Dasi *et al.*, 1981), in the horizons of a podsol soil (Goodfellow, 1969; Lowe and Gray, 1972), in several types of deciduous litter (Hissett and Gray, 1973), in the phylloplane (Goodfellow *et al.*, 1976a,b; Austin *et al.*, 1978; Ercolani, 1978, 1983), and in the rhizosphere of diverse plants (Rovira and Brisbane, 1967; Schofield and Whalley, 1978). Such surveys have also enabled bacterial populations in fresh water (Witzel *et al.*, 1981, 1982a,b), estuaries (Austin *et al.*, 1977a,b,c; Mallory *et al.*, 1977; Kaper *et al.*, 1983), and the sea (Floodgate and Hayes, 1963; Bousfield, 1978; Kaneko *et al.*, 1979, Hauxhurst *et al.*, 1980; Austin, 1982; Ventosa *et al.*, 1982) to be described and compared. It is also possible to distinguish between the bacterial communities of different environments (Goodfellow *et al.*, 1976a; Tables 2 and 3). Numerical taxonomy of randomly selected strains has also been used to examine the diversity of bacterial populations in the Beaufort Sea (Kaneko *et al.*, 1977).

In most taxonomic surveys of microbial communities a majority of the bacterial isolates have been recovered in a relatively small number of major clusters, many of which have been assigned to established genera (Tables 2 and 3). It is also interesting that little overlap has been found between species of *Bacillus* containing isolates from diverse habitats (Logan and Berkeley, 1981). In contrast, many workers have overlooked the numerous isolates that have been recovered either as single-member clusters or in minor clusters which contain relatively few strains. The latter may represent nuclei of novel groups, strains of established taxa lacking plasmids, or genetically unstable strains. Clearly single-member and minor clusters need to be taken more seriously if a balanced picture of the bacterial composition of ecologically complex communities is to be obtained. It has already been shown that acidophilic actinomycetes assigned to single-member clusters are novel on the basis of chemical criteria (Goodfellow *et al.*, unpublished data).

In many instances, marker or reference strains have proved to be poor guides to the identification of clusters when subjected to phenetic classification with further isolates. Rovira and Brisbane (1967) found that only 14 of 77 named strains fell into defined groups, and essentially similar results have been reported in many other studies (e.g., Goodfellow, 1969; Vanderzant *et al.*, 1972; Hagedorn and Holt, 1975a; Kaneko *et al.*, 1979; Austin, 1982; Seiler *et al.*, 1980; Seiler and Hennlich, 1983). These apparently disappointing results can be attributed to an inappropriate choice of marker strains, to the designated type strain not being typical of the taxon (Krieg and Lockhart, 1966), or to the unsuspected existence of numerous novel taxa (Williams *et al.*, 1984). It is not easy to choose a representative strain given the numerous genera found in natural environments (Fig. 1). Further, in the eighth edition of *Bergey's Manual of Determinative Bacteriology* (Buchanan and Gibbons, 1974), bacteria found in soil were assigned to 330 species of 71 genera. There is therefore considerable merit in selecting marker strains once the general nature of the isolates is known (Ercolani, 1978). However, as more becomes known about the species composition of microbial communities, the easier the choice of marker strains will become. There is also some evidence that single strains are not always adequate representatives of species and genera (Wilkinson and Jones, 1977; Goodfellow *et al.*, 1982c) and that prolonged laboratory culture might smooth out differences between marker strains and fresh isolates (Gyllenberg *et al.*, 1963).

Clusters that do not contain marker strains are sometimes identified using conventional keys and tables (e.g., Pfister and Burkholder, 1965; Bean and Everton, 1969; Bölter, 1977; Mallory *et al.*, 1977; Schofield and Whalley, 1978). Somewhat predictably, this expedient has been shown to be of limited value as two or more distinct clusters have keyed out to the same taxon (Goodfellow, 1969; Austin *et al.*, 1977a,b,c, 1978; Austin, 1982). There is also ample evidence to show that numerical taxonomic techniques can discriminate between isolates that would have been lumped together by *a priori* weighted features (Bean and Everton, 1969; Byrom, 1971; Goodfellow *et al.*, 1976a,b; Austin *et al.*, 1978).

Despite the shortcomings of numerical phenetic surveys, they have repeatedly been effective in detecting variation in natural populations of bacteria in both aquatic and terrestrial environments. Some of the more effective contributions have been in surveys of poorly studied habitats, such as the phylloplane, activated sludge, leaf litter, and solar salterns (Tables 2 and 3). Indeed, the lack of information on the bacterial flora of leaf surfaces has severely hampered ecological studies (Austin *et al.*, 1978; Ercolani, 1978). Austin *et al.* (1978) found that fluctuating populations of *Listeria grayi/murrayi, Pseudomonas fluorescens, Staphylococcus saprophyticus, Xanthomonas campestris,* and novel pink chromogenic bacteria predominated on the leaves of *Lolium perenne*. The xanthomonads and pink chromogens predominated in May, pseudomonads and xanthomonads in July, xanthomonads in September, and listeriae and staphylococci in October. Some of the pink chromogens were subsequently described as a new species, *Pseudomonas mesophilica* (Austin and Goodfellow, 1979), the type strain of which was shown to belong to a taxon accommodating pink-pigmented methylotrophic bacteria (Green and Bousfield, 1982).

Ercolani (1978) detected large populations of *Pseudomonas syringae* pv. *savastanoi,* the causative agent of olive knot disease, and smaller numbers of *Bacillus megaterium, Erwinia herbicola, Micrococcus luteus,* and *Xanthomonas campestris* on the phylloplane of *Olea europaea*. The predominance of *P. syringae* on healthy olive leaves in April and October supported a previous suggestion (Ercolani, 1970) that the phylloplane might be important in the epidemiology of the disease as a source of inoculum. A consistent pattern of fluctuation in the bacterial community was found over a 3-year period, which suggested that the dynamics of the different bacterial populations at any given time were governed by regularly recurrent events in this habitat. In additional studies Ercolani (1983) compared many strains of *P. syringae* pv. *savastanoi* from leaves of different ages at different times of the year and found a correlation between the phenotypic properties of the isolates and their origin. He attributed the phenotypic fluctuations to changes in the relative numbers of biovars on the phylloplane, rather than to any recurrent modification of a homogeneous but highly versatile population. The fluctuations were presumably due to the combined selective pressures of the host and the environment.

These studies on the bacterial flora of the phylloplane suggest that at any one time relatively few bacterial species predominate on leaf surfaces. This parallels the situation which has been described for the fungal community in this habitat (Dickinson and Wallace, 1976) and suggests that the phylloplane is a highly selective environment for microbial growth. This may follow from the extreme fluctuations in several of the leaves, which contrasts with the more 'buffered' environments experienced by those organisms inhabiting the rhizoplane or rhizosphere.

Most eco-taxonomic surveys have been restricted to an examination of only a few hundred of the millions of bacteria found in complex microbial communities. Given the problems involved, it is perhaps understandable that most workers have shown a marked reluctance to consider the statistical validity of procedures used to sample natural habitats. The use of unreplicated samples in the analysis of enormous microbial populations is statistically unsound, and it must be conceded that a few grams of soil or litres of water are unlikely to give a complete picture of microbial populations in a forest, a lake, or a sea. Obviously greater attention needs to be given to sampling regimes now that it is clear that bacterial populations can fluctuate on a diurnal (Holder-Franklin *et al.*, 1978; Bell *et al.*, 1982) and seasonal basis (Jones, 1977; Austin *et al.*, 1978; Ercolani, 1978, 1983). It has also been shown that bacteria isolated from a broad range of natural habitats can show a spectrum of forms as intermediate strains span the gaps between clusters that would have been recovered with less thorough sampling (Kapperud *et al.*, 1981). Another problem is that complex habitats, such as soil and compost, consist of an array of microhabitats. It seems likely that as these become recognised and analysed separately, then the number of distinct bacterial taxa will also increase. Obviously, sampling procedures will always represent a compromise between accuracy and feasibility (Williams and Gray, 1973). It has also to be acknowledged that the sampling problems faced by freshwater and marine microbiologists are even more acute than those encountered in studies of terrestrial environments.

Use of Numerical Taxonomic Data Bases

Numerical classification not only circumscribes taxa at selected levels of overall similarity, but it also provides quantitative data on the test reactions of strains within each of the defined clusters. This is usually presented as the percentage of strains within each cluster which show a positive state for each character used to build the classification. These data can be used to compare the physiological profiles of taxa from different habitats (Goodfellow, 1969; Hissett and Gray, 1973; Austin *et al.*, 1979b), and they can be usefully trawled for characters that might be used to construct dichotomous keys and diagnostic tables for the identification of unknown isolates. A numerical classification with a high information content also provides an excellent basis for the construction of computerised identification matrices; it facilitates the selection of strains for biochemical, genetic, physiological, taxonomic, ecological, and pathological studies, and it can also be used to improve the objectivity of procedures designed to isolate particular components of microbial communities.

It is well known that developments in microbial ecology have been severely hampered by the lack of good identificiation systems for bacteria. This situation contrasts markedly with that faced by the diagnostic medical or veterinary bacteriologist. In most broadly based numerical surveys, data bases have been scanned for possible diagnostic properties, but surprisingly few characters were found to be cluster specific. This does not imply an inherent weakness in the numerical taxonomy method, but rather it underlies the danger of establishing taxa on small numbers of properties. It is also surprising that few of the tests used to construct diagnostic schemes have been the subject of reproducibility studies. This failure to evaluate determinative schemes may explain why microbial ecologists have shown a distinct reluctance to use keys and tables recommended for the identification of unknown bacteria from soil (Goodfellow, 1969; Lowe and Gray, 1972; Hagedorn and Holt, 1975a), leaf surfaces (Austin *et al.*, 1978), food canneries (Bean and Everton, 1969; Byrom, 1971), food products (McMeekin *et al.*, 1972; Delarras *et al.*, 1979), freshwater habitats (Witzel *et al.*, 1982a,b), and marine habitats (Floodgate and Hayes, 1963; Ezura and Sakai, 1970; Singleton and Skerman, 1973; Austin *et al.*, 1977c, 1979a; Ventosa *et al.*, 1982). It has to be concluded that the problem of deriving determinative schemes for the rapid and accurate identification of isolates from large heterogeneous populations remains unsolved.

The renewed interest in computer-assisted identification of bacteria provides a possible way forward and builds upon ideas raised in the pioneering work of Gyllenberg (1963, 1965b). Computer-assisted schemes are to be preferred to conventional keys and tables, for not only are they quick and easy to use but identification can be achieved using incomplete results (Lapage *et al.*, 1970; Hill, 1974), while the chances of misidentification due to erroneous or unusual results are greatly reduced (Sneath, 1974). Numerical taxonomies are now being used to construct identification matrices (see Holmes and Hill, Chapter 10; Williams *et al.*, Chapter 11), and these contain the minimum number of characters needed to discriminate between taxa. At present, few numerical classifications of bacteria have been supported by probabilistic identification schemes, possibly because of the daunting problems associated with reproducibility studies on tests of presumptive diagnostic value (Wayne *et al.*, 1976). One of the few exceptions is the matrix produced by Wayne *et al.* (1980) for the identification of slow-growing mycobacteria of medical importance.

Probability matrices derived from numerical taxonomic data bases have been used to identify unknown streptomycetes (Williams *et al.*, 1983b; Williams *et al.*, Chapter 11) and vibrios (Dawson and Sneath, 1985) from natural habitats and bacteria from Alaskan outer continental shelf regions (Davis *et al.*, 1983). The preliminary results of these studies are encouraging, although it seems likely that criteria currently used to effect successful identifications will be tempered in the light of experience. A number of other probabilistic identification systems have been introduced, but they rest upon data which are less comprehensive than those found in sound numerical classifications. They include systems for Gram-negative fermentative rods (Schindler *et al.*, 1979; Schindler and Idlbek, 1982), anaerobic bacteria (Kelley and Kellogg, 1978), nitrogen-fixing soil bacteria (Rennie, 1980), and *Bacillus* species (Willemse-Collinet *et al.*, 1980). More elegant systems of numerical identification have been devised (Ercolani, 1978; Sielaff *et al.*, 1982), and it is possible to assign unknown strains to a particular region of a continuum (Sneath, 1979b) where bacteria show a spectrum of forms.

Improvements in classification and identification have also been used to devise selective isolation strategies for specific groups of bacteria (Williams *et al.*, 1984). Thus, the streptomycete data base of Williams *et al.* (1983a) has been used to formulate and evaluate media designed to isolate members of the streptomycete community other than those which flourish on conventional media based on colloidal chitin or starch casein (Goodfellow and Williams, 1983). The formulation of these media was assisted by the application of the DIACHAR program (Sneath, 1980) to the taxonomic data matrix to discover the most distinctive growth requirements and tolerances of selected groups (Vickers *et al.*, 1984). The isolates obtained were identified using the probabilistic system of Williams *et al.* (Chapter 11). A similar strategy has led to the isolation of several novel acidophilic actinomycetes (Goodfellow *et al.*, unpublished data). These studies underline the limitations of conventional isolation procedures and reinforce the view that even in intensively studied habitats many novel microbes still await discovery (Williams *et al.*, 1984).

Another important advantage of numerical taxonomy is that the data obtained may be used to effect an objective choice of representative isolates for ecological, taxonomic, or pathological studies. The strains chosen may show the highest average similarity to other members of the cluster (Silvestri *et al.*, 1962) or be deemed central to the taxon (Lowe and Gray, 1973a). Both of these procedures can be justified, especially where only one or two isolates from each cluster are studied. However, it is clearly preferable to carry out comparative ecological work on strains representing the normal limits of variability within taxa. To date, isolates representing numerically defined groups from soil (Lowe and Gray, 1973a,b), leaf surfaces (Austin *et al.*, 1977d; Al-Hadithi, 1979), and spruce litter (Dickinson *et al.*, 1981) have been the subject of growth and competitive interaction studies *in vitro* and *in vivo*.

Restricted Ecological Surveys

Conventional numerical taxonomy has generally proved to be most effective when it is used to clarify relationships within poorly studied taxa (see Wayne, Chapter 5; MacDonell and Colwell, Chapter 6) or in ecological surveys limited to a narrow range of cultures from well-defined habitats (Table 4). Such restricted studies are relatively easy to execute, as particular populations can be obtained using selective isolation procedures and characterised under conditions known to favour the growth of the bacteria in question. In such investigations the choice of marker strains also raises comparatively few problems, as authentic representatives, including type strains, can usually be obtained from public and private culture collections. Thus, it would appear that many advantages occur from the simple act of restricting the scope of the investigation in either a taxonomic or an ecological context.

In practice, however, most narrow ecological surveys have considered a few hundred aerobic, heterotrophic, mesophilic bacteria isolated from unreplicated samples and examined at temperatures well above those met *in situ*. In most of the restricted surveys data have been analysed using standard coefficients and clustering techniques (Table 4), and the results expressed in dendrograms and shaded diagrams. Good congruence has been found between numerical taxonomies based on different coefficients and clustering algorithms (Kersters *et al.*, 1973; Piot *et al.*, 1980; Lee *et al.*, 1981; Schofield and Schaal, 1981; Green and Bousfield, 1982; Banks and Board, 1983; West *et al.*, 1983; McCarthy and Cross, 1984). Many workers have evaluated the quality of their data by determining intra- and inter-group similarities (Melville, 1965; Colwell, 1970; Boeyé and Aerts, 1976; King *et al.*, 1979, Azad and Kado, 1980; Gillespie, 1981; Banks and Board, 1983), and test error (Goodfellow *et al.*, 1979, 1982a,b; Schofield and Schaal, 1981; Banks and Board, 1983; West *et al.*, 1983; McCarthy and Cross, 1984), but almost no attempt has been made to determine the hierarchicalness of data or the distinctness of defined clusters. However, in an analysis of obligately aerobic, Gram-negative bacteria from processed pork, Banks and Board (1983) noted that the cophenetic correlation coefficient was high (.8899) and that there was considerable overlap between three of the six major clusters. Nevertheless. in many instances the results of restricted studies have proved to be easy to interpret as most isolates have fallen into a small number of major clusters (Table 4), many of which were identified by marker strains (e.g., Barre, 1969; Schofield and Schaal, 1981; Barton and Hughes, 1982; Goodfellow *et al.*, 1982a; Green and Bousfield, 1982; Shaw and Latty, 1982; Banks and Board, 1983; McCarthy and Cross, 1984). However, as noted before, little attention has been paid to minor or single-member clusters.

Good agreement has usually been found between the results of restricted numerical phenetic surveys and data derived from chemical, genetic, and serological analyses of representatives of numerically defined clusters (Jones and Sackin, 1980; Goodfellow and Wayne, 1982). In particular, good concordance has been found between numerical phenetic and DNA pairing data as in the case of *Agrobacterium tumefaciens* (Kersters *et al.*, 1973), *Gardnerella vaginalis* (Piot *et al.*, 1980), *Klebsiella terrigena* (Gavini *et al.*, 1977; Izard *et al.*, 1981), and *Serratia fonticola* (Gavini *et al.*, 1979). The very high genetic relatedness and marked similarities in phenetic characters among *Erwinia rubrifaciens* strains, and the confinement of this phytopathogen to California, led Azad and Kado (1980) to the view that the organism originated from a single source. The results of numerical taxonomic investigations have also been supported by those from both qualitative and quantitative analyses of chemotaxonomic data (Izard *et al.*, 1981; Goodfellow *et al.*, 1982b, 1985). A sense of relief is usually felt when good agreement is found between taxonomies based upon different taxonomic criteria, but it is well to remember that lack of congruence should not be swept aside as discordant results may raise questions of considerable biological significance (see Sneath, Chapter 16).

Not surprisingly, detailed surveys of restricted groups of bacteria from defined habitats have frequently highlighted new centres of variation, which in some cases have been described as novel taxa. Thus, previously uncharacterised isolates have been recovered from bovine udders (Weckbach and Langlois, 1976), clinical material (Feltham, 1979; Piot *et al.*, 1980), fodder (Goodfellow *et al.*, 1979), frozen vegetables (Sheard and Priest, 1981), marine muds (Gunn *et al.*, 1983), oral cavities (Carlsson, 1968; Kuhn *et al.*, 1978), Scotch whisky distilleries (Barbour and Priest, 1983), soil (Pichinoty *et al.*, 1980), vacuum-packed meat (Shaw and Harding, 1984), and wine (Barre, 1969). Novel taxa accommodating isolates from natural habitats include *Alcaligenes faecalis* subsp. *humari* (Austin *et al., 1981b), Chromobacterium fluviatile* (Moss *et al.*, 1978), *Klebsiella trevisanii* (Ferragut *et al.*, 1983), *Mycobacterium fallax* (Lévy-Frébault *et al.*, 1983), *Pseudomonas mesophilica* (Austin and Goodfellow, 1979), *Rhodococcus coprophilus* (Rowbotham and Cross, 1977), and *Serratia fonticola* (Gavini *et al.*, 1979). Similarly, numerical phenetic analysis on *R. equi* strains from diverse sources led to the redefinition of the taxon (Goodfellow *et al.*, 1982).

Restricted numerical phenetic surveys have also helped to clarify relationships between poorly studied groups of bacteria found in a plethora of well-circumscribed habitats (Table 4). Such groups include pseudomonads from spoiled meats (Shaw and Latty, 1982), facultative methylotrophic bacteria (Green and Bousfield, 1982), and aerobic, endospore-forming bacilli from marine sediments (Boeyé and Aerts, 1976; Bonde, 1981), salt marshes (Logan and Berkeley, 1981), rhizosphere soil (Garcia *et al.*, 1982), and nonrhizosphere soil (Pichinoty *et al.*, 1980; Garcia *et al.*, 1982). Improved classification has also shed light on the ecology of many species including *Acinetobacter calcoaceticus* (Pagel and Seyfried, 1976), *Actinomyces israelii* (Schofield and Schaal, 1981), *Aeromonas punctata* (Popoff and Véron, 1976), *Agrobacterium tumefaciens* (Kersters *et al., 1973), Alteromonas putrefaciens* (Gillespie, 1981), *Bacillus sphaericus* (de Barjac *et al.*, 1980), *Corynebacterium pyogenes* (Roberts, 1968), *Erwinia rubrifaciens* (Azad and Kado, 1980), *Gardnerella vaginalis* (Piot *et al.*, 1980), *Mycobacterium africanum* (David *et al.*, 1978), *M. pulveris* (Tsukamura *et al.*, 1983), *Obesumbacterium proteus* (Priest *et al.*, 1973), *Pediococcus damnosus* (Lawrence and Priest, 1981), *Photobacterium phosphoreum* (Reichelt and Baumann, 1973), *Pseudomonas fragi* (Shaw and Latty, 1982; Banks and Board, 1983), *Pseudomonas pickettii* (King *et al.*, 1979), *Renibacterium salmoninarum* (Goodfellow, *et al.*, 1985), *Rhodococcus equi* (Barton and Hughes, 1982; Goodfellow *et al.*, 1982a), and *Vibrio metschnikovii* (Lee *et al.*, 1978).

Table 4. *Numerical phenetic analyses of specific bacteria from defined habitats*[a]

Taxa	Habitat/Source	Data matrix		Statistics		Strains assigned to major taxa (%)	References
		Strains	Tests	Coefficient	Algorithm		
Bacteria associated with humans							
Streptococcus spp.	Oral cavity	119	70	S_J, L	AL	100	Carlsson (1968)
Streptococcus spp.	Mainly of human origin	216	75	S_J, HM	SL	70	Colman (1968)
Streptococcus spp.	Oral cavity	298	144	S_J, HM	SL	NS	Drucker and Melville (1971)
Actinomyces spp.	Oral cavity	71	>50	S_{sm}	NS	100	Melville (1965)
Beneckea parahaemolytica	Gut, eye, and ear infection	204	110	S_{sm}	SL	96	Baumann *et al.* (1973)
Enterobacteriaceae	Human faeces, soil, and water	111	143	DC	AV, VC	96	Gavini *et al.* (1976)
Mycobacterium africanum	Sputum	98	39	S_{sm}	AL	100	David *et al.* (1978)
Micrococcus and *Staphylococcus* spp.	Clinical sources	277	61	S_{sm}	AL	100	Feltham (1979)
Pseudomonas cepacia, P. maltophila, P. pickettii	Blood, respiratory tract, and urine	174	64	S_G	AL, SL	100	King *et al.* (1979)
Gardnerella vaginalis	Vagina	116	149	S_J, S_p, S_{sm}	AL	72	Piot *et al.* (1980)
Actinomyces spp.	Pus, sinus discharge	222	124	S_J, S_p, S_{sm}	AL	73	Schofield and Schaal (1981)
Staphylococcus spp.	Clinical sources	264	64	S_J	AL	94	Gunn *et al.* (1983)
Mycobacterium fallax	Sputum, fresh water, and soil	22	47	S_{sm}	SL	100	Lévy-Frébault *et al.* (1983)

Bacteria associated with other animals							
Acinetobacter spp.	Poultry carcasses	195	83	S_G	SL	96	Thornley (1967)
Actinomyces (Corynebacterium) pyogenes	Abscesses of cattle, pigs, and sheep	100	62	S_J	SL	92	Roberts (1968)
Beneckea and *Photobacterium* spp.	Fish skins, intestines, and luminous organs	173	165	S_{sm}	SL	100	Reichelt and Baumann (1973)
Aeromonas hydrophila, A. sobria	Infected fish and frogs	68	59	TD	AV, VC	100	Popoff and Véron (1976)
Staphylococcus spp.	Bovine udders	303	54	S_{sm}	AL	87	Weckbach and Langlois (1976)
Rhodococcus rhodnii	Gut of *Rhodnius prolixus*	177	92	S_{sm}	AL	100	Goodfellow and Alderson (1977)
Simonsiella spp.	Oral cavities of warm-blooded animals	50	57	S_{sm}	AL	100	Kuhn *et al.* (1978)
Vibrio metschnikovii	Marine animals, sewage, and river water	85	118	S_G	AL, SL	100	Lee *et al.* (1978)
Bacillus sphaericus	Insects including mosquitoes	35	160	TD	AV, VC	97	de Barjac *et al.* (1980)
Alteromonas and *Pseudomonas* spp.	Marine fish and seawater	36	98	S_J, S_{sm}	AL	83	Gray and Stewart (1980)
Alcaligenes faecalis subsp. *homari*	Haemolymph of moribund lobsters	17	124	S_J, S_{sm}	AL	100	Austin *et al.* (1981b)
Alteromonas putrefaciens, Pseudomonas fragi	Spoilt freshwater and seawater fish	154	160	S_{sm}	CL	94	Gillespie (1981)
Yersinia enterocolitica, Y. kristensenii, Y. paratuberculosis	Fish, small rodents, water	332	46	ED	WC	100	Kapperud *et al.* (1981)

(continued)

Table 4 (*Continued*)

Taxa	Habitat/Source	Data matrix		Statistics		Strains assigned to major taxa (%)	References
		Strains	Tests	Coefficient	Algorithm		
Aeromonas hydrophila/punctata, Vibrio fluvialis, V. anguillarum	Marine animals and seawater	154	100	ED	AL, CL, SL, WC	97	Lee *et al.* (1981)
Rhodococcus equi	Foals with virulent pneumonia, dung, and soil	205	160	S_J, S_p, S_{sm}	AL	85	Goodfellow *et al.* (1982a)
Rhodococcus equi	Foals with virulent pneumonia, faeces, and soil	110	112	S_J, S_{sm}	AL	85	Barton and Hughes (1982)
Renibacterium salmoninarum	Infected kidneys of salmonid fish	56	86	S_J, S_p, S_{sm}	AL	100	Goodfellow *et al.* (1985)
Bacteria associated with plants							
Agrobacterium and *Rhizobium* spp.	Root nodules of legumes	59	191	S_J, S_{sm}	SL	100	Moffett and Colwell (1968)
Agrobacterium radiobacter, A. rhizogenes	Galls from diverse plants	79	92	S_p, S_{sm}	AL, SL	96	Kersters *et al.* (1973)
Pseudomonas mesophilica	Phylloplane of perennial rye grass	67	146	S_J, S_{sm}	AL	91	Austin and Goodfellow (1979)
'Actinomadura-like' clusters A and B	Grains and fodder	156	90	S_J, S_p, S_{sm}	AL, SL	37	Goodfellow *et al.* (1979)
Mycobacterium komossense	*Sphagnum* moss	50	52	S_{sm}	SL	100	Kazda and Müller (1979)

Mycobacterium sphagni	*Sphagnum* moss	50	52	S_{sm}	SL	100	Kazda (1980)
Erwinia rubrifaciens	Infected walnut trees	75	140	S_J, S_{sm}	SL	100	Azad and Kado (1980)
Acetobacter and *Gluconobacter* spp.	Flowers, fruits, and beverages	122	188	S_J, S_{sm}	AL	100	Gosselé *et al.* (1983b)
Thermomonospora spp.	Composts and fodder	113	101	S_J, S_p, S_{sm}	AL, SL	100	McCarthy and Cross (1984)
Bacteria from soil and water							
Streptomyces spp.	Pine and deciduous forest soil	18	46	S_{sm}	SL	95	Williams *et al.* (1969)
Anaerobic and microaerophilic taxa	Nonmethanogenic phase of anaerobic digester	93	101	S_{sm}	AL, SL	76	Toerien (1970b)
Bacillus spp.	Soil and other habitats	600	139	S_G	AL	100	Logan and Berkeley (1981)
Bacillus spp.	North Sea sediments	163	63	TD	AL	100	Boeyé and Aerts (1976)
Acinetobacter calcoaceticus, A. lwoffii	Fresh water	339	96	S_{sm}	AL	92	Pagel and Seyfried (1976)
Alteromonas and *Pseudomonas* spp.	Fresh water, seawater, and fish	38	165	S_J	CL	90	Lee *et al.* (1977)
Klebsiella spp.	Water and soil	122	95	DC	AV, VC	92	Gavini *et al.* (1977)
Rhodococcus coprophilus	Lake water and mud	36	78	S_{sm}	SL	92	Rowbotham and Cross (1977)
Chromobacterium spp.	Fresh water	186	32	ED	Various	98	Moss *et al.* (1978)
Serratia fonticola	Fresh water	92	117	DC	AV, VC	98	Gavini *et al.* (1979)
Bacillus spp.	Diverse soils	47	108	S_{sm}	SL	100	Pichinoty *et al.* (1980)

(*continued*)

Table 4 (*Continued*)

Taxa	Habitat/Source	Data matrix		Statistics		Strains assigned to major taxa (%)	References
		Strains	Tests	Coefficient	Algorithm		
Lactose-fermenting bacteria	Seawater and sediments	322	100	S_J, S_{sm}	AL	92	Austin *et al.* (1981a)
Mycobacterium agri	Soil	165	104	S_{sm}	SL	97	Tsukamura (1981)
Mycobacterium agri, M. chubuense, M. obuense, M. tokaiense	Soil	155	104	S_{sm}	SL	82	Tsukamura *et al.* (1981)
Bacillus coagulans and *B. stearothermophilus*	Rice field soil	74	116	TD	AV	75	Garcia *et al.* (1982)
Nocardia amarae	Abnormal foam from sewage treatment plants	123	92	S_J, S_p, S_{sm}	AL	100	Goodfellow *et al.* (1982b)
Methylobacterium spp.	Diverse habitats	194	140	S_J, S_{sm}	AL, CL, SL	86	Green and Bousfield (1982)
Klebsiella trevisanii	Water and soil	44	95	DC	AV, VC	98	Ferragut *et al.* (1983)
Staphylococcus spp.	Seawater	220	62	S_J	AL	100	Gunn and Colwell (1983)
Vibrio spp.	Fresh water, seawater, and diseased fish	237	148	ED, S_{sm}	AL, SL	95	West *et al.* (1983)

Bacteria associated with food							
Lactobacillus spp.	Cheddar cheese	68	109	S_J	SL	100	Hauser and Smith (1964)
Acinetobacter spp.	Poultry carcasses	195	83	S_G	SL	96	Thornley (1967)
Lactobacillus spp.	Wines	69	86	S_{sm}	AL	100	Barre (1969)
Lactobacillus spp.	French sausages	190	50	S_{sm}	SL	78	Laban *et al.* (1978)
Obesumbacterium proteus	Beers	37	50	S_p, S_{sm}	SL	100	Priest *et al.* (1973)
Pseudomonas fluorescens, P. fragi, Alteromonas putrefaciens	Meat	218	174	S_{sm}	AL	83	Molin and Ternstrom (1982)
Pediococcus damnosus	Beers	96	139	S_J, S_p, S_{sm}	AL	97	Lawrence and Priest (1981)
Bacillus spp.	Frozen vegetables	89	107	S_J, S_{sm}	AL	84	Sheard and Priest (1981)
Pseudomonas fragi	Spoiled meat	123	160	S_{sm}	AL	100	Shaw and Latty (1982)
Pseudomonas fragi	Fresh and processed pork	185	126	S_G, S_J, S_{sm}	AL, SL	86	Banks and Board (1983)
Lactobacillus spp.	Washback samples from whisky distilleries	178	NS	S_{sm}	CL	99	Barbour and Priest (1983)
Lactobacillus spp.	Vacuum-packed meat	123	79	S_{sm}	AL	98	Shaw and Harding (1984)
Leuconostoc mesenteroides	Grain mashes and fermentations	73	104	S_J, S_{sm}	CL	100	Priest and Barbour (see Chapter 7)

[a]Abbreviations: DC, Dice coefficient; ED, Euclidean distance; HM, Harrison method; S_G, Gower coefficient; S_J, Jaccard coefficient; S_p, pattern coefficient; S_{sm}, simple-matching coefficient; TD, taxonomic distance coefficient; AL, average-linkage coefficient; AV, aggregation according to variance coefficient; CL, complete-linkage coefficient; SL, single-linkage coefficient; VC, Veron's acuteness coefficient; WC, Ward's coefficient; NS, not stated.

Many workers have pinpointed properties of presumptive diagnostic value from data bases based on restricted numerical surveys and weighted them for the identification of unknown isolates. Diagnostic keys and tables have been proposed for the identification of aerobic, endospore-forming bacilli from marine muds (Boeyé and Aerts, 1976), brewery cocci (Lawrence and Priest, 1981), lactobacilli from wine (Barre, 1969), lactic acid bacteria from vacuum-packed meat (Shaw and Harding, 1984), pseudomonads from fresh and processed meat (Shaw and Latty, 1982; Banks and Board, 1983), and for *Acinetobacter* species (Thornley, 1967), *Chromobacterium* species (Moss *et al.*, 1978), *Rhodococcus* species (Goodfellow and Alderson, 1977; Rowbotham and Cross, 1977), *Agrobacterium* species (Kersters *et al.*, 1973), *Klebsiella* species (Gavini *et al.*, 1977; Ferragut *et al.*, 1983), *Staphylococcus* species (Gunn and Colwell, 1983; Gunn *et al.*, 1983), *Thermomonospora* species (McCarthy and Cross, 1984), *Vibrio* species (Colwell, 1970; Lee *et al.*, 1978, 1981), *Alcaligenes faecalis* subsp. *homari* (Austin *et al.*, 1981b), *Mycobacterium pulveris* (Tsukamura *et al.*, 1983), *Pseudomonas pickettii* (King *et al.*, 1979), and *Serratia fonticola* (Gavini *et al.*, 1979). In the absence of reproducibility studies it is debatable whether such identification schemes will work in practice, and it is also surprising that few, if any, of the schemes have been used to chart changes in bacterial populations due to diurnal, seasonal, or other environmental factors.

Application of Numerical Taxonomy to Other Groups of Organisms

Conventional numerical taxonomy is now widely used to determine relationships within closely related groups of bacteria. Thus, despite the enormous labour and computational inputs required for each numerical exercise they are now performed routinely in preference to more classical taxonomic studies. This situation has undoubtedly arisen because of a lack of confidence in traditional monothetic classification as it was applied to bacteria. By contrast almost all other groups of plants, animals, and micro-organisms are classified and identified according to time-honoured systems which are based on characteristics which are widely accepted as being important indicators of their natural affinities.

Such confidence in conventional taxonomic studies, however fragile or even misplaced it may be in particular instances, explains why so few attempts have been made to apply numerical taxonomy to groups other than the bacteria. The apparent simplicity and convenience of a system which enables decisions to be made on the basis of a handful of readily observed characters are powerful arguments for the continuation of the *status quo*. It is, however, of interest that numerical taxonomy has been employed on several occasions to provide additional confirmation of the validity of decisions taken on a conventional basis.

Foremost amongst such studies are those concerned with the Fungi Imperfecti and the yeasts. The former are natural candidates for an Adansonian classification as the conventional approach is based primarily on their method of spore formation, even though this emphasis frequently results in groupings which differ from those which emerge from a study of the teleomorph states, where these have been discovered (Whalley and Greenhalgh, 1973). Numerical taxonomy has, however, been used to reinforce the view that conidium ontogeny is a prime character in grouping the anamorph forms which constitute the Fungi Imperfecti. Dabinett and Wellman (1978) examined a collection of such fungi, and they found that a numerical classification, based on morphology, physiology, and conidium ontogeny, was in close agreement with the original subjective grouping which had been based purely on the detailed mechanisms of asexual spore formation.

The application of numerical taxonomy to the yeasts can be regarded as a relatively logical extension of the systems which are already in use in this group, where highly significant taxonomic characters are often at a premium. In many genera, species recognition is based on a range of biochemical and physiological tests, and the application of numerical taxonomy merely implies that a broader range of equally weighted characters is taken into account (Kockova-Kratochvilova *et al.*, 1978, 1981). In both these studies it is of interest that the clustering obtained using numerical taxonomy was supported by data concerning the G + C content of the type species of each grouping.

These studies, and others on *Rhizopus* (Dabinett and Wellman, 1973), *Conidiobolus* (King, 1976), and *Hypoxylon* (Whalley, 1976), have all been primarily concerned with reinforcing or amending existing classifications. All or most of the organisms entered into the programmes were selected on the basis of prior 'identification' according to a conventional scheme, and indeed the programmes were not designed to facilitate the classification of unknown isolates. A rather different approach was taken by Lourd *et al.* (1979), who examined 39 isolates of the ubiquitous plant pathogen *Colletotrichum gloeosporioides*. Their objective was to determine if valid subspecies or varieties could be distinguished, and using numerical taxonomy they were able to distinguish four subgroups, a discovery which may be most useful in future studies of this economically important pathogen. It is of interest that a similar use has been made of numerical taxonomy in respect of several genera of flowering plants. Here ecological or physiological subspecies or varieties have been identified and delimited using numerical taxonomy (e.g., Prentice, 1980).

Numerical methods have been applied to a number of microscopic algae, but again the emphasis has usually been on clarifying the systematics of particular genera or species (Da Silva and Gyllenberg, 1972; Komaromy, 1982). Of more interest in the present context is the study by Van Valkenburg *et al.* (1977), which was specifically designed to facilitate the identification of species of nannoplankton. These algae, which are mostly members of the families Chrysophyceae and Haptophyceae, are often difficult to identify on the basis of their superficial appearance, and hence a combination of biochemicial, morphological, physiological, and ultrastructural characters were employed to create a data matrix from which an identification key could be constructed. This was further refined to produce an on-line computer identification facility which will undoubtedly be valuable in ecological studies of these organisms.

Considering the diversity of the animal kingdom and the problems that are encountered in the identification of many of its members, it is notable that there have been relatively few attempts to employ numerical taxonomy in any context. As with the algae and the fungi, most studies have been concerned with the classification of particular genera or species (e.g., Smith and Hirshfield, 1975; Friesen and Bovee, 1976). Few zoologists have explored the possibility of using numerical taxonomy in either ecological or pathological studies, although it has been demonstrated that there is potential for its deployment in such work (Ready and Miles, 1981).

Conclusions

The isolation and classification of representatives of the bacterial flora found in complex microbial communities remains a difficult and laborious task. However, the problems involved in classifying and identifying specific groups of bacteria isolated from natural habitats are generally of the same order of magnitude as those experienced in the more familiar numerical phenetic analyses associated with improving the taxonomy of particular taxa. It is not surprising, therefore, that conventional numerical taxonomy has proved to be effective in the assignment of randomly chosen isolates to well-defined clusters. Numerical phenetic surveys have also led to the discovery of new taxa; they have shown that groups thought to be rare or insignificant are in fact widely distributed and common in nature, and they have enabled investigators to make an objective choice of strains for experimental studies. There is also an increasing awareness that the information in sorted data bases can be used for several purposes, notably for constructing computer probability identification matrices and for the formation of media designed to encourage the growth of specific fractions of the bacterial flora. Gains such as these are significant, but they do nevertheless fall short of the high expectations of some of those who pioneered the development of conventional numerical taxonomy.

In many numerical taxonomic surveys of bacterial populations both taxonomic and ecological principles have been either neglected or sacrificed completely. Little thought has been given to designing tests that have both taxonomic and ecological significance, to the mechanisation of data collection, to determining the quality of test data, to evaluating the results obtained by comparison with other modern taxonomic techniques, or to inter-locking numerical taxonomies derived from different studies. Indeed, in retrospect many studies seem to have been planned in a vacuum, with different workers seemingly beginning from scratch each time. It has also proved much more difficult than expected to devise simple, accurate, and practical schemes for the identification of specific groups of organisms. All too often the capacity of computers to handle large quantities of data has been abused by the somewhat aimless processing of poor-quality data in the forlorn belief that the groups obtained would be of taxonomic and ecological value. This blunderbuss approach has merely served to emphasize that numerical taxonomic surveys of the bacterial flora of natural habitats need to be carefully designed and executed.

In future much more attention needs to be given to the precise nature and limits of the habitat under study, to the extent and frequency of field sampling, and to the numbers of isolates which must be examined to determine population trends. It has long been recognised that habitats such as soil are markedly heterogeneous and contain a multiplicity of microhabitats that vary both in space and in time. Such heterogeneity casts doubt on the common practice of extrapolating from observations based on a few samples. A more precise definition of the habitats or processes under study should lead to the isolation of fewer types of organisms, which could then be examined more intensively than is possible for the innumerable taxa recovered from markedly heterogeneous samples. Further economy of effort would be possible if a better distinction could be drawn between alien and indigenous, and active and inactive organisms. It is astonishing that in many descriptive surveys ecological conclusions have been drawn from weekly or even longer interval sampling, given that such procedures bear little relevance to the generation times of natural populations of bacteria.

It still needs to be stressed that bacteria recovered on standard isolation media represent only a small fraction of the total bacterial flora of natural habitats. Evidence for the underestimation of bacterial numbers is provided by direct microscopic counts after fluorescent staining and by scanning electron microscopy, and the qualitative deficiencies of most general surveys are highlighted by the unending stream of publications describing novel taxa isolated from habitats which have been surveyed time and time again (Williams *et al.*, 1984). To date, the very real difficulty of isolating representative strains of all the bacterial taxa in ecologically complex communities has been compounded by the bias shown towards the aerobic, heterotrophic, mesophilic element. It needs to be appreciated that different selective procedures are required for the isolation of different fractions of the bacterial population. Indeed, it is possible that bacteria isolated on minimal media may represent a sizeable proportion of the actively metabolising bacterial flora in some habitats. Contingencies of anaerobic, microaerophilic, psychrophilic, and autotrophic bacteria may also constitute sizeable portions of the populations, especially in habitats which are subject to extremes of various environmental factors. It seems very likely that a more imaginative selection of isolation methods will profoundly alter our understanding of the types of bacteria occurring in most habitats.

Future descriptive studies of bacterial populations in ecologically complex communities should employ a three-phase strategy based on both ordination and clustering methods. The latter should not be considered as alternatives but as part of a coordinated approach designed to determine the factors and forces which govern the structure and function of the indigenous flora in natural habitats. The first aim would be to determine the relationship between the habitat and the population densities of the indigenous bacteria, for only then can a measure of confidence be placed in the ecological significance of the observed populations. The application of factor or principal components analysis provides a convenient way of determining the responses of bacterial populations to physico-chemical parameters. The development of selective isolation procedures for an increasing number of bacterial taxa should allow a much higher proportion of the bacterial flora to be examined than hitherto. Taxa shown to be alien to the habitat in question would not be considered further.

Next, large numbers of bacteria representing specific fractions of the bacterial flora could be described using a small number of carefully chosen properties and the data examined using factor or principal components analysis. At least some of the tests used in such studies should be of a functional nature, that is, they should relate to ecologically significant processes such as decomposition, competitive interactions, and nutrient transformations. The results of such investigations would not only yield an ecological fingerprint of populations but would also provide a framework for subsequent numerical phenetic surveys of groups of bacteria of particular interest. Information could then be extracted from the data bases derived from such analyses and used to develop improved selective isolation media and identification schemes. In light of past experience it needs to be stressed that numerical diagnostic systems depend critically on high-quality numerical taxonomies and careful test standardisation. The benefits which could then be derived by microbial ecologists from good-quality computer-assisted identification schemes are legion. Although this three-phase approach has been conceived with the bacteria in mind, it could be applied to other constituents of the microflora such as the microscopic algae, the fungi, and some protozoa.

Comprehensive programmes along the lines outlined will inevitably require extensive resources and can probably only be advanced through carefully planned and well-financed collaborative projects. The primary isolation of test strains will undoubtedly remain time-consuming, but rapid and automated techniques are available for collecting biochemical, nutritional, and physiological data. However, the data once assembled should be examined much more critically than in the past, and, where possible, an attempt should be made to read tests quantitatively so that effort can be maximised. New ways of handling the very large data bases will be needed, while access to them should be made easy. It is also not difficult to see that new applications of numerical taxonomy will be made in the ecological domain. At the very least, numerical analysis of phenotypic features derived from protein gel electrophoregrams (Gossele *et al.*, 1983a,b) or from other chemosystematic procedures (see Goodfellow and Minnikin, 1985) will be used to establish relationships between appropriate groups of bacteria.

It seems unlikely that microbial ecology will reach the level of precision current in plant and animal ecology until far more is known about the kinds, characteristics, and discriminating features of the microflora. Such descriptive studies remain difficult, but advances in selective isolation, data collection, and statistics provide a means of answering more of the problems which have hampered such investigations for so long. Numerical phenetic surveys of the microflora found in ecologically complex habitats require the cooperation of both ecologists and taxonomists if common objectives are to be realised. Ecology and taxonomy are subjects that advance through the synthesis of information derived from several of the more analytical disciplines and in the current scientific climate are perceived to be somewhat unfashionable. They are, however, basic biological disciplines which have led to new biological insights in the past and will undoubtedly do so in the future. However, in the final analysis let us not forget that, daunting though our problems may be, it is only syntheses that really count; all of the splendid snapshots in the world are but snapshots after all.

Acknowledgement

The authors are indebted to Dr A. G. O'Donnell for critically reading the manuscript.

References

Al-Hadithi, H. T. (1979). 'Bacteria Associated with *Alternaria brassicicola* Infections of Brassicas'. Ph. D. Thesis, University of Newcastle upon Tyne.

Austin, B. (1982). Taxonomy of bacteria isolated from a coastal, marine fish-rearing unit. *Journal of Applied Bacteriology* **53,** 253–268.

Austin, B., and Colwell, R. R. (1977). Evaluation of some coefficients for use in numerical taxonomy of micro-organisms. *International Journal of Systematic Bacteriology* **27,** 204–210.

Austin, B., and Goodfellow, M. (1979). *Pseudomonas mesophilica,* a new species of pink bacteria isolated from leaf surfaces. *International Journal of Systematic Bacteriology* **29,** 373–378.

Austin, B., Calomiris, J., Walker, J. D., and Colwell, R. R. (1977a). Numerical taxonomy and ecology of petroleum-degrading bacteria. *Applied and Environmental Microbiology* **34,** 60–68.

Austin, B., Colwell, R. R., Walker, J. D., and Calomiris, J. (1977b). The application of numerical taxonomy to the study of petroleum-degrading bacteria isolated from the aquatic environment. *Developments in Industrial Microbiology* **18,** 685–695.

Austin, B., Allen, D. A., Mills, A. L., and Colwell, R. R. (1977c). Numerical taxonomy of heavy metal-tolerant bacteria isolated from an estuary. *Canadian Journal of Microbiology* **10,** 1433–1447.

Austin, B., Dickinson, C. H., and Goodfellow, M. (1977d). Antagonistic interactions of phylloplane bacteria with *Drechslera dictyoides* (Drechsler) Shoemaker. *Canadian Journal of Microbiology* **23,** 710–715.

Austin, B., Goodfellow, M., and Dickinson, C. H. (1978). Numerical taxonomy of phylloplane bacteria isolated from *Lolium perenne*. *Journal of General Microbiology* **104,** 135–155.

Austin, B., Allen, D. A., Zachary, A., Belas, M. R., and Colwell, R. R. (1979a). Ecology and taxonomy of bacteria attaching to wood surfaces in a tropical habor. *Canadian Journal of Microbiology* **25,** 447–461.

Austin, B., Garges, S., Conrad, B., Harding, E. E., Colwell, R. R., Simidu, U., and Taga, N. (1979b). Comparative study of the aerobic heterotrophic bacterial flora of Chesapeake Bay and Tokyo Bay. *Applied and Environmental Microbiology* **37,** 704–714.

Austin, B., Hussong, D., Weiner, R. M., and Colwell, R. R. (1981a). Numerical taxonomy analysis of bacteria isolated from the completed 'most probable numbers' test for coliform bacilli. *Journal of Applied Bacteriology* **51,** 101–112.

Austin, B., Rodgers, C. J., Forns, J. M., and Colwell, R. R. (1981b). *Alcaligenes faecalis* subsp. *homari* subsp. nov., a new group of bacteria isolated from moribund lobsters. *International Journal of Systematic Bacteriology* **31,** 72–76.

Azad, H. R., and Kado, C. I. (1980). Numerical and DNA:DNA reassociation analyses of *Erwinia rubrifaciens* and other members of the Enterobacteriaceae. *Journal of General Microbiology* **120,** 117–129.

Banks, J. G., and Board, R. G. (1983). The classification of pseudomonads and other obligately aerobic Gram-negative bacteria from British pork sausage and ingredients. *Systematic and Applied Microbiology* **4,** 424–438.

Barbour, E. A., and Priest, F. G. (1983). Numerical classification of lactobacilli from Scotch whisky distilleries. *In* 'Current Developments in Malting, Brewing and Distilling' (Eds. F. G. Priest and I. Campbell), pp. 289–291. Institute of Brewing, London.

Barre, P. (1969). Taxonomie numerique du lactobacilles isoles du vin. *Archives für Mikrobiologie* **68,** 74–86.

Barton, M. D., and Hughes, K. L. (1982). Is *Rhodococcus equi* a soil organism? *Journal of Reproduction and Fertility, Supplement* **32,** 481–489.

Baumann, P., Baumann, L., and Mandel, M. (1971). Taxonomy of marine bacteria: the genus *Beneckea*. *Journal of Bacteriology* **107,** 268–294.

Baumann, L., Baumann, P., Mandel, M., and Allen, R. D. (1972). Taxonomy of aerobic marine eubacteria. *Journal of Bacteriology* **110,** 402–429.

Baumann, P., Baumann, L., and Reichert, J. L. (1973). Taxonomy of marine bacteria: *Beneckea parahaemolytica* and *Beneckea alginolytica*. *Journal of Bacteriology* **113,** 1144–1155.

Bean, P. G., and Everton, J. R. (1969). Observations on the taxonomy of chromogenic bacteria isolated from cannery environments. *Journal of Applied Bacteriology* **32,** 51–59.

Bell, C. R., Holder-Franklin, M. A., and Franklin, M. (1982). Correlations between predominant heterotrophic bacteria and physicochemical water quality parameters in two Canadian rivers. *Applied and Environmental Microbiology* **43,** 269–283.

Boeyé, A., and Aerts, M. (1976). Numerical taxonomy of *Bacillus* isolates from North Sea sediments. *International Journal of Systematic Bacteriology* **26,** 427–441.

Bölter, M. (1977). Numerical taxonomy and character analysis of saprophytic bacteria isolated from the Kiel Fjord and Kiel Bight. *In* 'Microbial Ecology of a Brackish Water Environment' (Ed. G. Rheinheimer), pp. 148–178. Springer-Verlag, Berlin.

Bonde, G. J. (1981). *Bacillus* from marine habitats: allocation to phena established by numerical techniques. *In* 'The Aerobic Endospore-Forming Bacteria' (Eds. R. C. W. Berkeley and M. Goodfellow), pp. 181–215. Academic Press, London.

Bousfield, I. J. (1978). The taxonomy of coryneform bacteria from the marine environment. *In* 'Coryneform Bacteria' (Eds. I. J. Bousfield and A. G. Callely), pp. 217–233. Academic Press, London.

Brisbane, P. G., and Rovira, A. D. (1961). A comparison of methods for classifying rhizosphere bacteria. *Journal of General Microbiology* **26,** 379–392.

Buchanan, R. E., and Gibbons, N. E., Eds. (1974). 'Bergey's Manual of Determinative Bacteriology', Eighth Edition. Williams & Wilkins, Baltimore, Maryland.

Burges, A. (1939). Soil fungi and root infection. *Broteria* **8,** 64–81.

Byrom, N. A. (1971). The Adansonian taxonomy of some cannery flavobacteria. *Journal of Applied Bacteriology* **34,** 339–346.

Carlsson, J. (1968). A numerical taxonomic study of human oral streptococci. *Odontologisk Revy* **18,** 55–74.

Carney, J. F., Wan, L., Lovelace, T. E., and Colwell, R. R. (1975). Numerical taxonomy study of *Vibrio* and *Spirillum* spp. *International Journal of Systematic Bacteriology* **25,** 38–46.

Colman, G. (1968). The application of computers to the classification of streptococci. *Journal of General Microbiology* **50,** 149–158.

Colwell, R. R. (1970). Polyphasic taxonomy of the genus *Vibrio:* numerical taxonomy of *Vibrio cholerae, Vibrio parahaemolyticus,* and related *Vibrio* species. *Journal of Bacteriology* **104,** 410–433.

Colwell, R. R., Litchfield, C. D., Vreeland, R. H., Kiefer, L. A., and Gibbons, N. E. (1979). Taxonomic studies of red halophilic bacteria. *International Journal of Systematic Bacteriology* **29,** 379–399.

Conn, H. J. (1917). Soil flora studies. 1. The general characteristics of the microscopic flora of soil. *New York Agricultural Experimental Station Technical Bullentin* **57.**

Conn, H. J. (1948). The most abundant groups of bacteria in soil. *Bacteriological Reviews* **12,** 257–273.

Cross, T., Rowbotham, T. J., Mishustin, E. N., Tepper, E. Z., Antoine-Portaels, F., Schaal, K. P., and Bickenbach, H. (1976). The ecology of nocardioform actinomycetes. *In* 'The Biology of the Nocardiae' (Eds. M. Goodfellow, G. H. Brownell, and J. A. Serrano), pp. 337–371. Academic Press, London.

Dabinett, P. E., and Wellman, A. M. (1973). Numerical taxonomy of the genus *Rhizopus* . *Canadian Journal of Botany* **51,** 2053–2064.

Dabinett, P. E., and Wellman, A. M. (1978). Numerical taxonomy of certain genera of Fungi Imperfecti and Ascomycotina. *Canadian Journal of Botany* **56,** 2031–2649.

Darland, G. (1975). Principal component analysis of infraspecific variation in bacteria. *Applied Microbiology* **30,** 282–289.

Dasi, W., Miaoying, C., Yitai, L., Daiwen, L., Huiling, Z., Sike, Z., Zhengfang, W., Junhua, H., and Tieshan, L. (1981). A study on numerical taxonomy of bacteria from

Toumuer peak of Tienshan mountain and other sources. *Acta Microbiologica Sinica* **21,** 385–401.

Da Silva, E. J., and Gyllenberg, H. G. (1972). A taxonomic treatment of the genus *Chlorella* by the technique of continuous classification. *Archiv für Mikrobiologie* **87,** 99–117.

David, H. L., Jahan, M.-T., Jumin, A., Grandry, J., and Lehman, E. H. (1978). Numerical taxonomy analysis of *Mycobacterium africanum. International Journal of Systematic Bacteriology* **28,** 467–482.

Davis, A. W., Atlas, R. M., and Krichevsky, M. I. (1983). Development of probability matrices for identification of Alaskan marine bacteria. *International Journal of Systematic Bacteriology* **33,** 803–810.

Dawson, C. A., and Sneath, P. H. A. (1985). A probability matrix for the identification of vibrios. *Journal of Applied Bacteriology* **58,** 407–423.

Deavin, A., Horsgood, R. K., and Rusch, V. (1981). Rhizosphere microflora in relation to soil conditions. Part 1: Comparison of bacteria in soil, rhizosphere and rhizoplane. *Zentralblatt für Bakteriologie, Parasitenkunde, Infektionskrankheiten und Hygiene, Abteilung 1, Originale, Reihe B* **136,** 613–618.

de Barjac, H., Véron, M., and Dumanoir, V. C. (1980). Caractérisation biochimique et sérologique de souches de *Bacillus sphaericus* pathogènes ou non pour les moustiques. *Annales de Microbiologie de l'Institut Pasteur* **131,** 191–202.

Debette, J., Losfeld, J., and Blundeau, R. (1975). Taxonomie numérique de bactéries telluriques non fermentantes à Gram-négatif. *Canadian Journal of Microbiology* **21,** 1322–1334.

Delabré, M., Bianchi, A., and Véron, M. (1973). Étude critique de méthodes de taxonomie numérique. Application a une classification de bactéries aquicoles. *Annales de Microbiologie de l'Institut Pasteur (Paris)* **124A,** 489–506.

Delarras, C., Laban, P., and Gayral, J. P. (1979). Micrococcaceae isolated from meat and dairy products (taxonomic study). *Zentralblatt für Bakteriologie, Mikrobiologie und Hygiene, Abteilung l, Originale B* **168,** 377–385.

De Leval, J., Houba, C., and Remacle, J. (1976). Les microorganismes en tant que bioindicateur de la qualité des eaux douces. *Mémoires Societe Royale de Botanique de Belgique* **7,** 129–140.

Dickinson, C. H., and Wallace, B. (1976). Effects of late applications of foliar fungicides on activity of micro-organisms on winter wheat flag leaves. *Transactions of the British Mycological Society* **76,** 103–112.

Dickinson, C. H., Dawson, D., and Goodfellow, M. (1981). Interactions between bacteria, streptomycetes and fungi from *Picea sitchensis* litter. *Soil Biology and Biochemistry* **13,** 65–71.

Drucker, D. B., and Melville, T. H. (1971). The classification of some oral streptococci of human and rat origin. *Archives of Oral Biology* **16,** 845–853.

Ercolani, G. L. (1970). Presenza epifitica di *Pseudomonas savastanoi* (E. F. Smith) Stevens sull'olivo, in Puglia. *Phytopathologia Mediterranea* **10,** 130–132.

Ercolani, G. L. (1978). *Pseudomonas savastanoi* and other bacteria colonizing the surface of olive leaves in the field. *Journal of General Microbiology* **109,** 245–257.

Ercolani, G. L. (1983). Variability among isolates of *Pseudomonas syringae* pv. *savastanoi* from the phylloplane of the olive. *Journal of General Microbiology* **129,** 901–916.

Erkenbrecher, C. W., and Stevenson, L. H. (1977). Factors related to the distribution of microbial biomass in saltmarsh-creeks. *Marine Biology* **40,** 121–125.

Ezura, Y., and Sakai, M. (1970). Numerical taxonomy of micrococci isolated from sea water. *Bulletin of the Faculty of Fisheries of Hokkaido University* **21,** 152–159.

Farris, J. S. (1969). On the cophenetic correlation coefficient. *Systematic Zoology* **18,** 279–285.

Feltham, R. K. A. (1979). A taxonomic study of the Micrococcaceae. *Journal of Applied Bacteriology* **47,** 243–254.

Ferragut, C., Izard, D., Gavini, F., Kersters, K., De Ley, J., and Leclerc, H. (1983). *Klebsiella trevisanii:* a new species from water and soil. *International Journal of Systematic Bacteriology* **33,** 133–142.

Floodgate, G. D., and Hayes, P. R. (1963). The Adansonian taxonomy of some yellow pigmented marine bacteria. *Journal of General Microbiology* **30,** 237–244.

Friesen, S., and Bovee, E. C. (1976). A preliminary numerical taxonomy for lobose amebas. *Journal of Protozoology* **23,** 11A.

Garcia, J. L., Roussos, S., Bensoussan, M., Bianchi, A., and Mandel, M. (1982). Taxonomie numérique de *Bacillus* thermophiles isolés de sols de rizière de l'afrique de l'ouest. *Annales de Microbiologie de l'Institut Pasteur (Paris).* **133A,** 471–488.

Garrett, S. D. (1956). 'Biology of Root-infecting Fungi'. Cambridge Univ. Press, Cambridge.

Gavini, F., Lefebvre, B., and Leclerc, H. (1976). Positions taxonomiques d'entérobactéries H_2S^- par rapport au genre *Citrobacter. Annales de Microbiologie (Paris)* **127A,** 275–295.

Gavini, F., Leclerc, H., Lefebvre, B., Ferragut, C., and Izard, D. (1977). Etude taxonomique d'enterobacteries appartenant au apparentées au genre *Klebsiella. Annales de Microbiologie de l'Institut Pasteur (Paris)* **128B,** 45–59.

Gavini, F., Ferragut, C., Izard, D., Trinel, P. A., Leclerc, H., Lefebvre, B., and Mossel, D. A. A. (1979). *Serratia fonticola* a new species from water. *International Journal of Systematic Bacteriology* **29,** 92–101.

Gillespie, N. C. (1981). A numerical taxonomic study of *Pseudomonas*-like bacteria isolated from fish in Southeastern Queensland and their association with spoilage. *Journal of Applied Bacteriology* **50,** 29–44.

Goodfellow, M. (1969). Numerical taxonomy of some heterotrophic bacteria isolated from a pine forest soil. *In* 'The Soil Ecosystem' (Ed. J. G. Sheals), pp. 83–104. The Systematics Association, London.

Goodfellow, M., and Alderson, G. (1977). The actinomycete genus *Rhodococcus:* a home for the *'rhodochrous'* complex. *Journal of General Microbiology* **100,** 99–122.

Goodfellow, M., and Minnikin, D. E., Eds. (1985). 'Chemical Methods in Bacterial Systematics'. Academic Press, London.

Goodfellow, M., and Pirouz, T. (1982). Numerical classification of sporoactinomycetes containing *meso*-diaminopimelic acid in the cell wall. *Journal of General Microbiology* **128,** 503–527.

Goodfellow, M., and Wayne, L. G. (1982). Taxonomy and nomenclature. *In* 'The Biology of Mycobacteria' (Eds. C. Ratledge and J. L. Stanford), pp. 471–521. Academic Press, London.

Goodfellow, M., and Williams, S. T. (1983). Ecology of actinomycetes. *Annual Review of Microbiology* **37,** 189–216.

Goodfellow, M., Austin, B., and Dawson, D. (1976a). Classification and identification of phylloplane bacteria using numerical taxonomy. *In* 'Microbiology of Aerial Plant Surfaces' (Eds. C. H. Dickinson and T. F. Preece), pp. 275–292. Academic Press, London.

Goodfellow, M., Austin, B., and Dickinson, C. H. (1976b). Numerical taxonomy of some yellow-pigmented bacteria isolated from plants. *Journal of General Microbiology* **97,** 219–233.

Goodfellow, M., Orlean, P. A. B., Collins, M. D., and Minnikin, D. E. (1978). Chemical and numerical taxonomy of strains received as *Gordona aurantiaca*. *Journal of General Microbiology* **109,** 57–68.

Goodfellow, M., Alderson, G., and Lacey, J. (1979). Numerical taxonomy of *Actinomadura* and related actinomycetes. Journal of General Microbiology **112,** 95–111.

Goodfellow, M., Beckham, A. R., and Barton, M. D. (1982a). Numerical classification of *Rhodococcus equi* and related actinomycetes. *Journal of Applied Bacteriology* **53,** 199–207.

Goodfellow, M., Minnikin, D. E., Todd, C., Alderson, G., Minnikin, S. M., and Collins, M. D. (1982b). Numerical and chemical classification of *Nocardia amarae*. *Journal of General Microbiology* **128,** 1283–1297.

Goodfellow, M., Weaver, C. R., and Minnikin, D. E. (1982c). Numerical classification of some rhodococci, corynebacteria and related organisms. *Journal of General Microbiology* **128,** 731–745.

Goodfellow, M., Embley, T. M., and Austin, B. (1985). Numerical taxonomy and emended description of *Renibacterium salmonnarum*. *Journal of General Microbiology* (in press).

Gosselé, F., Swings, J., Kersters, K., and De Ley, J. (1983a). Numerical analysis of phenotypic features and protein gel electropherograms of *Gluconobacter* Asai 1935 emend mut, char. Asai, Iizuka, and Komagata 1964. *International Journal of Systematic Bacteriology* **33,** 65–81.

Gosselé, F., Swings, J., Kersters, K., Pauwels, P., and De Ley, J. (1983b). Numerical analysis of phenotypic features and protein gel electropherograms of a wide variety of *Acetobacter* strains. Proposal for the improvement of the taxonomy of the genus *Acetobacter* Beijerinck 1898, 215. *Systematics and Applied Microbiology* **4,** 338–368.

Graham, P. H. (1964). The application of computer techniques to the taxonomy of root-nodule bacteria of legumes. *Journal of General Microbiology* **35,** 511–517.

Gray, P. A., and Stewart, D. J. (1980). Numerical taxonomy of some marine pseudomonads and alteromonads. *Journal of Applied Bacteriology* **49,** 375–383.

Green, P. N., and Bousfield, I. J. (1982). A taxonomic study of some Gram-negative facultatively methylotrophic bacteria. *Journal of General Microbiology*. **128,** 623–638.

Gunn, B. A., and Colwell, R. R. (1983). Numerical taxonomy of staphylococci isolated from the marine environment. *International Journal of Systematic Bacteriology* **33,** 751–759.

Gunn, B. A., Keiser, J. F., and Colwell, R. R. (1983). Numerical taxonomy of staphylococci isolated from clinical sources. *International Journal of Systematic Bacteriology* **33,** 738–750.

Gyllenberg, H. G. (1963). A general method for devising determination schemes for random collections of microbial isolates. *Annales Academiae Scientiarum Fennicae, Series A. IV. Biologica* **69,** 1–23.

Gyllenberg, H. G. (1964). An approach to numerical description of microbial populations. *Annales Academiae Scientiarum Fennica, Series A. IV. Biologica* **81,** 1–23.

Gyllenberg, H. G. (1965a). Character correlations in certain taxonomic and ecological groups of bacteria. *Annales Medicinae Experimentalis et Biologiae Fenniae* **43,** 82–90.

Gyllenberg, H. G. (1965b). A model for computer identification of micro-organisms. *Journal of General Microbiology* **39,** 401–405.

Gyllenberg, H. G. (1967). Significance of the Gram stain in the classification of soil bacteria. *In* 'The Ecology of Soil Bacteria' (Eds, T. R. G. Gray and D. Parkinson), pp. 351–359. Liverpool Univ. Press, Liverpool.

Gyllenberg, H. G., and Eklund, E. (1967). Application of factor analysis in microbiology. 2. Evaluation of character correlation patterns in psychrophilic pseudomonads. *Annales Academiae Scientiarum Fennicae, Series A. IV. Biologica* **113,** 1–19.

Gyllenberg, H. G., and Rauramaa, V. (1966). Taxometric models of bacterial soil populations. *Acta Agriculturae Scandinavica* **16,** 30–38.

Gyllenberg, H., Eklund, E., Antila, M., and Vartiovaara, U. (1963). Contamination and deterioration of market milk. V. Taxometric classification of pseudomonads. *Acta Agriculturae Scandinavica* **13,** 157–176.

Hagedorn, C., and Holt, J. G. (1975a). A nutritional and taxonomic survey of *Arthrobacter* soil isolates. *Canadian Journal of Microbiology* **21,** 353–361.

Hagedorn, C., and Holt, J. G. (1975b). Ecology of soil arthrobacters in Clarion–Webster toposequences of Iowa. *Applied Microbiology* **29,** 211–218.

Harman, R. (1960). 'Modern Factor Analysis'. Chicago Univ. Press, Chicago.

Håstein, T., and Smith, J. E. (1977). A study of *Vibrio anguillarum* from farmed and wild fish using principal components analysis. *Journal of Fish Biology* **11,** 69–75.

Hauser, M. M., and Smith, R. E. (1964). The characterisation of lactobacilli from Cheddar cheese. II. A numerical analysis of the data by means of an electronic computer. *Canadian Journal of Microbiology* **10,** 757–762.

Hauxhurst, J. D., Krichevsky, M. I., and Atlas, R. M. (1980). Numerical taxonomy of bacteria from the Gulf of Alaska. *Journal of General Microbiology,* **120,** 131–148.

Hill, I. R. (1974). Theoretical aspects of numerical identification. *International Journal of Systematic Bacteriology* **24,** 494–499.

Hissett, R., and Gray, T. R. G. (1973). Bacterial populations of litter and soil in deciduous woodland. 1. Qualitative studies. *Revue d'Écologie et de Biologie du Sol* **10,** 495–508.

Holder-Franklin, M. A. (1981). The development of biological and mathematical methods to study population shifts in aquatic bacteria in response to environmental change. Scientific series no. 124. Inland Water Directorate, Department of Environment, Ottawa.

Holder-Franklin, M. A., and Wuest, L. J. (1983). Population dynamics of aquatic bacteria in relation to environmental change as measured by factor analysis. *Journal of Microbiological Methods* **1,** 209–227.

Holder-Franklin, M. A., Franklin, M., Cashion, P., Cormier, C., and Wuest, L. (1978). Population shifts in heterotrophic bacteria in a tributary of the Saint John River as measured by taxometrics. *In* 'Microbial Ecology' (Eds. M. W. Loutit and J. A. R. Miles), pp. 44–50. Springer-Verlag, New York.

Hookey, J. V. (1983). Selective Isolation, Classification and Ecology of Nocardiae from Soil, Water and Biodeteriorating Rubber. Ph. D. Thesis, Univ. of Newcastle upon Tyne.

Hungate, R. E. (1966). 'The Rumen and Its Microbes'. Academic Press, New York.

Izard, D., Ferragut, C., Gavini, F., Kersters, K., De Ley, J., and Leclerc, H. (1981). *Klebsiella terrigena,* a new species from soil and water. *International Journal of Systematic Bacteriology* **31,** 116–127.

Jensen, V. (1967). The plate count technique. *In* 'The Biology of the Soil Bacteria' (Eds. T. R. G. Gray and D. Parkinson), pp. 158–170. Liverpool Univ. Press, Liverpool.

Jensen, V. (1971). The bacterial flora of beech leaves. *In* 'Ecology of Leaf Surface Micro-organisms' (Eds. T. F. Preece and C. H. Dickinson), pp. 463–469. Academic Press, London.

Johnson, R. M., Katarski, M. E., and Weisrock, W. P. (1968). Correlation of taxonomic criteria for a collection of marine bacteria. *Applied Microbiology* **16,** 708–713.

Jones, J. G. (1977). The effect of environmental factors on estimated viable and total populations of planktonic bacteria in lakes and experimental enclosures. *Freshwater Biology* **7,** 67–91.

Jones, D. (1978). An evaluation of the contributions of numerical taxonomic studies to the classification of coryneform bacteria. *In* 'Coryneform Bacteria' (Eds. I. J. Bousfield and A. G. Callely), pp. 13–46. Academic Press, London.

Jones, J. G. (1978). The distribution of some freshwater planktonic bacteria in two stratified eutrophic lakes. *Freshwater Biology* **8,** 127–140.

Jones, D., and Sackin, M. J. (1980). Numerical methods in the classification and identification of bacteria with especial reference to the Enterobacteriaceae. *In* 'Microbiological Classification and Identification' (Eds. M. Goodfellow and R. G. Board), pp. 73–106. Academic Press, London.

Jooste, A. (1979). The interrelationships between the biological and physicochemical parameters in the water phase of Seshego Dam, Northern Transvaal. *Journal of the Limnological Society of Southern Africa* **5,** 59–63.

Kaneko, T., Atlas, R. M., and Krichevsky, M. (1977). Diversity of bacterial populations in the Beaufort Sea. *Nature (London)* **270,** 596–599.

Kaneko, T., Krichevsky, M. I., and Atlas, R. M. (1979). Numerical taxonomy of bacteria from the Beaufort Sea. *Journal of General Microbiology* **110,** 111–125.

Kaper, J. B., Mills, A. L., and Colwell, R. R. (1978). Evaluation of the accuracy and precision of enumerating aerobic heterotrophs in water samples by the spread plate method. *Applied and Environmental Microbiology* **3,** 756–761.

Kaper, J. B., Lockman, H., Remmers, E. F., Kristensen, K., and Colwell, R. R. (1983). Numerical taxonomy of vibrios isolated from estuarine environments. *International Journal of Systematic Bacteriology* **33,** 229–255.

Kapperud, G., Bergan, T., and Lassen, J. (1981). Numerical taxonomy of *Yersinia enterocolitica* and *Yersinia enterocolitica*–like bacteria. *International Journal of Systematic Bacteriology* **31,** 401–419.

Kazda, J. (1980). *Mycobacterium sphagni* sp. nov. *International Journal of Systematic Bacteriology* **30,** 77–81.

Kazda, J., and Müller, K. (1979). *Mycobacterium komossense* sp. nov. *International Journal of Systematic Bacteriology* **29,** 361–365.

Kelley, R. W., and Kellogg, S. T. (1978). Computer-assisted identification of anaerobic bacteria. *Applied and Environmental Microbiology* **35,** 507–511.

Kersters, K., De Ley, J., Sneath, P. H. A., and Sackin, M. (1973). Numerical taxonomic analysis of *Agrobacterium. Journal of General Microbiology* **78,** 227–239.

King, D. S. (1976). Systematics of *Conidiobolus* (Entomophthorales) using numerical taxonomy. I. Biology and cluster analysis. *Canadian Journal of Botany* **54,** 45–65.

King, A., Holmes, B., Phillips, I., and Lapage, S. P. (1979). A taxonomic study of clinical isolates of *Pseudomonas pickettii, 'P. thomasii'* and 'Group 1Vd' bacteria. *Journal of General Microbiology* **114,** 137–147.

Kockova–Kratochvilova, A., Slavikova, E., and Jensen, V. (1978). Numerical taxonomy of the yeast genus *Debaryomyes* Lodder et Krejer van Rij. *Journal of General Microbiology* **104,** 257–268.

Kockova-Kratochvilova, A., Slavikova, E., Zemek, J., Augustin, J., Kuniak, L., and Dercova, K. (1981). Numerical taxonomy of the genus *Schwanniomyces* Klocker. *Biologia (Bratislava)* **36,** 693–701.

Komaromy, Z. P. (1982). Application of cluster analysis in the taxonomy of *Scotiella* species (Chlorophyceae). *Archiv für Hydrobiologies Supplementband* **60,** 432–438.

Krieg, R. E., and Lockhart, W. R. (1966). Classification of enterobacteria based on overall similarity. *Journal of Bacteriology* **92,** 1275–1280.

Kuhn, D. A., Gregory, D. A., Buchanan, G. E., Jr., Nyby, M. D., and Daly, K. R. (1978). Isolation, characterization, and numerical taxonomy of *Simonsiella* strains from the oral cavities of cats, dogs, sheep and humans. *Archives of Microbiology* **118,** 235–241.

Kuznetsov, S. I., Dubinina, G. A., and Lapteva, N. A. (1979). Biology of oligotrophic bacteria. *Annual Review of Microbiology* **33,** 377–387.

Laban. P., Tavre, C., Romet, F., and Larpent, J. P. (1978). Lactobacilli isolated from French Saucisson (Taxonomic study). *Zentralblatt für Bakteriologie Parasitenkunde, Infektionskrankheiten und Hygiene, Abteilung 1, Orginale, Reihe B* **166,** 105–111.

Lapage, S. P., Bascomb, S., Willcox, W. R., and Curtis, M. A. (1970). Computer identification of bacteria. *In* 'Automation, Mechanization and Data Handling in Microbiology (Eds. A. Baillie and R. J. Gilbert), pp. 1–22. Academic Press, London.

Lawrence, D. R., and Priest, F. G. (1981). Identification of brewery cocci. *In* 'Proceedings of the European Brewery Convention', Copenhagen, pp. 217–227. IRL Press Ltd., London.

Lee, J. V., Gibson, D. M., and Shewan, J. M. (1977). A numerical taxonomic study of some *Pseudomonas*-like marine bacteria. *Journal of General Microbiology* **98,** 439–451.

Lee, J. V., Donovan, T. J., and Furniss, A. L. (1978). Characterization, taxonomy, and emended description of *Vibrio metschnikovii. International Journal of Systematic Bacteriology* **28,** 99–111.

Lee, J. V., Shread, P., Furniss, A. L., and Bryant, T. N. (1981). Taxonomy and description of *Vibrio fluvialis* sp. nov. (synonym group F vibrios, group EF6). *Journal of Applied Bacteriology* **50,** 73–94.

Lee, C. Y., Fung, D. Y. C., and Kastner, C. L. (1982). Computer-assisted identification of bacteria on hot-boned and conventionally processed beef. *Journal of Food Science* **47,** 363–367.

Lévy-Frébault, V., Rafidinarivo, E., Prome, J.-C., Grandry, J., Biosvert, H., and David, H. L. (1983). *Mycobacterium fallax. International Journal of Systematic Bacteriology* **33,** 336–343.

Lighthart, B. (1974). A cluster analysis of some bacteria in the water column of Green Lake, Washington. *Canadian Journal of Microbiology* **21,** 392–394.

Liston, J., Wiebe, W., and Colwell, R. R. (1963). Quantitative approach to the study of bacterial species. *Journal of Bacteriology* **35,** 1061–1070.

Litchfield, C. D., Colwell, R. R., and Prescott, J. M. (1969). Numerical taxonomy of heterotrophic bacteria growing in association with continuous-culture *Chlorella sorokiniana. Applied Microbiology* **18,** 1044–1049.

Lochhead, A. G., and Chase, F. E. (1943). Qualitative studies of soil microorganisms. V. Nutritional requirements of the predominant bacterial flora. *Soil Science* **55,** 185–195.

Lochhead, A. G., and Taylor, C. B. (1938). Qualitative studies of soil micro-organisms. 1. General introduction. *Canadian Journal of Research, Section C* **16,** 152–161.

Logan, N. A., and Berkeley, R. C. W. (1981). Classification and identification of members of the genus *Bacillus* using API tests. *In* 'The Aerobic Endospore-forming Bacteria' (Eds. R. C. W. Berkeley and M. Goodfellow), pp. 105–140. Academic Press, London.

Lourd, M., Geiger, J. P., and Goujon, M. (1979). Les *Colletotrichum* agents d'anthracnoses en Côte-d'Ivoire. I.—Charactéristiques morphologiques et culturales d'isolates de *Colletotrichum gloeosporioides* Penz. *Annales Phytopathologie* **11,** 483–495.

Lowe, W. E., and Gray, T. R. G. (1972). Ecological studies on coccoid bacteria in a pine forest soil. I. Classification, *Soil Biology and Biochemistry* **4,** 459–467.

Lowe, W. E., and Gray, T. R.. (1973a). Ecological studies on coccoid bacteria in a pine forest soil. II. Growth of bacteria inoculated into soil. *Soil Biology and Biochemistry* **5,** 449–462.

Lowe, W. E., and Gray, T. R. G. (1973b). Ecological studies on coccoid bacteria in a pine forest. III. Competitive interactions between bacterial strains in soil. *Soil Biology and Biochemistry* **5,** 463–472.

McCarthy, A. J., and Cross, T. (1984). A taxonomic survey of *Thermomonospora* and other monosporic actinomycetes. *Journal of General Microbiology* **130,** 5–25.

McMeekin, T. A., Stewart, D. B., and Murray, J. G. (1972). The Adansonian taxonomy and the deoxyribonucleic acid base composition of some Gram-negative, yellow pigmented rods. *Journal of Applied Bacteriology* **35,** 129–137.

Mallory, L. M., Austin, B., and Colwell, R. R. (1977). Numerical taxonomy and ecology of oligotrophic bacteria isolated from the estuarine environment. *Canadian Journal of Microbiology* **23,** 733–750.

Melchiorri-Santolini, U. (1968). Numerical taxonomy of pelagic bacteria from the Ligurian sea. *Annaliai Microbiologiya* **18,** 67–83.

Melville, T. H. (1965). A study of the overall similarity of certain actinomycetes mainly of oral origin. *Jounal of General Microbiology* **40,** 309–315.

Moffett, M. L., and Colwell, R. R. (1968). Adansonian analysis of the Rhizobiaceae. *Journal of General Microbiology* **51,** 245–266.

Molin, G., and Ternstrom, A. (1982). Numerical taxonomy of psychrotrophic pseudomonads. *Journal of General Microbiology* **128,** 1249–1264.

Moss, M. O., Ryall, C., and Logan, N. A. (1978). The classification and characterization of chromobacteria from a lowland river. *Journal of General Microbiology* **105,** 11–21.

Niemelä, S., and Sundman, V. (1977). Effects of clear-cutting on the composition of bacterial populations of northern spruce forest soil. *Canadian Journal of Microbiology* **23,** 131–138.

Niemelä, S. I., Hopkins, J. W., and Quadling, C. (1968). Selecting an economical binary test battery for a set of microbial cultures. *Canadian Journal of Microbiology* **14,** 271–279.

Orchard, V. A., and Goodfellow, M. (1980). Numerical classification of some named strains of *Nocardia asteroides* and related isolates from soil. *Journal of General Microbiology* **118,** 295–312.

Orchard, V. A., Goodfellow, M., and Williams, S. T. (1977). Selective isolation and occurrence of nocardiae in soil. *Soil Biology and Biochemistry* **9,** 233–238.

Pagel, J. E., and Seyfried, P. L. (1976). Numerical taxonomy of aquatic *Acinetobacter* isolates. *Journal of General Microbiology* **95,** 220–232.

Persson, I. B., and Rosswall, T. (1978). Functional description of bacterial populations from lakes with varying degrees of eutrophication. SNV PM 1080, *Report to Swedish Environmental Protection Board (in Swedish).*

Pfister, R. M., and Burkholder, P. R. (1965). Numerical taxonomy of some bacteria isolated from Antarctic and tropical seawaters. *Journal of Bacteriology* **90,** 863–872.

Pichinoty, F., Garcia, J.-L., and Mandel, M. (1980). Taxonomie numérique de 46 souches dénitrifiantes et mésophiles de *Bacillus* isolées à partir du sol par culture élective en présence de nitrite. *Canadian Journal of Microbiology* **26,** 787–795.

Piot, P., Van Dyck, E., Goodfellow, M., and Falkow, S. (1980). A taxonomic study of *Gardnerella vaginalis* (*Haemophilus vaginalis*) Gardner and Dukes 1955. *Journal of General Microbiology* **119,** 373–396.

Pohja, M. S., and Gyllenberg, H. G. (1967). Application of factor analysis in microbiology. 5. Evaluation of the population development in cold stored meat. *Annales Academiae Scientiarum Fennicae, Series A, IV. Biologica* **116,** 1–8.

Popoff, M., and Véron, M. (1976). A taxonomic study of the *Aeromonas hydrophila–Aeromonas punctata* group. *Journal of General Microbiology* **94,** 11–22.

Prentice, H. C. (1980). Variation in *Silene dioica* (L.) Clairv.: numerical analysis of populations from Scotland. *Watsonia* **13,** 11–26.

Pretorius, W. A. (1964). An Ecological Study of the Aerobic Proteolytic Free-Living Nitrogen Fixing and Spiralform Bacteria in Stabilization Ponds. D. Sc. Thesis, University of Pretoria.

Priest, F. G., Somerville, H. J., Cole, J. A., and Hough, J. (1973). The taxonomic position of *Obesumbacterium proteus,* a common brewery contaminant. *Journal of General Microbiology* **75,** 295–307.

Quadling, C. (1967). Evaluation of tests and grouping of cultures by a two-state principal component method. *Canadian Journal of Microbiology* **13,** 1379–1400.

Quadling, S., and Colwell, R. R. (1964). The use of numerical methods in characterizing unknown isolates. *Developments in Industrial Microbiology* **5,** 151–161.

Quigley, M. M., and Colwell, R. R. (1968). Properties of bacteria isolated from deep-sea sediments. *Journal of Bacteriology* **95,** 211–230.

Ready, P. D., and Miles, M. A. (1981). Delimitation of *Trypanosoma cruzi* zymodemes by numerical taxonomy. *Systematic Parasitology* **2,** 207–211.

Reichelt, J. L., and Baumann, P. (1973). Taxonomy of the marine, luminous bacteria. *Archiv für Mikrobiologie* **94,** 283–330.

Rennie, R. J. (1980). Dinitrogen-fixing bacteria: computer assisted identification of soil isolates. *Canadian Journal of Microbiology* **26,** 1275–1283.

Roberts, R. J. (1968). A numerical taxonomic study of 100 isolates of *Corynebacterium pyogenes. Journal of General Microbiology* **53,** 299–303.

Rosswall, T. (1976). The need for rapid methods and automation in environmental microbiology. *In* 'Rapid Methods and Automation in Microbiology' (Eds. H. H. Johnson and S. W. B. Newsom), pp. 131–135. Learned Information (Europe) Ltd., Oxford.

Rosswall, T., and Kvillner, E. (1978). Principal components and factor analysis for the description of microbial populations. *Advances in Microbial Ecology* **2,** 1–48.

Rosswall, T., and Persson, I.-B. (1982). Functional description of bacterial populations from seven Swedish lakes. *Limnologica (Berlin)* **14,** 1–16.

Rouatt, J. W. (1967). Nutritional classifications of soil bacteria and their value in ecological studies. *In* 'The Ecology of Soil Bacteria' (Eds. T. R. G. Gray and D. Parkinson), pp. 360–370. Liverpool Univ. Press, Liverpool.

Rouatt, J. W., and Katznelson, H. (1961). A study of the bacteria in the root surface and in the rhizosphere soil of crop plants. *Journal of Applied Bacteriology* **24,** 164–171.

Rovira, A. D., and Brisbane, P. G. (1967). Numerical taxonomy and soil bacteria. *In* 'The Ecology of Soil Bacteria' (Eds. T. R. G. Gray and D Parkinson), pp. 337–350. Liverpool Univ. Press, Liverpool.

Rowbotham, T. J., and Cross, T. (1977). *Rhodococcus coprophilus* sp. nov.: an aerobic nocardioform actinomycete belonging to the '*rhodochrous*' complex. *Journal of General Microbiology* **100,** 123–138.

Rypka, E. W., Clapper, W. E., Bowen, I. G., and Babb, R. (1967). A model for the identification of bacteria. *Journal of General Microbiology* **46,** 407–424.

Sackin, M. J. (1981). Vigour and pattern as applied to multistate quantitative characters in taxonomy. *Journal of General Microbiology* **122,** 247–254.

Schindler, J., and Idlbek, J. (1982). A simplified strategy for the identification of Gram-negative fermenting rods using a desk-top computer. *Journal of Applied Bacteriology* **52,** 353–356.

Schindler, J., Duben, J., and Lysenko, O. (1979). Computer-aided numerical identification of Gram-negative fermentative rods on a desk-top computer. *Journal of Applied Bacteriology* **47,** 45–51.

Schofield, G. M., and Schaal, K. P. (1981). A numerical taxonomic study of the Actinomycetaceae and related taxa. *Journal of General Microbiology* **127,** 237–259.

Schofield, G. M., and Whalley, A. J. S. (1978). Numerical taxonomy of rhizosphere bacteria from sand dune grasses. *Annali Microbiologia ed Enzimologia* **28,** 111–125.

Seiler, H., and Busse, M. (1977). Taxonomic studies on Gram-positive coryneform bacteria from dairy waste water. *Milchwissenschaft* **32,** 525–530.

Seiler, H., and Hennlich, W. (1983). Characterization of coryneform bacteria in piggery wastes. *Systematics and Applied Microbiology* **4,** 132–140.

Seiler, H., Braat, R., and Ohmayer, G. (1980). Numerical cluster analysis of the coryneform bacteria from activated sludge. *Zentralblatt für Bakteriologie und Hygiene, Abteilung 1, Originale C* **1,** 357–375.

Shaw, B. G., and Harding, C. D. (1984). A numerical taxonomic study of lactic acid bacteria from vacuum-packed beef, pork, lamb and bacon. *Journal of Applied Bacteriology* **56,** 25–40.

Shaw, B. G., and Latty, J. B. (1982). A numerical taxonomic study of *Pseudomonas* strains from spoiled meat. *Journal of Applied Bacteriology* **52,** 219–228.

Sheard, M. A., and Priest, F. G. (1981). Numerical classification of some psychrotrophic bacilli isolated from frozen foods. *Journal of Applied Bacteriology* **51,** xxii–xxiii.

Sielaff, B. H., Matsen, J. M., and McKie, J. E. (1982). Novel approach to bacterial identification that uses the Autobac System. *Journal of Clinical Microbiology* **15,** 1103–1110.

Silvestri, L. G., Turri, M., Hill, L. R., and Gilardi, E. (1962). A quantitative approach to the systematics of actinomycetes based on overall similarity. *Symposia of the Society for General Microbiology* **12,** 333–360.

Singleton, R. J., and Skerman, T. M. (1973). A taxonomic study by computer analysis of marine bacteria from New Zealand waters. *Journal of the Royal Society for New Zealand* **3,** 129–140.

Skyring, G. W., and Quadling, C. (1969). Soil bacteria: comparisons of rhizosphere and non-rhizosphere populations. *Canadian Journal of Microbiology* **15,** 473–488.

Skyring, G. W., Jones, H. E., and Goodchild, D. (1977). The taxonomy of some new isolates of dissimilatory sulfate-reducing bacteria. *Canadian Journal of Microbiology* **23,** 1415–1425.

Smith, B.-H., and Hirshfield, H. I. (1975). Numerical taxonomy of *Blepharisma* based on the effects of selected antibodies. *Journal of Protozoology* **22,** 44A.

Sneath, P. H. A. (1968a). Vigour and pattern in taxonomy. *Journal of General Microbiology* **54,** 1–11.

Sneath, P. H. A. (1968b). The future outline of bacterial classification. *The Classification Society Bulletin* **1,** 28–45.

Sneath, P. H. A. (1974). Test reproducibility in relation to identification. *International Journal of Systematic Bacteriology* **24,** 508–523.

Sneath, P. H. A. (1977). A method of testing the distinctness of clusters: a test of the disjunction of two clusters in Euclidean space as measured by their overlap. *Journal of Mathematical Geology* **9,** 123–143.

Sneath, P. H. A. (1978). Classification of microorganisms. *In* 'Essays in Microbiology' (Eds. J. R. Norris and M. H. Richmond), No. 9, pp. 1–31. Wiley, London.

Sneath, P. H. A. (1979a). Basic program for a significance test for two clusters in Euclidean space as measured by their overlap. *Computers and Geosciences* **5,** 143–155.

Sneath, P. H. A. (1979b). BASIC programme for identification of an unknown with presence–absence data against an identification matrix of percent positive characters. *Computers and Geoscience* **5,** 195–213.

Sneath, P. H. A. (1980). BASIC program for the most diagnostic properties of groups from an identification matrix of percent positive characters. *Computers and Geosciences* **6,** 21–26.

Sneath, P. H. A., and Johnson, R. (1972). The influence on numerical taxonomic similarities of errors in microbiological tests. *Journal of General Microbiology* **72,** 377–392.

Soumare, S., Losfeld, J., and Blondeau, R. (1973). Apports de la taxonomie numérique a l'étude du spectre bactérien de la microflore des solo du nord de la France. *Annales de Microbiologie (Paris)* **124B,** 81–94.

Sperber, J. I., and Rovira, A. D. (1959). A study of bacteria associated with the roots of subterranean clover and Wimmera rye grass. *Journal of Applied Bacteriology* **22,** 85–95.

Splittstoesser, D. F., Wexler, M., White, J., and Colwell, R. R. (1967). Numerical taxonomy of Gram-positive and catalase-positive rods isolated from frozen vegetables. *Applied Microbiology* **15,** 158–162.

Stout, J. D. (1960). Bacteria of soil and pasture leaves at Claudelands showgrounds. *New Zealand Journal of Agricultural Research* **3,** 413–430.

Sundman, V. (1968). Characterization of bacterial populations by means of factor profiles. *Acta Agriculturae Scandinavica* **18,** 22–32.

Sundman, V. (1970). Four bacterial soil populations characterized and compared by a factor analytical method. *Canadian Journal of Microbiology* **16,** 455–464.

Sundman, V., and Carlberg, G. (1967). Application of factor analysis in microbiology. 4. The value of geometric parameters in the numerical description of bacterial soil populations. *Annales Academiae Scientiarum Fennica, Series A. IV. Biologica* **115,** 1–12.

Sundman, V., and Gyllenberg, H. G. (1967). Application of factor analysis in microbiology. 1. General aspects on the use of factor analysis in microbiology. *Annales Academiae Scientiarum Fennicae, Series A, IV. Biologica* **112,** 1–32.

Taylor, C. B. (1951). The nutritional requirements of the predominant bacterial flora of soil. *Journal of Applied Bacteriology* **14,** 101–111.

Taylor, C. B., and Lochhead, A. G. (1937). A study of *Bacterium globiforme* Conn in soils differing in fertility. *Canadian Journal of Research C,* **15,** 340–347.

Thornley, M. J. (1967). A taxonomic study of *Acinetobacter* and related genera. *Journal of General Microbiology* **49,** 211–257.

Toerien, D. F. (1970a). Population description of the non-methanogenic phase of anaerobic digestion. III. Non-hierachical classification of isolates by principal component analysis. *Water Research* **4,** 305–314.

Toerien, D. F. (1970b). Population description of the non-methanogenic phase of anaerobic digestion. II. Hierachical classification of isolates. *Water Research* **4,** 285–303.

Torstensson, N. T. L., and Rosswall, T. (1977). The effect of 20 years' application of 2,4-D and MCPA on the soil flora. *In* 'The Interaction of Soil Microflora and Environmental Pollutions' Vol. 1, pp. 170–176. Instytur Uprawg Nawozenia i Bleboznawsta, Polawg, Poland.

Tsukamura, M. (1981). Numerical analysis of rapidly growing, nonphotochromogenic mycobacteria, including *Mycobacterum agri* (Tsukamura 1972) Tsukamura sp. nov. nom. rev. *International Journal of Systematic Bacteriology* **31,** 247–258.

Tsukamura, M., Mizuno, S., and Tsukamura, S. (1981). Numerical analysis of rapidly growing, scotochromogenic mycobacteria, including *Mycobacterium obuense* sp. nov., nom. rev., *Mycobacterium rhodesiae* sp. nov. nom. rev., *Mycobacterium aichiense* sp. nov., nom. rev., *Mycobacterium chubuense* sp. nov., nom. rev., and *Mycobacterium tokaiense* sp. nov., nom. rev. *International Journal of Systematic Bacteriology* **31,** 263–275.

Tsukamura, M., Mizuno, S., and Toyama, H. (1983). *Mycobacterium pulveris* sp. nov., a nonphotochromogenic Mycobacterium with an intermediate growth rate. *International Journal of Systematic Bacteriology* **33,** 811–815.

Väätänen, P. (1980). Factor analysis of the impact of the environment on microbial communities in the Tvärminne Area, southern coast of Finland. *Applied and Environmental Microbiology* **40,** 55–61.

Vágnerová, K., Macura, J., and Čatská, V. (1960). Rhizosphere microflora of wheat. II. Composition and properties of bacterial flora during the vegetation period of wheat. *Folia Microbiologica* **5,** 311–319.

Van Valkenburg, S. D., Karlander, E. P., Patterson, G. W., and Colwell, R. R. (1977). Features for classifying photosynthetic aerobic nanoplankton by numerical taxonomy. *Taxon* **26,** 497–505.

Vanderzant, C., Judkins, P. W., Nickelson, R., and Fitzhugh, H. A., Jr. (1972). Numerical taxonomy of coryneform bacteria isolated from pond-reared shrimp (*Penaeus aztecus*) and pond water. *Applied Microbiology* **23,** 38–45.

Ventosa, A., Quesada, E., Rodriguez-Valera, F., Ruiz-Berraquerdo, F., and Ramos-Cormenzana, A. (1982). Numerical taxonomy of moderately halophilic Gram-negative rods. *Journal of General Microbiology* **128,** 1959–1968.

Vickers, J. C., Williams, S. T., and Ross, G. W. (1984). A taxonomic approach to selective isolation of streptomycetes. *In* 'Biological, Biochemical, and Biomedical Aspects of Actinomycetes' (Eds. L. Ortiz-Ortiz, L. F. Bojalil, and V. Yakoleff), pp. 553–561. Academic Press, Orlando.

Wayne, L. G., Engel, H. W. B., Grassi, C., Gross, W., Hawkins, J., Jenkins, P. A., Käppler, W., Kleeberg, H. H., Krasnow, I., Nel, E. E., Pattyn, S. R., Richards, P. A., Showalter, S., Slosarek, M., Szabo, I., Tárnok, I., Tsukamura, M., Vergmann, B., and Wolinsky, E. (1976). Highly reproducible techniques for use in systematic bacteriology in the genus *Mycobacterium:* Tests for niacin and catalase and for resistance to isoniazid, thiophene 2-carboxylic hydrazide, hydroxylamine and p-nitrobenzoate. *International Journal of Systematic Bacteriology* **26,** 311–318.

Wayne, L. G., Krichevsky, E. J., Love, L. L., Johnson, R., and Krichevsky, M. I. (1980). Taxonomic probability matrix for use with slowly growing mycobacteria. *International Journal of Systematic Bacteriology* **30,** 528–538.

Weckbach, L. S., and Langlois, B. E. (1976). Classification by numerical taxonomy of staphylococci isolated from the bovine udder. *Journal of Milk Food Technology* **39,** 246–249.

West, P. A., Lee, J. V., and Bryant, T. N. (1983). A numerical taxonomic study of species of *Vibrio* isolated from the aquatic environment and birds in Kent, England. *Journal of Applied Bacteriology* **55,** 263–282.

Whalley, A. J. S. (1976). Numerical taxonomy of some species of *Hypoxylon. Mycopathologia et Mycologia applicata* **59,** 155–161.

Whalley, A. J. S., and Greenhalgh, G. N. (1973). Numerical taxonomy of *Hypoxylon.* I. Comparison of classifications of the cultural and the perfect staes. *Transactions of the British Mycological Society* **61,** 435–454.

White, D. C. (1983). Analysis of microorganisms in terms of quantity and activity in natural environments. *In* 'Microbes in their Natural Environments' (Eds. J. H. Slater, R. Whittenbury, and J. W. T. Wimpenny), pp. 38–66. Cambridge Univ. Press, Cambridge.

Wilkinson, B. J., and Jones, D. (1977). A numerical taxonomic survey of *Listeria* and related bacteria. *Journal of General Microbiology* **98,** 399–421.
Willemse-Collinet, M. E., Tromp, T. F., and Huizinga, T. (1980). A simple and rapid computer-assisted technique for the identification of some selected *Bacillus* species using biochemical tests. *Journal of Applied Bacteriology* **49,** 385–394.
Williams, S. T., and Gray, T. R. G. (1973). General principles and problems of soil sampling. *In* 'Sampling: Microbiological Monitoring of Environments' (Eds. R. G. Board and D. W. Lovelock), pp. 111–122. Academic Press, New York.
Williams, S. T., Davies, F. L., and Hall, D. M. (1969). A practical approach to the taxonomy of actinomycetes isolated from soil. *In* 'The Soil Ecosystem' (Ed. J. G. Sheals), pp. 107–117. Systematics Association, London.
Williams, S. T., Goodfellow, M., Alderson, G., Wellington, E. N. H., Sneath, P. H. A., and Sackin, M. J. (1983a). Numerical classification of *Streptomyces* and related genera. *Journal of General Microbiology* **129,** 1743–1813.
Williams, S.T., Goodfellow, M., Wellington, E. M. H., Vickers, J. C., Alderson, G., Sneath, P. H. A., Sackin, M. J., and Mortimer, A. M. (1983b). A probability matrix for identification of some streptomycetes. *Journal of General Microbiology* **129,** 1815–1830.
Williams, S. T., Goodfellow, M., and Vickers, J. C. (1984). New microbes from old habitats? *In* 'The Microbe 1984, II: Prokaryotes and Eukaryotes' (Eds. D. P. Kelly and N. G. Carr), pp. 219–256. Cambridge Univ. Press, Cambridge.
Witzel, K.-P., Krambeck, H. J., and Overbeck, H. J. (1981). On the structure of bacterial communities in lakes and rivers—a comparison with numerical taxonomy of isolates. *Verhandlungen—Internationale Vereinigung für Theoretiqe und Angewandte Limnologie* **21,** 1365–1370.
Witzel, K.-P., Overbeck, H. J., and Moaledj, K. (1982a). Microbial communities in Lake Plussee—an analysis with numerical taxonomy of isolates. *Archiv für Hydrobiologie* **94,** 38–52.
Witzel, K.-P., Moaledj, K., and Overbeck, H. J. (1982b). A numerical taxonomic comparison of obligocarbophilic and saprophytic bacteria isolated from Lake Plussee. *Archiv für Mikrobiologie* **95,** 507–520.

9

The Application and Relevance of Nonhierarchic Methods in Bacterial Taxonomy

G. ALDERSON

School of Medical Sciences, University of Bradford, Bradford, UK

Introduction

Computer-assisted numerical methods used to group taxonomic units into taxa on the basis of their character states have been termed *numerical taxonomic* methods or *taxometrics*. In bacterial systematics, a literature search would yield more than 600 publications based on such methods (Sneath, 1962, 1972, 1976; Colwell, 1970; Sneath and Sokal, 1973; Jones and Sackin, 1980; and MacDonell and Colwell, Chapter 6; Priest and Barbour, Chapter 7; Goodfellow and Dickinson, Chapter 8). The vast majority of these publications have been concerned with hierarchic cluster analysis of bacteriological data. Thus, in bacteriology, the term numerical taxonomy is almost exclusively associated with cluster analysis, usually of *Q* matrix data (see Fig. 1), with results being presented in the now familiar form of the dendrogram (phenogram) and shaded diagram.

The primary aim of this contribution is to indicate some of the relationships between hierarchic and nonhierarchic methods and to familiarise readers with some of the nonhierarchic methods which may be useful in bacterial systematics. Finally, although it is impractical to mention all of the applications of nonhierarchic methods in bacteriology, some studies which have used nonhierarchic methods will be considered in an attempt to demonstrate successes and attendant problems. The planning and execution of a numerical taxonomic study will not be considered, but clearly as with any taxonomic task, both must be done with care. The reader is referred to the papers of Sneath (1972, 1978a) and Jones and Sackin (1980) for general overviews of the numerical taxonomic procedure.

Background

Bacterial taxonomists aim to find the natural pattern of the distribution of organisms (operational taxonomic units, OTUs) and taxa (groups of OTUs) in phe-

ISBN 0-12-289665-3

netic space. They wish to discover the taxonomic structure in their data. In numerical taxonomy information on such patterns is stored in resemblance (similarity or dissimiliarity) matrices (Fig. 1). In bacterial taxonomy the number of OTUs (t) and characters (test results, n) stored in such matrices is usually large and the pattern multidimensional. Methods of analysis of resemblance matrices aim to represent the pattern of relationships in a comprehensible form, with a much reduced number of dimensions. Many methods are available, but cluster analysis has been most commonly used in bacterial systematics. Ordinations and other nonhierarchic techniques are used infrequently and, although graphs and trees were introduced as early as 1951 (Florek *et al.*, 1951a,b), these methods are rarely found in the representation of bacterial taxonomies.

Both cluster analysis and ordination are branches of multivariate statistics. Cluster analysis includes many different techniques for finding compact clusters in phenetic hyperspace. However, all such techniques enable the user to produce a classification of OTUs *directly* from an explicit separation of OTUs into discrete clusters of similar entities. Clustering algorithms include those familiar to bacterial taxonomists, such as single-linkage and average-linkage techniques. The latter are agglomerative methods, which are easy to compute and which produce nested, strictly hierarchic groupings, often represented graphically in the form of a two-dimensional dendrogram. In ordination, however, OTUs are arranged into some order but are not split into distinct groups. A classification *as such* is not produced; clusters are not formed but may be made by eye later, from the indications of appropriate taxonomic structure found in two- or three-dimensional ordination diagrams. Ordination is the word used by ecologists for the same group of analyses termed multidimensional scaling by psychologists. There does not seem to be any specific statistical terminology, but the principle of ordination is to find the best representation, using a reduced number of dimensions, of distances in a many-dimensional phenetic space. It is because representation of any set of OTUs with respect to more than three characters is not possible that mathematical means have been devised to summarise the information about relationships implied by all the characters.

In biological systematics, a group of methods which includes single- and average-linkage techniques was given the acronym of SAHN (sequential, agglomerative, hierarchic, nonoverlapping clustering methods) by Sneath and Sokal (1973). These hierarchic methods are frequently employed to find taxonomic structure, but all clustering methods do not by nature produce hierarchic classifications. Nonhierarchic taxonomic procedures include ordinations as well as some nonhierarchic clustering techniques, such as those of Lance and Williams (1967) and Rogers and Tanimoto (1960). In practice, however, the structure of biological classifications is hierarchic. These classical, or traditional, hierarchies have their roots in the time of Linnaeus; the Linnaean system required mutually exclusive and hierarchically ordered classes. The classes are usually sharply defined with gaps in the spectrum of phenetic variation, and such classifications have the advantage that taxa may be compared at any desired level. The Linnaean system allows scientists to succumb to that often felt human desire to put things into neat little boxes. In this case the things are OTUs, more specifically, bacteria. It is familiarity with a classical system and a natural conservatism often found in the scientific community which has been partly responsible for maintaining the Linnaean system in biology.

Some biologists have broken free from the 'box' syndrome. Curtis (1959) and colleagues were among the forerunners in discussing the concept of a continuum of forms in plant ecology. These workers employed ordination methods. In bacterial ecology, Brisbane and Rovira (1961) considered that soil isolates formed a continuous spectrum rather than a series of groups, and other preliminary studies such as those of Gyllenberg and Raumamaa (1966) supported this view. Ordinations had been used by behavioural scientists (Hotelling, 1933a,b), but it was a botanist, D. W. Goodall (1953), who coined the term. Many new methods for ordination followed, but these were not widely applied in biological systematics.

Disenchantment with traditional hierarchic representations of relationships in biological systematics has come slowly and more recently. Growing criticisms have stemmed from the realisation that hierarchic classifications may be poor representations of actual phenetic relationships found in nature, and of the taxonomic structure stored in biological data. Techniques devised to measure stress—that is, how well a classification represents the data from which it was derived—have been instrumental in this process. Examples of such 'optimality criteria' (Sneath and Sokal, 1973) include cophenetic correlation (Sokal and Rohlf, 1962) and more recently the *W* test (Sneath, 1977).

Boyce (1969) concluded that ordination, using principal components analysis, gave a better representation of data on homonid skulls than did phenograms derived from cluster analyses. Phenograms have also been shown to be inadequate representations of phenetic analyses by Rohlf (1967, 1968). Jardine and Sibson (1968) have suggested that hierarchic but overlapping structures could be a classificatory compromise, but stated that their methods would require some ordination procedure to give an adequate pictorial representation of data. Sneath and Sokal (1973) commented that for data which could not be represented sufficiently well in a nonoverlapping hierarchy, ordination was sufficient to indicate relevant aspects of phenetic relationships between organisms. Thus, for some data sets far better representations may be obtained by summarising the data in an ordination. Some idea of the criticisms that can be levelled at hierarchic and nonhierarchic approaches in taxonomy are shown in Table 1. It is easy to see that both approaches have their drawbacks. Ordinations, however, are the antithesis of nested hierarchic classifications. If biologists were to produce major classifications based on ordinations it would not pass unnoticed, as these would not be

Table 1. *Relative merits of hierarchic and nonhierarchic methods of classification*[a,b]

Hierarchic (SAHN cluster analyses)	Nonhierarchic (ordinations)
Results of cluster analyses are usually expressed as dendrograms/phenograms. Clusters of OTUs are shown as tightly arranged twigs on long stems or branches.	Clusters are not formed but must be made by eye, by visual inspection of the ordination plots or models, often with the guide of previous knowledge or analysis.
Phenograms allow the comparison of taxa at any desired level. A classification may be directly produced.	Classifications are not produced directly.
Phenograms are useful for summarising taxonomic relationships.	Ordinations are a useful aid for understanding taxonomic structure and are of value particularly in ecology.
Loss of information to achieve the simplicity of a two-dimensional phenogram may obscure relationships or give a distorted view of data because of group size dependency.	Reduction in dimensionality in ordinations involves loss of information; distortion of relationships may result from inadequacies of projection. Relationships obscured in a two-dimensional phenogram *may* be revealed in the space of the first three dimensions of an ordination.
Phenograms are notorious for distorting distances between major groups, as much of the information on the metric properties of the object space may be lost in hierarchic classification.	Ordinations usually give faithful representation of distance between major groups.
Phenograms usually produce faithful representations of the distances between close neighbours.	Ordinations may falsify distances between close neighbours; the finer divisions thus obscured may have significance in classification and ecology. NMMS gives the best balance between inter- and intra-cluster distances.
Given a suitable clustering algorithm, a phenogram of a hierarchic classification can be produced even when the data do not warrant it. All data sets do not lend themselves to such an arrangement, and distortion of true phenetic relationships can result.	Data may be too complex in structure for display in the first three dimensions of an ordination to be of major value. However, ordinations can be an excellent means of confirming, or otherwise, whether suggested hierarchic groupings are really indicated by the observations.
Optimality criteria are readily available to assess the goodness of the classifica-	Classification relies heavily on interpretation; reduction in dimensionality may

Table 1. (*Continued*)

Hierarchic (SAHN cluster analyses)	Nonhierarchic (ordinations)
tion (cophenetic correlation, *W* test, etc.).	produce overlapping and the risk of not recognising clusters as distinct, especially in two-dimensional plots. Limited methodology for detection of overlap can make interpretation difficult.

[a]Based on information from Williams and Dale (1965); Gower (1966, 1969a,b); Rohlf (1967, 1968, 1970); Boyce (1969); Sneath (1972, 1978a, 1980, 1983); Sneath and Sokal (1973); Marriott (1974); Clifford and Stephenson (1975); Alderson *et al.* (1984).

[b]OTU, Operational taxonomic unit; NMMS, nonmetric multidimensional scaling; SAHN, sequential, agglomerative, hierarchic, nonoverlapping.

Linnaean. To reject altogether nested nonoverlapping hierarchic classification would require a consensus of the biological community.

Nevertheless, looking at biological data via a transformed space has become a legitimate scientific activity. Marriott (1974) suggested that the results of any cluster analysis, whatever the method used, should be published with scatter diagrams based on an ordination to illustrate how the clusters found related to the distribution of the OTUs. Such a strategy would leave the reader to judge how well the proposed clusters reflected the structure of the data. At present, there is no available method which permits one to look at a resemblance matrix and make the decision as to whether ordination or clustering would be better employed. Perhaps taxonomists should take the advice of Marriott and several others who considered clustering and ordination to be complementary procedures, rather than alternatives in classification (Gower, 1969a; Sneath, 1972, 1978a; Sneath and Sokal, 1973; Marriott, 1974; Clifford and Stephenson, 1975; Alderson *et al.*, 1984). A more experimental and less conservative approach to taxonomy is needed, for far too long the subject has been considered one of the more boring, though essential, biological disciplines.

Nonhierarchic Methods

Introduction

Ordinations are all nonhierarchic multidimensional scaling methods. They may be derived from either *Q* or *R* matrices (see Fig. 1). In all cases mathematical techniques are used to find the best representation, using a reduced number of dimensions, of distances in a many-dimensional phenetic (*A*) space. A number of

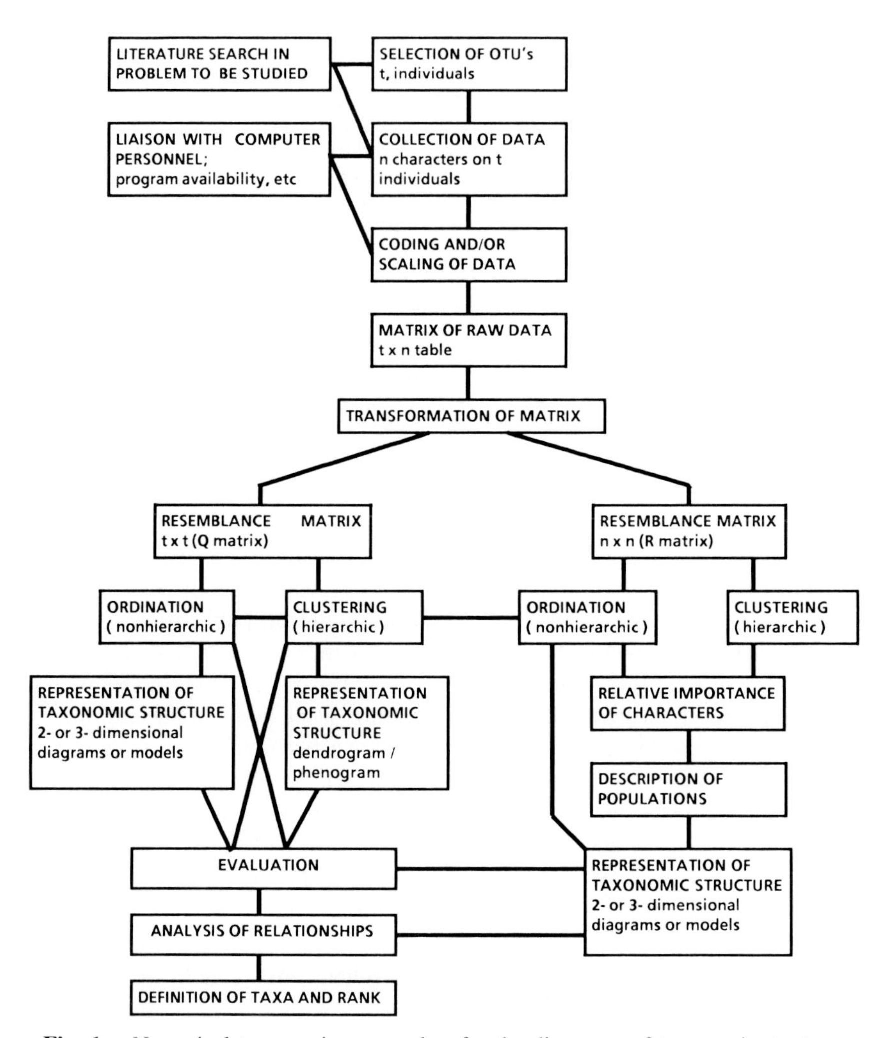

Fig. 1. Numerical taxonomic approaches for the discovery of taxonomic struture.

reviews are available for readers interested in mathematical derivations (Gower, 1966; Harman, 1967; Krzanowski, 1971; Sneath and Sokal, 1973; Marriott, 1974; Kendall, 1975; Clifford and Stephenson, 1975; Dunn and Everitt, 1982).

The results of ordinations may be displayed as taxonomic maps in two-dimensions (see Figs. 2, 3, 4, and 6) or as three-dimensional taxometric models (fish tank models, see Fig. 7), or as stereograms (Rohlf, 1968). Clusters are not formed and must be made by visual inspection, for example, within the context of the results of an earlier cluster analysis.

Duality of Q *and* R *Techniques*

To understand the wide-ranging application of nonhierarchic methods in microbial taxonomy it is necessary to understand the relationship between Q and R techniques. The origin of all numerical analysis is the table of test results of n characters on t individuals, or OTUs. This $t \times n$ table may be transformed in one of two ways: by forming correlations between all pairs of characters or by calculating the associations between all pairs of OTUs over all characters (see Fig. 1). The two kinds of transformations have been referred to as R and Q techniques, respectively (Cattell, 1952). Cattell's definitions originated in factor analysis, and since that time confusion has arisen as to the precise use of the terms, especially because factor analyses comparing characters have been used to produce groupings of OTUs (Williams and Dale, 1965). The latter suggested a resolution of the problem by considering an R technique as one which leads to a classification of characters and a Q technique as one which gave a classification of OTUs. The mathematical steps are the same in both cases, and it can be seen that the main emphasis in numerical taxonomy has been with Q techniques.

Gower (1966, 1967, 1969b) showed that identical ordinations could be obtained from either Q or R matrices if suitable mathematical transformations were used. Both Q and R techniques can give the coordinates of a set of t individuals in multidimensional space. Thus, in some cases Q and R techniques are dual to one another. Gower (1966) emphasised the duality of principal component analysis, operating on the R matrix, and principal coordinate analysis, based on the Q matrix. For every R analysis in common use there exists a corresponding Q analysis, but the converse does not apply. R techniques may sometimes be more convenient to compute than the corresponding Q techniques and are useful when there are fewer characters than OTUs; less computational space is required. Indeed, for the taxonomist or ecologist it is possible to choose whichever analysis is easier to compute given the particular data set.

Some of the earliest bacteriological work based on hierarchic and nonhierarchic methods was aimed at investigating relationships between character correlations and the taxonomic structure of the microbes under study. Pike (1965a,b) examined character correlations in the Micrococcaceae using cluster analysis, and Gyllenberg (1965a, 1967) used ordinations to examine characteristics of soil isolates. In such studies major R factors or R clusters (i.e., clusters of characters) often correspond with major Q factors or Q clusters, but fine detail is not usually apparent from R studies.

R analyses have been used in many other ways, for example, to reduce the size of a data base used for a subsequent cluster analysis. The early study of Hill *et al.* (1965) is a good example, but Sneath and Sokal (1973) pointed out some of the problems involved in estimating phenetic similarity on reduced space from factor scores extracted from a factor analysis. For many years R analyses have been found to be a valuable tool in bacterial ecology (Rosswall and Kvillner, 1978;

Table 2. *Multivariate statistics used in bacterial taxonomy*

Primarily for grouping of OTUs (classification)		Primarily for discriminating between groups and/or for identification of unknowns
A *Hierarchic* cluster analysis	**B** *Nonhierarchic* ordinations (multidimensional analyses)	**C** *Nonhierarchic* ordinations (multidimensional analyses requiring an imposed grouping of OTUs)
i Average linkage	**i** Principal component analysis	**i** Discriminant analysis
ii Complete linkage (furthest neighbour)	**ii** Principal factor analysis	**ii** Stepwise discriminant analysis
iii Single linkage	**iii** Multiple factor analysis	**iii** Multiple discriminant analysis (canonical variate analysis)
	iv Principal coordinate analysis	**iv** Taxon-radius models
	v Nonmetric multidimensional scaling	

Holder-Franklin and Wuest, 1983; Goodfellow and Dickinson, Chapter 8). They have also been used in a strictly taxonomic manner for both classification and identification. These applications, along with *Q* ordinations, are considered later. A summary of nonhierarchic methods used in bacterial systematics is shown in Table 2.

Some useful reading on computer packages which contain hierarchic and nonhierarchic techniques of analysis is included in SPSS. (1975), GENSTAT. (1980), and BMDP. (1981).

Analyses Used for the Grouping of Strains (Classification)

Principal Component Analysis. The technique of principal component (PCP) analysis was essentially intended for use on measurements made on a continuous scale, but Gower (1966) has shown that binary data may also be analysed. Analyses are usually carried out on standardised measurements and the components extracted from the correlation matrix. The covariance matrix is thought to be too dependent on scale, but it has been utilised. Commonly *R* analyses are ordinated in *A* space, but *Q* analyses have been published (Skyring *et al.*, 1977).

In PCP analysis an observed set of variables (test results/characters) are transformed to a new set, in which the first few principal components account for the majority of the variability in the original data. Each principal component is uncorrelated with preceding components, and the first dimension (principal axis), corresponding to the largest eigenvalue, is that which expresses the greatest scatter of spread of the OTUs, that is, accounts for the greatest amount of variance from the data. The OTUs are then arranged along this dimension; the next dimension which expresses the next greatest scatter of OTUs is then sought and the OTUs arranged along it. The procedure is repeated until a high proportion of the total variance (usually >75%) of the data is accounted for.

No assumptions are made concerning the original variables. This is a different and possibly more convenient way of expressing the same set of results. The first

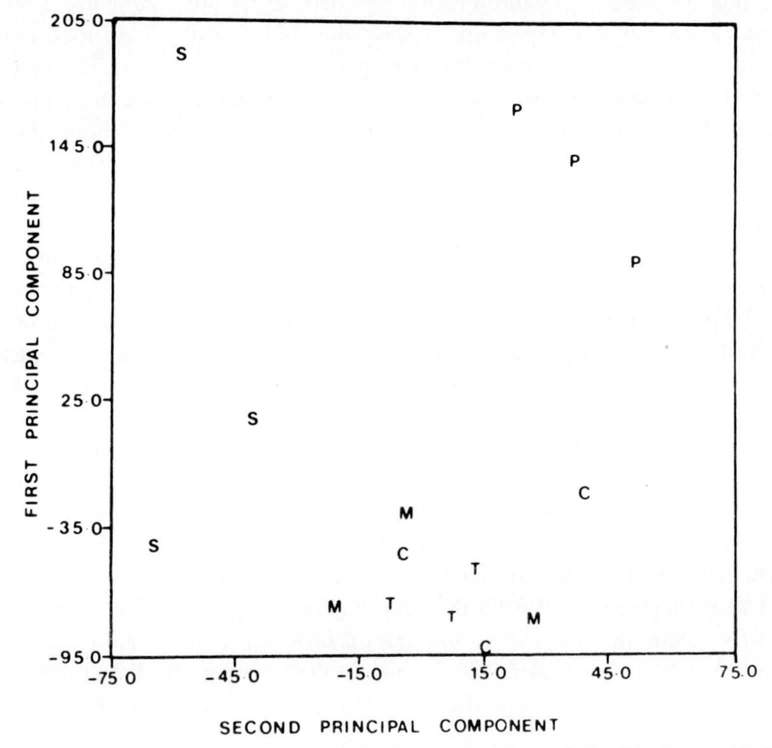

Fig. 2. Principal component analysis of strains of *Bacillus subtilis* (S), *B. pumilis* (P), *B. cereus* (C), *B. thuringiensis* (T), and *B. megaterium* (M). The two principal components represent 88.4% of the variation between samples. Data obtained from pyrolysis gas–liquid chromatography of whole cells. From O'Donnell and Norris (1981), with permission.

few principal components, if they account for the majority of the variance, may allow the remainder to be discarded. Thus, PCP scores that can be obtained for each OTU can be used to replace the original character values in the preparation of a Q matrix before cluster analysis. Clustering on reduced space, however, may be problematical from a theoretical standpoint (Sneath and Sokal, 1973).

Principal component scores may also be used to make two-dimensional scatter plots or three-dimensional models to view directly the relationships between OTUs, and thus to aid classification (Skyring *et al.*, 1977; Moss *et al.*, 1978; Kapperud *et al.*, 1981). They have also been used in the analysis of complex chemical data, as will be seen later (Fig. 2).

Principal Component Analysis and Factor Analysis. Both factor analysis and principal component analysis may be used for data reduction; that is, they are used to express information in a set of observations in a smaller number of components/factors. Unfortunately, the two terms are frequently confused. Thus, some workers, especially those in the United States, have used the term factor analysis as a synonym for principal component analysis. The confusion may have arisen because component analysis is one method used to extract factors for factor analysis while, for the taxonomist, there is often little difference between the results of a PCP and a principal factor (PF) analysis. Harman (1967), however, clearly made the distinction between PCP and PF analyses which has become generally accepted.

In PCP analysis the diagonals of the correlation matrix are unities, but with PF analysis these diagonals are reduced by their uniqueness to communalities, that is, the percentage of variation due to 'common factors'. This is because in PF analysis the variance of each attribute has two portions which are independent; one is unique variance and the other is common-factor variance, which is known in PF analysis as communality. The common factors generate the covariances among the observed variables, while the unique variances contribute only to variances of their particular variable. The objective of PF analysis is to find values of the communalities such that the correlation coefficients remaining in the matrix are reduced to zero, thus reducing the rank (the number of nonzero eigenvalues that can be extracted from a matrix is termed the rank of a matrix) and hence the space in which the data may be represented. Communalities are usually determined indirectly, and several different methods are available (see Harman, 1967). The factors are obtained as the square roots of the magnitude of each communality. The proportion of the variance for which they account is then known.

Dunn and Everitt (1982) commented that PCP analysis is simply a transformation of the coordinate axes of a multivariate system to new orientations, whereas factor analysis is a fundamental model for the covariance structure of observations. Like Gower (1966, 1969a,b), they dismissed factor analyses for biological work in general and for numerical taxonomy in particular. These authors have shown that alternative methods such as PCP analysis under suitable conditions are likely to give similar results to more complex factor analyses.

Multiple Factor Analysis. In multiple factor (MF) analysis, factors, once extracted, are rotated and may be allowed to depart from their orthogonal relationship by becoming oblique or correlated. This contrasts with PCP and PF analyses and other ordinations such as canonical variate analysis and discriminant analysis, in which components or factors usually remain orthogonal or uncorrelated.

The term factor analysis is sometimes used in a general sense and may refer to PCP, PF, or MF analysis. This lack of precision was noted by Harman (1972), but the confusion over the term remains unresolved. Rohlf and Sokal (1962) defined their MF technique as the 'complete centroid method of factor extraction, reextraction of the factors until estimates of the communalities stabilise, rotation of the centroid factor matrix to simple structure'. These authors were also the first to use MF analysis in numerical taxonomy of the Q matrix in a study of relationships amongst bees. Gyllenberg and co-workers (Gyllenberg and Eklund, 1967; Gyllenberg *et al.*, 1967; Pohja and Gyllenberg, 1967; Sundman and Carlberg, 1967; Sundman and Gyllenberg, 1967) and Harman (1972) have championed the use of MF analysis in microbial classification and in the description of microbial populations.

Work using MF analysis in taxonomy is usually carried out on suites of characters in populations of organisms, and as such is a form of R analysis. Similarly, ecological work using MF analysis usually reports R analyses, and it is in ecological work that MF and PF analyses have found a major application in microbiology (Holder-Franklin and Wuest, 1983; Goodfellow and Dickinson, Chapter 8). However, Skyring *et al.* (1977) used PCP analysis on a Q matrix derived from tests on dissimilatory sulphate-reducing bacteria. Clifford and Stephenson (1975) have suggested that there are occasions when factor analyses, PF and MF, would be preferable to PCP analysis, especially in ecological studies where data include a large proportion of attributes that are only weakly correlated. In practice, all three techniques have a drawback in that missing values in the raw data are not usually allowed. If no-comparison (NC) coding is available, PCP analysis results may be disturbed more by their inclusion than a principal coordinate analysis (PCO; Rohlf, 1972).

Principal Coordinate Analysis. This is a classical multidimensional scaling technique developed by Gower (1966), who was dissatisfied with the application of PCP and MF analyses to Q matrices in biological classification. Gower devised a sound basis for ordinations of the Q matrix in which eigenvalues and eigenvectors were calculated directly from a Q matrix of dissimilarity. Unlike

PCP analysis, which is only relevant when a Euclidean metric is considered suitable for the observations, PCO analysis is an ordination which is applicable to relationships between a set of OTUs in space whether the distances are Euclidean or not. If the observed proximities are Euclidean the results of a PCO analysis are equivalent to those of a PCP analysis (Gower, 1966; Dunn and Everitt, 1982; Alderson *et al.*, 1984). If the Q matrix has already been formed for a cluster analysis, and there is no corresponding R matrix, it is convenient to use PCO analysis as an ordination for taxonomic purposes (Alderson *et al.*, 1984). When R matrices exist, they may be smaller than the Q matrix, and in such cases an R matrix analysis is preferred, especially where large numbers of OTUs are involved. The availability of large-capacity, high-speed computers has removed most restrictions on the size of both data sets and Q matrices. Thus, when there are missing test results—a common phenomenon in bacteriology—similarity measures remain reliable and robust, whereas replacing missing values by estimates of values or by guesses is not satisfactory (Marriott, 1974).

Principal coordinate analyses have been used in bacterial classification (Logan and Berkeley, 1981; Bridge and Sneath, 1983; Alderson *et al.*, 1984; Figs. 4 and 5) and in the analysis of complex bacteriological data (Shute *et al.*, 1984). It also seems likely that PCP analysis could be useful in the examination of immunologic data where the matrix is by nature in distances (Sneath and Sokal, 1973), and it may be relevant in the production of identification schemes (Gower, 1968; Ross, 1975; Logan and Berkeley, 1981).

Nonmetric Multidimensional Scaling. This is a general ordination technique which has found some use in numerical taxonomy and has also been referred to as nonlinear mapping (Kruskal, 1964a,b). In this case *actual* numerical values of dissimilarity are not used; it is their rank order that is important. Again scatter plots of ordinations can be a powerful tool in summarising large amounts of data. Nonmetric multidimensional scaling (NMMS) may be used as an alternative to PCO analysis, and results may be very close to those obtained in such analyses. Dunn and Everitt (1982) have suggested that NMMS may be of more value than other ordinations in detecting relationships between close neighbours. Nonmetric multidimensional scaling/nonlinear mapping has been used in bacterial classification (see Fig. 3; Bonde, 1981) to display discriminant analyses of complex data, for example to visualise more than three discriminant functions (Wieten *et al.*, 1983).

Analyses Used for Discriminating between Groups or for the Identification of Unknowns

Discriminant Analysis. Discriminant analysis or discriminant function (DF) analysis may be used when two or more groups of OTUs are to be distinguished statistically. Many different analyses are available, and some may be viewed as

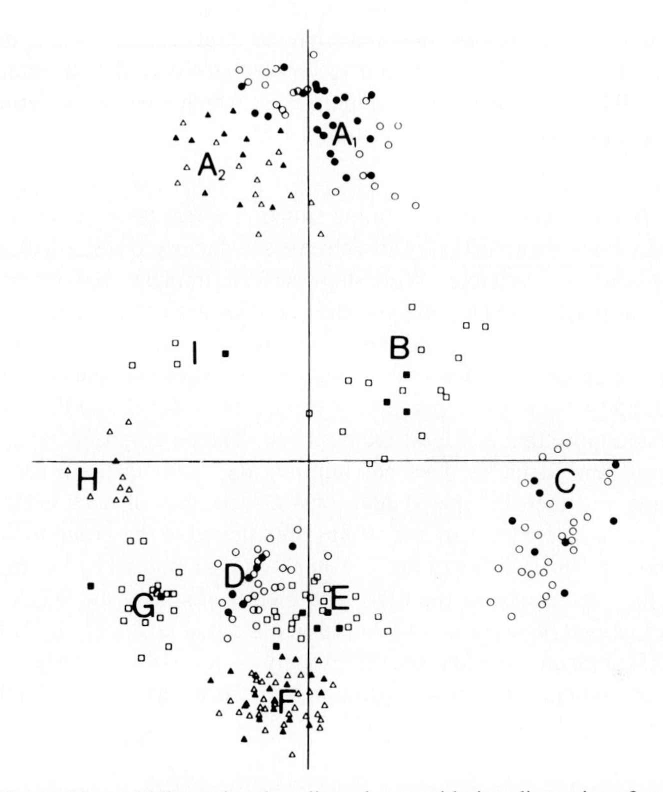

Fig. 3. Nonmetric multidimensional scaling plot considering dimensions 2 and 3. The analysis was of 231 selected strains of *Bacillus* including marine isolates which are indicated in black. Data were obtained using conventional phenetic bacteriological tests. Ten clusters are indicated. From Bonde (1981), with permission.

extensions of taxon-radius models in identification (e.g., the simultaneous keys of Gyllenberg, 1964, 1965b).

In classical DF analysis (Fisher, 1936) a linear discriminant function is formed, so that it has maximal variance between groups relative to the pooled variance within groups. Thus, a linear discriminant function of the characters describing the OTUs in one taxon will have high values compared with the low values for the OTUs in the second taxon. The discriminant function will usually serve as a much better discriminant of the two taxa than does any character taken singly. With DF analysis the data are reduced; scatter plots may be used to visualise the results of the first two discriminant functions, and distances in the transformed space are based on Mahalanobis distance D^2, where D is the distance between the two population means, after eliminating the effects of correlations. Discriminant functions have also been displayed as nonlinear maps, when more than three discriminant functions may be visualised (Weiten *et al.*, 1983). Classical DF analysis has been said to be best suited to the assignment of an unknown OTU to one of two overlapping taxa, which are usually defined on the basis of quantitative characters (Sneath, 1978b).

Stepwise Discriminant Analysis. In the situation where there are more discriminating variablies than necessary to achieve satisfactory discrimination, a stepwise procedure is available. With stepwise discriminant analysis or stepwise discriminant function (SDF) analysis the variables are entered into a 'classification function' one at a time, and thus the variables used in computing the linear functions are chosen in a stepwise manner. Selection of variables is made so that those that add most to the separation of groups are entered into the discriminant function and those that add least are removed. The process is stopped when the addition of new variables does not improve the classification. The computer programme will then list the Mahalanobis D^2 distance of each OTU from the centre of the *a priori* groups, and strains are allotted to the group to which they are the closest. The stability of the SDF analysis is monitored by leaving out each OTU in turn, recalculating the SDF, and then reallocating the OTU.

A discriminant function was first used in microbial taxonomy by Hill *et al.* in 1965. Discriminant function and SDF analyses are still regularly used in the analysis of complex data from pyrograms in the identification of micro-organisms (O'Donnell and Norris, 1981).

Multiple Discrimination Analysis. Multiple discrimination analysis or canonical variate (CV) analysis is another multivariate technique regularly used in the analysis of complex microbiological data. This technique again allows the investigation of relationships of OTUs in multidimensional space. In this case, as in SDF analysis, a grouping of OTUs is required prior to an analysis where distances are defined using the Mahalanobis D^2 statistic. As the method is applicable only to groups that may be considered to have a common within-group dispersion matrix, the taxonomic use of CV analysis is limited to discrimination, identification, or infra-specific taxonomy.

As with PCP analysis, transformed axes are sought, but in CV analysis the direction of the first axis is that of the greatest variability between the means of the different taxa. The second axis is chosen to be orthogonal to the first and in the direction of the next greatest variability, and so on. A set of canonical variate means for each taxon is found, and for each OTU requiring identification a set of canonical variate scores with which to determine its assignment to its closest taxon is determined. Plots of taxon means in CV space (usually two or three variates) can also be useful for visualising relationships between taxa (O'Donnell and Norris, 1981; Fig. 6). Bonde (1978, 1981) has shown that CV analysis may

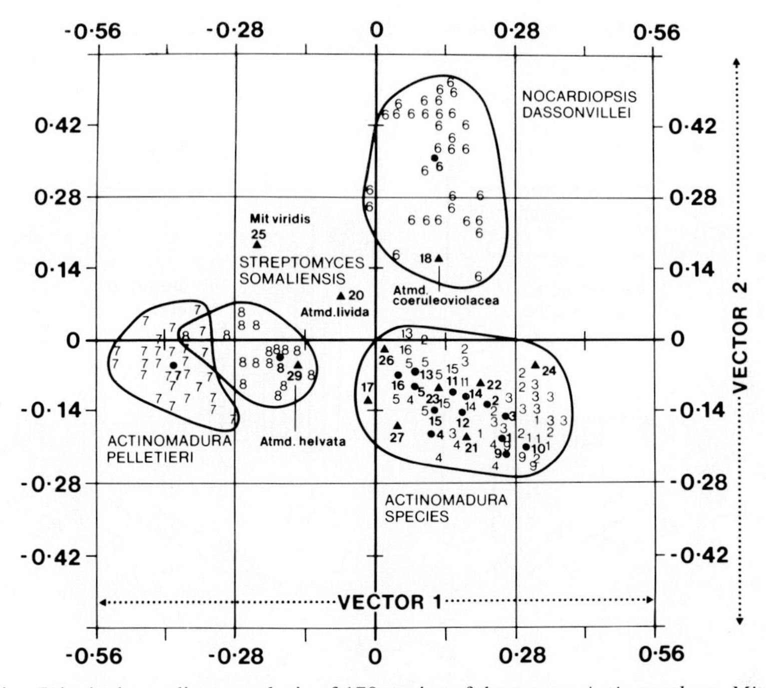

Fig. 4. Principal coordinate analysis of 170 strains of the genera *Actinomadura, Microbispora, Microtetraspora, Nocardiopsis,* and *Streptomyces.* Code numbers of individual strains (6, 7, etc.) are based on clustering from a previous analysis; ●, the mean of each cluster; ▲, a type strain. Data were obtained from conventional phenetic bacteriological tests. Note the overlap between strains labelled 7 and 8 in a plot based on the first two dimensions only. From Alderson *et al.* (1984), with permission.

be employed to aid the selection of diagnostic tests from numerical taxonomic classifications.

Taxon-radius Models. These are, like CV and DF analyses, probabilistic methods of identification which require an *a priori* group structure and computer assistance. An early model employing correlation coefficients and an *A* space reduced by PCP analysis was proposed by Gyllenberg (1964, 1965b). Sneath and Sokal (1973) suggested a more general scheme where distances between an unknown OTU and the centroid of each taxon hypersphere were measured. The taxon-radius model was fully explained by Sneath (1978b), and this and other computer-assisted identification models are dealt with by Holmes and Hill (Chapter 10). Taxon-radius models represent powerful identification systems which can be employed in conventional bacteriology. They are especially useful as they allow a reduction in the number of tests necessary to produce a result compared with conventional schemes (Sneath, 1978b).

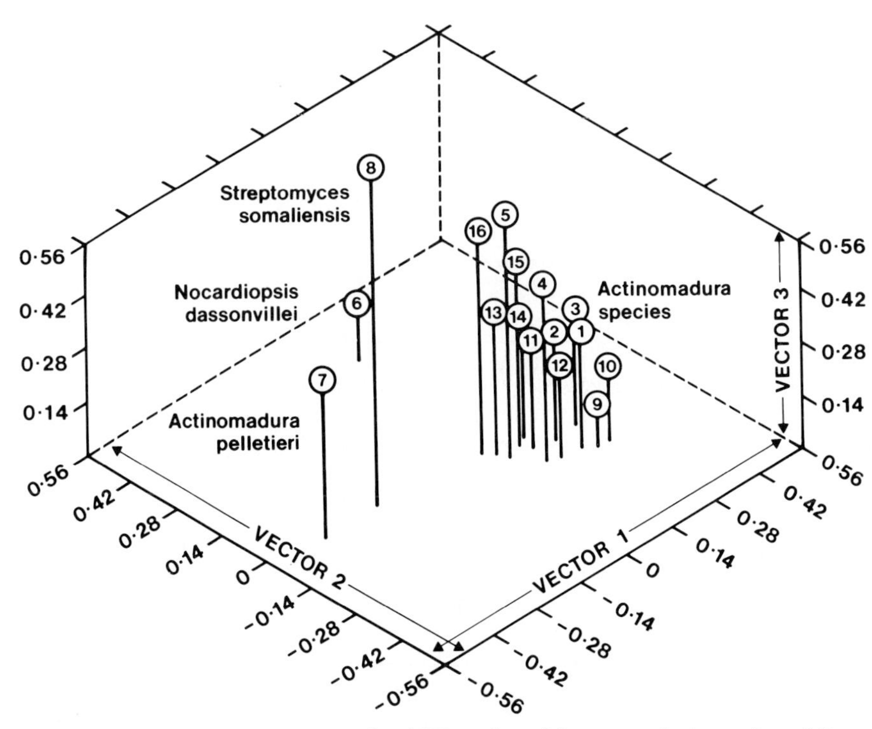

Fig. 5. Principal coordinate analysis of 170 strains of the genera *Actinomadura, Microbispora, Microtetraspora, Nocardiopsis,* and *Streptomyces*. The numbers (⑧, etc.) represent the position of the hypothetical mean organisms of clusters based on a previous hierarchic analysis. Data were obtained from conventional phenetic bacteriological tests. Note that a consideration of the third dimension resolves the overlap between clusters 7 and 8 seen in Fig. 4. From Alderson *et al.* (1984), with permission.

Interpreting Ordinations

The objective of taxonomic studies using ordinations is to display OTUs in a reduced space. Both ordination and clustering produce summaries of the variation in *A* space. In contrast to hierarchic cluster analyses, however, none of the techniques discussed in this chapter produce a classification *per se;* taxonomic grouping must be made by eye, either with the guidance of previous knowledge or with the help of a previous analysis. The dangers of establishing taxa simply by subjective inspection of ordination diagrams or maps have been pointed out many times (Sneath, 1972, 1980; Sneath and Sokal, 1973).

Problems of interpretation of ordinations include cluster overlap: clusters that are distinct in the full hyperspace may overlap in low-dimensional plots (Sneath and Sokal, 1973; Clifford and Stephenson, 1975; Sneath, 1980). Resolution of overlap in two dimensions was seen only by consideration of the third dimension in a PCO analysis of some sporoactinomycetes (Alderson *et al.*, 1984), but if clusters are numerous and dense, ordination may not give a simple low-dimensional result (Williams and Lance, 1968; Sneath, 1980). Simple techniques are not available to examine similarity matrices prior to ordination to see if ordination is an appropriate approach, but after the analyses have been performed evaluation is possible. Clifford and Stephenson (1975) suggested that if only a relatively small proportion of the variance was revealed in the first three axes of an ordination, then it was probable that the original data were too complex for the technique to be of great immediate value. Sneath (1980) showed how the risk of undetected overlap could be calculated in a PCO, PCP, or CV analysis.

Taxonomic relationships are multidimensional, and any pictorial representation, especially a two-dimensional plot, will involve the loss of information and the possibility of some relationships being distorted or obscured (Clifford and Stephenson, 1975). Both hierarchic cluster analyses represented as phenograms and low-dimensional ordination plots have their inadequacies (Table 1).

It is not unreasonable to expect that bacterial taxonomists involved in classification will use *both* ordinations and cluster analyses to investigate patterns of variation and taxonomic structure.

Application of Nonhierarchic Methods in Bacterial Taxonomy

Classification and Description of Bacterial Populations

Description of Populations in Bacterial Ecology. In the 1960s two independent groups were at the forefront in exploiting nonhierarchic methods in bacteriology; both placed particular emphasis on the description and general grouping of natural populations, rather than on strict taxonomic work. Gyllenberg and colleagues in Finland employed both PCP and MF analyses but added to confusion over terms by calling the PCP technique used by Hill *et al.* (1965) a factor analysis (Sundman and Gyllenberg, 1967). Gyllenberg (1964, 1965b) suggested that bacterial populations, could well be described by a geometric model. Correlation coefficients were calculated from raw data and PCP analysis used to reduce the *A* space to three relevant dimensions. This aided population description (Gyllenberg, 1964) and allowed the definition of a 'final identification space' in which the centre of gravity and the radius of each taxon was computed in order to produce an identification system (Gyllenberg, 1965b; Gyllenberg and Raumamaa, 1966). The system was related to DF analysis and to the taxon-radius model.

Factor analysis (MF) was attempted and PCP analysis used to extract factors in an effort to reveal character correlations in data on four very different groups of bacterial strains (Gyllenberg, 1965a). It was suggested that character correlations might provide a sound basis for natural classification and hence for reliable identification (Gyllenberg, 1965a; Harman, 1972). Sundman and Gyllenberg (1967) published the first of a series of papers which explored the use of factor analysis in microbiology (Gyllenberg and Eklund, 1967; Gyllenberg *et al.*, 1967; Pohja and Gyllenberg, 1967; Sundman and Carlberg, 1967), and this work continued into the 1970s (Sundman 1968, 1970, 1973; Gyllenberg, 1970, 1973).

Work of a similar nature, applying nonhierarchic techniques to the grouping of bacterial populations, was being carried out in Canada by a team centred around Quadling. A two-stage PCP procedure was evaluated using named cultures of Gram-positive and Gram-negative bacteria (Quadling and Hopkins, 1967). Separate PCP analyses were made on attribute complexes formed by applying cluster analysis to an *R* matrix. Forty-seven principal components vector scores were allotted to the relevant cultures and a second PCP analysis applied to the reduced space. The first four principal components of this analysis were plotted for each culture before a final cycle of clustering was used to detect clusters of strains in the further reduced space. Condensation of the data was said to retain only the features responsible for overall effective discrimination between strains. However, deficiencies were reported in minor but not in major groupings (Quadling and Hopkins, 1967). This work was continued using soil bacteria, named marker strains, and fresh isolates, and the ordinations were compared with conventional *Q* analyses using clustering techniques (Skyring and Quadling, 1969a,b, 1970; Skyring *et al.*, 1971).

Most of the work done by the two different groups had a greater impact on bacterial ecology (Goodfellow and Dickinson, Chapter 8) than on bacterial taxonomy. Subsequent work (Håstein and Smith, 1977; Skyring *et al.*, 1977; Rosswall and Kvillner, 1978) rekindled interest in the description of microbial populations using ordinations. The stress has been on understanding *how* ecological processes may be regulated and how ordination may aid this understanding (Rosswall and Kvillner, 1978). Håstein and Smith (1977) employed the ordination approach to detect subgroups of *Vibrio anguillarum* from diseased fish. An attempt was made to associate the subgroups with factors relevant to the disease.

Attempts to Classify. Some of the first publications in bacteriology which reported the use of nonhierarchic techniques in classification were on the Gram-positive cocci. Defayolle and Colobert (1962) compared results from an MF analysis of *R* matrix data with those from a *Q* hierarchic analysis of *Streptococcus faecalis* strains. The factor and the cluster analyses agreed with traditional results and showed that strains belonging to the species could be divided into three clear subgroups. Hill *et al.* (1965) examined data from 80 tests on 49 strains of the Micrococcaceae and found that results from various hierarchic analyses were substantially the same as those obtained from clusterings with the first five dimensions from PCP analysis. Thus, as with the work of Quadling and Hopkins (1967), ordination was used to reduce dimensionality before clustering. Hill *et al.* (1965) found that the projection of the first two axes of the PCP analysis was sufficient to separate the major groups. Principal component analysis also facilitated the recognition of taxonomically important characters.

Harman (1972) championed classical MF analysis for taxonomic purposes. He proposed that the technique would help in classification because the principle of objectivity would be better served by determining factors from the relationships among the characters. As was seen in Table 1, ordinations of all kinds have their drawbacks but in fact, apart from early work, it is nonhierarchic methods other than MF and PF analyses that have been applied with increasing regularity.

One of the first numerical taxonomic studies on the genus *Bacillus* employed MF analysis, and the results were presented as two- and three-dimensional diagrams of the positions of 18 clusters of strains (Defayolle *et al.*, 1968). The first three factors accounted for 70% of the total variance, and it was noted that *B. pumilus* and *B. subtilis* strains were not separated in the first three dimensions. The similarity of these two species had been noted by Gordon *et al.* (1973).

The genus *Bacillus* comprises a complex and heterogeneous group of organisms and as such has received considerable attention from those interested in the application of multivariate statistics to taxonomic problems. Bonde (1978, 1981) had problems in identifying marine isolates of *Bacillus* using conventional published schemes, and set about a numerical taxonomic study of more than 400 strains. These included many marker strains from the genus as well as new marine isolates. Results from 77 conventional tests were subjected to NMMS analysis (Kruskal, 1964a,b), and, since two-dimensional plots gave inconclusive results, 20 serial scalings were used to produce a numerical classification of 231 of the strains. More than 70% of the marine isolates clustered with other well-characterised *Bacillus* species (Bonde, 1981; Fig. 3), but phenon A_2 was recognised as a novel taxon within the genus. Canonical variates were used to select tests from the NMMS classification for identification (Bonde 1978, 1981). The first four eigenvalues accounted for 90% of the information in the complete data set, and 25 of the tests were finally selected for a dichotomous key. All of the original isolates were then identified and related to the source of the original marine sample. This work showed just how a range of multivariate statistics can be used to good effect in classification, identification, and ecology.

Logan and Berkeley (1981) also used ordinations on data collected from *Bacillus* strains. These workers used the API system in an attempt to improve the lot of those involved in the unenviable task of identifying 'aerobic spore-forming rods'. API 20E, 50E, API ZYM, ZYM II, AP1, AP2, and AP3 kits were used to provide reproducible tests, and the final data matrix contained results from 119 such tests, plus 20 morphological and physiological tests, on 600 cultures. The data were analysed in six separate runs on strains forming cluster groups in an initial hierarchic cluster analysis. The ordination selected was PCO analysis, but the first three vectors accounted for only 37 to 47% of the total variation in the six analyses. Data were presented in the form of a couple of two-dimensional plots for each cluster group. Cluster group IV included strains of *B. subtilis, B. licheniformis, B. pumilis, B. megaterium,* and '*B. amyloliquefaciens*', but ordination by PCO analysis showed that *B. subtilis* and '*B. amyloliquefaciens*' formed a single group related to *B. licheniformis* and the *B. pumilis* group. This result was in disagreement with that of O'Donnell *et al.* (1980), who argued for the recognition of '*B. amyloliquefaciens*' as a species separate from *B. subtilis* on the basis of CV analysis of data from pyrolysis products. Strains of *B. pumilis* and *B. subtilis,* however, were clearly separated in the study of Logan and Berkeley (1981). The latter also suggested that a further task would be to use the PCO analysis and the technique of Gower (1968) to aid identification of new isolates.

Not all of the experimental approaches to classification come from researchers in the taxonomy of the genus *Bacillus.* Moss *et al.* (1978) studied 186 chromobacteria from a lowland river and examined their phenetic data by clustering and PCP analysis based on a transformed distance matrix, as well as on the correlation matrix. In both ordinations strains of particular interest, previously unidentified diffuse purple colonial types, occupied a well-defined space in a two-dimensional plot. These isolates were also clearly separated by the hierarchic approach, and the grouping of other strains was also very similar. Results from these numerical studies, in conjunction with percentage guanine + cytosine content of DNA, allowed Moss and colleagues to describe a new species, *Chromobacterium fluviatile.*

Kapperud *et al.* (1978) used essentially the same techniques, cluster and PCP analyses, in a numerical phenetic study of *Yersinia enterocolitica* and related strains. The main emphasis in the interpretation of results from 46 tests on 332 strains was on the PCP data, and it was proposed that *Y. kristensenii* strains deserved species status whilst strains of *Y. enterocolitica,* '*Y. frederiksenii*', and '*Y. intermedia*' formed a phenotypic continuum in a large heterogeneous grouping.

Gyllenberg was a forerunner in the application of nonhierarchic techniques to the classification of sporoactinomycetes (Gyllenberg *et al.*, 1967; Gyllenberg, 1970). Factor analyses were used on a limited data base in an attempt to recognise significant character correlations for the description of species of *Streptomyces,* but the work led to the definition of a new group (Gyllenberg *et al.*, 1967). A minimal data set was also used in the work of Szulga (1978), who looked at results obtained on *Streptomyces* strains using a variety of numerical approaches. Principal components were said to 'aid the identification of individual strains', but only 11 characters were used, clearly limiting the value of the exercise.

Alderson *et al.* (1984) used PCO analysis as a means of solving a problem of taxonomic structure in the genus *Actinomadura.* Strains of *Streptomyces somaliensis* had been consistently linked with those of *A. pelletieri* in numerical phenetic analyses using clustering, whereas chemotaxonomic data allow a clear distinction to be drawn between the two taxa. The first two dimensions of the PCO analysis (Fig. 4) did not allow a sharp distinction to be made between the two taxa, but when a third dimension was considered (Fig. 5), additional information allowed a clear separation of *A. pelletieri* from *S. somaliensis* (Alderson *et al.*, 1984).

Coryneform bacteria were the subject of an extensive numerical taxonomic study (Seiler, 1983) which used a similar approach to that employed by Logan and Berkeley (1981). Six cluster groups produced in a hierarchic analysis of data on 557 strains were examined in detail using a nonhierarchic technique. Seiler employed a 'linkage-map procedure' and discriminant analysis; computational details were not given, but the program was most likely one for NMMS analysis. The linkage maps provided much greater detail on the structure of subclusters than any of seven different hierarchic techniques.

The Gram-positive cocci were the subject of early nonhierarchic analyses (Defayolle and Colobert, 1962; Hill *et al.*, 1965). Bridge and Sneath (1983) applied both hierarchic and nonhierarchic techniques in a comprehensive study of the genus *Streptococcus.* Principal coordinate analysis of the distances between centroids of the 28 phena found by average-linkage clustering allowed a general view of relationships to be seen (Bridge and Sneath, 1983). The variance on the first three axes was low (38.7%), but the main taxonomic structure was well represented in a three-dimensional model.

Discrimination and Identification of Bacteria

Analysis of Complex Bacteriological Data. The concept of using complex chemical techniques such as pyrolysis for microbial taxonomy first arose with the work of Reiner (1963, 1965) and Garner and Gennaro (1965). Since that time pyrolysis of whole microbial cells has been used increasingly as a tool to differentiate between strains of many different genera. Two approaches have been used to examine pyrolysed cells, gas–liquid chromatography (Py–gc) and mass spectrometry (Py–ms), and the advantages of each approach have been discussed (Quinn, 1976; Gutteridge and Norris, 1979; Irwin, 1982; Gutteridge *et al.*, Chapter 14).

Originally pyrograms or pyrolysis 'fingerprints' were simply evaluated visually. Qualitative differences were reported for bacteria (Cone and Lechowich,

1970; Emswiler and Kotula, 1978), but presence or absence of peaks is usually noticeable only with diverse organisms. For most applications pyrograms are basically similar, and differences are to be found in the quantity of characteristic peaks. When using multivariate statistics on pyrogram data it is the peak heights that become the variables (n), thus enabling attempts at an objective analysis. At present, there is no agreement on the best statistical approach to take with pyrolysis data; different groups have used different statistics. The group centred around Meuzelaar in the Netherlands (see Table 3) has favoured the examination of Py–ms fingerprints by DF analysis and nonlinear mapping. These workers followed the early work of Reiner and Kubica (1969), who successfully examined Py–gc fingerprints of mycobacteria visually, with several applications of Py–ms to mycobacteria (Meuzelaar *et al.*, 1976, 1978; Wieten *et al.*, 1979, 1981a,b, 1982, 1983). The mycobacterial pyrograms were evaluated and identification schemes set up. Multivariate statistical analyses were based on DF analysis and the work of Eshuis *et al.* (1977) in producing nonlinear maps of distance matrices. Mycobacteria from the 'tuberculosis complex' were to be differentiated from other mycobacteria, and a success rate of 92% was reported compared to classical identification procedures. The clinical mycobacterial data base was reported to be stable for more than 1 year (Wieten *et al.*, 1981a,b). Later Wieten *et al.* (1983) commented that species status was not justified for the group of African strains classified in the 'tuberculosis complex'. Nonlinear mapping was found to be a qualitative aid in both interpretation and evaluation of the pyrolysis data.

In contrast, Norris and co-workers concentrated on Py–gc. This group has been more adventurous in the application of multivariate statistics, using them for the critical evaluation of data as well as in the characterisation and identification of food spoilage bacteria such as those belonging to the genera *Bacillus* and *Clostridium*. Principal component analysis was extensively used as an exploratory technique to provide a pictorial representation of the major groups and to detect aberrant analyses (Fig. 2; MacFie and Gutteridge, 1978; MacFie *et al.*, 1978; Gutteridge *et al.*, 1979; O'Donnell and Norris, 1981). Hierarchic cluster analyses were also used (MacFie *et al.*, 1978; O'Donnell and Norris, 1981) but were dismissed as unsatisfactory, possibly because of the high proportion of redundant variables found in the data (MacFie *et al.*, 1978). Principal component analysis was also found to be less than satisfactory; MacFie *et al.* (1978) found it did not produce a consistent differentiation of food spoilage organisms. Gutteridge *et al.* (1979) noted that PCP analysis was not suitable for discriminating between species or genus groups. These latter workers did, however, find PCP analysis useful for assessing reproducibility and for detecting aberrant analyses and unsuspected trends in the data, as well as for providing information on relationships of pyrolysis peaks to clusters of OTUs. Identification of new isolates was found to be feasible using CV analysis (MacFie *et al.*, 1978), and the latter was found to be excellent for differentiating between pyrograms of very similar species (O'Donnell *et al.*, 1980; O'Donnell and Norris, 1981), often from only the first two canonical variates (Fig. 6).

Gutteridge and Puckey (1982) applied CV analysis to Py–ms data from 50 Gram-negative strains and found that a preprocessing data reduction step was necessary as more variables per sample were produced with Py–ms than Py–gc. This latter paper included a report on the use of SDF analysis and jacknifing, a technique first applied to pyrogram data by Gutteridge *et al.* (1980). Both CV and SDF analyses require a predetermined group structure, and the results of Gutteridge and colleagues, whilst not too encouraging, did suggest that the predetermined groups of toxin producers belonging to *Clostridium botulinum* could be discriminated by these techniques.

Shute *et al.* (1984) outlined a series of ordinations for the examination of complex Py–gc data from 53 strains of four closely related species of *Bacillus*. Principal component analysis was found to be valuable in detecting outliers and as an explanatory technique in the examination of intra- and inter-group relationships. An *a priori* group structure was imposed on the data before CV and

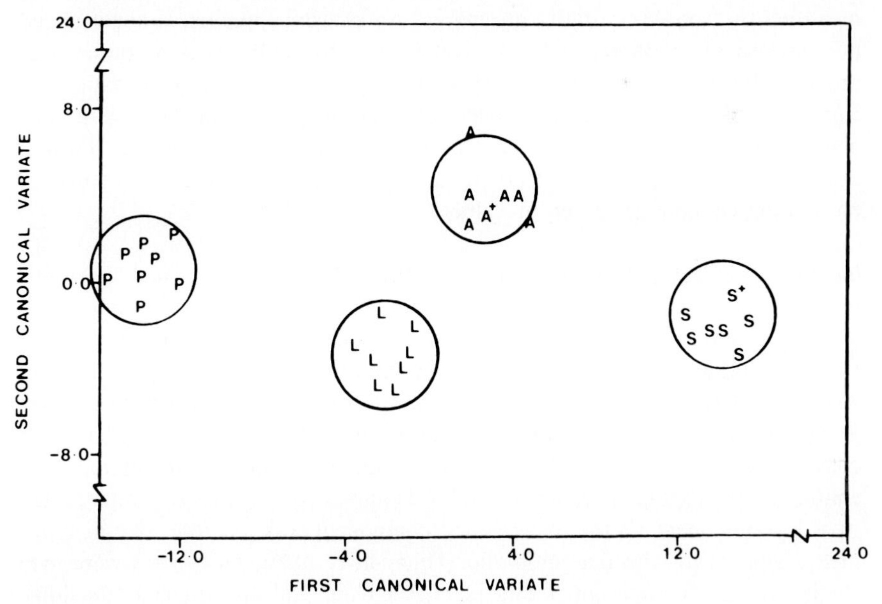

Fig. 6. Canonical variate analysis of the strain means of 32 nonsporing strains of *Bacillus: B. subtilis* (S), *B. pumilis* (P), *B. licheniformis* (L), and '*B. amyloliquefaciens*' (A). Points marked + represent the position of more than one strain. The first two canonical variates represent 98% of the variation between groups. Data obtained from pyrolysis gas–liquid chromatography of whole cells. From O'Donnell and Norris (1981), with permission.

SDF analyses were used; the structure was that based on API test results for the strains (O'Donnell *et al.*, 1980; Logan and Berkeley, 1981). The ordinations proved invaluable in the examination of results for reproducibility studies. It proved necessary to standardise culture conditions, sample preparation, and the growth phase of cultures before pyrolysis.

Other workers have found statistical analysis of pyrolysis data of potential in food bacteriology. Emswiler and Kotula (1978) and Stern *et al.* (1979, 1980) reported useful results from Py–gc data analysed with SDF analysis, but Stern (1982) could not differentiate food-borne strains of different genera using a similar combination of approaches. Certainly pyrolysis techniques in combination with multivariate nonhierarchic statistics offer objective ways of confirming and expanding classifications, as well as a new approach to the rapid identification of unknown strains.

Reports of analyses of other complex bacteriological data using the techniques discussed in this chapter have been sparse. Jenkins *et al.* (1977) examined data from gas–liquid chromatograms of fatty acid methyl esters of *Simonsiella* strains in SDF and CV analyses. The latter was said to demonstrate relationships between different strains, whilst the former allowed the strains to be correctly identified by 'source of origin' groupings. An unusual approach to analysis and presentation of taxonomic data derived from DNA hybridisation studies was provided by Moss and Bryant (1982), who used PCP analysis on data from *Chromobacterium fluviatile* and other Gram-negative bacteria. Only five variables were available and these all contributed to the first component. Further examples of the application of nonhierarchic multivariate statistics to complex bacteriological data are given in Table 3.

Identification Using Limited Data Sets. Gilardi (1971) suggested that antimicrobial susceptibility profiles, which are routinely collected on significant clinical bacterial isolates, might be used in the identification of isolates. In 1975 Darland used multivariate statistics on quantitative profiles, first applying PCO analysis to determine if his Enterobacteriaceae strains were representative of one or more populations; he then imposed an *a priori* eight-group structure based on conventional biochemical test results. Discriminant functions were derived from antibiotic sensitivities determined on 336 isolates; each isolate was represented by a 12-component vector, where each component was the zone size in millimetres associated with one antibiotic. Thirty-three unknown isolates were then studied for the 12 antibiotics and identified using the discriminant functions; agreement was 92% compared with conventional methods. Darland noted that the method was quick and economical.

Sielaff *et al.* (1976) reduced the identification time to a maximum of 6 hr by employing a semiautomated antimicrobial testing system (Autobac 1; Pfizer, Inc.), and Sielaff *et al.* (1982) reported further expansions and refinements. The

Table 3. *Taxonomic use of nonhierarchic methods in the analysis of complex bacteriological data*

Bacterial taxa/group	Number of strains	Origin of data[a]	Statistics employed[b]	Conclusions	Reference
Mycobacterium, 14 species	97	Py–ms of whole cells	NMMS	Automated differentiation of strains possible	Meuzelaar *et al.* (1976)
Listeria serotypes	20	Py–ms of whole cells	NMMS	Partition into serotypes possible	Eshuis *et al.* (1977)
Simonsiella strains	48	GLC of fatty acid methyl esters	DF	Discrimination possible, and identification to source of strain	Jenkins *et al.* (1977)
Streptococcus, oral strains	55	Py–gc of whole cells	CV, %S	Resolutions of position of intermediate strains allowed by CV	Stack *et al.* (1977)
Neisseria gonorrhoeae and related strains	15	Py–ms of whole cells	NMMS	Gonococci and closely related bacteria distinguished	Borst *et al.* (1978)
Salmonella serotypes	10	Py–gc of cell fragments	SDF	Differentiation of serotypes successful	Emswiler and Kotula (1978)
Bacillus spp.	32	Py–gc of whole cells	CV, SDF	Good congruence with the results of other classifications	O'Donnell (1978)
Clostridium botulinum and related strains	65	Py–gc of whole cells	CV, PCP	Classification possible by a combination of statistical methods	MacFie and Gutteridge (1978)
Aerobic meat spoilage isolates	25	Py–gc of whole cells	CA, CV, PCP	CA, PCP useful in exploratory analyses; CV discriminated groups of genera	MacFie *et al.* (1978)
Listeria serotypes and *Mycobacterium* spp.	20 11	Py–ms of whole cells	NMMS	Differentiation of strains successful	Meuzelaar *et al.* (1978)

(*continued*)

Table 3 (*Continued*)

Bacterial taxa/group	Number of strains	Origin of data[a]	Statistics employed[b]	Conclusions	Reference
Streptococcus, oral strains	19	Py–gc of whole cells	DF, %S	Adequate numerical treatment of data not provided by DF	Stack *et al*. (1978)
Miscellaneous Gram-negative rods, oral streptococci	NS	Py–gc of whole cells	PCP	PCP excellent for exploratory analysis, but not for identification or discrimination between established groups	Gutteridge *et al*. (1979)
Yersinia enterocolitica and related Gram-negative taxa	15	Py–gc of whole cells	CV, SDF	Discrimination successful	Stern *et al*. (1979)
Mycobacterium spp.	NS	Py–ms of whole cells	Analysis of variance, NMMS	Powerful tool for classification and identification	Wieten *et al*. (1979)
Acinetobacter and *Pseudomonas* spp.	49	Py–gc of whole cells	CV	Recognised species groups differentiated	French *et al*. (1980)
Clostridium botulinum and related strains	65	Py–gc of whole cells	CV, SDF	Discrimination of different toxin producers possible	Gutteridge *et al*. (1980)
Escherichia coli strains with/without K1 antigen	18	Py–ms of whole cells	NMMS	Fast screening for presence of specific properties possible	Haverkamp *et al*. (1980a)
Neisseria meningitidis serotypes	21	Py–ms of capsular polysaccharides	NMMS	Characterisation of polysaccharides possible	Haverkamp *et al*. (1980b)
Bacillus spp.	32	Py–gc of whole cells	CV	Phenotypically similar species discriminated	O'Donnell *et al*. (1980)

Yersinia enterocolitica, virulent and nonvirulent strains	14	Py–gc of whole cells	DF, SDF	Virulence prediction possible	Stern *et al.* (1980)
Streptococcus mutans plus strains of four diverse genera	10 4	Py–gc of whole cells	CV, DF	Promising for discrimination and identification	French *et al.* (1981)
Bacillus spp.	NS	Py–gc of whole cells	CA, CV, PCP, SDF	Useful in confirming and extending *Bacillus* taxonomy	O'Donnell and Norris (1981)
Mycobacterium tuberculosis complex and atypical mycobacteria	NS	Py–ms of whole cells	Analysis of variance, NMMS	NMMS allowed selection of discriminatory peaks used in successful identification	Wieten *et al.* (1981a)
Mycobacterium tuberculosis complex	91	Py–ms of whole cells	NMMS	Identification of constituents of complex achieved	Wieten *et al.* (1981b)
Streptococcus, oral strains	57	Py–gc of whole cells	DF	Identification not always successful	French and Phillips (1982)
Gram-negative bacteria	50	DP–ms of whole cells	CV, SDF	Discrimination and identification successful	Gutteridge and Puckey (1982)
Chromobacterium fluviatile	10	DNA–RNA hybridisation	PCP	PCP provided alternative presentation of data	Moss and Bryant (1982)
Endospore and non-endospore-forming food spoilage isolates	14	Py–gc of pre-inoculated can supernate	Df	Successful screen for identification of endospore-forming bacteria	Schafer *et al.* (1982)
Aerobic food-borne bacteria	18	Py–gc of whole cells	SDF	Differentiation not successful	Stern (1982)
Mycobacterium spp. including *M. leprae*	NS	Py–ms of whole cells	NMMS	Useful tool in classification of *M. leprae*	Wieten *et al.* (1982)

(*continued*)

Table 3 (*Continued*)

Bacterial taxa/group	Number of strains	Origin of data[a]	Statistics employed[b]	Conclusions	Reference
Mycobacterium tuberculosis complex	27	Py–ms of whole cells	Analysis of variance, DF, NMMS	Species status of '*M. africanum*' challenged	Wieten *et al.* (1983)
Escherichia coli strains with/without K1 antigen	18	Py–ms of whole cells	DF, PCP	K1 strains differentiated from non-K1	Windig *et al.* (1983)
Bacillus spp.	32	Py–ms of whole cells	CV, PCO, PCP, SDF	Differentiation successful using data from nonsporulated cultures	Shute *et al.* (1984)

[a]DP–ms, Direct-probe mass spectrometry; GLC, gas–liquid chromatography; Py–gc, pyrolysis gas–liquid chromatography; Py–ms, pyrolysis mass spectrometry.
[b]CA, Cluster analysis; CV, canonical variates; DF, discriminant function analysis; PCO, principal coordinates; PCP, principal components; NMMS, nonmetric multidimensional scaling; %S, percentage similarity; SDF, stepwise discriminant function analysis; NS, not stated.

Autobac was used to test susceptibility to 18 different agents, and a two-stage quadratic DF program was used for identification. Sielaff *et al.* (1982) suggested that the identification method was accurate, quick, and reliable enough to be feasible for routine use in a clinical laboratory, and this was confirmed in a collaborative evaluation of the scheme (Barry *et al.*, 1982). Boyd *et al.* (1978), however, pointed out that bacteria with atypical resistance patterns were likely to be misidentified and that such strains could form up to 17% of isolates, thereby limiting the use of the method and making it inferior to currently accepted biochemical methods available for the identification of medical bacteria. Nevertheless, General Diagnostics (Warner-Lambert Co.) have produced a system, the Autobac IDX, which incorporates a computer analysis of inhibition profiles based on an earlier data base for Gram-negative clinical isolates (Barry *et al.*, 1982). Two studies affirmed the convenience and utility of such a procedure to routine clinical laboratories, and 95% accuracy of the Autobac IDX result with traditional methods was recorded (Costigan and Hollick, 1984; Kelly *et al.*, 1984).

Identification Using Traditional Bacteriological Data. Many of the techniques discussed in this chapter can be used for identification with less specialised bacteriological methodology than mentioned above. Ordinations may be useful for selecting diagnostic markers: Bonde (1978, 1981) used CV analysis; Sundman and Gyllenberg (1967), and Catteau *et al.* (1973) used MF analysis; and Hill *et al.* (1965), Quadling and Hopkins (1967), and Skyring *et al.* (1971) used PCP analysis. Gower (1968) suggested that PCO analysis may offer one of the simplest means of identification, by adding results from new isolates as points to an existing analysis. The application of PCO analysis to identification was also reported by Gibbs *et al.* (1978), who were able to distinguish between *Staphylococcus aureus* biotype A and B strains isolated from poultry. The value of taxon-radius models and discriminant function analyses to identification has already been considered. General-purpose computer-assisted identification methods are discussed in detail by Holmes and Hill (Chapter 10).

Conclusions

Many different hierarchic and nonhierarchic techniques are available. However, it is as well to remember the old adage of computer scientists, ‘rubbish in, rubbish out’; the quality of the data to be computed remains paramount. Given the advent of faster, smaller, and cheaper computers, it is also as well to recall that computer programmes for taxonomic purposes are not only easy to use but are also easy to abuse. Cooperation with a biologically literate computer scientist is essential at an early stage and also subsequently, for help with interpretation and evaluation of results is important to the ultimate success of any exercise. Nonhierarchic multivariate statistics are an invaluable aid to the interpretation of classifications obtained with a hierarchic numerical taxonomic approach. They have also been demonstrably useful in the discrimination of bacteria using traditional and complex data, and in ecological studies of bacterial populations. In the final analysis, numerical methods, whether hierarchic or nonhierarchic, are tools in bacterial systematics and are not a replacement for the insight, skill, and careful thought it takes to provide classification and identification schemes.

Acknowledgements

Many thanks are due to H. J. H. Macfie and M. J. Sackin for reading the manuscript and to Dave Greatorex, Helena Larkin, and Dave Reardon for graphic, typographic, and photographic work.

References

Alderson, G., Athalye, M., and White, R. P. (1984). Numerical methods in the taxonomy of sporoactinomycetes. *In* 'Biological, Biochemical, and Biomedical Aspects of Actinomycetes' (Eds. L. Ortiz-Ortiz, L. F. Bajalil, and V. Yakoleff), pp. 597–615. Academic Press, Orlando.

Barry, A. L., Gavan, T. L., Smith, P. B., Matsen, J. M., Morello, J. A., and Sielaff, B. H. (1982). Accuracy and precision of the Autobac system for rapid identification of Gram-negative bacilli: a collaborative evaluation. *Journal of Clinical Microbiology* **15,** 1111–1119.

BMDP. (1981). Biomedical Computer Programs, P-Series. University of California, Los Angeles. Univ. of California Press, Berkeley.

Bonde, G. J. (1978). Application of canonical variates to clusters formed by multidimensional scaling. *Journal of Applied Bacteriology* **45,** xi–xii.

Bonde, G. J. (1981). *Bacillus* from marine habitats: allocation to phena established by numerical techniques. *In* 'The Aerobic Endospore-forming Bacteria: Classification and Identification' (Eds. R. C. W. Berkeley and M. Goodfellow), pp. 180–215. Academic Press, London.

Borst, J., van der Snee-Enkelaar, A. C., and Meuzelaar, H. C. L. (1978). Typing of *Neisseria gonorrhoeae* by pyrolysis mass spectrometry. *Antonie van Leeuwenhoek* **44,** 253.

Boyce, A. J. (1969). Mapping diversity: a comparative study of some numerical methods. *In* 'Numerical Taxonomy' (Ed. A. J. Cole), pp. 1–31. Academic Press, London.

Boyd, J. C., Lewis, J. W., Marr, J. J., Harper, A. M., and Kowalski, B. R. (1978). Effect of atypical antibiotic resistance on microorganism identification by pattern recognition. *Journal of Clinical Microbiology* **8,** 689–694.

Bridge, P. D., and Sneath, P. H. A. (1983). Numerical taxonomy of *Streptococcus. Journal of General Microbiology* **129,** 565–597.

Brisbane, P. G., and Rovira, A. D. (1961). A comparison of methods for classifying rhizosphere bacteria. *Journal of General Microbiology* **26,** 379–392.

Catteau. M., Poncelet, F., Beerens, H., and Losfeld, J. (1973). A rapid method for the identification of *Bifidobacterium* species using 50 characters. *In* 'Actinomycetales. Characteristics and Practical Importance' (Eds. G. Sykes and F. A. Skinner), pp. 301–310. Academic Press, London.

Cattell, R. B. (1952). 'Factor Analysis'. Harper, New York.

Clifford, H. T., and Stephenson, W. (1975). 'An Introduction to Numerical Classification'. Academic Press, London.

Colwell, R. R. (1970). Numerical analysis in microbial identification and classification. *Developments in Industrial Microbiology* **11,** 154–160.

Cone, R. D., and Lechowich, R. V. (1970). Differentiation of *Clostridium botulinum* types A, B and E by pyrolysis gas–liquid chromatography. *Applied Microbiology* **19,** 138–145.

Costigan, W. J., and Hollick, G. E. (1984). Use of Autobac IDX system for rapid identification of Enterobacteriaceae and nonfermentative Gram-negative bacilli. *Journal of Clinical Microbiology* **19,** 301–302.

Curtis, J. T. (1959). 'The Vegetation of Wisconsin: An Ordination of Plant Communities'. Univ. of Wisconsin Press, Madison.

Darland, G. (1975). Discriminant analysis of antibiotic susceptibility as a means of bacterial identification. *Journal of Clinical Microbiology* **2,** 391–396.

Defayolle, M., and Colobert, L. (1962). L'espece *Streptococcus faecalis* II.—Étude de l'homogénéité par l'analyse factorielle. *Annales de l'Institut Pasteur* **103,** 505–522.

Defayolle, M., Colobert, L., Poncet, P., Buissière, J., and Pontier, J. (1968). Application de l'analyse factorielle à la taxonomie des microorganisms. *Biometrie–Praximetrie* **9,** 14–51.

Dunn, G., and Everitt, B. S. (1982). 'An Introduction to Mathematical Taxonomy'. Cambridge Univ. Press, Cambridge.

Emswiler, B. S., and Kotula, A. W. (1978). Differentiation of *Salmonella* serotypes by pyrolysis–gas–liquid chromatography of cell fragments. *Applied and Environmental Microbiology* **35,** 97–104.

Eshuis, W., Kistemaker, P. G., and Meuzelaar, H. L. C. (1977). Some numerical aspects of reproducibility and specificity. *In* 'Analytical Pyrolysis' (Eds. C. E. Roland Jones and C. A. Cramer), pp. 151–166. Elsevier, Amsterdam.

Fisher, R. A. (1936). The use of multiple measurements in taxonomic problems. *Annals of Eugenics* **7,** 179–188.

Florek, K., Łukaszewicz, J., Perkal, J., Steinhaus, H., and Zubrzycki, S. (1951a). Sur la liason et la division des points d'un ensemble fini. *Colloquium Mathematicum* **2,** 282–285.

Florek, K., Łukaszewicz, J., Perkal, J., Steinhaus, H., and Zubrzycki, S. (1951b). Taksonomia Wrocławska. Przegļd Antropologiczny **17,** 193–211 (in Polish).

French, G. L., and Phillips, I. (1982). Discrimination and identification of oral streptococci by pyrolysis–gas chromatography. *In* 'Basic Concepts of Streptococci and Streptococcal Diseases' (Eds. S. E. Holm and P. Christensen), pp. 62–64. Academic Press, New York.

French, G. L., Gutteridge, C. S., and Phillips, I. (1980). Pyrolysis gas chromatography of *Pseudomonas* and *Acinetobacter* species. *Journal of Applied Bacteriology* **49,** 505–516.

French, G. L., Phillips, I., and Chinn, S. (1981). Reproducible pyrolysis–gas chromatography of micro-organisms with solid stationary phases and isothermal oven temperatures. *Journal of General Microbiology* **125,** 347–355.

Garner, W., and Gennaro, R. (1965). Gas chromatographic differentiation of closely related species of microorganisms. *150th Meeting of the American Chemical Society, Atlantic City New Jersey,* Abstract Q28.

GENSTAT. (1980). A general statistical program (16 authors). Rothamsted Experimental Station, Harpenden, England.

Gibbs. P. A., Patterson, J. T., and Harvey, J. (1978). Biochemical characteristics and enterotoxigenicity of *Staphylococcus aureus* strains isolated from poultry. *Journal of Applied Bacteriology* **44,** 57–74.

Gilardi, G. L. (1971). Antimicrobial susceptibility as a diagnostic aid in the identification of non-fermenting Gram-negative bacteria. *Applied Microbiology* **22,** 821–823.

Goodall, D. W. (1953). Objective methods for the classification of vegetation. I. The use of positive interspecific correlations. *Australian Journal of Botany* **1,** 39–63.

Gordon, R. E., Haynes, W. C., and Pang, C. H.-N. (1973). 'The Genus *Bacillus*'. U.S. Department of Agriculture, Washington, D.C.

Gower, J. C. (1966). Some distance properties of latent root and vector methods used in multivariate analysis. *Biometrika* **53,** 325–338.

Gower, J. C. (1967). Multivariate analysis and multidimensional geometry. *The Statistician* **17,** 13–28.

Gower, J. C. (1968). Adding a point to vector diagrams in multivariate analysis. *Biometrika* **55,** 582–585.

Gower, J. C. (1969a). A survey of numerical methods useful in taxonomy. *Acaralogia* **11,** 357–375.

Gower, J. C. (1969b). The basis of numerical methods in classification. *In* 'The Soil Ecosystem' (Ed. J. G. Sheals), London Systematics Association Publication No. 8, pp. 13–30. Systematics Association, London.

Gutteridge, C. S., and Norris, J. R. (1979). A review: the application of pyrolysis techniques to the identification of microorganisms. *Journal of Applied Bacteriology* **47,** 5–43.

Gutteridge, C. S., and Puckey, D. J. (1982). Discrimination of some Gram negative bacteria by direct probe mass spectrometry. *Journal of General Microbiology* **128,** 721–730.

Gutteridge, C. S., MacFie, H. J. H., and Norris, J. R. (1979). Use of principal components analysis for displaying variation between pyrograms of micro-organisms. *Journal of Analytical and Applied Pyrolysis* **1,** 67–76.

Gutteridge, C. S., Mackey, B. M., and Norris, J. R. (1980). A pyrolysis gas–liquid chromatography study of *Clostridium botulinum* and related organisms. *Journal of Applied Bacteriology* **49,** 165–174.

Gyllenberg, H. G. (1964). An approach to numerical description of microbiol populations. *Annales Academiae Scientarum Fennicae Series A, IV Biologica* **81,** 1–23.

Gyllenberg, H. G. (1965a). Character correlations in certain taxonomic and ecologic groups of bacteria. A study based on factor analysis. *Annales Medicinae Experimentalis e Biologiae Fennicae* **43,** 82–90.

Gyllenberg, H. G. (1965b). A model for computer identification of microorganisms. *Journal of General Microbiology* **39,** 401–405.

Gyllenberg, H. G. (1967). Significance of the Gram stain in the classification of soil bacteria. *In* 'The Ecology of Soil Bacteria. An International Symposium' (Eds. T. R. G. Gray and D. Parkinson), pp. 351–359. Liverpool Univ. Press, Liverpool.

Gyllenberg, H. G. (1970). Factor analytical evaluation of patterns of correlated characteristics in streptomycetes. *In* 'The Actinomycetales' (Ed. H. Prauser), pp. 101–105. Gustav Fischer Verlag, Jena.

Gyllenberg. H G. (1973). Numerical methods in automatic identification of microorganisms. *Bulletins from the Ecological Research Committee—NFR (Statens Naturvetenskapliga Forskingsråd)* **17,** 127–133.

Gyllenberg, H. G., and Eklund, E. (1967). Application of factor analysis in microbiology. II. Evaluation of character correlation patterns in psychrophilic pseudomonads. *Annales Academiae Scientiarum Fennicae Series A, IV Biologica* **113,** 1–16.

Gyllenberg, H. G., and Raumamaa, V. (1966). Taxometric models of bacterial soil populations. *Acta Agriculturae Scandinavica* **16,** 30–38.

Gyllenberg, H. G., Wóznicka, W., and Kuryłowicz, W. (1967). Application of factor analysis in microbiology III. A study of the 'yellow series' of streptomycetes. *Annales Academiae Scientiarum Fennicae Series A, IV Biologica* **114,** 1–15.

Harman, H. H. (1967). 'Modern Factor Analysis', 2nd Edition. Univ. of Chicago Press, Chicago.

Harman, H. H. (1972). How factor analysis can be used in classification. I. Mathematical part. *In* 'Yeasts Models in Science and Technics' (Eds. A. Kocková-Kratochivílová and E. Minarík), pp. 273–295. Publishing House of the Slavac Academy of Sciences, Bratislava, Czechoslovakia.

Håstein, T., and Smith, J. E. (1977). A study of *Vibrio anguillarum* from farmed and wild fish using principal components analysis. *Journal of Fish Biology* **11,** 69–75.

Haverkamp, J., Eshuis, W., Boerboom, A. J. H., and Guinée, P. A. M. (1980a). Pyrolysis mass spectrometry as a rapid screening method of biological materials. *In* 'Advances in Mass Spectrometry' (Ed. N. R. Daly), Vol. 8, pp. 983–989. Heyden, London.

Haverkamp, J., Meuzelaar, H. L. C., Beuvery, E. C., Boonkamp, P. M., and Tiesjema, R. H. (1980b). Characterisation of *Neisseria meningitidis* capsular polysaccharides containing sialic acid by pyrolysis mass spectrometry. *Analytical Biochemistry* **104,** 407–418.

Hill, L. R., Silverstri, L. G., Ihm, P., Farchi, G., and Lanciani, P. (1965). Automatic classification of staphylococci by principal component analysis and a gradient method. *Journal of Bacteriology* **89,** 1393–1401.

Holder-Franklin, M. A., and Wuest, L. J. (1983). Factor analysis as an analytical method in microbiology. *In* 'Mathematics in Microbiology' (Ed. M. Bazin), pp. 139–169. Academic Press, London.

Hotelling, H. (1933a). Analysis of a complex of statistical variables into principal components. *Journal of Educational Psychology* **24,** 417–441.

Hotelling H. (1933b). Analysis of a complex of statistical variables into principal components. *Journal of Educational Psychology* **24,** 498–520.

Irwin, W. J. (1982). 'Analytical Pyrolysis: A Comprehensive Guide'. Dekker, New York.

Jardine, N., and Sibson, R. (1968). The construction of hierarchic and non-hierarchic classifications. *Computer Journal* **11,** 177–184.

Jenkins, C. L., Kuhn, D. A., and Daly, K. R. (1977). Fatty acid composition of *Simonsiella* strains. *Archives of Microbiology* **113,** 209–214.

Jones, D., and Sackin, M. J. (1980). Numerical methods in the classification and identification of bacteria with especial reference to the Enterobacteriaceae. In 'Microbiological Classification and Identification' (Eds. M. Goodfellow and R. G. Board), pp. 73–106. Academic Press, London.

Kapperud, G., Bergan, T., and Lassen, J. (1981). Numerical taxonomy of *Yersinia enterocolitica* and *Yersinia enterocolitica*-like bacteria. *International Journal of Systematic Bacteriology* **31,** 401–419.

Kelly, M. T., Matsen, J. M., Morello, J. A., Smith, P. B., and Tilton, R. C. (1984). Collaborative clinical evaluation of the Autobac IDX system for identification of Gram-negative bacilli. *Journal of Clinical Microbiology* **19,** 529–533.

Kendall, J. (1975). 'Mulivariate Analysis'. Griffin, London.
Kruskal, J. B. (1964a). Multidimensional scaling by optimizing goodness of fit to a nonmetric hypothesis. *Psychometrika* **29,** 1–27.
Kruskal, J. B. (1964b). Nonmetric multidimensional scaling: a numerical method. *Psychometrika* **29,** 115–129.
Krzanowski, W. J. (1971). The algebraic basis of classical multivariate methods. *The Statistician* **20,** 51–61.
Lance, G. N., and Williams, W. T. (1967). A general theory of classificatory sorting strategies. II. Clustering systems. *Computer Journal* **10,** 271–277.
Logan, N. A., and Berkeley, R. C. W. (1981). A classification of the genus *Bacillus* based upon the API system. *In* 'The Aerobic Endospore-forming Bacteria: Classification and Identification' (Eds. R. C. W. Berkeley and M. Goodfellow), pp. 105–140. Academic Press, London.
MacFie, H. J. H., and Gutteridge, C. S. (1978). Analysis of pyrolysis gas–liquid chromatography data using multivariate statistical techniques. *Journal of Applied Bacteriology* **45** iv–v.
MacFie, H. J. H., Gutteridge, C. S., and Norris, J. R. (1978). Use of canonical variates analysis in differentiation of bacteria by pyrolysis gas–liquid chromatography. *Journal of General Microbiology* **104,** 67–74.
Marriott, F. H. C. (1974). 'The Interpretation of Multiple Observations'. Academic Press, London.
Meuzelaar, H. L. C., Kistemaker, P. G., Eshuis, W., and Engel, H. W. B. (1976). Progress in automated and computerised characterization of microorganisms by pyrolysis mass spectometry. *In* 'Rapid Methods and Automation in Microbiology' (Eds. H. H. Johnston and S. W. B. Newsom), 2nd Edition, pp. 225–229. Learned Information (Europe) Ltd., Oxford.
Meuzelaar, H. L. C., Kistemaker, P. G., Eshuis, W., and Boerboom, H. A. J. (1978). Automated pyrolysis–mass spectrometry: application to the differentiation of microorganisms. *In* 'Advances in Mass Spectrometry' (Ed. N. R. Daly), Vol. 7B, 1452–1456. Heyden, London.
Moss, M. O., and Bryant, T. N. (1982). DNA ribosomal RNA hybridization studies of *Chromobacterium fluviatile. Journal of General Microbiology* **128,** 829–834.
Moss, M. O., Ryall, C., and Logan, N. A. (1978). The classification and characterisation of chromobacteria from a lowland river. *Journal of General Microbiology* **105,** 11–21.
O'Donnell, A. G. (1978). The application of pyrolysis gas–liquid chromatography to some aerobic sporeformers. *Journal of Applied Bacteriology* **45,** v.
O'Donnell, A. G., and Norris, J. R. (1981). Pyrolysis gas–liquid chromatographic studies. *In* 'The Aerobic Endospore-forming Bacteria: Classification and Identification' (Eds. R. C. W. Berkeley and M. Goodfellow), pp. 141–179. Academic Press, London.
O'Donnell, A. G., Norris, J. R., Berkeley, R. C. W., Claus, D., Kaneko, T., Logan, N. A., and Nozaki, R. (1980). Characterization of *Bacillus subtilis, Bacillus pumilus, Bacillus licheniformis,* and *Bacillus amyloliquefaciens* by pyrolysis gas–liquid chromatography, deoxyribonucleic acid–deoxyribonucleic acid hybridization, biochemical tests, and API systems. *International Journal of Systematic Bacteriology* **30,** 448–459.
Pike, E. B. (1965a). A trial of association methods for selecting determinative characters from a collection of Micrococcaceae isolates. *Journal of General Microbiology* **41,** xix.
Pike, E. B. (1965b). A trial of statistical methods for selection of determinative characters from Micrococcaceae isolates. *Spisy Prirodovedecke Fakulty University J. E. Purkyne —Brne Series K* **35,** 316–317.

Pohja, M. S., and Gyllenberg, H. G. (1967). Application of factor analysis in microbiology 5. Evaluation of the population development in cold-stored meat. *Annales Academiae Scientarum Fennicae Series A, IV Biologica* **116,** 1–8.

Quadling, C., and Hopkins, J. W. (1967). Evaluation of tests and grouping of cultures by a two-stage principal component method. *Canadian Journal of Microbiology* **13,** 1379–1400.

Quinn, P. (1976). Identification of microorganisms by pyrolysis; the state of the art. *In* 'Rapid Methods and Automation in Microbiology' (Eds. H. H. Johnston and S. W. B. Newsom), 2nd Edition, pp. 178–186. Learned Information (Europe) Ltd., Oxford.

Reiner, E. (1963). Identification of bacterial strains by pyrolysis—gas–liquid chromatography. *Nature (London)* **200,** 1058–1059.

Reiner, E. (1965). Identification of bacterial strains by pyrolysis—gas–liquid chromatography. *Nature (London)* **206,** 1272–1274.

Reiner, E., and Kubica, G. P. (1969). Predictive value of pyrolysis—gas–liquid chromatography in the differentiation of mycobacteria. *American Review of Respiratory Diseases* **99,** 42–249.

Rogers, D. J., and Tanimoto, T. T. (1960). A computer program for classifying plants. *Science* **132,** 1115–1118.

Rohlf, F. J. (1967). Correlated characters in numerical taxonomy. *Systematic Zoology* **16,** 109–126.

Rohlf, F. J. (1968). Stereograms in numerical taxonomy. *Systematic Zoology* **17,** 246–255.

Rohlf, F. J. (1970). Adaptive hierarchical clustering schemes. *Systematic Zoology* **19,** 58–82.

Rohlf, F. J. (1972). An empirical comparison of three ordination techniques in numerical taxonomy. *Systematic Zoology* **21,** 271–280.

Rohlf, F. J., and Sokal, R. R. (1962). The description of taxonomic relationships by factor analysis. *Systematic Zoology* **11,** 1–16.

Ross, G. J. S. (1975). Rapid techniques for automatic identification. *In* 'Biological Identification with Computers' (Ed. R. J. Pankhurst), pp. 93–102. Academic Press, London and New York.

Rosswall, T., and Kvillner, E. (1978). Principal components and factor analysis for the description of bacterial populations. *In* 'Advances in Microbial Ecology' (Ed. M. Alexander), Vol. 2, 1–48. Plenum, New York.

Schafer, M. L., Peeler, J. T., Bradshaw, J. G., Hamilton, C. H., and Carver, R. B. (1982). A rapid gas chromatographic method for the identification of sporeformers and nonsporeformers in swollen cans of low-acid foods. *Journal of Food Science* **47,** 2033–2037.

Seiler, H. (1983). Identification key for coryneform bacteria derived by numerical taxonomic studies. *Journal of General Microbiology* **129,** 1433–1471.

Shute, L. A., Gutteridge, C. S., Norris, J. R., and Berkeley, R. C. W. (1984). Curie-point pyrolysis mass spectrometry applied to characterization and identification of selected *Bacillus* species. *Journal of General Microbiology* **130,** 343–355.

Sielaff, B. H., Johnson, E. A. , and Matsen, J. M. (1976). Computer-assisted bacterial identification utilizing antimicrobial susceptibility profiles generated by Autobac. *Journal of Clinical Microbiology* **3,** 105–109.

Sielaff, B. H., Matsen, J. M., and McKie, J. E. (1982). Novel approach to bacterial identification that uses the Autobac system. *Journal of Clinical Microbiology* **15,** 1103–1110.

Skyring, G. W., and Quadling, C. (1969a). Soil bacteria: principal component analysis of descriptions of named cultures. *Canadian Journal of Microbiology* **15,** 141–158.

Skyring, G. W., and Quadling, C. (1969b). Soil bacteria: comparisons of rhizosphere and non-rhizosphere populations. *Canadian Journal of Microbiology* **15,** 473–488.

Skyring, G. W., and Quadling, C. (1970). Soil bacteria: a principal component analysis and guanine–cytosine contents of some arthrobacter–coryneform soil isolates and some named cultures. *Canadian Journal of Microbiology* **16,** 95–106.

Skyring, G. W., Quadling, C., and Rouatt, J. W. (1971). Soil bacteria: principal component analysis of physiological descriptions of some named cultures of *Agrobacterium, Arthrobacter* and *Rhizobium. Canadian Journal of Microbiology* **17,** 1299–1311.

Skyring, G. W., Jones, H. E., and Goodchild, D. (1977). The taxonomy of some new isolates of dissimilatory sulfate-reducing bacteria. *Canadian Journal of Microbiology* **23,** 1415–1425.

Sneath, P. H. A. (1962). The construction of taxonomic groups. *In* 'Microbial Classification' (Eds. G. E. Ainsworth and P. H. A. Sneath), pp. 289–332. Cambridge Univ. Press, Cambridge.

Sneath, P. H. A. (1972). Computer taxonomy. *In* 'Methods in Microbiology' (Eds. J. R. Norris and D. W. Ribbons), Vol. 4, pp. 29–98. Academic Press, London.

Sneath, P. H. A. (1976). An evaluation of numerical taxonomic techniques in the taxonomy of *Nocardia* and allied taxa. *In* 'The Biology of the Nocardiae' (Eds. M. Goodfellow. G. H. Brownell, and J. A. Serrano), pp. 74–101. Academic Press, London.

Sneath, P. H. A. (1977). A method for testing the distinctness of clusters: a test of the disjunction of two clusters in Euclidean space as measured by their overlap. *Journal of the International Association for Mathematical Geology* **9,** 123–143.

Sneath, P. H. A. (1978a). Classification of microorganisms. *In* 'Essays in Microbiology' (Eds. J. R. Norris and M. R. Richmond). Wiley, Chichester.

Sneath, P. H. A. (1978b). Identification of microorganisms. *In* Essays in Microbiology' (Eds. J. R. Norris and M. R. Richmond). Wiley, Chichester.

Sneath, P. H. A. (1980). The probability that distinct clusters will be unrecognised in low dimensional ordinations. *The Classification Society Bulletin* **4,** 22–43.

Sneath, P. H. A. (1983). Distortions of taxonomic structure from incomplete data on a restricted set of reference strains. *Journal of General Microbiology* **129,** 1045–1073.

Sneath, P. H. A., and Sokal, R. R. (1973). 'Numerical Taxonomy: The Principles and Practice of Numerical Classification'. Freeman, San Francisco.

Sokal, R. R., and Rohlf, F. J. (1962). The comparison of dendrograms by objective methods. *Taxon* **11,** 33–40.

SPSS. (1975). 'Statistical Package for the Social Sciences', 2nd Edition. McGraw-Hill, New York.

Stack, M. V., Donoghue, H. D., Tyler, J. E., and Marshall, M. (1977). Comparison of oral streptococci by pyrolysis GLC. *In* 'Analytical Pyrolysis' (Eds. C. E. Roland Jones and C. A. Cramer), pp. 57–68. Elsevier, Amsterdam.

Stack, M. V., Donoghue, H. D., and Tyler, J. E. (1978). Discrimination between oral streptococci by pyrolysis gas–liquid chromatography. *Applied and Environmental Microbiology* **35,** 45–50.

Stern, N. J. (1982). The inability of pyrolysis gas–liquid chromatography to differentiate selected foodborne bacteria. *Journal of Food Protection* **45,** 229–234.

Stern, N. J., Kotula, A. W., and Pierson, M. D. (1979). Differentiation of selected Enterobacteriaceae by pyrolysis—gas–liquid chromatography. *Applied and Environmental Microbiology* **38,** 1098–1102.

Stern, N. J., Kotula, A. W., and Pierson, M. D. (1980). Virulence prediction of *Yersinia enterocolitica* by pyrolysis gas–liquid chromatography. *Applied and Environmental Microbiology* **40,** 646–651.

Sundman, V. (1968). Characterisation of bacterial populations by means of factor profiles. *Acta Agriculturae Scandinavica* **18,** 22–26.

Sundman, V. (1970). Four bacterial soil populations characterised and compared by a factor analytical method. *Canadian Journal of Microbiology* **16,** 455–464.

Sundman, V. (1973). Description and comparison of microbial populations in ecological studies with the aid of factor analysis. *Bulletins from the Ecological Research Committee—NFR (Statens Naturvetenskapliga Forskingsråd)* **17,** 135–140.

Sundman, V., and Carlberg, G. (1967). Application of factor analysis in microbiology. IV. The value of geometric parameters in the numerical description of bacterial soil populations. *Annales Academiae Scientiarum Fennicae Series A, IV Biologica* **115,** 1–12.

Sundman, V., and Gyllenberg, H. V. (1967). Application of factor analysis in microbiology. 1. General aspects on the use of factor analysis in microbiology. *Annales Academiae Scientiarum Fennicae Series A, IV Biologica* **112,** 1–32.

Szulga, T. (1978). A critical evaluation of taxonomic procedures applied in *Streptomyces. Zentralblatt fur Bakteriologie, Parasitenkunde, Infektionskrankheiten und Hygiene, Abteilung 1, Supplement* **6,** 31–42.

Wieten, G., Haverkamp, J., Engel, H. W. B., and Tárnok, I. (1979). Pyrolysis mass spectrometry in mycobacterial taxonomy and identification. *In* 'Twenty-five Years of Mycobacterial Taxonomy' (Eds. K. P. Kubica, L. G. Wayne, and L. S. Good), pp. 171–189. C. D. C. Press, Atlanta, Georgia.

Wieten, G., Haverkamp, J., Engel, H. W. B., and Berwald, L. G. (1981a). Application of pyrolysis mass spectrometry to the classification and identification of mycobacteria. *Reviews of Infectious Diseases* **3,** 871–877.

Wieten, G., Haverkamp, J., Meuzelaar, H. L. C., Engel, H. W. B., and Berwald, L. G. (1981b). Pyrolysis mass spectrometry: a new method to differentiate between the mycobacterium of the "tuberculosis complex" and other mycobacteria. *Journal of General Microbiology* **122,** 109–118.

Wieten, G., Haverkamp, J., Berwald, L. G., Groothuis, D. G., and Draper, P. (1982). Pyrolysis mass spectrometry: its applicability to mycobacteriology, including *Mycobacterium leprae. Annales de Microbiologie* **133B,** 15–27.

Wieten, G., Haverkamp, J., Groothuis, D. G., Berwald, L. G., and David H. L. (1983). Classification and identification of *Mycobacterium africanum* by pyrolysis mass spectrometry. *Journal of General Microbiology* **129,** 3679–3688.

Williams, W. T., and Dale, M. B. (1965). Fundamental problems in numerical taxonomy. *Advances in Botanical Research* **2,** 35–68.

Williams, W. T., and Lance, G. N. (1968). The choice of strategy in the analysis of complex data. *Statistician* **18,** 31–43.

Windig, W., Haverkamp, J., and Kistemaker, P. G. (1983). Interpretation of sets of pyrolysis mass spectra by discriminant analysis and graphical rotation. *Analytical Chemistry* **55,** 81–88.

10

Computers in Diagnostic Bacteriology, Including Identification

B. HOLMES AND L. R. HILL

National Collection of Type Cultures, Central Public Health Laboratory, London, UK

Introduction

Beers and Lockhart (1962) were the first to suggest that a mathematical model could be used for bacterial identification. These authors outlined three possible lines of approach. The first, to carry out a numerical phenetic classification for each new isolate, would be impracticable for single strains but might be useful for batch processing of large numbers of organisms. For example, in a survey of a hospital environment, newly isolated strains would be characterised using the same range of tests as a set of reference strains of known taxa. Unknown strains would be identified after a numerical classification only if they joined a cluster containing one or more reference strains. Methods for identification by comparison of unidentified strains with taxonomic groups previously constructed using numerical clustering methods have been described by Quadling and Colwell (1964) and Gyllenberg (1965); these are related to the third approach described below.

The second approach was the mathematical construction of keys for use in place of conventional identification keys. The construction of identification keys based on mathematical evaluation of the information content of each test was suggested by Maccacaro (1958). Hill and Silvestri (1962), for example, constructed a probability key for the identification of a number of actinomycete taxa; the mathematical basis of that key was described by Möller (1962). A simple mathematical model for estimating the differentiating power of diagnostic tests was described by Gyllenberg (1963), who proposed a formula for calculating the minimum number of two-state tests that would be necessary for separating given numbers of taxa. Gyllenberg's methods were subsequently developed by Rypka *et al.* (1967) and Rypka and Babb (1970), who described a numerical approach for estimating the differentiating power of a test in relation to the other tests considered for the construction of the set. A method for test selection based on

ISBN 0-12-289665-3

information theory was also described by Niemelä *et al.* (1968). In all these approaches a mathematical model is used once only to determine the tests with the highest differentiating power, and unknowns are then identified by conventional comparison of their test results with those expected for each taxon. Although these approaches share the advantage that a reduction in the number of tests necessary for identification can be expected, they also have the same disadvantage as conventional identification, namely the problem of identifying isolates that are atypical in one or more characteristics.

The third approach was to estimate the probability of strains of each taxon giving positive results in a series of tests and then to use these probabilities to yield a statistical estimate of the reliability of the diagnosis. Payne (1963) was the first to describe the use of a computer for identification in this way. He suggested that identification should be performed sequentially with the choice of additional tests with which to further the identification also being made by computer. Dybowski and Franklin (1968) described the use of a computer-assisted, conditional probability method for the identification of enterobacteria. Lapage *et al.* (1970) used a similar probability method for the identification of nonfastidious Gram-negative bacteria of clinical importance (predominantly enterobacteria). They also described a method for selecting tests with the highest discriminating power between the taxa suggested as possible identifications by the probability calculations. These methods were used from 1967 to 1970 for the identification of 1595 strains of bacteria. Details of the methods and the problems encountered were given in three publications. The first paper (Lapage *et al.*, 1973) dealt with the general problems of such a project, the second (Bascomb *et al.*, 1973) gave a bacteriological account, and in the third (Willcox *et al.*, 1973) the mathematical aspects were considered. At the same time, Friedman *et al.* (1973) described a probabilistic method for computer-assisted identification of bacteria using biochemical tests. This method was applied also to antimicrobial susceptibility patterns (Friedman and MacLowry, 1973). The use of probability matrices was also described by some other investigators (Robertson and MacLowry, 1974; Gyllenberg and Niemelä, 1975a,b; and see API, 1983).

Numerical identification has been the subject of several reviews (Sneath and Sokal, 1973; Pankhurst, 1974; Gyllenberg and Niemelä, 1975a,b; Gyllenberg, 1976; Sneath, 1978, 1979; Willcox *et al.*, 1980). Although techniques of numerical identification, more advanced than those suggested by Beers and Lockhart (1962), have been described (e.g., Euclidean distances and correlation coefficients), the probabilistic approach has been the most popular. Nevertheless, probability matrices are rather few in number. Bascomb *et al.* (1973) published a matrix for the identification of Gram-negative rods of clinical importance, and similar matrices have been published for *Bacteroides* (Johnson and Ault, 1978), slowly growing mycobacteria (Wayne *et al.*, 1980), and Micrococcaceae and streptococci (Feltham and Sneath, 1982). Probability matrices have been developed also for anaerobes of clinical interest (Kelley and Kellogg, 1978) and for coryneform bacteria (Hill *et al.*, 1978), but these have yet to be published. However, taxonomic papers often contain information in the form required for the construction of probability matrices, that is, they contain data on the percentage of strains of a given taxon giving positive results in a given test. Similar tables are also found in standard texts such as *Bergey's Manual of Determinative Bacteriology* (Buchanan and Gibbons, 1974). In theory at least, probability matrices can be devised from these sources. In practice, however, difficulties arise mainly as a result of the inadequate standardization of test methods and the poor classification of certain bacterial groups. There can also be computing problems as data banks increase in size. Ideally there will be a tendency to improve standardization of tests and to increase the publication of comprehensive tables of percentages of strains of various taxa giving positive results in a given test. It is likely that the identification manuals of the future will contain detailed information on the results of selected tests for the majority of bacterial species readily handled by conventional cultural techniques.

The majority of published probability matrices are based on a large number of conventional tests (often >50). Hospital and other routine laboratories cannot stock media for so many tests or easily ensure standardization of such tests; consequently these matrices have been used most successfully in reference laboratories. The two problems have, however, been tackled and largely overcome by the manufacturers of commercial identification systems, thereby bringing probabilistic identification into the hospital laboratory.

Commercial Identification Systems

Commercial identification systems, produced and quality controlled at a central source, have the potential for overcoming the problem of inadequate standardization of conventional tests. Within a laboratory the average probability of errors for conventional biochemical tests is usually 2–4%, whilst between laboratories the corresponding values are normally 6–10% (Lapage *et al.*, 1973; Sneath, 1974). Within-laboratory studies of the test reproducibility of commercial identification systems have yielded average probability of error values of ≤2% (Butler *et al.*, 1975; Holmes *et al.*, 1977c, 1979), which correspond very favourably with the range obtained with conventional tests. Between-laboratory studies of the test reproducibility of commercial identification systems have yet to be published with precise figures (but see Logan and Berkeley, 1981); it is hoped that the values obtained will be better than the corresponding values for between-laboratory conventional tests.

To be commercially viable, the number of tests in these identification systems is limited and seldom exceeds 20. With so few tests, identification rates are often poor but can be improved by adopting the probabilistic approach using a computer (Holmes *et al.*, 1977c). Even with a single commercially available system, such as the API 20E system, successive identification schemes have been produced by the manufacturer over the years and the highest identification rates are obtained with those schemes using the computer-assisted probabilistic approach (Holmes *et al.*, 1978b).

Pattern Matching

To achieve satisfactory identification rates, most manufacturers of identification systems have combined two concepts: identification by exact pattern matching and identification using probability matrices. The first concept was one of the earliest techniques tried by the manufacturers once they had recognized the shortcomings of diagnostic tables. Using a register, usually called 'profile register', identification was achieved by exact match of a pattern of test results for an unknown strain with the pattern ('profile') known to correspond to strains of a particular taxon. This approach is ideal for an identification system where the set of tests is fixed. A given taxon will generally be listed more than once, that is, it will have two or more patterns listed; thus, atypical strains can be correctly identified. If the number of tests in the identification system is low ($\leq$10), then it should be feasible to list all possible patterns. Alternatively, when the number of tests is high ($\sim$20), it will only be possible to list commonly observed patterns. A strain will not be identified if a pattern is not listed in the register. A major problem in the development of registers was to find a way of representing patterns of test results so that comparison between the unknown and the reference material would be simplified. This problem was overcome by converting the pattern of test results to a numerical code which was unique for that pattern. Such coding schemes were not commercial innovations; they had been proposed originally for use with conventional tests but did not come into common use (e.g., see Fey, 1959). Numerical coding schemes were reviewed extensively by D'Amato *et al.* (1981). Methods of coding may be found to vary slightly between identification systems, but this is due to different test-ordering conventions adopted by the various manufacturers.

Probabilistic Identification

Identification by exact match of test patterns, as used in registers, was later combined with the probabilistic approach in the form of an index, usually called 'profile index'. Reference laboratories such as the National Collection of Type Cultures (NCTC), which use the probabilistic approach in conjunction with a large number of tests ($\geq$50), tend to compute the likelihoods for the test results obtained for each new isolate. Manufacturers of identification systems, where the number of tests is much smaller, can (depending on the number of tests) process all or the commoner patterns of results possible with each system through a computer to generate printouts of the likelihoods. The pattern of results that yielded that printout can then be listed, as in a register, with the accompanying likelihoods, so that the index becomes in effect a collection of compressed computer printouts. As with registers, if the number of tests is about 10, then all possible combinations can be listed ($2^{10} = 1024$ possible patterns), and further compression may be obtained by printing the likelihoods only when a particular pattern does not achieve identification to a single taxon. With about 20 tests, it will not be feasible to list all possible patterns and the index will contain only a selection of these. The 21 tests of the API 20E system, for example, give 2^{21} (= 2,097,952) possibilities, which would fill about 10 thick volumes. Nevertheless, access to the full data base may often be obtained via a telephone service provided by the manufacturer, or the full data base may be available on a floppy disk sold by the manufacturer for use on the laboratory's own microcomputer.

During the calculation of the likelihoods from a probability matrix (Holmes, 1982a), two values are obtained: the *absolute likelihood* and the *relative likelihood.* The multiplication of the probabilities obtained from the matrix of the individual test results is the absolute likelihood. This is a measure of how well the characters of an unknown correspond to the characters of the taxon as a whole. The normalization of the absolute likelihood yields the relative likelihood and is a measure of the share of a particular taxon of the total sum of probabilities for all the taxa considered. Ideally, to identify a strain as a member of a particular taxon requires that the absolute likelihood of belonging to the taxon be sufficiently high, and also that the relative likelihood for the unknown strain be close to 1. The absolute and relative likelihoods may be expressed mathematically in different forms, for example as percentages or ratios. Both or only one of the two likelihood values may be displayed by the computer programme. Whether displayed or not, an identification decision may not be made by the computer programme from either likelihood value (Dybowski and Franklin, 1968), or an identification decision may be made on a combination of both absolute and relative likelihood values (Gyllenberg and Niemelä, 1975a). In our laboratory, the identification decision is based solely on the relative likelihood, and the identification threshold level is .999 (Lapage *et al.*, 1973). This has the slight disadvantage that a strain belonging to a taxon not included in the matrix could be misidentified as a taxon in the matrix if the strain resembled that taxon much more closely than it did any other taxon in the matrix. The problem can be overcome by all computer-suggested identifications being checked by an experienced microbiologist with whom the final identification decision rests. To overcome such difficulties, and to cope with the additional problem that differentiation between taxa will sometimes be inadequate if only a comparatively small number of tests are available, the manufacturers of identification systems have had to derive as much information as possible from the likelihood calculations. Most therefore provide both absolute and relative likelihood values, and some introduce a Euclidean distance element by expressing the absolute likelihood of the unknown relative to the absolute likelihood of a typical strain of each taxon. Some manufacturers, unfortunately, neither give in their index for each pattern of test results a statement as to the acceptability of the suggested identification (e.g., 'acceptable identification', 'very good identification'), nor indicate the threshold identification level (see next paragraph). Since the likelihood values only are printed, the onus of accepting as the correct identification that taxon with the highest (but not necessarily significant) likelihood values, is placed on the user rather than the manufacturer. In more advanced systems, the program makes an identification decision from a combination of the absolute and relative likelihood values, and comments are given as to the level of acceptability of the suggested identification. D'Amato *et al.* (1981) gave a summary of some of the identification methods using likelihoods. Although the programs employed by the manufacturers of identification systems are not usually published, the principles involved in arriving at a statement as to the level of acceptability of a suggested identification were illustrated by D'Amato *et al.* (1981).

The threshold levels that the likelihood values must exceed for the various identification decisions made in the more sophisticated commercial systems are largely unknown. However, this information has been made available by the manufacturer of the API 20E system (API, 1983). All manufacturers face the same dilemma: a high identification threshold level reduces the risk of misidentification but lowers the rate of identification; conversely, a low identification threshold level increases the rate of identification but carries a higher risk of misidentification (Lapage *et al.*, 1973). Because of the limitation on the number of tests in commercial systems, the identification threshold level will generally be lower than that acceptable to a reference laboratory. However, various features, such as recommending the serological confirmation of certain suggested identifications, may be incorporated into the manufacturers' identification schemes in order to reduce the risk of misidentification (see Holmes *et al.*, 1978b).

Various instruments for bacterial identification also appear to incorporate a probability matrix, for example, the Auto-Microbic System and Abbott MS-2. The manufacturer of the Autobac System has, however, adopted a multivariate statistical technique termed quadratic discriminant function (Sielaff *et al.*, 1976).

No one has yet pursued the lines followed by the manufacturers of identification systems and produced indices or complete data bases on floppy disks for the laboratory that wishes to continue using conventional tests. The few probability matrices that have been published contain too many tests for the routine laboratory and too many taxa, several of the latter being of little or no clinical importance. The person using conventional tests may thus turn to a published matrix and abstract from it the probability figures for a limited range of taxa and tests. The choice will vary according to the individual. Given that there will also be variation according to the individual writing the identification programme, it can be seen that there is much potential for duplication of effort and lack of standardization in the efforts of individuals to apply probabilistic identification in their own laboratories using conventional tests. Even then, probabilistic identification is only available to the laboratory with its own microcomputer; no indices are available to bring probabilistic identification to the laboratory which is without a microcomputer.

The NCTC Computer Identification Laboratory

Most of what is described in the pages that follow is based on our own experiences at the NCTC. The early developmental work, including the probability matrix, was described by Bascomb *et al.* (1973), Lapage *et al.* (1973), and Willcox *et al.* (1973). The matrix (Bascomb *et al.*, 1973) comprised 56 fermentative and 14 nonfermentative taxa, and proved successful for the identification of fermenting bacteria (90.8%) but less so for the non-fermenting bacteria (82.1%). The reason for the lack of success with non-fermenting strains was the unsuitability of certain tests in the matrix. Thus, methyl red and Voges–Proskauer tests are invariably negative with non-fermenting strains and in peptone–water–sugar media, saccharolytic non-fermenting bacteria generally give negative results as any acid produced from the carbohydrate is masked by parallel alkali production from the peptone. Since these tests accounted for nearly half those in the matrix, the number of tests available for discrimination was very much reduced for the non-fermenting strains.

It was decided to improve the identification of the non-fermenting bacteria by incorporating the probabilities for more suitable tests into a separate matrix for these organisms. For example, ammonium salt sugars, originally used for differentiation of *Bacillus* species, gave clear reactions when used to detect acid production from carbohydrates by non-fermenting bacteria. The full range of tests and taxa used has been described (Willcox *et al.*, 1980). The separate matrix worked well for saccharolytic non-fermenting bacteria, but again there was poor differentiation between the non-saccharolytic taxa, principally the alkali-producing pseudomonads, *Branhamella, Kingella, Moraxella,* and *Neisseria.* Although constructed over a number of years and used for routine identification purposes, the matrix for the non-fermenting strains has not been published. In the development of the matrix, several taxonomic studies proved necessary, either to establish new species or to give revised descriptions of species previously not known to occur with any regularity in clinical specimens (see Holmes and Roberts, 1981; Holmes *et al.*, 1977a,b, 1978a, 1981, 1982; King *et al.*, 1979).

Rather than construct a third matrix with tests suitable for discrimination between the non-saccharolytic, non-fermenting strains, it would seem more appropriate to construct a single new matrix based on test methods which would be equally applicable to all the taxa in the two existing matrices, for example, specific enzyme tests, or computer-stored profiles of protein patterns derived by electrophoretic techniques (see Kersters, Chapter 13).

Inauguration of Identification Service

Until 1971, a few laboratories submitted strains to NCTC for identification. From 1972 on it was decided to offer a service to more laboratories, and approximately 800 strains were received each year. Several of the sending laboratories are abroad, so bacteria rarely encountered in clinical material in the United Kingdom are received, for example *Pseudomonas pseudomallei* from Malaysia. The service is offered for the identification of so-called difficult organisms, and strains fall into this category for a variety of reasons:

1. Incorrect test results may have been obtained by the sending laboratory.
2. The culture might be mixed.
3. The unknown strain may belong to a taxon which is rare or unfamiliar to the sender.
4. The unknown may be an atypical strain of a well-known taxon and may not be recognized as such by the sender, or the sender may suspect that this is the case but would like confirmation.
5. The strain may belong to a taxon in a poorly classified group.
6. The strain may belong to a new taxon.

The way in which the service operates has been described fully elsewhere (Willcox *et al.*, 1980). With the original 1000 strains used to develop the published matrix, and the 9000 received during the operation of the identification service to date, test results are available for some 10,000 strains. Each strain examined in our laboratory usually undergoes several computer ‘runs’ during which the test results obtained up to that point are processed through the probability matrix for calculation of the likelihoods. On the first such run, certain alpha-numeric data are also entered in the computer. These data comprise our reference number for the strain, the matrix to be used for the probability calculations, the type of run, the run number, and a code number specific to the sending laboratory. With this code, the full address of the sender can be located in the appropriate computer file. Also entered at this time is the sender’s reference number for the strain and the name of the patient. This information is stored in the computer on the current file, together with the test results that are overwritten on successive computer runs. The identification score resulting from the probability calculations is also stored. When all the tests on a strain have been carried out, all runs completed, and the identification reported to the sender, the strain is given a final designation code. These codes are allotted to each taxon in the matrix and to new taxa possibly to be included at some future revision of the matrix. Periodically the data on completed strains are transferred from the current file to an archive file, and at this point the final designation code is also stored in the computer. At this time the taxon to which the sender thought the strain might belong is also stored; this is achieved using the same code numbers as those used for final designations. Following an archive run, a printout is obtained of the current file, listing the strains in numerical order. In this way a check can be made on the progress of strains, and any that appear to have been in the current file for an abnormally long period can be readily picked out and investigated. As mentioned above, the archive file now contains data on some 10,000 strains.

Production of Data Summaries

When the separate matrix on the non-fermenting bacteria had been compiled, 10 reference strains of each taxon, usually a species, were obtained and tested in order to derive the probability figures for the matrix. Although most strains of, for example, *Pseudomonas maltophilia* were successfully identified, occasional strains displayed characteristics not observed in the 10 strains of this species used to compile the original matrix. There was no way of knowing accurately just how commonly strains with 'unusual' characteristics were being encountered. Since the archive file by now contained the test results on 281 strains of *P. maltophilia,* it became desirable to retrieve data from the archive file in the form of printed summaries. These data were then used to revise the existing matrices so that for *P. maltophilia,* for example, the probability figures were now based on 281 strains, not the original 10. At the time of such revisions, new taxa can be added to derive revised matrices, which then need thorough evaluation before being adopted for the routine identification service or being published. By use of the final designation codes, computer programmes can sort the archived strain data by taxa, and print for each taxon, five separate summary pages which are headed, respectively:

1. Strain data
2. Computer results
3. Cumulative results (A)
4. Delta diagrams
5. Cumulative results (B)

These pages are described fully below:

1. Strain Data. This page lists the alphanumeric data entered for each strain and records the laboratory from which the strain was sent, the sender's reference number for the strain, the source or patient's name, and the taxon to which the sender thought the unknown strain might belong.

2. Computer Results. This page indicates for each strain the type of run on which the identification was reached, how many computer runs it had undergone, on which version of the matrix the likelihoods were calculated, and how many tests were carried out. Then the name of the taxon which received the highest identification score is listed, followed by the actual score. If a strain failed to reach identification level, the name of the taxon which received the second highest identification score is also listed, followed by its actual score. If these two taxa are closely related and the sum of their identification scores exceeds .999, they may be linked by the computer program and allowed to identify as a composite group (usually where a strain possesses characteristics intermediate between the two taxa). On this page it will also be stated when such composite group identifications occur, and there will also be an indication of the number of unexpected test results shown by any one strain. This page thereby permits the rapid determination of the identification rate for each taxon. Strains not reaching identification level and displaying several characters unexpected for the taxon receiving the highest identification score, may not belong to the taxon under consideration and may in some cases have received an inappropriate final designation. For example, the page for *Acinetobacter calcoaceticus* shows that, of 240 strains tested only 2 failed to reach identification level, even though some of these were processed only through the original matrix of Bascomb *et al.* (1973) and not through the later, separate matrix for nonfermenting bacteria. A different picture is seen for *A. lwoffii.* Of the 317 strains examined only 16% (10/61) reached identification level on the original matrix of Bascomb *et al.* (1973), whereas with the separate matrix for nonfermenting bacteria, the identification rate was 96% (245/256).

3. Cumulative Results (A). This page lists the individual results obtained for each strain in all the tests in which it was examined.

4. Delta Diagrams. This page lists the tests in which all strains of the taxon gave positive results followed by the tests in which all strains gave negative results. It also lists which strains gave identical results to each other in the range of tests carried out. Finally, the computer prints for the strains that differ from each other, all the tests in which different strains gave different results. In summary pages 1–3, all the strains of a particular taxon and the accompanying data are listed in strict numerical order. On summary page 4, however, the strains and test results are ordered according to the method of Lapage and Willcox (1974). This has the effect of placing the most typical strains on the right and the least typical to the left. Thus, if one wishes to evaluate a revised matrix by simply running through the new matrix the test results for strains identified on an earlier matrix, this page is most useful. It allows strains to be chosen showing varying degrees of atypicality, and by listing strains with identical results, inadvertent duplication can be avoided. Finally, since the most atypical strains appear at the left, it is easy to see if particular strains possess several characters in which they differ from all other strains in the taxon and therefore possibly warrant exclusion from the taxon.

5. Cumulative Results (B). This page summarises the test results obtained for each taxon. It states how many strains of the taxon have been received and then for each test how many strains were tested, how many of those gave positive results, and the percentage of strains giving positive results in each test. Next to each of these percentage figures appears the corresponding probability figure in the latest version of the matrix. It can be seen that this page is the most important from the point of view of revising the two current matrices in routine use. We know the current probability figure, in the case of *Pseudomonas maltophilia* based on 10 strains, and can see at a glance if that probability figure needs altering now that it can be seen what proportion of the 281 strains of *P. maltophilia* gave positive results in a particular test.

Use of Data Summaries to Revise Probability Matrices

It can be seen that the five summary pages are interrelated and any change on one requires alteration of the other four. It is therefore necessary for them to be re-run periodically. Armed with the summaries, it at first seemed a relatively straightforward task to revise the matrices then in current use. Because of the large number of strains on which the revised probability figures were to be based, it was decided to use the actual percentage figure of strains positive in a given test rather than rounding them to the nearest 5 as was the case with the current matrices. This approach meant scrutinizing percentage figures like 2 or 98, as these represented minority results for the taxon. Each was checked in case the figure was incorrect, because a strain was given the wrong final designation code or an incorrect test result had been stored in the computer. As corrections took place, many of the percentage figures had to be recalculated.

Revision of the Matrix for the Identification of Non-fermenting Strains

For the non-fermenting strains the task appeared easier, since there were no plans to add new taxa but merely to revise the matrix figures for the existing taxa. As previously mentioned, a number of taxonomic studies on various taxa included in the matrix had been undertaken so that many of the original taxonomic problems had been resolved. Thus, in the case of *Agrobacterium* the various species had originally been defined largely according to their phytopathogenic properties, but subsequently classification based on phenotypic characters was proposed (Holmes and Roberts, 1981). Following this study the final designation codes for the *Agrobacterium* strains were amended accordingly, and so in the summaries the cumulated percentage figures were correct. In other cases, for example *Flavobacterium* taxa, certain taxonomic studies had yet to be published (Holmes, 1983; Holmes *et al.*, 1983) and the final designation codes had not been amended prior to printing the summaries. In these cases the percentage figures were derived directly from the results of the unpublished studies. A continual problem was the inability to differentiate between the reference strains of *Achromobacter* species biotypes 1 and 2 using the differential tests described in the literature (Tatum *et al.*, 1974). In addition, there were 56 strains that did not conform to either biotype. The reference strains of *Achromobacter* species biotypes 1 and 2 plus these 56 *Achromobacter*-like strains have now been the subject of a numerical taxonomic analysis designed to clarify the taxonomy of these organisms (Holmes and Dawson, 1983).

Alcaligenes was also a problem; attempts to distinguish between *A. faecalis* and *A. odorans* were not successful. Later publications showed that *A. odorans* is a synonym of *A. faecalis* (Kiredjian *et al.*, 1981; Rüger and Tan, 1983). Consequently, the percentage figures for *A. faecalis* and *A. odorans* in the summaries were combined. A further difficulty lay with members of nonreactive genera such as *Branhamella, Moraxella,* and *Neisseria.* We had not conducted numerical taxonomic studies on these bacteria and so could not base our taxa in the matrix on the clusters formed. We therefore selected those species which we thought could be identified satisfactorily on the data available, for example *Moraxella osloensis, M. phenylpyruvica,* and *M. urethralis. Moraxella bovis, M. equi, M. lacunata,* and *M. liquefaciens,* most strains of which produce gelatinase, were placed together in the '*Moraxella* proteolytic group'. Similarly, particularly non-reactive species of the genera *Branhamella* and *Neisseria,* together with *Moraxella nonliquefaciens,* were placed in a single '*Branhamella–M. nonliquefaciens–Neisseria*' taxon. For many of these taxa the printed summaries proved of little value and the percentage figures had to be derived manually. As part of the revision exercise, names were revived for *Flavobacterium breve* (Holmes and Owen, 1982) and *Pseudomonas cepacia* (Palleroni and Holmes, 1981), which were included in the matrix but did not appear on the *Approved Lists of Bacterial Names* (Skerman *et al.*, 1980).

In its present form, the revised matrix on the non-fermenting bacteria comprises 59 taxa and 68 tests, compared with the 14 taxa and 50 tests in the original matrix of Bascomb *et al.* (1973). Six further taxa will be added now our *Achromobacter* study has been completed (Holmes and Dawson, 1983).

Revision of the Matrix for the Identification of Fermenting Strains

Many new taxa described since the original matrix was published had to be included in the revised matrix. As new species descriptions were published, suitable reference strains were obtained and characterised using our own range of tests. Probability figures were readily available for many such taxa from the summary pages. For strains of other taxa received since the summaries were last printed, the test results and percentages were compiled manually. Many of the new taxa, however, had been regarded formerly as atypical strains of existing species. In some cases, these new taxa were recognized before the summaries were printed and were given a new final designation code. Therefore, when the summaries were printed the figures were correct both for the new taxon and for the taxon from which the strains of the new species had been removed. In other cases, a new species was recognized only after the summaries were printed, and this meant compiling manually the probability figures for both the new taxon and the revised old taxon. The matrix of Bascomb *et al.* (1973) contained strains designated '*Citrobacter* matrix biotype 1' and '*Citrobacter* matrix biotype 2' distinguished largely on the results for H_2S production in triple sugar iron agar and indole production. Ewing and Davis (1972) showed that the strains of these two taxa should have been classified as those tolerating KCN but not fermenting adonitol (*C. freundii*) and those giving the converse results in these two tests (*C. koseri,* synonym *C. diversus*). The final designation codes for these strains were revised accordingly and so, when the summaries were printed, there were separate pages for *C. freundii* and for *C. koseri.* The next problem was the later recognition that several of the *C. freundii* strains (those that produced indole and failed to form H_2S) were members of the more recently described species *C. amalonaticus* (formerly *Levinea amalonatica;* see Farmer, 1981). It seemed a relatively simple matter to sort the *C. freundii* strains according to these characters and derive new probability figures either manually or by re-printing the relevant summary pages, but unfortunately several strains were found to produce indole as well as H_2S. It is not clear if the latter should be regarded as atypical *C. freundii* strains. A numerical taxonomic analysis would help resolve the problem, but in the meantime we took 10 reference strains of *C. amalonaticus* and 10 of *C. freundii* and based the probability figures for the revised matrix on them rather than on the 287 strains listed as *C. freundii* in the summaries.

Enterobacter sakazakii strains, which form yellow-pigmented colonies and fail to ferment sorbitol, were relatively easily removed from *E. cloacae* where they were formerly placed. Non-pigmented strains failing to ferment sorbitol were left in *E. cloacae,* as it was known that some strains of this species are negative in those tests. Pending further taxonomic study, strains of Group EF-4 (Holmes and Ahmed, 1981) were regarded as a single taxon when the summaries were printed. However, more recent DNA base composition data suggest that strains producing arginine dihydrolase should be regarded as a taxon separate from those not having an arginine dihydrolase. It was decided to establish two separate biovars of Group EF-4 in the revised matrix, and again this meant arriving at the respective probability figures largely by manual methods. In the original matrix of Bascomb *et al.* (1973), there was a combined taxon for *Klebsiella aerogenes* and *K. oxytoca*. Subsequently, Jain *et al.* (1974) showed by DNA–DNA hybridization that strains of *K. oxytoca* were so different from other *Klebsiella* strains that they should be placed in a separate genus. Since strains of *K. oxytoca* differ from those of *K. aerogenes* in producing indole, strains of the former could be picked out easily and given a separate final designation, so that they all appeared on a single page when the summaries were printed.

A more recent development was the recognition that three biogroups of *Proteus vulgaris* represent separate species. The name *P. penneri* was proposed for biogroup 1 (Hickman *et al.*, 1982b), whilst the type strain of *P. vulgaris* belonged to biogroup 3. No name was proposed for biogroup 2. The summaries contained all the test results for the three biogroups as a single page for *P. vulgaris,* so the probability figures for each of the three biogroups had to be derived manually. The same problem arose for strains formerly classified as *Vibrio cholerae* but atypical in failing to ferment sucrose. Such strains are now recognized as a separate species, *V. mimicus* (Davis *et al.*, 1981). Similarly, *Yersinia enterocolitica* contained strains that have been assigned to three separate species: *Y. frederiksenii* (Ursing *et al.*, 1980), *Y. intermedia* (Brenner *et al.*, 1980), and *Y. kristensenii* (Bercovier *et al.*, 1980). Even *Y. enterocolitica sensu stricto* poses problems. Five biovars are recognized, one of which (biovar 5) differs from the other four in failing both to reduce nitrate and to ferment trehalose. Since there was only one strain conforming to biovar 5, it was not included in the revised matrix as a separate taxon. Subsequently it was found that this strain identified as *Y. enterocolitica* with a score of 1.000 in the revised matrix, so our strategy had been correct. There was also the possibility that some *Y. enterocolitica* strains which fermented raffinose could have been strains of *Y. intermedia*. In fact, the strains in question proved to be plasmid-bearing *Y. enterocolitica* strains capable of fermenting both lactose and raffinose as described by Cornelis *et al.* (1978).

A similar problem arose with *Vibrio cholerae*. The original matrix of Bascomb *et al.* (1973) contained only two *Vibrio* species: *V. cholerae* and *V. parahaemolyticus*. The former obviously contained the non-O1 serovar strains or non-cholera vibrios as well as the pathogenic variety of *V. cholerae*. However, because of fears of unnecessary panic in sending laboratories if they received computer printouts showing *V. cholerae* as the most likely taxon to which their unknown strain belonged, the name of the taxon was changed to '*Vibrio* spp.—not *parahaemolyticus*'. This removed possible confusion from sending laboratories but caused some difficulty within our own laboratory! Although all identifications are checked by an experienced microbiologist, it is left to the junior staff to enter the appropriate final designation codes. Pathogenic *V. cholerae* strains, as well s non-O1 strains biochemically identical to them, all received the final designation code for '*Vibrio* spp.—not *parahaemolyticus*.' The choice of this name proved most unfortunate, as successive junior staff members who were not fully aware of the situation, also gave the final designation code for this name to strains of other *Vibrio* species such as *V. anguillarum* and *V. vulnificus*. The summary page for '*Vibrio* spp.—not *parahaemolyticus*' thus contained the cumulative figures for several *Vibrio* species, including *V. mimicus,* and again probability figures for the various species had to be derived manually. Only 23 of the 42 strains on that summary page were finally included in *V. cholerae*.

In some cases taxa were merged in the revised matrix, although they had been separate in the original one of Bascomb *et al.* (1973). Strains formerly classified as *Aeromonas formicans* are now regarded as atypical *A. hydrophila. Escherichia coli* and *Shigella* clearly constitute a single species (Brenner *et al.,* 1972a, 1973), and although they are kept separate for practical and historical reasons, it is not surprising that intermediate strains occur. To allow the correct identification of these intermediate strains in the matrix, Bascomb *et al.* (1973) had a separate taxon for them: '*Escherichia:* Alkalescens–Dispar group'. However, the Alkalescens–Dispar strains are now regarded as atypical *E. coli* (Edwards and Ewing, 1972). In order to reflect current taxonomic thinking, and yet to allow for the identification of these strains, probability figures were derived for a single taxon *E. coli,* redefined to include the Alkalescens–Dispar strains. For this taxon, and for the two *Shigella* taxa in the matrix (*S. sonnei* and *Shigella* spp.—not *sonnei*) only, we tested strains and derived probability figures for three new tests in which *E. coli* strains are generally positive and *Shigella* strains generally negative: alkali production on Christensen's citrate, acetate utilization, and mucate fermentation. Although these three tests are included in the matrix, there are only probability figures for the three taxa above; probability figures are not currently available for the remaining taxa. A similar problem arose with *Klebsiella terrigena.* The test results for strains of all the more recently recognized taxa were processed through the un-revised matrix. As expected, most failed to achieve significant identification scores, which suggested that when the recently recognized taxa were included in the matrix, strains of these taxa should identify successfully. Strains of *K. terrigena,* however, misidentified as *K. aerogenes* and *K. oxytoca* without unusual results. Therefore, to include *K. terrigena* in the revised matrix, we examined strains of *K. terrigena* and of the other *Klebsiella* species, and added probability figures for two new tests (Izard *et al.*, 1981)—growth at 5°C and at 42°C—with which it was hoped to attain satisfactory differentiation of the *Klebsiella* species in the revised matrix.

For the routine identification service, use is made of the sender's test results, which are run through the appropriate matrix. Strains of *Acinetobacter cal-*

coaceticus do not produce oxidase but produce acid from glucose and often from arabinose and xylose in peptone–water medium. Unless the sender has also carried out the Hugh and Leifson O–F test (oxidative) or nitrate reduction test (negative), such strains may initially be taken as fermenting strains and their results processed through the apparently appropriate matrix. To allow for this possibility, *A. calcoaceticus* is included in both the matrix for the fermenting organisms and that for the nonfermenting strains. Similarly, other non-fermenting organisms, such as *Pseudomonas pseudomallei,* can also be identified using either of the two matrices.

Nomenclature also proved to be a difficulty in revising the matrix for fermenting bacteria. A major problem is that clinicians in the United Kingdom refer to typical klebsiellas as *K. aerogenes* and think of *K. pneumoniae* as biochemically atypical klebsiellas from the respiratory tract. In the United States, however, all of these klebsiellas are referred to as *K. pneumoniae,* the type species of the genus. *Klebsiella aerogenes* does not appear on the *Approved Lists of Bacterial Names* (Skerman *et al.*, 1980) and is currently without standing in nomenclature. To complicate the picture further, *K. pneumoniae, K. ozaenae,* and *K. rhinoscleromatis* should be regarded as a single species (Brenner *et al.*, 1972b). In order to resolve this problem, the revised matrix for the fermenting bacteria contains entries for the three species *K. oxytoca, K. pneumoniae,* and *K. terrigena,* but *K. pneumoniae* is divided into four subspecies—*aerogenes, ozaenae, pneumoniae,* and *rhinoscleromatis*—making six *Klebsiella* taxa in all. In its present form the revised matrix comprises 110 taxa compared with 56 in the original model of Bascomb *et al.* (1973). Other new taxa have been described such as *K. planticola* (Bagley *et al.*, 1981) and *Vibrio hollisae* (Hickman *et al.*, 1982a). Reference strains of these species will have to be examined before the taxa can be included in a further revision of the matrix.

Evaluation of Revised Matrices

Both of the revised matrices were initially evaluated by examining the test results for the type strain of each species or a typical strain in the case of unnamed taxa. There was almost 100% success with the matrix for the fermenting bacteria. Typical strains of Group EF-4 (arginine-negative biovar) and *Neisseria denitrificans* did not reach the identification level to their respective taxa, as both these organisms are largely unreactive in the tests and so cannot be differentiated adequately from each other. More importantly, the type strains of *Hafnia alvei* and *Pasteurella multocida* also failed to reach the identification level, but in the set of tests both strains were atypical in their biochemical reactions. It is of interest to note that even on the original matrix of Bascomb *et al.* (1973) the type strains of these two taxa failed to reach identification level, so the performance of the revised matrix was no worse in this respect. More typical strains of these two taxa did reach the acceptable identification level.

The initial evaluation of the revised matrix for the non-fermenting strains was also encouraging, but the success rate was, not unexpectedly, lower than for the other matrix. The difficulty of distinguishing the largely unreactive taxa of the genera *Branhamella, Brucella, Kingella, Moraxella,* and *Neisseria* was the most serious problem. In the revised matrix the various *Brucella* species were combined in a single taxon, '*Brucella* spp.' The type strains of *B. abortus* and *B. neotomae* failed to reach the identification level, although the type strains of the other *Brucella* species included in this taxon were identified. A biochemically atypical *B. abortus* did reach identification level. There was no alternative but to accept that only strains of certain *Brucella* species would reach identification level and possibly only atypical strains of *B. abortus* and *B. neotomae* would do so. The revised matrix contained the three species of *Kingella* as separate taxa, but only the type strain of *K. indologenes* identified satisfactorily. On merging *K. denitrificans* and *K. kingae* into a single taxon, '*Kingella* spp. other than *indologenes*', the type strain of *K. kingae* would still not identify to the combined taxon. It was thought that this change offered no real advantage, so the original position with the three species as separate taxa was taken. Although only *K. indologenes* strains would identify, the identification scores were so high for the type strains of *K. denitrificans* and *K. kingae* that there should be no problem for an experienced microbiologist to determine the identity of such strains. The taxon '*Branhamella–Moraxella nonliquefaciens–Neisseria*' comprised the most unreactive taxa in the matrix, and several species were contained within it. The type strains of the 10 constituent species all failed to reach identification level. This was not unexpected given the tests used in the revised matrix. As stated earlier, a third matrix with still different tests is needed for the differentiation of the nonsaccharolytic, non-fermenting strains. Similarly, with the *Moraxella* proteolytic group, which is composed of several species, the type strains of *M. bovis* and *M. lacunata* failed to reach identification level. It is not possible to resolve the problem at present.

The initial evaluation of the matrix on the fermenting strains revealed that when certain taxa had been later recognized as comprising more than one species, and these species had been represented in the matrix as separate taxa, differentiation between them was sometimes inadequate. To overcome this problem more composite groups were allowed, so that strains could reach identification level when the sum of the scores for two biochemically similar taxa (e.g., *Citrobacter amalonaticus* and *C. freundii*) exceeded the identification threshold level of .999. Some such composite groups were also allowed in the revised matrix for the non-fermenting strains.

Further evaluation of both matrices was undertaken with the least typical strain given that final designation code. This approach also caused problems. Atypical strains, which previously identified to a taxon but with unusual results, generally identified to the same taxon in the revised matrix but now without unusual results because allowance had been made for these in the revised probability figures. For other strains, however, that still reached identification level with unusual results, or failed to reach identification level, careful assessment was needed to determine if such strains should have been given that final designation code originally. Depending on whether or not the strain was a definite member of the taxon, it could have correctly or incorrectly reached identification level, or if it failed to identify then it might have been correct for it not to do so. Consequently, it was difficult to judge the performance of the revised matrices using strains that were too atypical. It is possible that some of these atypical strains may represent new species. Indeed, over the years a number of strains which could not be identified at the time were given the final designation code 'Irregular Enterobacteriaceae'. Amongst these strains there may be representatives of new species, but several could belong to the newly described taxa added to the revised matrix for the fermenting strains. Re-running the test results for such strains through the revised matrix should indicate if any of the strains received in the past fall into the 'new' taxa.

Further evaluation of the matrices is under way by rerunning the test results of additional strains of each taxon up to a maximum of 10 strains per taxon. Strains being identified routinely are being processed in parallel through the original routine matrix and the revised one. When all the test strains have been re-run through the revised matrix, the summaries will be re-printed. Using the revised summaries it will be possible to determine the overall identification rate of the revised matrices, compare the rates with those of the original matrix of Bascomb *et al.* (1973), and also determine the identification rate for each taxon. Provisional identification rates for the 21 species of *Pseudomonas* in the revised matrix for the non-fermenting strains were very encouraging (≥88% for most taxa), although not unexpectedly the rates were lower for the largely non-saccharolytic, alkali-producing pseudomonads (67–100%). Programmes for the statistical analysis of probability matrices have been described. One programme calculates the best identification score that an entirely typical strain of each taxon could achieve (Sneath, 1980a); another determines the extent to which pairs of taxa overlap (Sneath, 1980b). Such programmes should be of great value to us in the further evaluation of our matrices.

The Future of Computers in Bacterial Identification

Clinical microbiology has lagged behind other medically important disciplines such as clinical chemistry. Even the majority of the various commercial identification systems for bacteria are not a major advance over traditional biochemical tests performed in test tubes. For such tests, which mostly yield binary data (presence or absence characters), the probabilistic approach described above is satisfactory and is not likely to be further developed. Other approaches such as

Euclidean distances have not been widely used with traditional test methods, as they seem to offer little advantage over the probabilistic approach. Close agreement has been reported between likelihood and Euclidean distance methods in the identification of 223 isolates of bacteria (Gyllenberg and Niemelä, 1975a). Euclidean distance and other taxon-radius identification models may prove more suitable for the analysis of data derived from other taxonomic methods such as electrophoresis. The latter technique is a promising new approach to bacterial identification. An unknown strain could be identified by comparison of its profile of protein patterns with those obtained for reference strains and stored in a computer (see Kersters, Chapter 13). Other more recent approaches to microbial identification, such as specific-enzyme tests, pyrolysis–gas chromotography, or pyrolysis–mass spectrometry (see Gutteridge *et al.,* Chapter 14; Holmes, 1982b), attempt to discriminate between taxa using quantitative data, as results obtained with these techniques tend to be qualitatively similar. New approaches are being explored for the interpretation of quantitative data, and these include discriminant or canonical variates analysis (see Gutteridge *et al.,* Chapter 14; Macfie *et al.,* 1978).

References

API. (1983). API 20E Analytical Profile Index. API System, La Balme Les Grottes.

Bagley, S. T., Seidler, R. J., and Brenner, D. J. (1981). *Klebsiella planticola* sp. nov.: a new species of Enterobacteriaceae found primarily in nonclinical environments. *Current Microbiology* **6,** 105–109.

Bascomb, S., Lapage, S. P., Curtis, M. A., and Willcox, W. R. (1973). Identification of bacteria by computer: identification of reference strains. *Journal of General Microbiology* **77,** 291–315.

Beers, R. J., and Lockhart, W. R. (1962). Experimental methods in computer taxonomy. *Journal of General Microbiology* **28,** 633–640.

Bercovier, H., Ursing, J., Brenner, D. J., Steigerwalt, A. G., Fanning, G. R., Carter, G. P., and Mollaret, H. H. (1980). *Yersinia kristensenii:* a new species of Enterobacteriaceae composed of sucrose-negative strains (formerly called atypical *Yersinia enterocolitica* or *Yersinia enterocolitica*–like). *Current Microbiology* **4,** 219–224.

Brenner, D. J., Fanning, G. R., Skerman, F. J., and Falkow, S. (1972a). Polynucleotide sequence divergence among strains of *Escherichia coli* and closely related organisms. *Journal of Bacteriology* **109,** 953–965.

Brenner, D. J., Steigerwalt, A. G., and Fanning, G. R. (1972b). Differentiation of *Enterobacter aerogenes* from klebsiellae by deoxyribonucleic acid reassociation. *International Journal of Systematic Bacteriology* **22,** 193–200.

Brenner, D. J., Fanning, G. R., Miklos, G. V., and Steigerwalt, A. G. (1973). Polynucleotide sequence relatedness among *Shigella* species. *International Journal of Systematic Bacteriology* **23,** 1–7.

Brenner, D. J., Bercovier, H., Ursing, J., Alonso, J. M., Steigerwalt, A. G., Fanning, G. R., Carter, G. P., and Mollaret, H. H. (1980). *Yersinia intermedia:* a new species of Enterobacteriaceae composed of rhamnose-positive, melibiose-positive, raffinose-positive strains (formerly called *Yersinia enterocolitica* or *Yersinia enterocolitica*-like). *Current Microbiology* **4,** 207–212.

Buchanan, R. E., and Gibbons, N. E. (eds.) (1974). 'Bergey's Manual of Determinative Bacteriology', 8th edition. Williams & Wilkins, Baltimore, Maryland.

Butler, D. A., Lobregat, C. M., and Gavan, T. L. (1975). Reproducibility of the Analytab (API 20E) system. *Journal of Clinical Microbiology* **2,** 322–326.

Cornelis, G., Luke, R. K. J., and Richmond, M. H. (1978). Fermentation of raffinose by lactose-fermenting strains of *Yersinia enterocolitica* and by sucrose-fermenting strains of *Escherichia coli. Journal of Clinical Microbiology* **7,** 180–183.

D'Amato, R. F., Holmes, B., and Bottone, E. J. (1981). The systems approach to diagnostic microbiology. *CRC Critical Reviews in Microbiology* **9,** 1–44.

Davis, B. R., Fanning, G. R., Madden, J. M., Steigerwalt, A. G., Bradford, H. B., Jr., Smith, H. L., Jr., and Brenner, D. J. (1981). Characterization of biochemically atypical *Vibrio cholerae* strains and designation of a new pathogenic species, *Vibrio mimicus. Journal of Clinical Microbiology* **14,** 631–639.

Dybowski, W., and Franklin, D. A. (1968). Conditional probability and the identification of bacteria: a pilot study. *Journal of General Microbiology* **54,** 215–229.

Edwards, P. R., and Ewing, W. H. (1972). 'Identification of Enterobacteriaceae'. Burgess, Minneapolis, Minnesota.

Ewing, W. H., and Davis, B. R. (1972). Biochemical characterization of *Citrobacter diversus* (Burkey) Werkman and Gillen and designation of the neotype strain. *International Journal of Systematic Bacteriology* **22,** 12–18.

Farmer, J. J., III. (1981). The genus *Citrobacter. In* 'The Prokaryotes, a Handbook on Habitats, Isolation, and Identification of Bacteria' (eds. M. P. Starr, H. Stolp, H. G. Trüper, A. Balows, and H. G. Schlegel), pp. 1140–1147, Springer-Verlag, Berlin.

Feltham, R. K. A., and Sneath, P. H. A. (1982). Construction of matrices for computer-assisted identification of aerobic Gram-positive cocci. *Journal of General Microbiology* **128,** 713–720.

Fey, H. (1959). Differenzierungsschema für gramnegative aerobe Stäbchen. *Schweizerische Zeitschrift für Allgemeine Pathologie und Bakteriologie* **22,** 641–652.

Friedman, R., and MacLowry, J. (1973). Computer identification of bacteria on the basis of their antibiotic susceptibility patterns. *Applied Microbiology* **26,** 314–317.

Friedman, R. B., Bruce, D., MacLowry, J., and Brenner, V. (1973). Computer-assisted identification of bacteria. *American Journal of Clinical Pathology* **60,** 395–403.

Gyllenberg, H. G. (1963). A general method for deriving determination schemes for random collections of microbial isolates. *Annales Academiae Scientarum Fennicae A, IV Biologica* **69,** 1–23.

Gyllenberg, H. G. (1965). A model for computer identification of micro-organisms. *Journal of General Microbiology* **39,** 401–405.

Gyllenberg, H. G. (1976). Development of reference systems for automatic identification of clinical isolates of bacteria. *Archivum Immunologiae et Therapiae Experimentalis* **24,** 1–19.

Gyllenberg, H. G., and Niemelä, T. K. (1975a). Basic principles in computer-assisted identification of microorganisms. *In* 'New Approaches to the Identification of Microorganisms' (Eds. C.-G. Hedén and T. Illeni), pp. 201–223. Wiley, New York.

Gyllenberg, H. G., and Niemelä, T. K. (1975b). New approaches to automatic identification of microorganisms. *In* 'Biological Identification with Computers' (Ed. R. J. Pankhurst), pp. 121–136. Academic Press, London.

Hickman, F. W., Farmer, J. J., III, Hollis, D. G., Fanning, G. R., Steigerwalt, A. G., Weaver, R. E., and Brenner, D. J. (1982a). Identification of *Vibrio hollisae* sp. nov. from patients with diarrhea. *Journal of Clinical Microbiology* **15,** 395–401.

Hickman, F. W., Steigerwalt, A. G., Farmer, J. J., III, and Brenner, D. J. (1982b). Identification of *Proteus penneri* sp. nov., formerly known as *Proteus vulgaris* indole negative or as *Proteus vulgaris* biogroup 1. *Journal of Clinical Microbiology* **15,** 1097–1102.

Hill, L. R., and Silvestri, L. G. (1962). Quantitative methods in the systematics of Actinomycetales. III. The taxonomic significance of physiological–biochemical characters and the construction of a diagnostic key. *Giornale di Microbiologia* **10,** 1–28.

Hill, L. R., Lapage, S. P., and Bowie, I. S. (the late). (1978). Computer assisted identification of coryneform bacteria. *In* 'Coryneform Bacteria' (Eds. I. J. Bousfield and A. G. Callely), pp. 181–215. Academic Press, London.

Holmes, B. (1982a). Computer-assisted microbial identification as applied to conventional tests and identification kits. *Medical Laboratory World* (Jan.) 21–23.

Holmes, B. (1982b). New approaches to rapid microbial identification. *In* 'Rapid Methods and Automation in Microbiology (Proceedings of the Third International Symposium on Rapid Methods and Automation in Microbiology, Washington D.C., May 1981)' (Ed. R. C. Tilton), pp. 220–223. American Society for Microbiology, Washington, D.C.

Holmes, B. (1983). The taxonomy of the genus *Flavobacterium. In* 'Gram Negative Bacteria of Medical and Public Health Importance: Taxonomy–Identification–Applications'. *Les Editions INSERM* **114,** 273–294.

Holmes, B., and Ahmed, M. S. (1981). Group EF-4: A *Pasteurella*-like organism. *In* 'Haemophilus, Pasteurella and Actinobacillus' (Eds. M. Kilian, W. Frederiksen, and E. L. Biberstein), pp. 161–174. Academic Press, London.

Holmes, B. and Dawson, C. A. (1983). Numerical taxonomic studies on *Achromobacter* isolates from clinical material. *In* 'Gram Negative Bacteria of Medical and Public Health Importance: Taxonomy—Identification—Applications'. *Les Editions INSERM* **114,** 331–341.

Holmes, B., and Owen, R. J. (1982) *Flavobacterium breve* sp. nov., nom. rev. *International Journal of Systematic Bacteriology* **32,** 233–234.

Holmes, B., and Roberts, P. (1981). The classification, identification and nomenclature of agrobacteria. Incorporating revised descriptions for each of *Agrobacterium tumefaciens* (Smith & Townsend) Conn 1942, *Agrobacterium rhizogenes* (Riker *et al.*) Conn 1942, and *Agrobacterium rubi* (Hildebrand) Starr & Weiss 1943. *Journal of Applied Bacteriology* **50,** 443–467.

Holmes, B., Owen, R. J., Evans, A., Malnick, H., and Willcox, W. R. (1977a). *Pseudomonas paucimobilis,* a new species isolated from human clinical specimens, the hospital environment, and other sources. *International Journal of Systematic Bacteriology* **27,** 133–146.

Holmes, B., Snell, J. J. S., and Lapage, S. P. (1977b). Revised description from clinical isolates of *Flavobacterium odoratum* Stutzer and Kwaschnina 1929, and designation of the neotype strain. *International Journal of Systematic Bacteriology* **27,** 330–336.

Holmes, B., Willcox, W. R., Lapage, S. P., and Malnick, H. (1977c). Test reproducibility of the API (20E), Enterotube, and Pathotec systems. *Journal of Clinical Pathology* **30,** 381–387.

Holmes, B., Snell, J. J. S., and Lapage, S. P. (1978a). Revised description from clinical strains, of *Flavobacterium breve* (Lustig) Bergey *et al.* 1923 and proposal of the neotype strain. *International Journal of Systematic Bacteriology* **28,** 201–208.

Holmes, B., Willcox, W. R., and Lapage, S. P. (1978b). Identification of Enterobacteriaceae by the API 20E system. *Journal of Clinical Pathology* **31,** 22–30.

Holmes, B., Dowling, J., and Lapage, S. P. (1979). Identification of Gram-negative non-fermenters and oxidase-positive fermenters by the Oxi/Ferm Tube. *Journal of Clinical Pathology* **32,** 78–85.

Holmes, B., Owen, R. J., and Weaver, R. E. (1981). *Flavobacterium multivorum,* a new species isolated from human clinical specimens and previously known as Group IIk, biotype 2. *International Journal of Systematic Bacteriology* **31,** 21–34.

Holmes, B., Owen, R. J., and Hollis, D. G. (1982). *Flavobacterium spiritivorum,* a new species isolated from human clinical specimens. *International Journal of Systematic Bacteriology* **32,** 157–165.

Holmes, B., Hollis, D. G., Steigerwalt, A. G., Pickett, M. J., and Brenner, D. J. (1983). *Flavobacterium thalpophilum,* a new species recovered from human clinical material. *International Journal of Systematic Bacteriology* **33,** 677–682.

Izard, D., Ferragut, C., Gavini, F., Kersters, K., De Ley, J., and Leclerc, H. (1981). *Klebsiella terrigena,* a new species from soil and water. *International Journal of Systematic Bacteriology* **31,** 116–127.

Jain, K., Radsak, K., and Mannheim, W. (1974). Differentiation of the *Oxytocum* group from *Klebsiella* by deoxyribonucleic acid–deoxyribonucleic acid hybridization. *International Journal of Systematic Bacteriology* **24,** 402–407.

Johnson, J. L., and Ault, D. A. (1978). Taxonomy of the *Bacteroides* II. Correlation of phenotypic characteristics with deoxyribonucleic acid homology groupings for *Bacteroides fragilis* and other saccharolytic *Bacteroides* species. *International Journal of Systematic Bacteriology* **28,** 257–268.

Kelley, R. W., and Kellogg, S. T. (1978). Computer-assisted identification of anaerobic bacteria. *Applied and Environmental Microbiology* **35,** 507–511.

King, A., Holmes, B., Phillips, I., and Lapage, S. P. (1979). A taxonomic study of clinical isolates of *Pseudomonas pickettii, 'P. thomasii'* and 'Group IVd' bacteria. *Journal of General Microbiology* **114,** 137–147.

Kiredjian, M., Popoff, M., Coynault, C., Lefèvre, M., and Lemelin, M. (1981). Taxonomie du genre *Alcaligenes. Annales de Microbiologie (Paris)* **132B,** 337–374.

Lapage, S. P., and Willcox, W. R. (1974). A simple method for analysing binary data. *Journal of General Microbiology* **85,** 376–380.

Lapage, S. P., Bascomb, S., Willcox, W. R., and Curtis, M. A. (1970). Computer identification of bacteria. *In* 'Automation, Mechanization and Data Handling in Microbiology' (Eds. A. Baillie and R. J. Gilbert), pp. 1–22. Academic Press, London.

Lapage, S. P., Bascomb, S., Willcox, W. R., and Curtis, M. A. (1973). Identification of bacteria by computer: general aspects and perspectives. *Journal of General Microbiology* **77,** 273–290.

Logan, N. A., and Berkeley, R. C. W. (1981). Classification and identification of members of the genus *Bacillus* using API tests. *In* 'The Aerobic Endospore-forming Bacteria: Classification and Identification' (Eds. R. C. W. Berkeley and M. Goodfellow), pp. 105–140. Academic Press, London.

Maccacaro, G. A. (1958). La misura della informazione contenuta nei criteri di classificazione. *Annali di Microbiologia ed Enzimologia* **8,** 231–239.

Macfie, H. J. H., Gutteridge, C. S., and Norris, J. R. (1978). Use of canonical variates analysis in differentiation of bacteria by pyrolysis gas–liquid chromatography. *Journal of General Microbiology* **104,** 67–74.

Möller, F. (1962). Quantitative methods in the systematics of Actinomycetales. IV. The theory and application of a probabilistic identification key. *Giornale di Microbiologia* **10,** 29–47.

Niemelä, S. I., Hopkins, J. W., and Quadling, C. (1968). Selecting an economical binary test battery for a set of microbial cultures. *Canadian Journal of Microbiology* **14,** 271–279.

Palleroni, N. J., and Holmes, B. (1981). *Pseudomonas cepacia* sp. nov., nom. rev. *International Journal of Systematic Bacteriology* **31,** 479–481.

Pankhurst, R. J. (1974). Automated identification in systematics. *Taxon* **23,** 45–51.

Payne, L. C. (1963). Towards medical automation. *World Medical Electronics* **2,** 6–11.

Quadling, C., and Colwell, R. R. (1964). The use of numerical methods in characterizing unknown isolates. *Developments in Industrial Microbiology* **5,** 151–161.

Robertson, E. A., and MacLowry, J. D. (1974). Mathematical analysis of the API enteric 20 profile register using a computer diagnostic model. *Applied Microbiology* **28,** 691–695.

Rüger, H.-J., and Tan, T. L. (1983). Separation of *Alcaligenes denitrificans* sp. nov., nom. rev. from *Alcaligenes faecalis* on the basis of DNA base composition, DNA homology, and nitrate reduction. *International Journal of Systematic Bacteriology* **33,** 85–89.

Rypka, E. W., and Babb, R. (1970). Automatic construction and use of an identification scheme. *Medical Research Engineering* **9,** 9–19.

Rypka, E. W., Clapper, W. E., Bowen, I. G., and Babb, R. (1967). A model for the identification of bacteria. *Journal of General Microbiology* **46,** 407–424.

Sielaff, B. H., Johnson, E. A., and Matsen, J. M. (1976). Computer-assisted bacterial identification utilizing antimicrobial susceptibility profiles generated by Autobac 1. *Journal of Clinical Microbiology* **3,** 105–109.

Skerman, V. B. D., McGowan, V., and Sneath, P. H. A. (1980). Approved lists of bacterial names. *International Journal of Systematic Bacteriology* **30,** 225–420.

Sneath, P. H. A. (1974). Test reproducibility in relation to identification. *International Journal of Systematic Bacteriology* **24,** 508–523.

Sneath, P. H. A. (1978). Identification of microorganisms. *In* 'Essays in Microbiology' (Eds. J. R. Norris and M. H. Richmond), pp. 10/1–10/32. Wiley, New York.

Sneath, P. H. A. (1979). Numerical taxonomy and automated identification: some implications for geology. *Computers and Geosciences* **5,** 41–46.

Sneath, P. H. A. (1980a). BASIC program for determining the best identification scores possible from the most typical examples when compared with an identification matrix of percent positive characters. *Computers and Geosciences* **6,** 27–34.

Sneath, P. H. A. (1980b). BASIC program for determining overlap between groups in an identification matrix of percent positive characters. *Computers and Geosciences* **6,** 267–278.

Sneath, P. H. A., and Sokal, R. R. (1973). 'Numerical Taxonomy'. Freeman, San Francisco.

Tatum, W. H., Ewing, W. H., and Weaver, R. E. (1974). Miscellaneous Gram-negative bacteria. *In* 'Manual of Clinical Microbiology' (Eds. E. H. Lennette, E. H. Spaulding, and J. P. Truant), 2nd edition, pp. 270–294. American Society for Microbiology, Washington, D.C.

Ursing, J., Brenner, D. J., Bercovier, H., Fanning, G. R., Steigerwalt, A. G., Brault, J., and Mollaret, H. H. (1980). *Yersinia frederiksenii:* a new species of Enterobacteriaceae composed of rhamnose-positive strains (formerly called atypical *Yersinia enterocolitica* or *Yersinia enterocolitica*-like). *Current Microbiology* **4,** 213–217.

Wayne, L. G., Krichevsky, E. J., Love, L. L., Johnson, R., and Krichevsky, M. I. (1980). Taxonomic probability matrix for use with slowly growing mycobacteria. *International Journal of Systematic Bacteriology* **30,** 528–538.

Willcox, W. R., Lapage, S. P., Bascomb, S., and Curtis, M. A. (1973). Identification of bacteria by computer: theory and programming. *Journal of General Microbiology* **77,** 317–330.

Willcox, W. R. (the late), Lapage, S. P., and Holmes, B. (1980). A review of numerical taxonomic methods in bacterial identification. *Antonie van Leeuwenhoek* **46,** 233–299.

11

Application of New Theoretical Concepts to the Identification of Streptomycetes

S. T. WILLIAMS

Department of Botany, University of Liverpool, Liverpool, UK

J. C. VICKERS

Department of Microbial Biochemistry, Glaxo Group Research Ltd., Greenford, Middlesex, UK

M. GOODFELLOW

Department of Microbiology, The Medical School, Newcastle upon Tyne, UK

Introduction

The definition and recognition of *Streptomyces* species have provided taxonomists with a major problem for many years. Hundreds of species have been legitimately described (Shirling and Gottlieb, 1967, 1968a,b, 1969, 1972; Pridham and Tresner, 1974) and included in the *Approved Lists of Bacterial Names* (Skerman *et al.*, 1980). Still more have been cited in the patent literature (Trejo, 1970). Many attempts have been made since the 1950s to allocate the numerous species to groups (or series), thereby facilitating their identification (see Williams *et al.*, 1981, 1983a,b, for detailed discussion). The vast majority of such groupings were based on a few subjectively chosen morphological and pigmentation properties, and hence the species groups were artificial and inclusion could be ruled out by one aberrant character state. Biochemical, nutritional, and physiological characters have been used in many species descriptions, but there has been little standardisation in the test selection, in test procedures, or in the range of species to which they have been applied. One notable exception was the use of standardised carbon source utilization tests in the International *Streptomyces* Project (Shirling and Gottlieb, 1966) and hence in the last edition of *Bergey's Manual of Determinative Bacteriology* (Pridham and Tresner, 1974).

The first and most comprehensive attempt to construct a numerical classification of streptomycetes using a wide range of characters was by Silvestri *et al.*

ISBN 0-12-289665-3

(1962). Twenty-five centres of variation were recognised, and the data were used to construct a probabilistic identification key (Möller, 1962; Hill and Silvestri, 1962). Subsequently there were a number of attempts to construct both numerical classification and identification systems, but they were based on a relatively small number of characters (Gyllenberg *et al.*, 1967, 1975; Gyllenberg, 1970; Kuryłowicz *et al.*, 1975). Results of factor analysis also suggested that many characters used to describe *Streptomyces* species were highly variable and prone to errors in interpretation (Gyllenberg, 1970).

Subsequently a more comprehensive numerical classification of streptomycetes and related genera was constructed (Williams *et al.*, 1981, 1983a). The data from this study provided a basis for the construction of a probabilistic identification matrix for streptomycetes (Williams *et al.*, 1983b), using various new programs devised by Professor P. H. A. Sneath. The results obtained are summarised and evaluated here.

The Numerical Classification

Strains and Characters

The study included 475 strains and 44 duplicate cultures, type cultures being selected whenever possible. Emphasis was placed on *Streptomyces* species which included 394 ISP cultures (Shirling and Gottlieb, 1967, 1968a,b, 1969, 1972); marker strains of 14 other genera were also studied. After preliminary evaluation of the reproducibility of tests, 162 unit characters were determined for all strains. The characters included those used traditionally for streptomycetes as well as newly applied tests. They were categorised as morphological, pigmentation, antimicrobial activity, biochemical properties, degradative ability, antibiotic resistance, growth requirements, and utilization of carbon and nitrogen sources.

Computation and Analysis

Most characters existed in one of two mutually exclusive states and were scored plus or minus. Qualitative multistate characters, such as pigmentation and spore chain morphology, were coded as several independent characters, and each was scored plus for the character state shown and minus for the alternatives. Quantitative multistate characters, such as tolerance to inhibitors, were coded by the additive method (Sneath and Sokal, 1973). Data were computed to determine both the simple-matching coefficient S_{sm} (Sokal and Michener, 1958), which includes both positive and negative matches, and the Jaccard coefficient S_J (Sneath, 1957) including only positive matches. Clustering was achieved using unweighted pair-group arithmetic average clustering of Sneath and Sokal (1973).

The distinctness of the major clusters defined was assessed by calculation of the degree of cluster overlap and the overlap statistics between all cluster pairs using the OVCLUST program (Sneath, 1979a).

Test reproducibility was assessed by examining the determinations of the 162 unit characters for the 44 duplicate cultures and their partners. Test variance S_i^2 was calculated (Sneath and Johnson, 1972), and the average test variance S^2 was used to calculate the average probability p of an erroneous test result (Sneath and Johnson, 1972). Similarity between duplicates was calculated by computation of the S_{sm} coefficient.

Results of the Numerical Classification

Test Error. Analysis of the results obtained with the duplicate cultures showed that the average probability p of an erroneous test result was 3.36%, which was well within the limit of 10% suggested by Sneath and Johnson (1972). The 44 pairs of duplicate cultures showed a mean similarity of 93.1% S_{sm}, indicating that test error was clearly within acceptable limits. The majority of tests gave S_i^2 values below .05, but 7 showed a variance >0.1 and a further 16 provided little or no separation value and were therefore deleted from the data matrix before computation of overall similarities. The final classification matrix therefore contained 139 unit characters.

Composition of Cluster Groups. Of the cluster groups defined at 70.1% S_{sm}, by far the largest was group A, which contained most (73%) of the *Streptomyces* strains (Table 1). Several other genera sharing a wall chemotype I (Lechevalier and Lechevalier, 1970) with *Streptomyces* were also included. *Actinopycnidium* and *Actinosporangium* were clearly synonyms of *Streptomyces; Chainia, Elytrosporangium,* and *Microellobosporia* are morphologically distinguishable from *Streptomyces* but fell within the genus on the basis of overall similarity. *Nocardioides albus* was on the fringe of this cluster group and was excluded from it by the S_J coefficient analysis, as was *Saccharopolyspora hirsuta* (wall chemotype IV). The main anomaly was *Nocardiopsis dassonvillei* (wall chemotype III), which fell clearly into this cluster group with both coefficients. The genera with a wall chemotype I excluded from cluster group A were *Intrasporangium, Kitasatoa,* and *Streptoverticillium,* as were *Actinomadura* spp. and *Microtetraspora glauca* (wall chemotype III), together with *Nocardia asteroides* and *'Nocardia' mediterranei* (wall chemotype IV). Some *Streptomyces* species (27%) clearly fell outside the '*Streptomyces*' cluster group A, including those in the major clusters *S. rimosus* and *S. lavendulae*. The former constituted cluster group B, whilst the latter joined with *Kitasatoa* and *Streptoverticillium* species to form cluster group F. Thus, the overall generic status of the cluster groups is uncertain.

Table 1. *Composition of cluster groups defined by the* S_{sm} *coefficient at 70.1%*

Cluster group	Number of strains	Number of clusters defined at 77.5% S_{sm}[a]	Major components	Number of *Streptomyces* strains
A	340	48	*Streptomyces* spp. *Actinopycnidium caeruleum* *Actinosporangium violaceum* *Chainia* spp. *Elytrosporangium* spp. *Microellobosporia* spp. *Nocardiopsis dassonvillei* *Saccharopolyspora hirsuta* *Nocardioides albus*	218
B	8	2	*Streptomyces rimosus*	8
C	16	8	*Streptomyces* spp.	16
D	2	2	*Microellobosporia flavea* *Streptomyces massasporeus*	1
E	25	8	*Actinomadura* spp. *'Nocardia' mediterranei* *Streptomyces* spp.	8
F	60	19	*Kitasatoa* spp. *Streptoverticillium* spp. *Streptomyces* spp.	32
G	3	2	*Streptomyces fradiae*	3
H	2	2	*Streptomyces* spp.	2
I	3	2	*Nocardia* spp.	1
J	15	7	*Actinomadura pelletieri* *Intrasporangium calvum* Acidophilic *Streptomyces* spp.	11

[a]Includes single-member clusters.

The Major Clusters. Groups defined at the 77.5% S_{sm} level consisted of 22 major clusters containing six or more strains (Table 2) and 51 minor clusters. The major clusters contained 307 strains (64.6%), the minor ones 140 (29.5%), with 28 strains (5.9%) being recovered as single-member clusters. Clusters were named, where possible, after the earliest validly described species which they contained.

Some details of the major clusters are given in Table 2. These fell into the '*Streptomyces*' cluster group A, with the exception of *Streptomyces rimosus,*

Table 2. *Major clusters defined at 77.5% by the S_{sm} coefficient and at 63% by the S_J coefficient*

Cluster name	Number of strains		Predominant characteristic features[a,b]			
	77.5% S_{sm}	63% S_J	Spore surface	Spore chain	Spore colour	Melanin pigment
Streptomyces albidoflavus	71	72	Sm	RF	Y-Gy	−
Streptomyces atroolivaceus	9	6	Sm	RF	W-Gy	−
Streptomyces exfoliatus	18	16 (2)[c]	Sm	RF	R-Gy	+/−
Streptomyces violaceus	8	9	Sm/Spy	RF/RA/S	V-Y	+
Streptomyces fulvissimus	9	9	Sm	RF/RA/S	R	+/−
Streptomyces rochei	26	23	Sm/Spy/Hy	S/RA	Gy	−
Streptomyces chromofuscus	9	7 (3)	Sm/Spy/Hy	RF/S	Y-W-Gy	+/−
Streptomyces albus	6	6	Sm	S	W	−
Streptomyces griseoviridis	6	6	Sm	S	R	−
Streptomyces cyaneus	38	37	Sm/Spy	S	B-R-Gy	+
Streptomyces diastaticus	20	22 (5)	Sm	RF/RA/S	W-R-Gy	+/−
Streptomyces olivaceoviridis	7	8 (2)	Sm	S	Gy	−
Streptomyces griseoruber	8	9	Sm	S	Gy	−
Streptomyces lydicus	11	10 (2)	Sm	S	Gy	−
Streptomyces violaceoniger	6	6	Rug	S	Gy	−
Streptomyces griseoflavus	6	5	Spy/Hy	RA	Gn	−
Streptomyces phaeochromogenes	6	6	Sm	RA/S	?	+/−
Streptomyces rimosus	7	7	Sm	RF/S	W-Y	−
Actinomadura spp.	6	6 (3)	Sm	RF/RA/S	?	−
'Nocardia' mediterranei	8	8	Sm	RF	?	−
Streptoverticillium griseocarneum	9	9 (3)	Sm	BV	R-Y-Gy	+
Streptomyces lavendulae	12	11 (4)	Sm	RF	R	+

[a]Features used to define species groups in *Bergey's Manual of Determinative Bacteriology* (Pridham and Tresner, 1974).
[b]Abbreviations: Hy, hairy; Rug, rugose; Sm, smooth; Spy, spiny; BV, Biverticillati; RA, Retinaculiaperti; RF, Rectiflexibiles; S, Spirales; B, blue; Gn, green; Gy, grey; R, red; V, violet; W, white; Y, yellow.
[c]Figures in parentheses indicate the number of subclusters obtained with the S_J coefficient.

Streptomyces lavendulae, Streptoverticillium griseocarneum, Actinomadura spp., and *'Nocardia' mediterranei.* As four characters were used to group *Streptomyces* species in the last edition of *Bergey's Manual of Determinative Bacteriology* (Pridham and Tresner, 1974), the predominant states of these characters within each cluster are given (Table 2). Some clusters, such as *Streptomyces albidoflavus* [which approximates to the '*griseus*' group of Hütter (1963) and other workers], *Streptomyces albus,* and *Streptomyces violaceoniger,* were reasonably consistent in their character states. Others, such as *Streptomyces chromofuscus* and *Streptomyces diastaticus,* showed considerable variation in these characters. It is not surprising that polythetic groups defined using 139 characters do not always show concordance with those constructed with 4 subjectively chosen characters. A detailed discussion of the relationships of these clusters to previous groupings of streptomycetes was given by Williams *et al.* (1983a).

All of the major clusters and most of the minor ones defined by the S_{sm} coefficient were recovered using the S_J coefficient, although in some cases cluster composition was changed. Of the 22 major clusters, 14 remained intact and 8 split into 2 or more sub-clusters (Table 2). This suggested that the classification was quite robust. It was also encouraging that most of the major clusters defined in the S_{sm} analysis showed little significant overlap (taking 5% as expected overlap), especially in view of the difficulties experienced in distinguishing streptomycete taxa in earlier studies (Gyllenberg *et al.*, 1967; Gyllenberg, 1970). The value of 5% is not stringent but is less than that of about 8.3% which corresponds to continuous variation (Sneath, 1977). The pattern of groups defined may sometimes represent overlapping variation rather than entirely sharply defined, well-separated species. It is therefore appropriate to regard them as either species or species groups pending further evaluation of their taxonomic status.

Construction of the Probabilistic Identification Matrix

Numerical classification results in the definition of phena at selected levels of similarity. It also provides quantitative data on the test reactions within each group defined, this being expressed as the percentage of strains showing a positive state for each character studied. Such data are in a form that is ideal for the construction of an identification matrix (Hill, 1974; Sneath, 1978), which contains the minimum number of selected characters required for discrimination between the groups previously defined by numerical classification. The matrix can then be used for the probabilistic identification of unknown strains, which is the logical end product of a numerical taxonomic study. Therefore, the classification test data were used to construct an identification matrix for the major clusters defined by the numerical classification (Williams *et al.*, 1983b).

Selection of Tests

A total of 23 clusters were selected for the matrix, consisting of all the major clusters (Table 2) together with *Streptomyces fradiae,* a well-known source of antibiotics. The characters most diagnostic for these clusters were selected from the 139 tests used in the classification matrix.

The first step in the selection procedure was the determination of the number of clusters in which each test was predominantly positive or negative; a good test showed a consistent state within as many clusters as possible and ideally gave a good balance between positive and negative reactions between clusters. The product of these values gives the S_i separation index of Gyllenberg (1963). A

Table 3. *Diagnostic value of characters selected for the identification matrix using the CHARSEP and DIACHAR programmes*[a]

	Number of clusters in which character is predominantly			
Characters[b]	Present	Absent	S_i index[c]	VSP index[d]
Morphology				
1. Spore surface smooth	15	2	30	27.0
2. Spore surface rugose	1	22	22	3.5
3. Spore chain BV	2	21	42	10.5
4. Spore chain RA	2	18	36	16.8
5. Spore chain RF	3	14	42	38.3
6. Spore chain S	6	7	42	54.9
7. Fragmentation of mycelium	1	22	22	3.6
Pigmentation				
8. Melanin	3	11	33	38.6
9. Substrate yellow-brown	14	2	28	22.1
10. Substrate red-orange	1	18	18	12.1
11. Spore mass grey	3	12	36	40.5
12. Spore mass red	2	13	26	34.2
13. Spore mass green	0	21	0	2.4
Carbon source utilization				
14. Adonitol	4	10	40	47.3
15. Cellobiose	18	2	36	19.7
16. D-Fructose	15	1	15	10.1
17. *meso*-Inositol	9	3	27	39.0
18. Inulin	1	13	13	18.6
19. D-Mannitol	17	4	68	42.1
20. Raffinose	6	2	12	39.3
21. L-Rhamnose	8	2	16	43.0
22. D-Xylose	14	2	28	26.5

(*continued*)

Table 3 (*Continued*)

Characters[b]	Number of clusters in which character is predominantly		S_i index[c]	VSP index[d]
	Present	Absent		
Nitrogen source utilization				
23. DL-α-Aminobutyric acid	3	8	24	33.4
24. L-Histidine	8	2	16	33.0
25. L-Hydroxyproline	2	6	12	35.1
Degradation				
26. Allantoin	3	5	15	39.9
27. Arbutin	13	4	52	40.6
28. Xanthine	9	5	45	53.8
Enzyme production				
29. Lecithinase	4	13	52	48.6
30. Pectinase	3	10	30	38.5
31. H_2S production	13	3	39	35.7
32. NO_3 reduction	4	6	24	42.6
Antibiosis				
33. *Aspergillus niger*	3	11	33	39.7
34. *Bacillus subtilis*	5	4	20	44.4
35. *Streptomyces murinus*	6	5	30	45.8
Antibiotic resistance				
36. Neomycin (50 μg ml^{-1})	4	14	56	42.0
37. Rifampicin (50 μg ml^{-1})	8	4	32	43.0
Growth				
38. 45°C	4	8	32	48.1
39. Sodium azide (0.01% w/v)	2	9	18	36.9
40. Sodium chloride (7.0% w/v)	3	5	15	44.6
41. Phenol (0.1% w/v)	8	6	48	55.8

[a]CHARSEP (Sneath, 1979c); DIACHAR (Sneath, 1980a).
[b]See footnote to Table 2 for abbreviations.
[c]See Gyllenberg (1963).
[d]See Sneath (1979c).

further selection of tests was achieved using the CHARSEP programme (Sneath, 1979c), which includes five different separation indices for assessing the diagnostic value of characters, including the VSP index, which gives higher scores for the more useful characters (Table 3). The next step was to apply the DIACHAR programme (Sneath, 1980a), which selects the most diagnostic tests for each group in an identification matrix. In a well-constructed matrix there should be several strongly diagnostic characters for each group. Results were

Table 4. *Some examples of tests with poor diagnostic values for the major clusters*

Characters	Number of clusters in which character is predominantly		S_i index[a]	VSP index[b]
	Present	Absent		
1. Spore surface hairy	0	21	0	0.3
2. Spore mass blue	0	22	0	0.2
3. Utilization of nitrate	20	0	0	2.8
4. Utilization of L-arginine	19	0	0	1.3
5. Utilization of D-mannose	21	0	0	1.3
6. Proteolysis	14	0	0	3.0
7. Degradation of RNA	20	0	0	0.9
8. Degradation of aesculin	20	0	0	0.9
9. Resistance to cephaloridine (100 μg ml^{-1})	23	0	0	0.02
10. Resistance to phenyl ethanol (0.3% v/v)	17	0	0	1.1

[a]See Gyllenberg (1963).
[b]See Sneath (1979c).

satisfactory for the majority of clusters in the matrix. It could also be seen by eye that a few characters were strongly diagnostic for one particular cluster, for example, rugose spore surface for *Streptomyces violaceoniger*, and these too were included in the final matrix although their overall separation values (S_i and VSP) were low. The final matrix therefore consisted of 23 clusters and 41 tests, the latter covering a wide range of characters (Table 3). The percentage positives for each test and cluster were stored in a computer for subsequent testing and use. Some examples of tests rejected because of their poor diagnostic values are given in Table 4.

Theoretical Evaluation of the Matrix

The importance of evaluating identification matrices was stressed by Sneath and Sokal (1973) and Sneath (1978). The quality of the matrix was therefore assessed both theoretically and practically.

A programme (OVERMAT; Sneath, 1980c) for determining overlap between groups in an identification matrix was applied to the percentage positive values for characters in the matrix; it is not possible to assess overlap in a large matrix by simple inspection. If there is much overlap between groups, unknowns may not identify well to any one of them. OVERMAT determines the *disjunction index W* for each pair of groups and the corresponding *nominal overlap* V_G; the significance of the determined overlap can also be assessed against a selected

Table 5. *Examples of identification scores for hypothetical median organisms*

Cluster name	Number of strains in cluster	Identification scores: Willcox probability	Taxonomic distance	Standard error of taxonomic distance
Streptomyces lavendulae	12	1.0	.23	−2.76
Streptomyces fulvissimus	9	1.0	.19	−3.62
Streptomyces griseoflavus	6	1.0	.18	−3.35
Streptomyces atroolivaceus	9	0.999	.20	−3.17
Streptomyces albus	6	1.0	.14	−3.97
Streptoverticillium griseocarneum	9	1.0	.15	−4.02
Streptomyces albidoflavus	71	0.999	.24	−2.70
Streptomyces exfoliatus	18	0.999	.24	−2.74
Streptomyces olivaceoviridis	7	0.999	.19	−3.25

[a]MOSTTYP programme (Sneath, 1980b).

critical overlap value V_0. In this case the chosen critical value was 5% (cf. OVCLUST programme for cluster overlap). No significant overlap between any of the clusters in the matrix was detected.

All subsequent assessments of the matrix, both theoretical and practical, involved use of the MATIDEN programme (Sneath, 1979b) to obtain the best identification scores for known or unknown strains against the groups in the matrix. Three of the identification coefficients included in this programme were used:

1. *Willcox probability* (Willcox *et al.*, 1973). This is the likelihood of unknown character-state values against a particular group divided by the sum of the likelihoods against all groups; the closer the score is to 1.0, the better is the fit.
2. *The taxonomic distance*. This expresses the distance of an unknown from the centroid of the group with which it is being compared; a low score indicates relatedness to the group, and ideally it is less than about .15.
3. *Standard error of the taxonomic distance*. This assumes that the groups are in hyperspherical normal clusters. An acceptable score is less than about 2.0 to 3.0, and about half the members of a taxon will have negative scores, that is, they are closer to the centroid than average.

Identification is achieved if the best scores are good *and* sufficiently better than the next best two alternatives against other groups. The output also lists atypical properties of the unknown against its best group, which should be few for a good identification.

Table 6. *Examples of identification scores for cluster representatives using classification test data*

Cluster name	Cluster representative	Identification scores		
		Willcox probability	Taxonomic distance	Standard error of taxonomic distance
Streptomyces lavendulae	*S. lavendulae*	1.0	.35	0.75
Streptomyces fulvissimus	*Streptomyces spectabilis*	1.0	.29	−0.82
Streptomyces griseoflavus	*Streptomyces hirsutus*	1.0	.28	0.04
Streptomyces atroolivaceus	*'Streptomyces scabies'*	0.999	.29	−0.76
Streptomyces albus	*Streptomyces albus*	0.999	.24	−0.06
Streptoverticillium griseocarneum	*Streptoverticillium cinnamoneum*	0.998	.29	0.55
Streptomyces albidoflavus	*Streptomyces griseus*	0.988	.30	−0.91
Streptomyces exfoliatus	*Streptomyces umbrinus*	0.924	.35	−0.03
Streptomyces olivaceoviridis	*Elytrosporangium brasiliense*	0.879	.34	1.27

The first identification scores using the matrix were determined by the MOSTTYP programme (Sneath, 1980b), which evaluates matrices by calculating the best scores which the 'hypothetical median organism' (HMO) of each group could achieve. Results obtained with 9 of the 23 clusters, selected to illustrate the range of response, are given in Table 5. It is clear that the matrix withstood this theoretical test. All Willcox probabilities were .999 or 1.0; taxonomic distances were low, ranging from .14 to .24, and standard errors of taxonomic distance were all negative.

The next step in evaluation of the matrix was to feed in the test results, obtained from the classification data, of a randomly chosen strain from each cluster. Results obtained for the nine representative clusters are given in Table 6. All strains identified to the correct cluster. Willcox probabilities were high, but those for the representative strains of the *Streptomyces exfoliatus* and *Streptomyces olivaceoviridis* clusters were lower than the values for the corresponding HMOs; standard errors had low or negative values, with that for the *S. olivaceoviridis* cluster being highest. These results were regarded as very satisfactory, some deterioration of the scores compared with the HMO being inevitable.

Practical Evaluation of the Matrix

The identification matrix appeared to be theoretically sound, so the next logical step was to assess it by feeding in data obtained from the independent determination of the character states of both known and unknown strains.

The character states of the same cluster representatives used in the theoretical evaluation (examples given in Table 6) were independently redetermined and the new identification scores obtained (Table 7). Generally there was little deterioration in the scores compared with those obtained using the original classification data (Table 6), the notable exceptions being the reduced Willcox probabilities for the strains representing clusters *Streptomyces albidoflavus, S. exfoliatus,* and *S. olivaceoviridis*. Such changes were clearly due to discrepancies between some of the classification test results and their redeterminations. However, overall test agreement for all tests and clusters was high; total discrepancies were 55 out of 943 (5.8%), and the average test variance S_i^2 was .029. This is equivalent to a probability of error of 3%, from the formula of Sneath and Johnson (1972), and well below the acceptable limit of 5% for test error within the same laboratory.

Finally, the 41 characters were determined for unknown isolates from soil,

Table 7. *Examples of identification scores for cluster representatives from the independent redetermination of character states*

Cluster name	Cluster representative	Identification scores		
		Willcox probability	Taxonomic distance	Standard error of taxonomic distance
Streptomyces lavendulae	*S. lavendulae*	1.0	.37	1.20
Streptomyces fulvissimus	*Streptomyces spectabilis*	1.0	.27	−1.39
Streptomyces griseoflavus	*Streptomyces hirsutus*	0.999	.31	0.92
Streptomyces atroolivaceus	*'Streptomyces scabies'*	0.997	.27	−1.33
Streptomyces albus	*Streptomyces albus*	0.999	.29	1.59
Streptoverticillium griseocarneum	*Streptoverticillium cinnamoneum*	0.999	.29	0.84
Streptomyces albidoflavus	*Streptomyces griseus*	0.856	.31	−0.67
Streptomyces exfoliatus	*Streptomyces umbrinus*	0.641	.36	0.36
Streptomyces olivaceoviridis	*Elytrosporangium brasiliense*	0.240	.35	1.43

Table 8. *Examples of identification scores for unknown isolates*

Isolate number	Cluster identification	Identification scores: Willcox probability	Taxonomic distance	Standard error of taxonomic distance
1	*Streptomyces albidoflavus*	.999	.34	−0.04
2	*Streptomyces albidoflavus*	.999	.32	−0.58
3	*Streptomyces rochei*	.992	.35	0.27
4	*Streptomyces diastaticus*	.986	.40	0.93
5	*Streptomyces griseoruber*	.913	.32	1.90
6	*Streptomyces chromofuscus*	.907	.35	0.45
7	Not identified	.840	.41	3.35
8	Not identified	.790	.39	0.78
9	Not identified	.640	.41	1.80
10	Not identified	.520	.40	1.41
11	Not identified	.480	.43	3.54

water, and other habitats, and their identification scores were calculated. The criteria for a successful identification, based on the output from the MATIDEN programme, were as follows:

1. A Willcox probability greater than .850, with low scores for taxonomic distance and its standard error
2. All first scores significantly better than those for the next best two alternative groups
3. A small number of characters of the unknown listed as being atypical of those of the group in which it is placed

Table 9. *Summary of the identification of isolates from different habitats*

Origin of isolates	Total number of isolates	Isolates identified (%)
Seawater	55	65.5
Fresh water	9	89.0
Salt marsh	12	100.0
Sand dune	24	75.0
Indian soil	20	75.0
Garden soil	10	80.0
Pasture soil	9	77.5
Miscellaneous sources	14	78.5

Table 10. *Summary of cluster identifications of unknown isolates: genus* Streptomyces

Clusters	Number of isolates identified	Percentage of total identified strains
S. albidoflavus	59	49.5
S. rochei	15	12.6
S. cyaneus	13	11.0
S. chromofuscus	9	7.6
S. diastaticus	8	6.7
S. atroolivaceus	5	4.2
S. lydicus	5	4.2
S. exofoliatus	2	1.7
S. griseoflavus } *S. griseoruber* } *S. violaceoniger* }	1	0.8

A diversity of information is provided in the output, and the user must decide what scores are acceptable from experience (Sneath, 1979b). Examples of scores for identified and nonidentified isolates are given in Table 8. As values for the Willcox probability decreased, those for taxonomic distance, and in particular its standard error generally increased. Of the strains which identified, 60% did so at Willcox probability levels of .990 or above.

The results of attempts to identify 153 isolates from various natural sources are summarised in Tables 9 and 10. Overall, 72.5% of the isolates identified to one of the major clusters in the matrix, the success rate varying somewhat between isolates from different habitats. Almost half of the identified isolates fell into the *Streptomyces albidoflavus* cluster. This was by far the largest cluster (71 strains) defined in the numerical classification and is broadly equivalent to the '*griseus*' group, which is widely distributed in soil and water. The remaining identified isolates were distributed amongst 10 other clusters.

Conclusions

Numerical taxonomy is of proven value for both the classification and identification of bacteria. However, most numerical classifications have not been supported by probabilistic identification systems, one of the few exceptions being the probability matrix for identification of slowly growing mycobacteria (Wayne *et al.*, 1980). Conversely, most probabilistic identification schemes—for example, those of Lapage *et al.* (1973), Gyllenberg *et al.* (1975), and Hill *et al.* (1978)—have been constructed using data less comprehensive than those provided by numerical classification. A numerical classification of streptomycetes was therefore used to devise a probabilistic identification scheme for the major clusters.

The classification results illustrated the wide range of variation in *Streptomyces*. Although 22 major clusters were defined, 34.3% of the type strains studied fell into minor or single-member clusters at the 77.5% S_{sm} level. This should represent real variation, as a sufficient number of tests were used to avoid creation of artificial discontinuities. Some clusters were homogeneous with respect to the 'traditional' morphological and pigmentation characters and thus were comparable with species groups of earlier workers such as Hütter (1967) and Pridham and Tresner (1974). However, these characters were not generally cluster-specific, proving that a meaningful sub-generic classification of streptomycetes cannot be based on a few subjectively chosen characters. The results vindicate the view of Sneath (1970) that numerical analysis was the only practical way of dealing with the overspeciation in this genus. Some of the clusters defined represent species, while others probably represent species groups. Studies of DNA–DNA and DNA–RNA pairing, chemotaxonomy, serology, and genetic exchange will help to clarify this problem.

The identification matrix, constructed using recently devised programmes, is the most comprehensive and fully tested of any published to date. Nevertheless it has some imperfections. It was not practically feasible at this stage to include all clusters, and the boundaries of a few of the major clusters were rather indistinct. Also, the minimum number of tests needed (41) to distinguish amongst the clusters was quite large, reflecting the variation within clusters and the necessity to have at least as many tests as taxa in a matrix (Sneath and Chater, 1978). Nevertheless, the matrix has proved its practicality by identifying isolates from a variety of habitats. The mean frequency of successful identification (72.5%) compares favourably with those of other matrices applied to field strains (Lapage *et al.*, 1973; Hill *et al.*, 1978; Willcox *et al.*, 1980). The application of the Willcox probability at a less stringent level than that selected by these workers can be justified by the use of the additional identification data provided by the MATIDEN programme and the likelihood that at least some of the clusters are species groups. Thus, the application of both well-established and recently devised procedures of numerical taxonomy has provided a more objective means of dealing with the genus *Streptomyces*. The results should serve as a sound basis for further improvements in streptomycete systematics, whilst also assisting workers studying other aspects of this important genus.

Acknowledgement

We gratefully acknowledge the support of the Science and Engineering Research Council (Grants GR/A/04309; GR/A/8552).

References

Gyllenberg, H. G. (1963). A general method for deriving determination schemes for random collections of microbial isolates. *Annales Academiae Scientiarum Fennicae Series A, IV Biologica* **69,** 1–23.

Gyllenberg, H. G. (1970). Factor analytical evolution of patterns of correlated characteristics in streptomycetes. *In* 'The Actinomycetales' (Ed. H. Prauser), 101–105. Gustav Fischer, Jena.

Gyllenberg, H. G., Woznicka, W., and Kuryłowicz, W. (1967). Application of factor analysis in microbiology. 3. A study of the ''yellow series'' of streptomycetes. *Annales Academiae Scientiarium Fennicae Series A, IV Biologica* **114,** 3–15.

Gyllenberg, H. G., Niemelä, T. K., and Niemi, J. S. (1975). A model for automatic identification of streptomycetes. *Postepy Higieny i Medycyny Doswialdczalnej* **29,** 357–383.

Hill, L. R. (1974). Theoretical aspects of numerical identification. *International Journal of Systematic Bacteriology* **24,** 494–499.

Hill, L. R., and Silvestri, L. G. (1962). Quantitative methods in the systematics of Actinomycetales. III. The taxonomic significance of physiological–biochemical characters and the construction of a diagnostic key. *Giornale di Microbiologia* **10,** 1–27.

Hill, L. R., Lapage, S. P., and Bowie, I. S. (1978). Computer assisted identification of coryneform bacteria. *In* 'Coryneform Bacteria' (Eds. I. G. Bousfield and A. G. Callely), pp. 181–215. Academic Press, London.

Hütter, R. (1963). Zur Systematik der Actinomyceten. 10. Streptomyceten mit griseus-Luftmycel. *Giornale di Microbiologia* **11,** 191–246.

Hütter, R. (1967). 'Systematik der Streptomyceten'. Karger, Basel.

Kuryłowicz, W., Paszkiewicz, A., Woznicka, W., Kurzatowski, W., and Szulga, T. (1975). Classification of *Streptomyces* by different numerical methods. *Postepy Higieny i Medycyny Doswialdczalnej* **29,** 281–355.

Lapage, S. P., Bascomb, S., Willcox, W. R., and Curtis, M. A. (1973). Identification of bacteria by computer: general aspects and perspectives. *Journal of General Microbiology* **77,** 273–290.

Lechevalier, M. P., and Lechevalier, H. (1970). Chemical composition as a criterion in the classification of aerobic actinomycetes. *International Journal of Systematic Bacteriology* **20,** 435–443.

Möller, F. (1962). Quantitative methods in the systematics of Actinomycetales. IV. The theory and application of a probabilistic identification key. *Giornale di Microbiologia* **10,** 29–47.

Pridham, T. G., and Tresner, H. D. (1974). Family VII Streptomycetaceae Waksman and Henrici 1943. *In* 'Bergey's Manual of Determinative Bacteriology' (Eds. R. E. Buchanan and N. E. Gibbons), pp. 747–845. Williams & Wilkins, Baltimore, Maryland.

Shirling, E. B., and Gottlieb, D. (1966). Methods for characterisation of *Streptomyces* species. *International Journal of Systematic Bacteriology* **16,** 313–340.

Shirling, E. B., and Gottlieb, D. (1967). Cooperative description of type cultures of *Streptomyces*. I. The International *Streptomyces* Project. *International Journal of Systematic Bacteriology* **17,** 315–322.

Shirling, E. B., and Gottlieb, D. (1968a). Cooperative description of type cultures of *Streptomyces*. II. Species descriptions from first study. *International Journal of Systematic Bacteriology* **18,** 69–189.

Shirling, E. B., and Gottlieb, D. (1968b). Cooperative description of type cultures of *Streptomyces*. III. Additional species descriptions from first and second studies. *International Journal of Systematic Bacteriology* **18,** 279–391.
Shirling, E. B., and Gottlieb, D. (1969). Cooperative description of type cultures of *Streptomyces*. IV. Species descriptions from the second, third and fourth studies. *International Journal of Systematic Bacteriology* **19,** 391–512.
Shirling, E. B., and Gottlieb, D. (1972). Cooperative description of type strains of *Streptomyces*. V. Additional descriptions. *International Journal of Systematic Bacteriology* **22,** 265–394.
Silvestri, L. G., Turri, M., Hill, L. R., and Gilardi, E. (1962). A quantitative approach to the systematics of actinomycetes based on overall similarity. *Symposium of the Society of General Microbiology* **12,** 333–360.
Skerman, V. B. D., McGowan, V., and Sneath, P. H. A. (1980). Approved lists of bacterial names. *International Journal of Systematic Bacteriology* **30,** 225–420.
Sneath, P. H. A. (1957). The application of computers to taxonomy. *Journal of General Microbiology* **17,** 201–226.
Sneath, P. H. A. (1970). Application of numerical taxonomy to Actinomycetales: problems and prospects. *In* 'The Actinomycetales' (Ed. H. Prauser), pp. 371–377. Gustav Fischer, Jena.
Sneath, P. H. A. (1977). A method for testing the distinctness of clusters: a test of the disjunction of two clusters in Euclidean space as measured by their overlap. *Journal of Mathematical Geology* **9,** 123–143.
Sneath, P. H. A. (1978). Identification of micro-organisms. *In* 'Essays in Microbiology' (Eds. J. R. Norris and M. H. Richmond), pp. 10/1–10/32. Wiley, Chichester.
Sneath, P. H. A. (1979a). BASIC program for a significance test for two clusters in Euclidean space as measured by their overlap. *Computers and Geosciences* **5,** 143–155.
Sneath, P. H. A. (1979b). BASIC program for identification of an unknown with presence–absence data against an identification matrix of percent positive characters. *Computers and Geosciences* **5,** 195–213.
Sneath, P. H. A. (1979c). BASIC program for character separation indices from an identification matrix of percent positive characters. *Computers and Geosciences* **5,** 349–357.
Sneath, P. H. A. (1980a). BASIC program for the most diagnostic properties of groups from an identification matrix of percent positive characters. *Computers and Geosciences* **6,** 21–26.
Sneath, P. H. A. (1980b). BASIC program for determining the best identification scores possible for the most typical example when compared with an identification matrix of percent positive characters. *Computers and Geosciences* **6,** 27–34.
Sneath, P. H. A. (1980c). BASIC program for determining overlap between groups in an identification matrix of percent positive characters. *Computers and Geosciences* **6,** 267–278.
Sneath, P. H. A., and Chater, A. O. (1978). Information content of keys for identification. *In* 'Essays in Plant Taxonomy' (Ed. H. E. Street), pp. 79–95. Academic Press, London.
Sneath, P. H. A., and Johnson, R. (1972). The influence on numerical taxonomic similarities of errors in microbiological tests. *Journal of General Microbiology* **72,** 377–392.
Sneath, P. H. A., and Sokal, R. R. (1973). 'Numerical Taxonomy. The Principles and Practice of Numerical Classification'. Freeman, San Francisco.

Sokal, R. R., and Michener, C. D. (1958). A statistical method for evaluating systematic relationships. *Kansas University Science Bulletin* **38,** 1409–1438.

Trejo, W. H. (1970). An evaluation of some concepts and criteria used in the speciation of streptomycetes. *Transactions of the New York Academy of Sciences* **32,** 989–997.

Wayne, L. G., Krichevsky, E. J., Love, L. L., Johnson, R., and Krichevsky, M. I. (1980). Taxonomic probability matrix for use with slowly growing mycobacteria. *International Journal of Systematic Bacteriology* **30,** 528–538.

Willcox, W. B., Lapage, S. P., Bascomb, S., and Curtis, M. A. (1973). Identification of bacteria by computer: theory and programming. *Journal of General Microbiology* **77,** 317–330.

Willcox, W. R., Lapage, S. P., and Holmes, B. (1980). A review of numerical methods in bacterial identification. *Antonie van Leeuwenhoek* **46,** 233–299.

Williams, S. T., Wellington, E. M. H., Goodfellow, M., Alderson, G., Sackin, M., and Sneath, P. H. A. (1981). The genus *Streptomyces*—a taxonomic enigma. *Zentralblatt für Bakteriologie Mikrobiologie und Hygiene, Abteilung 1, Supplement* **11,** 45–57.

Williams, S. T., Goodfellow, M., Alderson, G., Wellington, E. M. H., Sneath, P. H. A., and Sackin, M. J. (1983a). Numerical classification of *Streptomyces* and related genera. *Journal of General Microbiology* **129,** 1743–1813.

Williams, S. T., Goodfellow, M., Wellington, E. M. H., Vickers, J. C., Alderson, G., Sneath, P. H. A., Sackin, M. J., and Mortimer, A. M. (1983b). A probability matrix for identification of streptomycetes. *Journal of General Microbiology* **129,** 1815–1830.

12

Protein Sequencing and Taxonomy

R. P. AMBLER

Department of Molecular Biology, University of Edinburgh, Edinburgh, UK

Introduction

Protein sequencing is one of a number of molecular methods (Fig. 1) that can be used for detecting and estimating the amount of similarity between genomes, which is the essential process in taxonomy. With all approaches there are three stages in forming a classification for a group of organisms; the acquisition of the data, the processing of the data to derive relationships, and the assessment of the scheme proposed with respect to other information or speculation. Sequence information, whether it is from proteins or nucleic acids, differs from other biological information by being 'digital' rather than 'analogue' in nature (Ambler, 1976), and by being directly related to the germ line. The beginning of the use of sequence information for taxonomic purposes was coeval with the origin of numerical taxonomy, and the subjects have grown up together. Crick (1958) said, 'Biologists should realize that before long we shall have a subject which might be called "protein taxonomy"—the study of the amino acid sequences of the proteins of an organism and the comparison of them between species'. Peter Sneath has been one of the foremost theorists about the use of sequence information in taxonomy (Sokal and Sneath, 1963; Sackin and Sneath, 1965; Sneath and Sokal, 1973; Sneath, 1974; Sneath *et al.*, 1975), and the value of his personal contribution to the subject is very great.

Proteins had been recognized as being species-specific long before there was any detailed knowledge of their chemical structure, and before there was any idea at all as to how they were synthesized.

Attempts were made to use inter-species differences for taxonomic purposes, but these were limited by the difficulties in giving a quantitative value to such differences. A classic study was that of Reichert and Brown (1909), who found that the shape and angles of the haemoglobin crystals were characteristic for each species. Their methods were still being used more than 70 years later in a pioneer study of prehistoric proteins (Loy, 1983), where haemoglobin was isolated from

ISBN 0-12-289665-3

DNA	Protein
Base composition	Physical properties, such as size or electrophoretic mobility
DNA–DNA hybridization	Peptide maps
DNA–rRNA hybridization	Immunological cross-reaction
Non-coding DNA sequences	Amino acid compositions
rRNA nucleotide catalogues	Protein sequences
DNA sequences of selected genes	Tertiary structures
DNA sequences of complete genomes	

Fig. 1. Macromolecular methods for comparing genomes. The methods vary in the evolutionary distance over which they can detect similarity, and the objectivity with which difference can be quantified. The sequence methods are the only 'digital' ones. The amount of information in long sequences such as a complete viral or mitochondrial genome is so great that estimation of similarity becomes computationally very demanding. Collins and Coulson (1984) have described methods for handling such large amounts of sequence data.

blood stains on tools used by Stone Age humans, and the victim species identified crystallographically. In the same period, Nuttall (1904) studied by immunology the serum proteins of a large number of species, by testing them with antisera made in rabbits, and attempting to quantify the strengths of the precipitin reactions. Both these methods suggested that the amount of protein difference was directly related to the evolutionary separation of the species compared.

The primary structure of a protein, which determines its other properties, is a unique sequence made up of the '20' universal α-amino acids, all in α-peptide linkage. This was demonstrated by the determination of the order of the amino acids in a specific protein, insulin (Sanger and Thompson, 1953), and by the demonstration in 1969 by synthesis that the biological activity was quantitatively associated with a particular molecular structure (by Gutte and Merrifield, and by a totally independent method, by Hirschmann and his associates; see Richards and Wyckoff, 1971). Techniques for the determination of amino acid sequences have developed considerably since then (Allen, 1981), but the procedures remain time-consuming, and demanding of both care and material, and for many purposes are now being superseded by direct DNA sequencing.

Early results confirmed that species specificity existed at the sequence level—for instance in the different N-terminal residues of horse (Porter and Sanger, 1948) and whale (Schmid, 1949) myoglobins, and in the complete sequences of insulins from different species (Brown *et al.*, 1955; Harris *et al.*, 1956)—and it was at this stage of knowledge that Crick made his 1958 affirmation. The dialogue between molecular biologists and systematists did not yet start, as the priorities of the former were to understand the relationship of proteins and genes, and to make the elucidation of protein sequences easier and more sensitive. The latter felt that the necessary techniques were too difficult, slow, and expensive for them to use independently, and that, for the time being, admittedly less satisfactory methods, such as the comparison of whole-protein electrophoresis patterns (Sibley, 1960), were the best they could hope to apply on a usefully wide scale. The excitement and thinking of these years was summarized in *The Molecular Basis of Evolution* (Anfinsen, 1959).

After 1960, more sequence information became available, as instruments and techniques for rapid and sensitive sequencing were developed, and other laboratories became involved now that Sanger, Moore, and Stein (Hirs *et al.*, 1959) and Anfinsen (Potts *et al.*, 1962) had shown that protein sequencing was feasible. In this period the first examples of several classes of protein became known, including that of the first cytochrome *c* (Margoliash *et al.*, 1961), and the first bacterial (Ambler, 1963) and viral (Anderer *et al.*, 1960; Tsugita *et al.*, 1960) proteins. As sequence data accumulated, problems of collating the results soon arose. Sorm and Keil (1962) made an early statistical survey of regularities in sequences, in an approach that was criticised by Williams *et al.* (1961) in the first application of computer analysis to sequences. Sackin and Sneath (1965) developed a computer program for comparing sequences in protein chains, and in the same year the first edition of the *Atlas of Protein Sequence and Structure* (Dayhoff *et al.*, 1965) appeared, containing 65 sequences. [It is appropriate to record the great personal contribution that Margaret Dayhoff (ob. 5 February 1983) made to the subject of this chapter, right up until the day of her sudden and unexpected death, through her enthusiasm, energy, and imagination.]

'Evolving Genes and Proteins'

A valuable symposium with this title was held (Bryson and Vogel, 1965), setting the stage for the application of the technique to evolutionary and taxonomic problems, and in the succeeding years many protein chemists have felt the lure of molecular phylogeny. The sequence determination of few 'new' proteins has been carried out primarily for such purposes, but once that of the first member of a class is known there is a strong temptation to study proteins from other sources to acquire comparative information, and a high proportion of sequence publications mention the study of evolution as part of the justification for the work (Ambler, 1976). Criteria of suitability of proteins for taxonomic study can be made (Ambler, 1971a), which mostly relate to making the collection of sequence information sufficiently efficient that a useful quantity is attainable. Important criteria are therefore small protein size, high yield, and ease of purification by a standard method, while other essential factors are the distribution of the protein among organisms, and its apparent rate of evolution (see below). Different criteria apply for the choice of genes to study by direct DNA sequencing. In practice, most of the information that has provided the fuel for the arguments in molecular evolution has come from a very small number of protein systems, and in particular from mitochondrial cytochromes *c*.

The first and most important question to be asked was whether phylogenetic trees derived from the study of a single-protein gene product were concordant (or, indeed, bore any resemblance) to trees derived from the sum of morphological and palaeontological evidence for the whole organisms. Concordance might apply only to topology, or also to branch lengths and so to rates of evolution. The early results with mitochondrial cytochromes *c* (Margoliash, 1963; Fitch and Margoliash, 1967; Dayhoff and Eck, 1968) and fibrinopeptides (Doolittle and Blombäck, 1964) showed that the concordance was good enough to give some credence to sequence-based phylogenies that extended beyond the limits of classical evidence, and gave hope that a 'natural classification' for such organisms as the bacteria (Van Niel, 1946) might be attainable.

The study of homologous proteins from different organisms suggested that single amino acid substitutions (Brown *et al.*, 1955; Ingram, 1957) were the predominant type of change in protein evolution, an impression that has been confirmed by subsequent work. Insertions or deletions of one or a small number of amino acids do occur, particularly at the ends of polypeptide chains, although internal events may be rarer in bacteria than in eukaryotes. It has been suggested that the majority of single substitutions have no selective effect in evolution (Kimura, 1968; King and Jukes, 1969), and Wilson *et al.* (1977) have argued that gene translocation events are more important in speciation than the background changes through amino acid alterations.

The elucidation of the three-dimensional structure of proteins by X-ray crystallography has demonstrated that families of proteins exist that possess the same 'fold', such as the globins from vertebrates, invertebrates, and legumes, and many but not all of the various types of cytochrome *c* (Dickerson *et al.*, 1976). Although the three-dimensional structure is completely specified by the amino acid sequence, proteins with the same 'fold' may have diverged so far that no sequence similarities can be detected (Rossmann *et al.*, 1974). Tertiary structure may therefore provide a way to recognize very distant genetic relationships.

Comparison of Classical and Protein-derived Phylogenies

Although the early sequence results from fibrinopeptides (Doolittle and Blombäck, 1964) and cytochrome *c* (Fitch and Margoliash, 1967) showed that there was general concordance with classical phylogenetic trees, different proteins appeared to be evolving at very different rates (Dayhoff, 1972; Wilson *et al.*, 1977). These observations have generally been explained as being the result of different functional constraints on the structures of different sorts of proteins (Zuckerkandl and Pauling, 1962). For each set of proteins, the evolutionary rate (the rate of acceptance of mutations) seemed to be approximately constant in each line of descent, and so protein trees could be used to make estimates of times of divergence of lineages, quite independently of any fossil record, and formed the basis of a 'biological clock'.

The effort since 1970 has mainly been involved with the apparent exceptions to these rules, and evaluating whether these exceptions are genuine. Many of the anomalies have been shown to be due to the comparison of paralogous rather than orthologous genes (Fitch and Margoliash, 1970). It was early recognized that some classes of proteins were coded for by multiple non-identical genes (e.g., the α- and β-globins in vertebrates), presumably arising from an ancestral gene duplication, followed by functional divergence. Discordance would be observed if one paralogous sequence, such as a β-haemoglobin, was compared with a set of orthologous α-haemoglobins. An important case occurs with the egg white lysozymes of birds, where two very different proteins have been characterized. Most orders of birds produce only one or the other of the two types (Prager *et al.*, 1974), although some species of geese produce both. The proteins are so different that the existence of any sequence homology is dubious, whilst the tertiary structure equivalence is incomplete (Grütter *et al.*, 1983).

Other anomalies have been explained by errors in the original sequence determination. The technical standard of some of the 'second-generation' amino acid sequencing was inadequate (Ambler, 1976), and although many errors have since been detected and corrected, more must still exist in the data used for constructing phylogenetic trees. Mistakes have been found because the original methodology seemed inadequate (*Chromatium* 'RHP'; Kennel *et al.*, 1972; Ambler *et al.*, 1979b), by X-ray crystallography (papain; Drenth *et al.*, 1968), and by DNA sequencing (*lac* repressor; Farabaugh, 1978). It is to be hoped that the threat of independent checking of protein sequences by DNA sequencing now makes workers more careful.

The large amount of sequence information now existing, and the difficulty of the mathematics involved, make the calculation of protein phylogenetic trees a specialized field of activity, largely monopolized by the groups of Dayhoff (1972), Goodman (1982), and Fitch (Maeda and Fitch, 1981), and it is difficult for an outsider to experiment to see the effect of adding a new sequence (or a corrected old one) to an existing data set.

Mitochondrial Cytochrome c

The sequence of mitochondrial cytochrome *c* has been studied from a wider range of organisms than any other protein, and nearly 100 sequences are now available for analysis. These include proteins from fungi, protozoa, algae, and several invertebrate groups, as well as many sequences from vertebrates and higher plants. Whereas the protein is located in the mitochondrion, it is coded for by a nuclear gene in all cases where this has been checked for (Sherman *et al.*,

1966; Anderson *et al.*, 1981), although it seems quite possible that in some protists it might still be coded for by the mitochondrial DNA. The protein has been evolving slowly, so sequences from mammals in the same order generally only differ in one or two residues out of about 100. Proteins which by sequence criteria are 'mitochondrial cytochromes *c*' have been found in several bacteria, including the nonsulphur purple photosynthetic bacteria (Ambler *et al.*, 1976), *Agrobacterium* (Van Beeumen *et al.*, 1980), and *Nitrobacter* (Yamanaka *et al.*, 1982).

For many years the mitochondrial cytochromes *c* were considered to be an excellent example of a simple set of orthologous proteins, although some yeasts were known to contain an isocytochrome (Stewart *et al.*, 1966). However, Hennig (1975) found that mouse testis contained an isocytochrome that differed from the protein of adult differentiated tissues at 13 positions, and a larval form has been found in the housefly (Yamanaka *et al.*, 1980), differing from the form from adults in at least five positions. These reports have made little impression on molecular evolutionists, and the results have apparently been ignored in several analyses of cytochrome *c* evolution (Baba *et al.*, 1981; Goodman *et al.*, 1982).

It has been shown that the cytochrome *c* gene system in mammals is in fact very complicated. Scarpulla *et al.* (1981, 1982) have isolated and determined the sequence of a rat cytochrome *c* gene, which would translate to give the same amino acid sequence as that of the adult mouse (Hennig, 1975) or rat (Carlson *et al.*, 1977) protein. However, in addition, the rat genome contains 20 to 30 further different DNA sequences that hybridize with the sequenced gene at high stringency. The genomes of other mammals contain a different but equally complicated set of hybridizing sequences. These results suggest that there may be several as yet unidentified tissue- and developmental-specific cytochromes *c*, although some of the hybridizing sequences are likely to be pseudogenes (Proudfoot, 1980).

The principal anomalies that Baba *et al.* (1981) found in 'fitting the gene phylogeny to the species phylogeny' for cytochrome *c* were with the prawn (*Macrobranchium malcolmsonii;* Lyddiatt and Boulter, 1976) and the rattlesnake sequences. In deriving the lowest nucleotide replacement tree by their maximum parsimony method, the prawn lineage was joined to that of the horse and donkey, caused by 'apparently a few fortuitous convergent amino acid substitutions'. Such substitutions were not apparent in a visual comparison of the sequences, and there seemed to be no more 'convergent' identities between prawn and horse than between prawn and dog or rat.

Rattlesnake Cytochrome c

The rattlesnake cytochrome *c* sequence (Bahl and Smith, 1965) was one of the earlier structures to be reported, and it was fairly soon recognized as being

Table 1. *Amino acid sequence of rattlesnake heart cytochrome* c[a]

	Species			
		Rattlesnake		
Residue	Human	*Crotalus adamanteus*[b]	*Crotalus vividis* and *Crotalus atrox*[c]	Turtle
11[d]	Ile	Thr[e]	Ser[f]	Val
12[d]	Met	Met	Met	Gln
15[d]	Ser	Ser	Gly[f]	Ala
16	Gln	Gln	Thr[f]	Gln
22	Lys	Lys	Glu[f]	Lys
33	His	His	His	Asn
36	Phe	Phe	Phe	Ile
44	Pro	Val	Val	Glu
46[d]	Tyr	Tyr	Tyr	Phe
50[d]	Ala	Ala	Ala	Glu
58[d]	Ile	Ile	Ile	Thr
61	Glu	Asp	Asp	Glu
62	Asp	Asp	Asp	Glu
81	Ile	Val	Val	Ile
83[d]	Val	Thr[e]	Thr[f]	Ile
85	Ile	Leu	Leu	Ile
86	Lys	Ser[e]	Lys	Lys
87	Lys	Lys	Ser[f]	Lys
89[d]	Glu	Lys[e]	Lys[f]	Ala
92	Ala	Thr[e]	Thr[f]	Ala
93	Asp	Asn[e]	Asp	Asp
100	Lys	Glu	Glu	Asp
101	Ala	Lys[e]	Ala	Ala
103	Asn	Ala	Ala	Ser
104	Glu	Ala[e]	Lys	Lys

[a]Positions where there is believed to be an identical residue in all four sequences are not shown. The haem is attached to Cys-14 and Cys-17.
[b]As reported by Bahl and Smith (1965).
[c]Combined results for the two species; no species differences detected in this investigation.
[d]Residues characteristic of primate cytochrome *c* sequences.
[e]Residues unique to the rattlesnake sequence originally proposed.
[f]Residues now believed to be unique to the rattlesnake sequence.

anomalous. Bahl and Smith (1965) noted that there were several positions where the sequence resembled that of the human protein (Matsubara and Smith, 1963), a finding that became odder as more concordant sequences were added to the data set (Crowson, 1972). Jukes and Holmquist (1972) explained the anomaly as being due to a species-specific acceleration in evolutionary rates in the snake lineage, whilst Fitch (1973) suggested that the determination of the rattlesnake sequence might be wrong. The only other available reptile cytochrome *c* sequence, from turtle, has no positions in which it is identical to snake and where snake is different to primate. Goodman (1981) mentioned that he used the sequence of the cytochrome *c* from the lizard *Varanus varanus* in his calculations, and ascribed it to 'Borden, D. (1980), personal communication of published data' but did not give the actual sequence. Goodman *et al.* (1982) gave a cytochrome *c* tree that includes 'lizard', with branch lengths that imply that rapid evolution has not taken place in the lineage of this organism.

The rattlesnake sequence has been reinvestigated (Table 1), and it now seems that the sequence differs in nine positions from that which has been used as the basis for evolutionary theorizing since 1965. The amount of 'evolution' is comparable to that between the most divergent of mammals. Four of the differences are near the haem attachment site, in a region that was only analysed for amino acid composition (and not sequenced) in the original investigation. Bahl and Smith (1965) placed these residues in sequence by 'homology' with the already known cytochrome *c* sequences, as was recorded by them and in the *Atlas of Protein Sequence and Structure* (Dayhoff, 1972). The other five differences are towards the C-terminus of the molecule, and are explicable as being due to the wrong ordering of amino acids within peptides that had been satisfactorily purified. Such mistakes were easy and very understandable in 1965, when sensitive methods for determining the N-terminal residues of small peptides were not in common use.

It is not yet known how the revised sequence will affect positions of the snake branch on phylogenetic trees, but it is clear that, in terms of mutational events, it will remain a long branch. Interestingly, the revisions do not affect most of the positions in which the snake resembled the primate sequences. The proposed sequence around the haem attachment site is unprecedented in mitochondrial cytochromes *c*. It will be worthwhile reexamining some other published sequences, such as that from bullfrog, for which this region was not completely sequenced.

DNA Sequences of Cytochrome c *Genes*

It should now be feasible to isolate and sequence cytochrome *c* genes, at least from any vertebrate, by probing a genomic library with a rat cytochrome *c* gene (Scarpulla *et al.*, 1982). Much smaller amounts of tissue may suffice than were needed for protein isolation, so the range of organisms that can be looked at is greatly increased. However, the complexity of the cytochrome *c* genomic situation that has been revealed by Scarpulla *et al.* (1982) means that great care will be necessary to ensure that any gene that is isolated and sequenced is the true orthologous one. It is possible that the rattlesnake cytochrome *c* will prove to be coded for by a paralogous gene.

Evolutionary Rate Anomalies

The rattlesnake cytochrome *c* case was only the first example of a protein appearing to have evolved in one line much more rapidly than in others. Another is insulin in the hystricomorphs, the South American rodents. The first insulin sequences to be examined were from cattle and pig, the animals available from slaughterhouses, and used as commercial sources of insulin for human medication. All these insulins were very similar to each other, the only difference being in a three-residue region in the A chain, now known to be exposed on the surface of the native hexamer. Human insulin was also found to be nearly identical to pig. However, guinea pig insulin was known to be immunologically distinctive (Moloney and Coval, 1955), so Smith (1966) determined the sequence and found it to be different from the pig in 17 of 51 positions, whilst the insulin from rat differed from that of the pig in only four positions. Smith (1972) also showed that coypu insulin was also very different, although that from the chinchilla, another hystricomorph, was similar to pig. These important but unexpected results attracted very little attention until Blundell and Wood (1975) explained them in terms of a change in a structural constraint. A mutation had become accepted in the guinea pig insulin gene which resulted in the protein no longer binding zinc to form a protease-resistant hexamer, and so selection was no longer acting to conserve the sequences for the regions of the subunit contacts. Hystricomorph ribonucleases have been examined and appear to have evolved fairly normally (Beintema and Lenstra, 1982), and cytochrome *c* from the guinea pig is identical to that from the rat and the mouse (Carlson *et al.*, 1977). Wriston (1981) has pointed out that there are many unusual biochemical features present in the guinea pig.

Inter- and Intra-species Sequence Variation

From theoretical considerations, Kimura (1968) suggested that a high proportion of the amino acid substitutions that occur during molecular evolution, both between species, and within species as polymorphisms, was selectively neutral. These ideas were developed by King and Jukes (1969) as 'non-Darwinian evolution'. The hypothesis gained a lot of support from experiments that were unable to distinguish functionally between mitochondrial cytochromes *c* that differed considerably in sequence (Margoliash *et al.*, 1972). Subsequently, alterations in assay conditions have allowed kinetic differences to be detected between cytochromes *c* that only differ from each other in one or two positions (Margoliash *et al.*, 1976), although the physiological validity of the assay differences has not yet been established (Kamen *et al.*, 1978).

Some proteins appear to be evolving rapidly in all lines. For example, the fibrinopeptides (Doolittle and Blombäck, 1964), which are released from fibrinogen during the clotting process, seem to be evolving at least 10 times as fast as cytochrome *c*, so differences occur even between species as similar as dog and fox. The rapid evolution has been explained as being the result of the absence of structural constraints on changes in these peptides. The isolated peptides do possess some biological activities (Colman *et al.*, 1967; Kay *et al.*, 1974), so selection is clearly possible, although I know of no studies of species variation in activity. On the Kimura (1968) hypothesis, a high level of intra-species polymorphism would also be expected, but no sequence variants were detected in a survey of 125 individual humans (Doolittle *et al.*, 1970).

A low level of intra-species sequence variation was also found in a survey of cytochromes c_{551} from a collection of strains of the bacterium *Pseudomonas aeruginosa* from very different habitats (Ambler, 1973; and see Table 2).

In *Drosophila melanogaster,* two enzymically and electrophoretically distinct forms of the alcohol dehydrogenase occur in all natural populations, but with geographic and habitat differences in the relative proportions of each allele in each population. The differences correlate with a single amino acid substitution (Retzios and Thatcher, 1979). DNA sequencing has now shown that for 11 genes from five different natural populations, this single substitution is the only amino acid difference, but that 42 silent polymorphisms affecting the DNA as changes between synonymous codons were detected in these structural genes (Kreitman, 1983).

Hence, the present consensus of evidence favours the view that neutral mutation plays only a relatively minor role in the fixation of amino acid substitutions in evolving proteins, although it does not appear that anyone has found a logical flaw in Kimura's arguments.

Gene Conversion

In higher organisms, many genes occur as members of multigene families. Protein and DNA sequence studies demonstrate the relatedness of the genes, which will have arisen through successive duplications. They may be functionally distinct, and be switched on and off at distinct times in the development of the organism. The genomes may also contain defective pseudogenes (Proudfoot, 1980). The multigene families have characteristic clusters of related genes, and the arrangement and spacing of the genes within a cluster is as distinctive as the sequence of bases in a gene or of amino acids in a protein. Comparison of trees deduced from the differences in gene cluster structure and those from DNA or protein sequences of particular genes in these clusters show anomalies, generally by duplicated genes not diverging in DNA sequence as rapidly as expected. It is believed that this effect is caused by recombination between the two similar and adjacent genes, and divergence is slowed by repeated recombination and sequence homogenisation. The process appears to be fairly common with globin genes, and as many vertebrate genes are members of multigene families, gene conversion may have an important effect on perturbing protein-based phylogenies, even for cytochrome *c* (Jeffreys, 1982). Multigene families and pseudogenes do not appear to occur in prokaryotes.

Molecular Phylogeny without Fossil Evidence

Protein Similarities between Kingdoms

The basic belief of molecular phylogeny is that if paralogous sequences are known for three or more organisms, then the two organisms with the most similar sequences are likely to be the most nearly related. Such statements must be covered by many caveats, and as we have seen in the last section, anomalies do occur in sets of organisms for which there is a good fossil record. However, these anomalies seem to be sufficiently rare that careful extrapolation of the method beyond the fossil evidence is useful. One of the areas of most interest is the relationship between the kingdoms, and the evaluation of theories about the origin of chloroplasts and mitochondria (Margulis, 1970).

The first need was to see over what evolutionary distance sequence homology could be detected. The early cytochrome *c* results showed that the mitochondrial proteins from animals, plants, and fungi were all clearly homologous, but that the functionally equivalent protein from an aerobic bacterium was so different that the similarity in the sequence might have arisen by convergence (Ambler, 1963). The first serine proteases from bacteria were convincing structural analogues of the pancreatic enzymes, but examples of homologous proteins have now been found (cytochrome c_2, Dus *et al.*, 1968; Ambler *et al.*, 1976; *Streptomyces griseus* trypsin, Olafson *et al.*, 1975; and see Hartley, 1970), and three-dimensional structures (Dickerson *et al.*, 1976) now show that many more of the bacterial cytochromes *c* are structurally homologous to the mitochondrial proteins.

The endosymbiosis theory maintains that chloroplasts arose from cyanobacteria, and the mitochondria from aerobic bacteria. Cytochromes *c*, plastocyanins, ferredoxins, and 5 S rRNAs have been sequenced from cyanobacteria and from chloroplasts, and clearly demonstrate a close relationship between the organelles and the modern free-living organisms. The evidence has been summarized by Schwartz and Dayhoff (1978, 1981), and it supports but does not prove the theory. The same evidence has also been interpreted by Uzzell and Spolsky (1981) as being consistent with the autogenous origin theory (and see other papers in the same symposium). Dayhoff (1983) has also summarized the sequence evidence for connexions between the (five) kingdoms.

Sequence Variation amongst Bacteria

Sequence taxonomy has been seen as the way to the 'natural classification' for bacteria envisaged by Van Niel (1946), and, indeed, Woese and associates (Fox *et al.*, 1980; Stackebrandt and Woese, 1981) believe that this has already been achieved through their use of partial-sequence evidence from 16 S rRNA. Results from protein sequencing do not cover the wide range of organisms studied by the rRNA method, but overlap for some groups of organisms. The two approaches in general agree, but there are some points of discordance.

Much less comparative sequence work has been done with bacterial proteins than with those from higher organisms, but enough is now available to give an idea of the range of sequence difference that may correspond to different taxa. There is no evidence for a time scale for speciation in bacteria, and no *a priori* reason the 'biological clock' should keep the same time in such different organisms as bacteria and vertebrates, even for functionally equivalent proteins. However, most of the comparative results for bacteria have used cytochromes *c*, and in discussing them the amounts of sequence difference will be compared to the taxonomic distance this corresponds to for mitochondrial cytochrome *c*.

The sequences of cytochromes *c* from two or more different strains of several well-established bacterial species have been determined (Table 2). Species vary in their 'tightness', and *Pseudomonas aeruginosa* and probably *Rhodopseudomonas sphaeroides* appear to be very homogeneous, whilst other species are more varied. The 'tightness' is generally in agreement with the intra-species DNA–DNA hybridization results (e.g., see Champion *et al.*, 1980). On the mitochondrial cytochrome *c* scale, tight species would have as much variation as there is between different orders of mammals, whilst in one of the looser species the amount of difference could be as much as among the whole span of vertebrates. A few inter-species differences are at this level, but most are larger. It seems likely that in some cases, such as *Rhodospirillum molischianum* and *Rhodospirillum fulvum* (Table 2), a closer examination of the phenotypic properties will allow the species to be merged. In conventionally close species like *P. aeruginosa* and *P. fluorescens* (Rhodes, 1961), the differences in sequence are as big as the mammal–insect span in mitochondrial cytochrome *c*, whilst between the more closely similar sequences from different members of the Rhodospirillaceae (Ambler *et al.*, 1979c; the 'M' sequences of Dickerson, 1980a) the range is as great as for the whole range of mitochondrial cytochromes *c* from human to protozoa. The distribution of sequences seems to be discontinuous, with a 'new' sequence being either quite closely related to a known one, or completely different. If homologous proteins are found in members of different currently recognized genera, they may be closer in sequence than are some proteins from organisms from the same genus. Cases of such similarity are between the cytochromes c_2 of *Rhodopseudomonas capsulata* and *Paracoccus*

Table 2. *Intra- and Inter-species differences in bacterial cytochromes* c

Species	Number of strains compared	Range of difference (residues)	Size of protein (residues)
Pseudomonas aeruginosa	11	0–1	82
Pseudomonas fluorescens biotype C	7	0–7	82
Pseudomonas stutzeri (sensu lato)	11	0–13	82
Pseudomonas stutzeri (sensu stricto)	7	0–5	82
Pseudomonas 'stanieri'	4	0–1	82
Rhodopseudomonas capsulata	3	2–3	116
Rhodopseudomonas palustris	4	12–20	114
Rhodopseudomonas sphaeroides	3	0	124
Rhodopseudomonas gelatinosa	2	2	85
Rhodospirillum tenue	2	20	92
Rhodospirillum salexigens	2	3	110
Desulfovibrio vulgaris	2	14	107
Agrobacterium tumefaciens	2	12	110
Inter-species			
Pseudomonas stutzeri/Pseudomonas 'stanieri'	7/4	10–13	82
Rhodospirillum tenue/Rhodocyclus purpureus	2/1	29–30	91
Rhodospirillum fulvum/Rhodospirillum molischianum	1/1	12–13[a]	96–100
Rhodospirillum rubrum/Rhodospirillum photometricum	1/1	50	113

[a]Two isocytochromes *c*.

denitrificans (Ambler *et al.*, 1981a), the cytochromes *c′* of *Chromatium vinosum* and *Rhodospirillum tenue* (Ambler *et al.*, 1979a, 1981b), and the ferredoxins from *Rhodospirillum rubrum* and *Clostridium pasteurianum* (Matsubara *et al.*, 1983).

Gene Distribution

The bacterial genes whose products have been studied by protein sequencing have erratic distributions, which follow neither traditional nor Woese 16 S rRNA classifications (Fig. 2), although it is possible that some of the proteins shown in this figure have a wider distribution but are often expressed at only a low level. It is difficult to prove the absence of a protein or a gene from an organism, particularly as sequence differences can be so great that hybridization would not be detected using the genomic probe from another organism. Anomalously large differences in sequence between organisms that had been thought to be closely

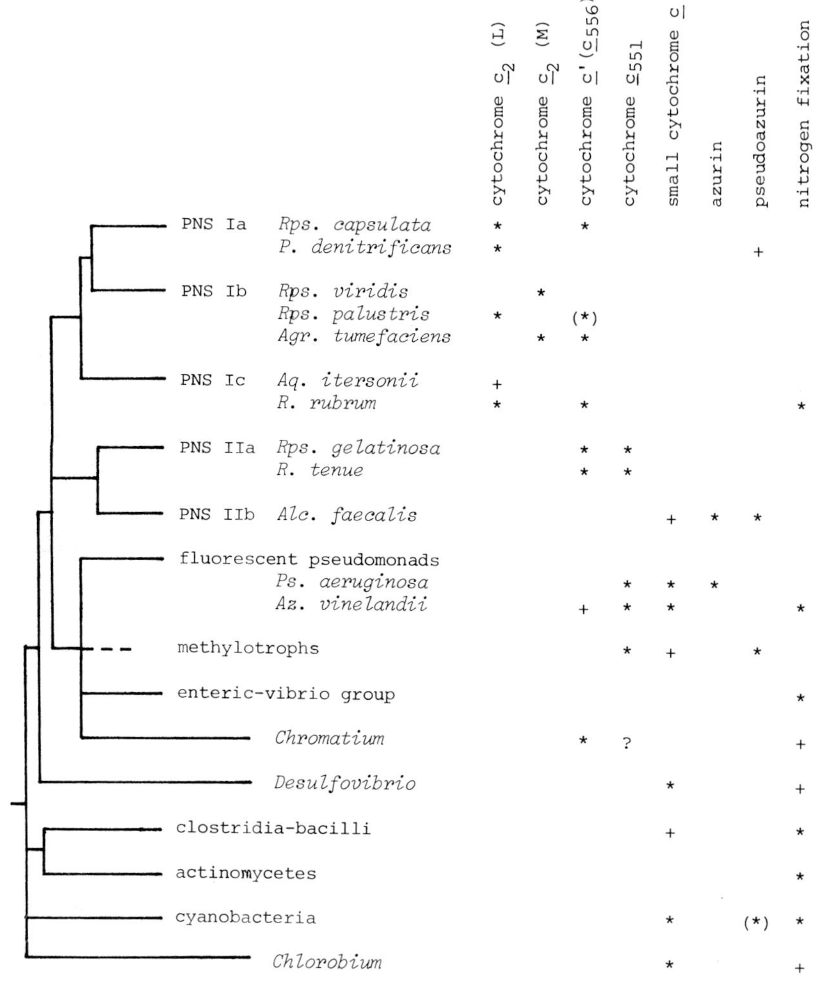

Fig. 2. Distribution of electron-transport proteins among eubacteria. Cytochromes *c* L (long) and M (medium) refer to the cytochrome c_2 types of Dickerson (1980a); the M type is that found in mitochondria. Many small cytochromes *c* have been characterized, but insufficient intermediate forms to relate them to each other have yet been found. The pseudoazurins are a group of plastocyanin-like proteins; sequence information for one from *Achromobacter cycloclastes* is available (Ambler, 1977). The information for nitrogen fixation comes mainly from DNA studies (Ruvkun and Ausubel, 1980). The connexions between groups are from the 16 S rRNA classification of Woese (Fox *et al.*, 1980; Stackebrandt and Woese, 1981). PNS, Purple nonsulphur; *Rps.*, *Rhodopseudomonas; P.*, *Paracoccus; Agr.*, *Agrobacterium; Aq.*, *Aquaspirillum; R.*, *Rhodospirillum; Alc.*, *Alcaligenes; Ps.*, *Pseudomonas; Az.*, *Azotobacter.* ★, Protein present and sequenced; +, protein present; ( ), protein present only in some strains; ?, some evidence for presence of protein.

related can be explained away as being paralogous gene comparison. Unexpected similarities cannot be dismissed in this way and must imply a genetic connexion (Ambler *et al.*, 1979a).

Other examples of widespread but erratic distributions of functions in bacteria are the production of β-lactamases and the ability to fix molecular nitrogen. There are several structurally unrelated types of β-lactamase (Ambler, 1980; Jaurin and Grundström, 1981), but one clearly homologous sequence type is produced by the Gram-positive bacteria *Staphylococcus aureus, Bacillus licheniformis,* and *B. cereus,* but also by many strains of the Gram-negative *Escherichia coli.* In *S. aureus* and *E. coli,* the gene for the enzyme is normally carried on a plasmid, and the most likely explanation for the apparent close genetic relationship between such diverse organisms is that plasmid exchange has occurred.

Genetic Exchange between Species

The introduction of foreign DNA into bacteria, and its subsequent replication and expression have constituted a basic part of experimental molecular genetics for many years (Avery *et al.*, 1944). At first this transfer could only be brought about with DNA from an organism very similar to the recipient, but the taxonomic distances across which transfer has been observed or demonstrated has greatly increased, and now the techniques of genetic engineering allow virtually any segment of DNA to be introduced and expressed in a recipient cell. The ways in which exchange can take place naturally between closely related genomes are now well understood (Shapiro, 1983), and the rapid spread of *R* factors all over the world is an indication of their efficiency. The possibility that genetic exchange might have long-term evolutionary importance has been part of molecular biological speculation for some time (Pollock, 1966). It was considered a possible explanation for similarities in β-lactamases from Gram-positive and Gram-negative organisms (Ambler and Meadway, 1969), whilst Anderson (1970) suggested that virus-mediated DNA transfer might even have a role in higher organism evolution, and Hartley (1970) thought that a cow might have infected a bacterium with the information that enabled it to make a mammalian-type serine protease.

Jones and Sneath (1970) reviewed the then known range of bacteria between which genetic exchange had been demonstrated, and discussed what taxonomic information could be gained by such knowledge. Hedges (1972) considered the pattern of evolutionary change in bacteria that could be brought about by genetic exchange, and the consequence it would have for phylogenetic and taxonomic studies. He predicted that a reticulate network rather than a tree would be necessary to represent the evolution of a bacterial genome.

If such lateral gene transfer does play a significant part in bacterial evolution, how would it manifest itself? It would show as organisms and characters that did not fit a 'natural' tree-based classification, and there would be discordance in topologies between phylogenetic trees constructed for different characters, and in particular for characters controlled by single genes. One would also find characters erratically distributed among disparate organisms, and several different permutations of homologous characters in a set of related organisms. Are such effects observed?

Apparent Cases of Lateral Gene Transfer

Some possible examples of lateral gene transfer are listed in Table 3. In some of these cases, alternate explanations are possible, such as contaminated cultures or immunological convergence, but in others lateral transfer seems the only reasonable explanation.

One of the most spectacular examples is that of *Photobacterium* superoxide dismutase (Martin and Fridovich, 1981). Superoxide dismutase (SOD) is an enzyme that protects organisms from the toxic effects of the superoxide ion. Several different types occur, all of which are metalloenzymes, and are distinguished by the metal they contain: iron, manganese, or copper and zinc. The Mn form is found in mitochondria and prokaryotes, the Fe form mostly in prokaryotes, whereas the CuZn form is characteristic of the cytosol of eukaryotes. The only known example of a CuZn SOD in prokaryotes is in *P. leiognathi* (Puget and Michelson, 1974), which also contains a normal bacterial Fe SOD. The CuZn SOD has been isolated from *Photobacterium* cells grown in pure culture, and compared in amino acid composition with SOD from other sources. By this criterion it could not be distinguished from CuZn SOD from teleost fish (six different species), but was clearly different from mammal and bird CuZn SOD and from all the Mn SOD and Fe SOD preparations analysed. The organism had been isolated from the luminous glands of a teleost fish, and there has clearly been a symbiosis between the bacterium and the pony fish for a considerable period. Some divergence of the bacterial SOD from that of the host fish has already taken place, as the proteins do not cross-react immunologically.

The fixation of molecular nitrogen is a property that is erratically but very widely distributed among bacteria, and representatives of four out of five of the eubacterial groups of Fox *et al.* (1980) have the capability. Ruvkun and Ausubel (1980) have used isolated nitrogen-fixation (*nif*) genes from *Klebsiella pneumoniae* to look for homologous sequences in other bacterial genomes. They found them in all of 19 widely disparate nitrogen-fixing strains, including clostridia, actinomycetes, and cyanobacteria, but not in any of the 10 non-fixing strains that they examined. The great sequence similarity in the structural genes for the enzymes of nitrogen fixation has been confirmed by Mevarech *et al.* (1980), by direct sequencing of the *nifH* gene in *Anabaena* 7120, and by comparison of the predicted amino acid sequence with that determined directly for

Table 3. *Possible cases of lateral gene transfer*[a]

Protein	Kingdom	Organisms	Reference
CuZn Superoxide dismutase	A to M	*Photobacterium*, teleost fish	Martin and Fridovich (1981)
Nitrogen reductase	Within M	Wide exchange among bacteria; Gram-positive, Gram-negative, and cyanobacteria	Ruvkun and Ausubel (1980) Mevarech *et al.* (1980)
Cytochrome *c′*	Within M	*Rhodospirillum tenue* and *Chromatium vinosum*	Ambler *et al.* (1979a)
Leghaemoglobin	A to P	Leguminous plants and (?) some animals	Hyldig-Nielsen *et al.* (1982)
Opine oxidase, etc.	M to P	*Agrobacterium tumefaciens* and dicotyledonous plants	Drummond (1979)
Histones H3 and H4	Within A	Two distantly related echinoderms	Busslinger *et al.* (1982)
β-Lactamase	Within A	Gram-positive and Gram-negative bacteria	Ambler and Meadway (1969); Ambler (1980)
Azurin	Within A	*Pseudomonas fluorescens* species cluster	Champion *et al.* (1980)
Elongation factor Tu	Within M	Gram-positive and Gram-negative bacteria and cyanobacteria	Filer and Furano (1980)
Tropomyosin	A to M	*Streptococcus* and mammal (?)	Hosein *et al.* (1979)
Serine protease	A to M	*Myxobacter* and vertebrate	Olafson *et al.* (1975); Hartley (1970)
Cytochromes	Within M	Halobacteria and eubacteria	Bayley (1982)
Ferredoxin	Within M	*Rhodospirillum rubrum* and *Clostridium pasteurianum*	Matsubara *et al.* (1983)
Alkaline phosphatase	Within A	Precambrian species with mineral skeletons	Mourant (1971)
Insulin and other hormones	A to M A to Pr	*Escherichia coli*, *'Progenitor cryptocides'*, *Tetrahymena*, some vertebrate cell (?)	LeRoith *et al.* (1981)

[a]The cases for which the evidence is strongest are towards the top of the list. The kingdoms that are most likely to be involved in each case are indicated: A, Animalia; P, Plantae; Pr, Protoctista; M, Monera.

the nitrogen reductase from *Clostridium pasteurianum* and *Azotobacter vinelandii*. If nitrogen fixation was a primitive character possessed by the common ancestor of these organisms, the sequences would have been evolving as slowly as histone H4 is supposed to have done. *Nif* genes are known to be encoded on plasmids in some species. Nitrogen fixation in *Klebsiella* appears to be regulated by elements similar to those that control other types of nitrogen metabolism in *Escherichia coli* (Ow *et al.*, 1983).

An example of tree discordance occurs between cytochromes *c* and *c′* among the *Rhodospirillaceae* (Ambler *et al.*, 1979a, 1981b; and Fig. 3). The cytochrome *c* from *Rhodospirillum tenue* and *Rhodopseudomonas gelatinosa* are similar in sequence, and resemble sequences from organisms like *Pseudomonas aeruginosa* and *Azotobacter vinelandii*, but the cytochrome *c′* from *Rhodospirillum tenue* resembles that from *Chromatium vinosum*, whereas that from *Rhodopseudomonas gelatinosa* is most like one from an *Alcaligenes* species. Dayhoff (1983) considered this to be an inconsistency 'quite beyond the lack of precision of the data or method', and was most likely to indicate a gene transfer event involving *Rhodospirillum tenue*.

Frequency of Lateral Gene Transfer in Bacteria

Champion *et al.* (1980) compared 93 strains of the *Pseudomonas fluorescens* species cluster by DNA–DNA hybridization, by a battery of 150 phenotypic tests, and by quantitative microcomplement fixation studies using six reference strains with antibodies directed against the protein azurin (Champion *et al.*, 1975). For 2 of the 93 strains the azurin results are inconsistent with the hybridization and phenotypic test results, suggesting a possible level of lateral gene transfer at species-cluster level.

A False Case of Apparent Lateral Gene Transfer

'*Chloropseudomonas ethylica*' was thought to be an organism that combined features of the electron-transport chains of sulphate reducers and of green photosynthetic bacteria (Ambler, 1971b; Van Beeumen *et al.*, 1976). The 'organism' was subsequently shown to be a consortium of a *Chlorobium* strain (Gray *et al.*, 1973) with a new type of sulphur reducer, *Desulfuromonas acetoxidans* (Pfennig and Biebl, 1976).

Conclusion

Phylogenetic Achievements of Protein Sequencing

In every single-gene phylogeny there appear to be some organisms for which anomalous evolution has occurred, for instance in the snake line for cytochrome *c* and the hystricomorphs for insulin. Classical taxonomists (Cronquist, 1976) understandably will not accept phylogenetic novelties based on the variation of a single gene.

The most sustained attempt to use protein sequences for specifically taxonomic purposes has been that of Boulter and colleagues at Durham, with their work on the flowering plants using mitochondrial cytochrome *c* (Boulter *et al.*, 1972) and chloroplast plastocyanin (Boulter *et al.*, 1979). Boulter interpreted his results cautiously and with full awareness of the mathematical and conceptual difficulties, but asserted that his evidence did modify the classical theories in some important ways, although Cronquist (1976) rather ponderously maintained that the available information is inadequate even as the basis for hypotheses. Boulter's studies were unfortunate in that the evolutionary rates for plastocyanins and cytochromes *c* in higher plants were sufficiently different that the comparison of trees was unsatisfactory. However, much more data from a third protein, ribulose-1,5-bisphosphate carboxylase, is now becoming available (Martin and Jennings, 1983), and we can look forward to some exciting and well-substantiated taxonomic predictions for the flowering plants.

A book edited by Goodman (1982) summarized the results achieved by protein sequencing in evolutionary studies, principally with reference to vertebrates, and looked ahead at the prospects of evolutionary investigations through genomic DNA sequences. There are now several proteins (globins, cytochromes *c*, lens α-crystallins, ribonucleases) which have been sequenced from a wide range of vertebrates, and phylogenetic trees have been constructed based on the composite evidence from several genes (Goodman *et al.*, 1982); this smooths away the anomalies. However, the major radiation of the eutherian mammals to give the orders that we can now study all took place ~60 million years ago, and comparative sequence studies are not very effective at distinguishing the order of nearly contemporary events a long time ago. De Jong (1982) made a number of suggestions about mammalian relationships on the basis of α-crystallin sequences, such as that the edentates are early eutherian offshoots, that the aardvark is related to the paenungulates (elephants, hyraxes, and manatees), and that the pholidotes (pangolins) are not close to the edentates. The consensus phylogeny (Goodman *et al.*, 1982) suggests a close relationship of the lagomorphs (rabbits) and tree shrews with the primates, a conclusion that is neither supported nor contradicted by palaeontology, but does not find strong evidence for ordinal positions for the carnivora, insectivora, or chiroptera.

If the biological clock exists for vertebrate proteins, it must be calibrated from the fossil record. Molecular biologists were slow to realize the difficulty palaeontologists have in making estimates of the times of divergence of different lineages, but this imprecision has been well emphasized by Wilson *et al.* (1977). Novacek (1982) reviewed evidence for the times of divergence of mammalian orders for the benefit of molecular biologists.

Protein or DNA for Taxonomic Sequencing?

The direct sequencing of DNA is now both more rapid and more accurate than protein sequencing, although it is extremely expensive in isotopes and reagents, but protein-sequencing methodology is again improving and becoming more sensitive. The isolation of the protein or the gene remains the critical step in any taxonomic sequence study. Isolation of genes by hybridization with a heterologous probe is simple and effective, but will only work if the sequences are quite similar. For bacterial cytochromes *c* (Table 2) it could not be relied on even between species of the same genus, and the absence of hybridization will not prove the absence of the gene. In cases where heterologous probes hybridize with DNA from very distant organisms, such as with *nif* (Ruvkun and Ausubel, 1980) or *tuf* (elongation factor Tu; Filer and Furano, 1980), lateral gene transfer is a likely explanation for the sequence conservation.

Gene Transfer and Natural Classification

If the rate of lateral gene transfer among a set of organisms was great enough, there could be no natural classification for them, and each organism would be just a selection of genes from one large gene pool. Gene transfer does seem to occur among the eukaryotes (Table 3), but not to an extent that disturbs the natural classification that certainly exists for large parts of the animal and plant kingdoms, if not for the fungi and protozoa. For the bacteria, the question is still open. The 'natural classification' of Fox *et al.* (1980) has certainly a better success rate than the eighth edition of *Bergey's Manual of Determinative Bacteriology* (Buchanan and Gibbons, 1974) at predicting genetic relationships amongst bacteria, but there is no evidence either for the stability of bacterial genomes through geological time or for any time scale in bacterial speciation. De Ley (1968) has emphasized that although there is evidence that bacteria existed in Precambrian times, there has been abundant time since for total change in their genomes, and it is quite possible that the presently living bacteria are quite different from those living in other geological periods, particularly in the Precambrian seas.

It is possible that the bacterial genome has an inviolate core of genes that code for essential functions, and which are so well matched to each other that no segregation is possible without making them function inadequately. It is more likely that there is a hierarchy for genes based on the ease with which they can be assimilated into a foreign genome. Factors which would affect this order would include the degree of structural interaction of the gene product with other molecules, the accuracy of control necessary for effective functioning, and the innovativeness of the function. The β-lactamases are transferred rapidly and widely because they confer a new function that has not been necessary until recently, and they are assimilated rapidly even though they normally need to be exported to a periplasmic location to function. Both cytochromes *c* and 16 S rRNA are integrated into complex structures, so may be expected to be assimilated only with difficulty into a foreign cell and genome.

There is no need for lateral transfer to involve a complete gene, and there is evidence for partial transfer in the case of the *tuf* gene of *Chromatium vinosum* (Filer and Furano, 1980). Another case may be between *Paracoccus denitrificans* and *Rhodopseudomonas capsulata* for part of a cytochrome *c* gene. When the sequences are compared, the first one-third of the molecules have 86% of their residues identical, whereas for the C-terminal two-thirds there is only 36% identity. I predict that when a number of complete 16 S rRNA sequences have been determined, examples of such chimaeras will be found among them.

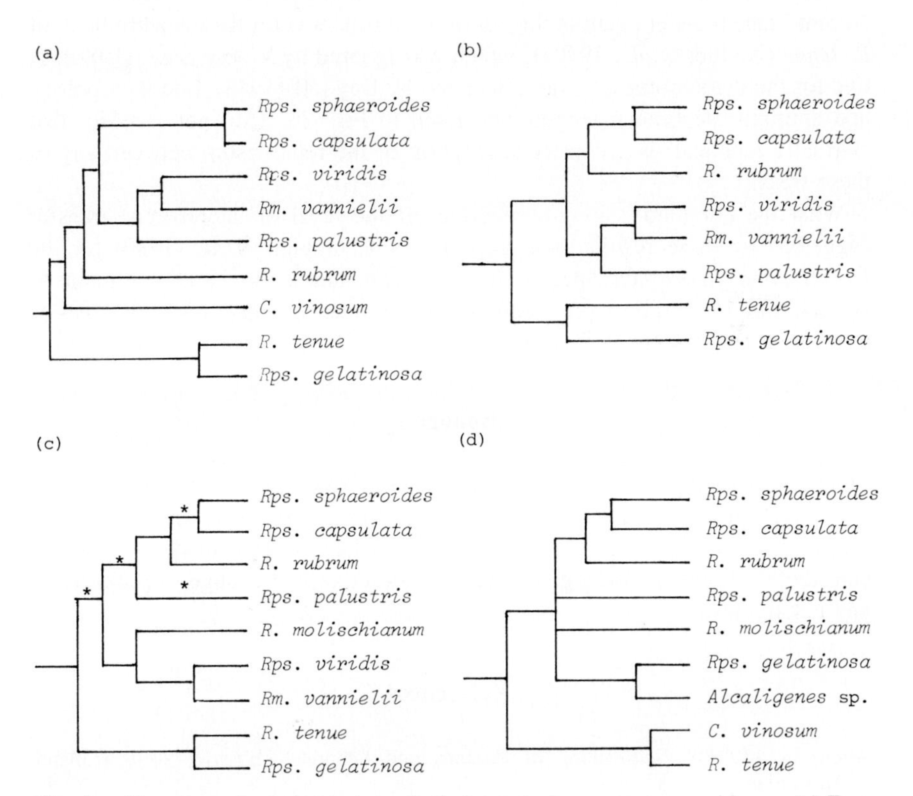

Fig. 3. Grouping of purple photosynthetic bacteria from sequence evidence. (a) From 16 S rRNA catalogues (Woese *et al.*, 1980); (b) from cytochrome c_2 sequences ignoring insertions and deletions (Woese *et al.*, 1980); (c) from cytochrome c_2 considering chain insertions, indicated by asterisks (Dickerson, 1980b); (d) from cytochrome c' (Ambler *et al.*, 1979a, 198lb; calculated by Dayhoff, 1983). Compare the connexions rather than the lengths of branches. *Rps.*, *Rhodopseudomonas; Rm.*, *Rhodomicrobium; R.*, *Rhodospirillum; C.*, *Chromatium.*

Stackebrandt and Woese (1981) said that 'the groupings [of the purple non-sulphur photosynthetic bacteria] generated by ribosomal RNA typing are virtually identical to those generated by comparative analyses of cytochrome *c* sequences. This effectively rules out inter-specific gene transfer as the cause of either phylogeny'. The results on which this statement is based are summarized in Fig. 3a and 3b (Woese *et al.*, 1980). The connexions of *Rhodospirillum rubrum* are significantly different between Fig. 3a and 3b. The grouping of cytochromes *c* in Fig. 3b was on the basis of numbers of identities when the sequences were optimally aligned. This grouping totally ignored the chain insertions, which Dickerson (1980b), in the adjoining paper, regarded as critical structural events in the divergence of the sequences. The topology of Dickerson's tree is shown in Fig. 3c. I believe that the most compelling evidence for a 'recent' gene transfer event in this set of organisms was for the cytochrome *c′* of *R. tenue* (Ambler *et al.*, 1979a), which was ignored by Woese *et al.* (1980). A tree for the cytochromes *c′* was calculated by Dayhoff (1983), and its topology and approximate branch lengths are given in Fig. 3d. I do not consider that 'virtually identical' is a correct description of the relationship between any of these trees.

What the cumulative long-term effect of the level of lateral gene transfer suggested by these results, and by those of Champion *et al.* (1980) for the *Pseudomonas fluorescens* species cluster, has on natural classification remains to be seen, but it is certainly premature to rule out inter-species transfer of genes as a factor in prokaryotic evolution.

Epilogue

Protein sequencing for taxonomic purposes has kept many scientists busy and used much ninhydrin during the years since 1960. The results have not had a radical effect on any areas of systematics, but have instead made an essential contribution to our knowledge of genomic structure and change in prokaryotes and eukaryotes.

References

Allen, G. (1981). 'Sequencing of Proteins and Peptides'. Elsevier-North Holland, Amsterdam.

Ambler, R. P. (1963). The amino acid sequence of *Pseudomonas* cytochrome *c*-551. *Biochemical Journal* **89,** 349–378.

Ambler, R. P. (1971a). Sequence data acquisition for the study of phylogeny. *In* 'Developpements Recents dans L'Etude Chimique de la Structure des Proteines' (Eds. A. Previero, J.-F. Pechère, and M.-A. Coletti-Previero), pp. 289–305. INSERM, Paris.

Ambler, R. P. (1971b). The amino acid sequence of cytochrome *c*-551.5 from the green photosynthetic bacterium *Chloropseudomonas ethylica*. *FEBS Letters* **18,** 347–350.

Ambler, R. P. (1973). The evolutionary stability of cytochrome *c*-551 in *Pseudomonas aeruginosa* and *Pseudomonas fluorescens* biotype C. *Biochemical Journal* **137,** 3–14.

Ambler, R. P. (1976). Standards and accuracy in amino acid sequence determination. *In* 'Structure–Function Relationships of Proteins' (Eds. R. Markham and R. W. Horne), pp. 1–14. Elsevier-North Holland, Amsterdam.

Ambler, R. P. (1977). Cytochrome *c* and copper protein evolution in prokaryotes. *In* 'Evolution of Metalloproteins' (Ed. G. J. Leigh), pp. 100–118. Symposium Press, London.

Ambler, R. P. (1980). The structure of β-lactamases. *Philosophical Transactions of the Royal Society of London, Series B* **289,** 321–331.

Ambler, R. P., and Meadway, R. J. (1969). Chemical structure of bacterial penicillinases. *Nature (London)* **222,** 24–26.

Ambler, R. P., Meyer, T. E., and Kamen, M. D. (1976). Primary structure determination of two cytochromes c_2: close similarity to functionally unrelated mitochondrial cytochrome *c*. *Proceedings of the National Academy of Sciences of the United States of America* **73,** 472–475.

Ambler, R. P., Meyer, T. E., and Kamen, M. D. (1979a). Anomalies in amino acid sequences of small cytochromes *c* and cytochromes *c′* from two species of purple photosynthetic bacteria. *Nature (London)* **278,** 661–662.

Ambler, R. P., Daniel, M., Meyer, T. E., Bartsch, R. G., and Kamen, M. D. (1979b). The amino acid sequence of cytochrome *c′* from the purple sulphur bacterium *Chromatium vinosum*. *Biochemical Journal* **177,** 819–823.

Ambler, R. P., Daniel, M., Hermoso, J., Meyer, T. E., Bartsch, R. G., and Kamen, M. D. (1979c). Cytochromes c_2 sequence variation among the recognized species of purple nonsulphur photosynthetic bacteria. *Nature (London)* **278**, 659–660.

Ambler, R. P., Meyer, T. E., Kamen, M. D., Schichman, S. A., and Sawyer, L. (1981a). A reassessment of the structure of *Paracoccus* cytochrome *c*-550. *Journal of Molecular Biology* **147,** 351–356.

Ambler, R. P., Bartsch, R. G., Daniel, M., Kamen, M. D., McLellan, L., Meyer, T. E., and Van Beeumen, J. (1981b). Amino acid sequences of bacterial cytochromes *c′* and *c* -556. *Proceedings of the National Academy of Sciences of the United States of America* **78,** 6854–6857.

Anderer, F. A., Uhlig, H., Weber, E., and Schramm, G. (1960). Primary structure of the protein from tobacco mosaic virus. *Nature (London)* **186,** 922–925.

Anderson, N. G. (1970). Evolutionary significance of virus infection. *Nature (London)* **227,** 1346–1347.

Anderson, S., Bankier, A. T., Barrell, B. G., de Bruijn, M. H. L., Coulson, A. R., Drouin, J., Eperon, I. C., Nierlich, D. P., Roe, B. A., Sanger, F., Schreier, P. H., Smith, A. J. H., Staden, R., and Young, I. G. (1981). Sequence and organization of the human mitochondrial genome. *Nature (London)* **290,** 457–465.

Anfinsen, C. B. (1959). 'The Molecular Basis of Evolution.' Wiley, New York.

Avery, O. T., MacLeod, C. M., and McCarty, M. (1944). Studies on the chemical nature of the substances inducing transformation of pneumococcal types. I. Induction of transformation by a desoxyribonucleic acid fraction isolated from pneumococcus type III. *Journal of Experimental Medicine* **79,** 137–157.

Baba, M. L., Darga, L. L., Goodman, M., and Czelusniak, J. (1981). Evolution of cytochrome *c* investigated by the maximum parsimony method. *Journal of Molecular Evolution* **17,** 197–213.

Bahl, O. P., and Smith, E. L. (1965). Amino acid sequence of rattlesnake heart cytochrome *c*. *Journal of Biological Chemistry* **240,** 3585–3593.
Bayley, S. T. (1982). Problems in tracing the early evolution of cells as illustrated by the archaebacteria and particularly by the halobacteria. *Zentralblatt für Bakteriologie, Mikrobiologie und Hygiene, Series C* **3,** 65–68.
Beintema, J. J., and Lenstra, J. A. (1982). Evolution of pancreatic ribonucleases. *In* 'Macromolecular Sequences in Systematic and Evolutionary Biology' (Ed. M. Goodman), pp. 43–73. Plenum, New York.
Blundell, T. L., and Wood, S. P. (1975). Is the evolution of insulin Darwinian or due to selectively neutral mutation? *Nature (London)* **257,** 197–203.
Boulter, D., Ramshaw, J. A. M., Thompson, E. W., Richardson, M., and Brown, R. H. (1972). A phylogeny of higher plants based on the amino acid sequences of cytochrome *c*, and its biological implications. *Proceedings of the Royal Society of London Series B* **181,** 441–455.
Boulter, D., Peacock, D., Guise, A., Gleaves, J. T., and Estabrook, G. (1979). Relationship between the partial amino acid sequences of plastocyanins from members of the families of flowering plants. *Phytochemistry* **18,** 603–608.
Brown, H., Sanger, F., and Kitai, R. (1955). The structure of pig and sheep insulins. *Biochemical Journal* **60,** 556–565.
Bryson, H., and Vogel, H. J. (1965). 'Evolving Genes and Proteins'. Academic Press, New York.
Buchanan, R. E., and Gibbons, N. E. (1974). 'Bergey's Manual of Determinative Bacteriology', 8th Edition. Williams & Wilkins, Baltimore, Maryland.
Busslinger, M., Rusconi, S., and Birnsteil, M. L. (1982). An unusual evolutionary behaviour of a seaurchin histone gene cluster. *EMBO Journal* **1,** 27–33.
Carlson, S. S., Mross, G. A., Wilson, A. C., Mead, R. T., Wolin, D., Bowen, S. F., Foley, N. T., Muijsens, A. O., and Margoliash, E. (1977). Primary structure of mouse, rat and guinea pig cytochromes *c*. *Biochemistry* **16,** 1437–1442.
Champion, A. B., Soderberg, K. L., Wilson, A. C., and Ambler, R. P. (1975). Immunological comparison of azurins of known amino acid sequence. *Journal of Molecular Evolution* **5,** 291–305.
Champion, A. B., Barrett, E. L., Palleroni, N. J., Soderberg, K. L., Kunisawa, R., Contopoulou, R., Wilson, A. C., and Doudoroff, M. (1980). Evolution in *Pseudomonas fluorescens*. *Journal of General Microbiology* **120,** 485–511.
Collins, J. F., and Coulson, A. F. W. (1984). Applications of parallel processing algorithms for DNA sequence analysis. *Nucleic Acids Research* **12,** 181–192.
Colman, R. W., Osbahr, A. J., and Morris, R. E. (1967). New vasoconstrictor, bovine peptide B released during blood coagulation. *Nature (London)* **215,** 292–293.
Crick, F. H. C. (1958). On protein synthesis. *Symposium of the Society for Experimental Biology* **12,** 138–163.
Cronquist, A. (1976). The taxonomic significance of the structure of plant proteins: a classical taxonomist's view. *Brittonia* **28,** 1–27.
Crowson, R. A. (1972). A systematist looks at cytochrome *c*. Journal of Molecular Evolution **2,** 28–37.
Dayhoff, M. O. (1972). 'Atlas of Protein Sequence and Structure', 5th Edition. National Biomedical Research Foundation, Washington, D.C.
Dayhoff, M. O. (1983). Evolutionary connections of biological kingdoms based on protein and nucleic acid sequence evidence. *Precambrian Research* **20,** 299–318.
Dayhoff, M. O., and Eck, R. V. (1968). 'Atlas of Protein Sequence and Structure 1967–68. National Biomedical Research Foundation, Silver Spring, Maryland.

Dayhoff, M. O., Eck, R. V., Chang, M., and Sochard, M. R. (1965). 'Atlas of Protein Sequence and Structure'. National Biomedical Research Foundation, Silver Spring, Maryland.

De Ley, J. (1968). Molecular biology and bacterial phylogeny. *In* 'Evolutionary Biology' (Eds. T. Dobzhansky, M. K. Hecht, and W. C. Steere), pp. 103–156. Appleton, New York.

De Jong, W. W. (1982). Eye lens proteins and vertebrate phylogeny. *In* 'Macromolecular Sequences in Systematic and Evolutionary Biology' (Ed. M. Goodman), pp. 75–114. Plenum, New York.

Dickerson, R. E. (1980a). Cytochrome *c* and the evolution of energy metabolism. *Scientific American* **242,** 98–110.

Dickerson, R. E. (1980b). Evolution and gene transfer in purple photosynthetic bacteria. *Nature (London)* **283,** 210–212.

Dickerson, R. E., Timkovich, R., and Almassy, R. J. (1976). The cytochrome fold and the evolution of bacterial energy metabolism. *Journal of Molecular Biology* **100,** 473–491.

Doolittle, R. F., and Blombäck, B. (1964). Amino acid sequence investigation of fibrinopeptides from various mammals; evolutionary implications. *Nature (London)* **202,** 147–152.

Doolittle, R. F., Chen, R., Glasgow, C., Mross, G., and Weinstein, M. (1970). The molecular constancy of fibrinopeptides A and B from 125 individual humans. *Humangenetik* **10,** 15–29.

Drenth, J., Jansonius, J. W., Koekoek, R., Swen, H. M., and Wolthers, B. G. (1968). Structure of papain. *Nature (London)* **218,** 929–932.

Drummond, M. (1979). Crown gall disease. *Nature (London)* **281,** 343–347.

Dus, K., Sletten, K., and Kamen, M. D. (1968). Cytochrome c_2 of *Rhodospirillum rubrum*. II. Complete amino acid sequence and phylogenetic relationships. *Journal of Biological Chemistry* **243,** 5507–5518.

Farabaugh, P. J. (1978). Sequence of the *lacI* gene. *Nature (London)* **218,** 929–932.

Filer, D., and Furano, A. V. (1980). Portions of the gene encoding elongation factor Tu are highly conserved in prokaryotes. *Journal of Biological Chemistry* **255,** 728–734.

Fitch, W. M. (1973). Aspects of molecular evolution. *Annual Review of Genetics* **7,** 343–380.

Fitch, W. M., and Margoliash, E. (1967). Construction of phylogenetic trees. *Science* **155,** 279–284.

Fitch, W. M., and Margoliash, E. (1970). The usefullness of amino acid and nucleotides sequences in evolutionary studies. *Evolutionary Biology* **4,** 67–109.

Fox, G. E., Stackebrandt, E., Hespell, R. B., Gibson, J., Maniloff, J., Dyer, T., Wolfe, R. S., Balch, W., Tanner, R., Magrum, L. J., Zablen, L. B., Blakemore, R., Gupta, R., Luehrsen, K. R., Bonen, L., Lewis, B. J., Chen, K. N., and Woese, C. R. (1980). The phylogeny of prokaryotes. *Science* **209,** 457–463.

Goodman, M. (1981). Decoding the pattern of protein evolution. *Progress in Biophysics and Molecular Biology* **38,** 105–164.

Goodman, M., Ed. (1982). 'Macromolecular Sequences in Systematic and Evolutionary Biology'. Plenum, New York.

Goodman, M., Romero-Herrera, A. E., Dene, H., Czelusniak, J., and Tashian, R. E. (1982). Amino acid sequence evidence on the phylogeny of primates and other eutherians. *In* 'Macromolecular Sequences in Systematic and Evolutionary Biology' (Ed. M. Goodman), pp. 115–191. Plenum, New York.

Gray, B. H., Fowler, C. F., Nugent, N. A., Rigopoulos, N., and Fuller, R. C. (1973).

Reevaluation of *Chloropseudomonas ethylica* strain 2K. *International Journal of Systematic Bacteriology* **23,** 256–264.

Grütter, M. G., Weaver, L. H., and Matthews, B. W. (1983). Goose lysozyme structure: an evolutionary link between hen and bacteriophage lysozymes? *Nature (London)* **303,** 828–831.

Harris, J. I., Sanger, F., and Naughton, M. A. (1956). Species differences in insulin. *Archives of Biochemistry and Biophysics* **65,** 427–438.

Hartley, B. S. (1970). Homologies in serine proteases. *Philosophical Transactions of the Royal Society of London Series B* **257,** 77–87.

Hedges, R. W. (1972). The pattern of evolutionary change in bacteria. *Heredity* **28,** 39–48.

Hennig, B. (1975). Change of cytochrome *c* structure during development of the mouse. *European Journal of Biochemistry* **55,** 167–183.

Hirs, C. H. W., Moore, S., and Stein, W. H. (1959). The sequence of amino acids in performic acid-oxidized ribonuclease. *Journal of Biological Chemistry* **235,** 633–647.

Hosein, B., McCarty, M., and Fischetti, V. (1979). Amino acid sequence and physicochemical similarities between streptococcal M protein and mammalian tropomyosin. *Proceedings of the National Academy of Sciences of the United States of America* **76,** 3765–3768.

Hyldig-Nielsen, J. J., Jensen, H. Ø, Paludan, K., Wiborg, O., Garrett, R., Jørgensen, P., and Marcker, K. A. (1982). The primary structure of two leghemoglobin genes from soybean. *Nucleic Acids Research* **10,** 689–701.

Ingram, V. M. (1957). Gene mutation in human haemoglobin: the chemical difference between normal and sickle cell haemoglobin. *Nature (London)* **180,** 326–328.

Jaurin, B., and Grundström, T. (1981). The *ampC* cephalosporinase of *E. coli* K12 has a different evolutionary origin from that of β-lactamases of the penicillinase type. *Proceedings of the National Academy of Sciences of the United States of America* **78,** 4897–4901.

Jeffreys, A. J. (1982). Evolution of globin genes. *In* 'Genome Evolution' (Eds. G. A. Dover and R. B. Flavell), pp. 157–176. Academic Press, London.

Jones, D., and Sneath, P. H. A. (1970). Genetic transfer and bacterial taxonomy. *Bacteriological Reviews* **34,** 40–81.

Jukes, T. H., and Holmquist, R. (1972). Evolutionary clock: non constancy of rate in different species. *Science* **177,** 530–532.

Kamen, M. D., Errede, B. J., and Meyer, T. E. (1978). Comparative studies of cytochromes *c*. *In* 'Evolution of Protein Molecules' (Eds. H. Matsubara and T. Yamanaka), pp. 373–385. Japan Scientific Societies Press, Tokyo.

Kay, A. B., Pepper, D. S., and McKenzie, R. (1974). The identification of fibrinopeptide B as a chemotactic agent derived from human fibrinogen. *British Journal of Haematology* **27,** 669–677.

Kennel, S. J., Meyer, T. E., Kamen, M. D., and Bartsch, R. G. (1972). On the monoheme character of cytochromes *c'*. *Proceedings of the National Academy of Sciences of the United States of America* **69,** 3432–3435.

Kimura, M. (1968). Evolutionary rate at the molecular level. *Nature (London)* **217,** 624–626.

King, J. L., and Jukes, T. H. (1969). Non-Darwinian evolution. *Science* **164,** 788–798.

Kreitman, M. (1983). Nucleotide polymorphism at the alcohol dehydrogenase locus of *Drosophila melanogaster*. *Nature (London)* **304,** 412–419.

LeRoith, D., Shiloach, J., Roth, J., and Lesniak, M. (1981). Insulin or a closely related molecule is native to *Escherichia coli*. *Journal of Biological Chemistry* **256,** 6533–6536.

Loy, T. H. (1983). Prehistoric blood residues: detection on tool surfaces and identification of species of origin. *Science* **220,** 1269–1271.

Lyddiatt, A., and Boulter, D. (1976). A comparison of cytochrome *c* from *Macrobrachium malcolmsonii* with other invertebrate cytochromes *c*. *Comparative Biochemistry and Physiology* **55B,** 337–342.

Maeda, N., and Fitch, W. M. (1981). Amino acid sequence of a myoglobin from lace monitor lizard, *Varanus varanus,* and its evolutionary implications. *Journal of Biological Chemistry* **256,** 4301–4309.

Margoliash, E. (1963). Primary structure and evolution of cytochrome *c*. *Proceedings of the National Academy of Sciences of the United States of America* **50,** 672–679.

Margoliash, E., Smith, E. L., Kreil, G., and Tuppy, H. (1961). Amino acid sequence of horse heart cytochrome *c*. *Nature (London)* **192,** 1121–1127.

Margoliash, E., Fitch, W. M., Markowitz, E., and Dickerson, R. E. (1972). Functional limits of cytochrome *c* variability. *In* 'Oxidation–Reduction Enzymes' (Eds. A. Akeson and A. Ehrenburg), pp. 5–17. Pergamon, Oxford.

Margoliash, E., Ferguson-Miller, S., Brautigan, D. L., and Chaviano, A. H. (1976). Functional basis for evolutionary change in cytochrome *c* structure. *In* 'Structure–Function Relationships of Proteins' (Eds. R. Markham and R. W. Horne), pp. 145–165. Elsevier-North Holland, Amsterdam.

Margulis, L. (1970). 'Origin of Eukaryotic Cells'. Yale Univ. Press, New Haven.

Martin, J. P., and Fridovich, I. (1981). Evidence for a natural gene transfer from the pony fish to its bioluminescent bacterial symbiont, *Photobacter leiognathi*. *Journal of Biological Chemistry* **256,** 6080–6089.

Martin, P. G., and Jennings, A. C. (1983). The study of plant phylogeny using amino acid sequences of ribulose-1,5-biphosphate carboxylase. I. Biochemical methods and the pattern of variability. *Australian Journal of Botany* **31,** 395–409.

Matsubara, H., and Smith, E. L. (1963). Human heart cytochrome *c*. *Journal of Biological Chemistry* **238,** 2732–2753.

Matsubara, H., Inoue, K., Hase, T., Hiura, H., Kakuno, T., Yamashita, J., and Horio, T. (1983). Structure of the extracellular ferredoxin from *Rhodospirillum rubrum:* close similarity to clostridial ferredoxins. *Journal of Biochemistry (Tokyo)* **93,** 1385–1390.

Mevarech, M., Rich, D., and Haselkorn, R. (1980). Nucleotide sequence of a cyanobacterial *nifH* gene coding for nitrogenase reductase. *Proceedings of the National Academy of Sciences of the United States of America* **77,** 6476–6480.

Moloney, P. J., and Coval, M. (1955). Antigenicity of insulin: diabetes induced by specific antibodies. *Biochemical Journal* **59,** 179–185.

Mourant, A. E. (1971). Transduction and skeletal evolution. *Nature (London)* **231,** 466–467.

Novecek, M. J. (1982). Information for molecular studies for anatomical and fossil evidence on higher eutherian phylogeny. *In* 'Macromolecular Sequences in Systematical Evolutionary Biology' (Ed. M. Goodman), pp. 3–41. Plenum, New York.

Nuttall, G. H. F. (1904). 'Blood Immunity and Blood Relationship'. Cambridge Univ. Press, London.

Olafson, R. W., Juvasek, L., Carpenter, M. R., and Smillie, L. B. (1975). Amino acid sequence of *Streptomyces griseus* trypsin: cyanogen bromide fragments and complete sequence. *Biochemistry* **14,** 1168–1177.

Ow, D. W., Sundaresan, V., Rothstein, D. M., Brown, S. E., and Ausubel, F. M. (1983). Promoters regulated by the *glnG* (*ntrC*) and *nifA* gene products share a heptameric consensus sequence in the −15 region. *Proceedings of the National Academy of Sciences of the United States of America* **80,** 2524–2528.

Pfennig, N., and Biebl, H. (1976). *Desulfuromonas acetoxidans* gen. nov. and sp. nov., a new anaerobic, sulfur reducing, acetate-oxidizing bacterium. *Archives of Microbiology* **110,** 3–12.

Pollock, M. R. (1966). 'What is Molecular Biology?, Inaugural Lecture No. 30. Univ. of Edinburgh, Edinburgh.

Porter, R. R., and Sanger, F. (1948). The free amino groups of haemoglobin. *Biochemical Journal* **42,** 287–294.

Potts, J. T., Berger, A., Cooke, J., and Anfinsen, C. B. (1962). A reinvestigation of the sequence of residues 11–18 in bovine pancreatic ribonuclease. *Journal of Biological Chemistry* **237,** 1851–1855.

Prager, E. M., Wilson, A. C., and Arnheim, N. (1974). Widespread distribution of lysozyme g in birds. *Journal of Biological Chemistry* **249,** 7295–7297.

Proudfoot, N. (1980). Pseudogenes. *Nature (London)* **286,** 840–841.

Puget, K., and Michelson, A. M. (1974). Iron containing superoxide dismutases from luminous bacteria. *Biochimie* **56,** 1255–1267.

Reichart, E. T., and Brown, A. P. (1909). 'The Crystallography of Haemoglobins', Publication No. 16. Carnegie Institute, Washington, D.C.

Retzios, A. D., and Thatcher, D. R. (1979). Chemical basis of the electrophoretic variation at the alcohol dehydrogenase locus of *Drosophila melanogaster. Biochimie* **61,** 701–704.

Rhodes, M. E. (1961). The characterization of *Pseudomonas fluorescens* with the aid of an electronic computer. *Journal of General Microbiology* **25,** 331–345.

Richards, F. M., and Wyckoff, H. W. (1971). Bovine pancreatic ribonuclease. *In* 'The Enzymes' (Ed. P. D. Boyer), 3rd Edition, Vol. 4, pp. 647–806. Academic Press, New York.

Rossmann, M. G., Moras, D., and Olsen, K. W. (1974). Chemical and biological evolution of a nucleotide-binding protein. *Nature (London)* **250,** 194–199.

Ruvkun, C. B., and Ausubel, F. M. (1980). Interspecies homology of nitrogenase genes. *Proceedings of the National Academy of Sciences of the United States of America* **77,** 191–195.

Sackin, M. J., and Sneath, P. H. A. (1965). Amino acid sequences in proteins: a computer study. *Biochemical Journal* **96,** 70P–71P.

Sanger, F., and Thompson, E. O. P. (1953). The amino acid sequence in the glycyl chain of insulin. *Biochemical Journal* **53,** 353–366, 366–374.

Scarpulla, R. C., Agne, K. M., and Wu, R. (1981). Isolation and structure of a rat cytochrome *c* gene. *Journal of Biological Chemistry* **256,** 6480–6486.

Scarpulla, R. C., Agne, K. M., and Wu, R. (1982). Cytochrome *c* gene-related sequences in mammalian genomes. *Proceedings of the National Academy of Sciences of the United States of America* **79,** 739–743.

Schmid, K. (1949). Untersuchungen über das Wal-myoglobin. *Helvetica Chimica Acta* **32,** 105–114.

Schwartz, R. M., and Dayhoff, M. O. (1978). Origins of prokaryotes, eukaryotes, mitochondria and chloroplasts. *Science* **199,** 395–403.

Schwartz, R. M., and Dayhoff, M. O. (1981). Chloroplast origins: inferences from protein and nucleic acid sequences. *Annals of the New York Academy of Sciences* **361,** 260–269.

Shapiro, J. A. (1983). 'Mobile Genetic Elements'. Academic Press, New York.

Sherman, F., Stewart, J. W., Margoliash, E., Parker, J., and Campbell, W. (1966). The structural gene for yeast cytochrome *c*. *Proceedings of the National Academy of Sciences of the United States of America* **55,** 1498–1504.

Sibley, C. G. (1960). The electrophoretic patterns of avian egg white proteins as taxonomic characters. *Ibis* **102,** 215–284.

Smith, L. F. (1966). Species variation in the amino acid sequence of insulin. *American Journal of Medicine* **40,** 662–666.

Smith, L. F. (1972). Amino acid sequences of insulins. *Diabetes, Supplement 2,* **21,** 457–460.

Sneath, P. H. A. (1974). Phylogeny of microorganisms. *Symposium of the Society for General Microbiology* **24,** 1–39.

Sneath, P. H. A., and Sokal, R. R. (1973). 'Numerical Taxonomy'. Freeman, San Francisco.

Sneath, P. H. A., Sackin, M. J., and Ambler, R. P. (1975). Detecting evolutionary incompatabilities from protein sequences. *Systematic Zoology* **24,** 311–332.

Sokal, R. R., and Sneath, P. H. A. (1963). 'Principles of Numerical Taxonomy'. Freeman, San Francisco.

Sorm, F., and Keil, B. (1962). Regularities in the primary structures of proteins. *Advances in Protein Chemistry* **17,** 167–207.

Stackebrandt, E., and Woese, C. R. (1981). The evolution of prokaryotes. *Symposium of the Society for General Microbiology* **32,** 1–31.

Stewart, J. W., Margoliash, E., and Sherman, F. (1966). Location of histidyl residue in yeast iso-2-cytochrome *c* and the structure of the cytochrome *c* hemochrome. *Federation Proceedings, Federation of American Societies for Experimental Biology* **25,** 647 (Abstract 2587).

Tsugita, A., Gish, D. T., Young, J., Fraenkel-Conrat, H., Knight, C. A., and Stanley, W. M. (1960). The complete amino acid sequence of the protein of tobacco mosaic virus. *Proceedings of the National Academy of Sciences of the United States of America* **46,** 1463–1469.

Uzzell, T., and Spolsky, C. (1981). Two data sets: alternative explanations and interpretations. *Annals of the New York Academy of Sciences* **361,** 481–499.

Van Beeumen, J., Ambler, R. P., Meyer, T. E., Kamen, M. D., Olson, J. M., and Shaw, E. K. (1976). The amino acid sequences of the cytochrome *c*-555 from two green sulphur bacteria of the genus *Chlorobium. Biochemical Journal* **159,** 757–774.

Van Beeumen, J., Tempst, P., Stevens, P., Bral, D., Van Dammen, J., and De Ley, J. (1980). Cytochromes *c* of two different sequence classes in *Agrobacterium tumefaciens. Protides of the Biological Fluids* **28,** 69–74.

Van Niel, C. B. (1946). The classification and natural relationships of bacteria. *Cold Spring Harbor Symposia on Quantitative Biology* **11,** 285–301.

Williams, J., Clegg, J. B., and Mutch, M. O. (1961). Coincidence and protein structure. *Journal of Molecular Biology* **3,** 532–540.

Wilson, A. C., Carlson, S. S., and White, T. J. (1977). Biochemical evolution. *Annual Review of Biochemistry* **46,** 573–639.

Woese, C. R., Gibson, J., and Fox, G. E. (1980). Do genealogical patterns in purple photosynthetic bacteria reflect interspecific gene transfer? *Nature (London)* **283,** 210–212.

Wriston, J. C. (1981). Biochemical peculiarities of the guinea pig and some possible examples of convergent evolution. *Journal of Molecular Evolution* **17,** 1–9.

Yamanaka, T., Inoue, S., and Hiroyoshi, T. (1980). Structural differences between larval and adult cytochromes *c* of the housefly *Musca domestica. Journal of Biochemistry (Tokyo)* **88,** 601–604.

Yamanaka, T., Tanaka, Y., and Fukumori, Y. (1982). *Nitrobacter agilis* cytochrome *c*-550: isolation, physicochemical and enzymatic properties, and primary structure. *Plant and Cellular Physiology* **23,** 441–449.

Zuckerkandl, E., and Pauling, L. (1962). Molecular disease, evolution and genetic heterogeneity. *In* 'Horizons in Biochemistry' (Eds. M. Kasha and B. Pullman), pp. 189–225. Academic Press, New York.

13

Numerical Methods in the Classification of Bacteria by Protein Electrophoresis

K. KERSTERS

Laboratorium voor Microbiologie en Microbiële Genetica, Faculteit Wetenschappen, Rijksuniversiteit, Ghent, Belgium

---

Introduction

The information contained in the microbial genome finds its expression at different levels. The base sequence of DNA is the primary level of information, which is expressed at the second level in the structure of protein molecules. The chemical structure of cellular components (the chemical composition of cell walls, lipids, etc.), and the phenotypic behaviour of cells are, respectively, the third and fourth levels of expression (Norris, 1980). Techniques such as DNA–DNA and DNA–ribosomal (r) RNA hybridization, cataloguing of 16 S rRNA (Fox *et al.*, 1980), electrophoresis of DNA fragments generated by restriction endonucleases, and the determination of the DNA base composition (mol % G + C) are methods of investigating the first level of genetic information. Amino acid sequencing techniques (see Ambler, Chapter 12), gel electrophoresis of proteins, and immunodiffusion and immunoelectrophoresis are techniques suitable for studying the second level of expression of genetic information.

The use of polyacrylamide gel electrophoresis in microbial systematics has been established for many years (for reviews see Garber and Ribbon, 1968; Kersters and De Ley, 1980; Jackman, 1985a). Proteins form a primary information source of enormous potential for the identification and differentiation of micro-organisms. The electrophoretic separation of cellular proteins is a very sensitive technique, mainly providing valuable information on the similarity of strains within species and subspecies. Different protein electrophoretic techniques are being used increasingly in taxonomic studies. For instance, the last six issues of the *International Journal of Systematic Bacteriology* contain at least 20 publications dealing with the use of polyacrylamide gel electrophoresis of proteins in the systematic studies of various genera: *Acetobacter* (Gillis *et al.*, 1983), *Bacteroides* (Cato *et al.*, 1982b; Holdeman and Johnson, 1982; Holdeman *et al.*, 1982; Johnson and Holdeman, 1983), *Bifidobacterium* (Biavati *et*

ISBN 0-12-289665-3

al., 1982), *Campylobacter* (Hanna *et al.*, 1983), *Clostridium* (Cato *et al.*, 1982a), *Corynebacterium* (Carlson and Vidaver, 1982), *Gluconobacter* (Gosselé *et al.*, 1983a), *Klebsiella* (Ferragut *et al.*, 1983), *Lactobacillus* (Cato *et al.*, 1983b), *Mycoplasma* (Hill, 1983), *Peptostreptococcus* (Cato *et al.*, 1983a), *Pseudomonas* (Nakajima *et al.*, 1983), *Rickettsia* (Philip *et al.*, 1983), *Rochalimaea* (Weiss and Dasch, 1982), *Ureaplasma* (Howard and Gourlay, 1982), and *Veillonella* (Mays *et al.*, 1982).

Electrophoresis of bacterial proteins is usually carried out by one of the following two procedures:

1. A mixture of proteins of a bacterial strain [e.g., soluble proteins from a cell-free extract, or proteins solubilized by treatment with sodium dodecylsulphate (SDS)] is submitted to electrophoresis in a polyacrylamide gel and stained; then the whole banding pattern is examined and compared with similar protein electrophoregrams from other bacterial strains, without any attempt to characterize each band.
2. The native proteins of bacterial strains are submitted to electrophoresis in a suitable stabilizing medium (e.g., starch gel, agar gel, polyacrylamide gel, or polyacrylamide–agarose gel mixtures) and then stained for specific enzymes such as dehydrogenases, esterases, phosphatases, β-lactamase, and so on (for a review, see Williams and Shah, 1980). In contrast to the first technique, zymogram patterns yield a small number of bands. Therefore, they can be analysed more easily and may be assessed visually. However, comparative zymogram analysis of only one or two enzymes generally does not yield results of value for bacterial systematics, because too few features are involved. To increase the taxonomic value of the comparisons of zymograms, a much greater number of different and specific enzymes have to be detected per strain (Baptist *et al.*, 1969, 1971; Seidler *et al.*, 1972).

The present contribution is concerned with the application of the first technique, where the entire electrophoretic protein-banding pattern is considered. The resemblance between electrophoregrams is often determined by visual comparison of the stained gels or their photographs. Although our eyes are very sensitive instruments, resemblance between protein patterns cannot be quantified visually, and large numbers of different patterns are hard to remember. Assistance by computers is therefore required. Developments made in objective, computerized analysis of protein electrophoregrams and their impact on the classification and identification of several bacterial groups will be summarized here.

Rationale for the Application of Protein Electrophoresis in Taxonomy

Proteins of the bacterial cell can be considered as an indirect copy of the genome. The primary structure (the amino acid sequence, molecular weight, and net electrical charge) of each protein species is a reflection of the DNA sequences in the corresponding cistron. The amount of each protein is likewise genetically determined. Zone electrophoresis of whole-cell proteins in well-defined standardized conditions produces a complex protein banding pattern (called a protein electrophoregram), which can be considered as a 'fingerprint' of the strain investigated. A bacterial strain, grown in identical conditions, will always produce the same set of proteins and hence the same electrophoretic protein pattern. In theory, loss or gain of a plasmid can affect the protein electrophoregram of a bacterial strain, provided the plasmid codes for proteins which are synthesized in such quantities that they can be detected by the techniques employed. However, no differences in protein patterns were observed between plant pathogenic *Agrobacterium* strains containing a large tumour-inducing plasmid (Ti plasmid, molecular mass at least 100×10^6) and their nonpathogenic mutants cured of the Ti plasmid (K. Kersters, J. Schell, and J. De Ley, unpublished results).

In one-dimensional gels each protein band will usually consist of a number of structurally different protein species with identical electrophoretic mobility. However, identical electrophoretic mobility of protein bands from different bacteria does not necessarily imply that these proteins are structurally related. Further, it should be realized that even the finest techniques available detect only a fraction of the proteins in the bacterial extract, because proteins present in too small a quantity will not be detected by, for example, Coomassie blue staining.

The taxonomic information yielded by the different electrophoretic techniques depends on the conditions of protein sample preparation and on the principle of the electrophoretic separation. Under nondenaturing conditions, ribosomal, membrane, and nucleic acid-bound proteins are usually removed from the final extract by high-speed centrifugation. Polyacrylamide gel electrophoretic patterns of material prepared in such nondenaturing conditions will reflect size as well as charge differences of the proteins, whereas patterns obtained in denaturing conditions using SDS are based on size differences only. As the molecular size of homologous proteins is more conserved than their net charge or isoelectric point, patterns based on separation by size only (e.g., SDS gel electrophoresis) should in theory detect broader taxonomic relationships than those relying on both charge and size parameters.

In a number of cases it is now well proven that one-dimensional whole-cell protein electrophoregrams discriminate at much the same level as information derived from DNA–DNA hybridization data (Kersters and De Ley, 1975; Swings *et al.*, 1976; Izard *et al.*, 1981; Owen and Jackman, 1982; Ferragut *et al.*, 1983; and this chapter). Bacterial strains with 90 to 100% of DNA base sequence homology usually show almost identical protein patterns, and strains with at least 70% of DNA homology tend to have similar protein patterns. These observations are the major pillars on which the application of protein electrophoresis in microbial systematics is based. Comparison of electrophoretic patterns is a powerful technique with a fairly fine taxonomic resolution, that is, at the level of species, subspecies, or biotypes. Consequently, more remote relationships, for example those between different genera or between different species of heterogeneous genera, usually cannot be detected by protein electrophoresis. In an examination of strains labelled *Pseudomonas paucimobilis*, Owen and Jackman (1982) found a high degree of congruence ($r = .92$) between the percentage protein similarity and the percentage DNA–DNA hybridization. Strains with zero DNA homology, however, showed a calculated protein pattern similarity of 30 to 40%, because electrophoretic traces of completely unrelated bacteria usually display an inherent similarity in their general shape. It remains to be investigated whether such correlations also exist for other groups of bacteria. Protein electrophoregrams can be used in microbial systematics for screening large numbers of isolates and detecting genetically highly related strains. Genetic relatedness between different electrophoretic groups can then be measured by DNA–DNA hybridizations of a few representative strains of each electrophoretic group (see the example on *Xanthomonas*).

Principles and Different Types of Electrophoretic Separations of Proteins

The technique of polyacrylamide gel electrophoresis (PAGE) for the separation of proteins was pioneered by Ornstein (1964) and Davis (1964). Currently, the following types of electrophoretic separation of proteins are being used in microbial systematics: (i) homogeneous PAGE, (ii) gradient PAGE, (iii) isoelectric focussing, and (iv) two-dimensional PAGE.

Homogeneous Polyacrylamide Gel Electrophoresis

The size of the pores of the polyacrylamide gel depends on the relative concentrations of acrylamide and bisacrylamide. Most frequently alkaline buffers are used for the separation of native proteins, migrating through the gel at a rate determined by their net electrical charge and molecular size. Several workers investigating representatives of the Mycoplasmatales have used acetic acid and urea-containing gel and buffer systems for the electrophoretic separation of proteins solubilized by a mixture of phenol–acetic acid–water (Razin and Rottem, 1967; Rosendal, 1973). Proteins solubilized with SDS are electrophoretically separated on the basis of size only (Weber and Osborn, 1969).

Gradient Polyacrylamide Gel Electrophoresis

This is a high-resolution technique, where a gradient of polyacrylamide concentration is formed, producing a gradient of pore size decreasing in the direction of protein migration. Gradient polyacrylamide gel electrophoresis has not yet been extensively applied in taxonomic studies, probably because difficulties may be encountered in the preparation of highly standardized and reproducible gradient gels, and because commercially available gradient gels are expensive. Gradient gel electrophoresis and numerical analysis of the resulting protein patterns has been applied in taxonomic studies of corynebacteria from axillary skin (Jackman, 1981, 1982) and of *Pseudomonas paucimobilis* (Owen and Jackman, 1982).

Isoelectric Focussing

This is also a high-resolution technique, where a gradient of pH is created in a low-concentration polyacrylamide gel by a mixture of carrier ampholytes (synthetic polyaminocarboxylic acids). During electrophoresis, protein molecules migrate through the gel until they reach a pH corresponding to their isoelectric point; there they concentrate (focus) in very fine bands. To date, isoelectric focussing has not been used very widely for the classification and identification of bacteria. Matthew and Harris (1976) successfully separated the isoenzymes of chromosomal β-lactamases of almost 240 strains representing 5 Gram-positive and 16 Gram-negative genera. These authors claimed that the patterns obtained were genus-, species-, and subspecies-specific.

Two-dimensional Polyacrylamide Gel Electrophoresis

In two-dimensional polyacrylamide gel electrophoresis, proteins are first separated in a tube gel by isoelectric focussing and then by electrophoresis in an SDS homogeneous slab gel or an SDS-containing gradient gel. The technique results in a very high resolution. More than 1000 proteins have been separated in a whole-cell extract of *Escherichia coli* using autoradiographic detection (O'Farrell, 1975). As the interpretation of the similarities between such protein maps is fairly complex, two-dimensional polyacrylamide gel electrophoresis has found limited application in bacteriological classification up to now. The technique has been applied for taxonomic purposes in the following genera: *Mycoplasma* (Rodwell and Rodwell, 1978), *Rhizobium* (Leps *et al.*, 1980; Roberts *et al.*, 1980), *Spiroplasma* (Mouches *et al.*, 1979), and *Ureaplasma* (Swenson *et al.*, 1983). Rodwell and Rodwell (1978) concluded that, at least for mycoplasmas, comparison of such protein fingerprints would seem a suitable method for distinguishing between closely related strains or between mutants of the same strain. The scanning, recording, and computer analysis of such two-dimensional patterns is much more complex than for one-dimensional patterns.

The Quantitative Comparison of Protein Electrophoregrams by Computer-assisted Techniques

Comparison of protein electrophoregrams can be done visually when only a few strains are compared and when no quantification of any differences is required—

for instance, to differentiate between colony variants and contaminants in a bacterial culture (see Kersters and De Ley, 1980). However, in most circumstances, it is necessary and advantageous to record objectively the protein pattern in a form suitable for computing, so that similarities and differences between patterns can be quantified. Some investigators have recorded the optical densities of just a few homologous protein bands present in all the strains (Hudson *et al.*, 1976) and compared the data by calculating the correlation coefficient. Other investigators have compared the entire banding pattern by computer (Kersters

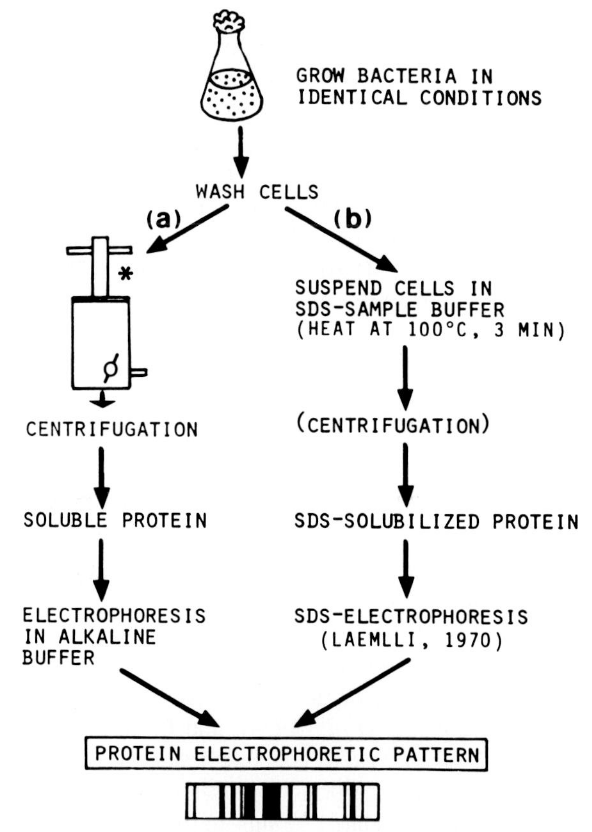

Fig. 1. Main steps in the preparation of protein samples for the characterization of bacteria by polyacrylamide gel electrophoresis. See text for explanation. (a) The method of Kersters and De Ley (1975) for the preparation of soluble native proteins with cell disruption by French pressure cell (marked by an asterisk). Electrophoresis is performed in a homogeneous, continuous, and alkaline gel and buffer system. (b) The method of Jackman (1985a,b) for the preparation of whole-cell proteins by disrupting the cells in sodium dodecyl sulphate (SDS) -containing sample buffer. Electrophoresis is performed in a homogeneous, discontinuous, SDS-containing alkaline gel and buffer system according to Laemmli (1970).

and De Ley, 1975, 1980; Swings *et al.*, 1976; Feltham and Sneath, 1979; Jackman, 1985a,b). With such techniques the relative mobility, the sharpness and relative protein concentration of the peaks, and the valleys between the peaks are taken into account. Quantitative comparison of electrophoretic protein patterns by computer is a typical problem of pattern analysis. The techniques used for comparison of total fatty acid profiles (Bousfield *et al.*, 1983) and pyrolysis gas–liquid chromatographic data (Gutteridge and Norris, 1979) may prove to be useful for the comparison and grouping of protein electrophoretic patterns.

Because the computer methods of Kersters and De Ley (1975) and Jackman (1985a,b) have been used in the classification of several bacterial genera (see Table 1), both approaches will be described briefly here (Figs. 2 and 3). Full details can be found in Kersters and De Ley (1975), Swings *et al.* (1976), and Jackman (1985a,b). Microcomputer-adapted programs in BASIC for the numerical analysis of electrophoretic protein patterns have been published (Jackman, 1983; Jackman *et al.*, 1983). The latter programs are based largely on the work of Feltham and Sneath (1979). In order to be amenable to computer treatment, highly reproducible protein electrophoregrams should be obtained. Therefore, techniques for protein sample preparation and electrophoresis will be summarized first (Fig. 1).

Culture of Bacteria and Sample Preparation

A great number of bacterial enzymes are inducible; therefore, the medium chosen should allow a good growth rate for the range of organisms under study. Liquid media can be used, but it is often preferable to grow bacteria on solid media in Roux flasks or in large Petri dishes. On solid media, contaminants are detected easily and the growth yield of aerobic bacteria is usually high.

Protein samples can be obtained in different ways. The left side of Fig. 1 (a) summarizes the methods of Kersters and De Ley (1975). These have been used for the study of more than 2000 bacterial strains (Swings *et al.*, 1977; Kersters, 1978; van Vuuren, 1978; Jarvis and Wolff, 1979; Kersters and De Ley, 1980; Izard *et al.*, 1981; Lambert *et al.*, 1981; van Vuuren *et al.*, 1981; Vantomme *et al.*, 1982; Ferragut, 1983; Ferragut *et al.*, 1983; Gillis *et al.*, 1983; Gosselé *et al.*, 1983a,b; see also Table 1).

The right side of Fig. 1 (b) represents schematically the procedures followed by Jackman (1982, 1985a,b) for taxonomic studies on *Corynebacterium* (Jackman, 1981, 1982), *Pseudomonas paucimobilis* (Owen and Jackman, 1982), and oral streptococci (Whiley *et al.*, 1982). The use of SDS in the sample preparation (Jackman, 1985a,b) results in the killing and disruption of small amounts of cells (approximately 0.1 g) in a single step. This is useful, particularly for handling pathogenic bacteria. The composition of the SDS–sample buffer is as follows: 0.0625 *M* Tris-HCl buffer, pH 6.8–2% (w/v) SDS–5%

(w/v) mercaptoethanol–10% (w/v) glycerol. The pellet of cells is suspended in 0.2 ml SDS–sample buffer and heated at 100°C for 3 min. After centrifugation for 20 min at 38,000 *g* the SDS-solubilized proteins are stored at −20°C until required (Jackman, 1985a,b).

Electrophoresis

Kersters and De Ley (1975) adapted the original methods of Ornstein (1964) and Davis (1964), and used 7% homogeneous tube gels with an alkaline continuous-buffer system (64 m*M* Tris-HCl buffer, pH 8.7). To compare protein patterns obtained from different electrophoretic experiments, protein extracts were studied in duplicate in the same electrophoretic run. One tube was loaded with the same protein sample to which two reference proteins (thyroglobulin and ovalbumin) were added. These reference proteins enabled the patterns to be normalized for computer treatment. Jackman (1985a,b) used a modification of the SDS-containing slab gel technique of Laemmli (1970). To my knowledge, no systematic comparison has been made of the two approaches with respect to their discriminatory value in bacterial taxonomy. It is therefore difficult to draw generalized conclusions on their relative merits. The two techniques are not strictly comparable. The SDS technique separates proteins by the more conserved parameter of molecular weight. Therefore, in theory separations by SDS techniques should detect broader microbial relationships than those which also include charge parameters.

Numerical Analysis of Electrophoretic Patterns

The stained gels are scanned by a densitometer, preferably linked to a microcomputer or computer, allowing the recording and initial processing of the traces.

The left side of Fig. 2 (a) shows our somewhat outdated method, where the densitometer is not on-line with a computer. The trace is normalized by dividing the distance between both reference proteins (ovalbumin and thyroglobulin) into 90 equal parts. The trace is then converted into a sequence of 110 (sometimes up to 130) numbers, representing the optical density (expressed as height in millimetres) of each position on the scan. These normalized data are punched via a computer terminal, and a data file of normalized traces is created (Fig. 2). A digitizing tablet (Fig. 2) can be used to eliminate manual reading of the traces.

More elegantly, a densitometer equipped with an analogue-to-digital converter (ADC) and controlled by a microcomputer can be used, as in Fig. 2 (b). Full details are given by Jackman (1985a,b) and Jackman *et al.* (1983). The starting and end points of the gel are automatically determined, and interpolation procedures are used to reduce the number of data per gel to 100. Background

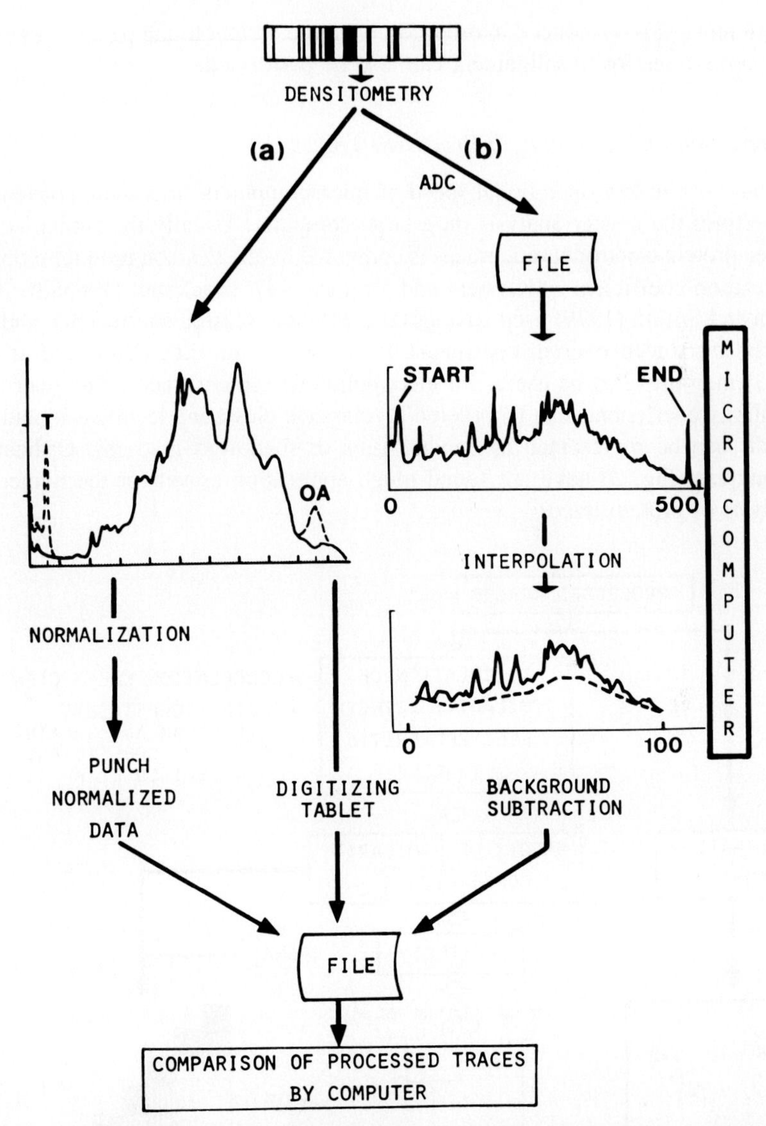

Fig. 2. Main steps in the recording, digitizing, and processing of the electrophoretic patterns. See text for explanation. (a) The semimanual method of Kersters and De Ley (1975); reference proteins: T, thyroglobulin; OA, ovalbumin. (b) The microcomputer-controlled methods of Jackman (1985a,b). ADC, Analogue-to-digital converter.

interference can be reduced in order to give more weight to the pattern of peaks, and corrections for misalignment can also be performed.

Comparison and Grouping of Processed Traces

Because of the low operational speed of microcomputers, it is more convenient to perform the cluster analysis on a large computer. Usually the similarity between protein electrophoretic traces is computed by the Pearson product-moment correlation coefficient *r* (Kersters and De Ley, 1975; Jackman, 1985a,b). Feltham and Sneath (1979) used an angular coefficient (cosine coefficient), and the newly introduced overlap coefficient for fatty acid profiles (Bousfield *et al.*, 1983) may prove to be useful for electrophoretic protein traces. The matrix of similarity coefficients can be ordered by classical clustering techniques, and the results can be represented as dendrograms or shadowed matrices. Ordination techniques (Fig. 3) have not found much application as yet for the numerical analysis of protein traces.

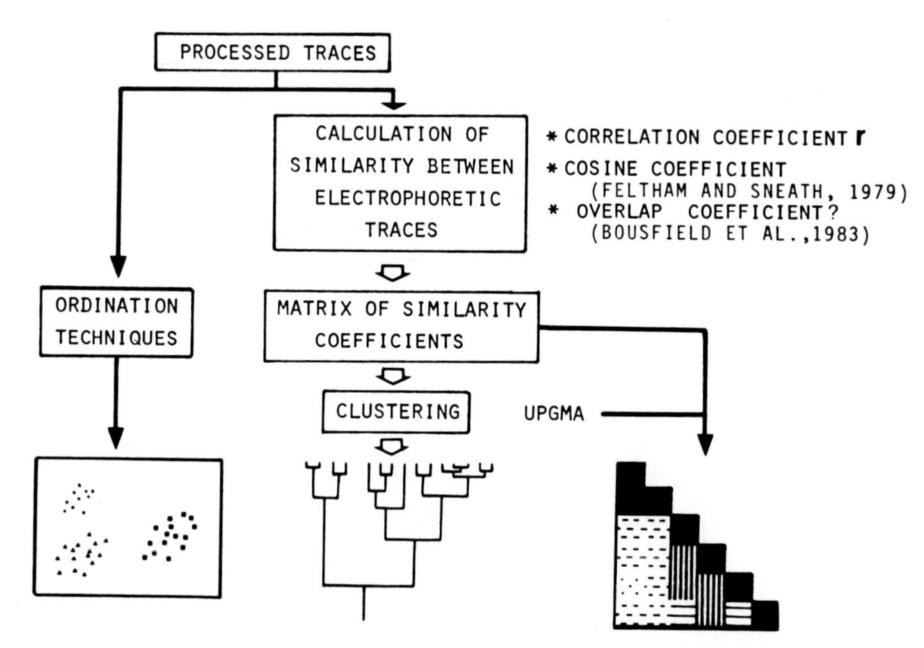

Fig. 3. Main steps and different methodologies for the numerical analysis of digitized and processed protein electrophoretic traces. See text for explanation. UPGMA, Unweighted pair-group method with averages algorithm. The asterisks denote the different types of similarity coefficients which can be used for calculating the similarity between each pair of traces.

Reproducibility and Resolution of Electrophoretic Protein Patterns

When examining large numbers of bacteria by numerical analysis of their electrophoretic protein patterns, reproducibility of the protein band mobilities is of critical importance and should be carefully checked. The reproducibility of the gel electrophoretic technique itself should be measured by examining a single protein sample on different occasions on several gels. When such experiments are performed properly, the mean of the matrix of correlation coefficients r is generally above .90, and frequently above .95. Small variations can occur between different samples of the same strain. The effect of differences in sample preparation and minor differences in culture conditions were investigated by Kersters and De Ley (1975) and Jackman (1981). Correlation coefficients between patterns of independently cultured and prepared samples were found to be in approximately the same range as the reproducibility of the electrophoretic technique itself. The protein patterns of some groups of bacteria (e.g., *Bordetella bronchiseptica*) were not much influenced by variation in culture conditions. On the other hand, patterns of *Agrobacterium* were found to differ considerably when grown on a minimal basal medium instead of a complex medium (Kersters and De Ley, 1975). Generally, it is advisable to perform all experiments under conditions as standardized as possible. Variations in the total protein concentration affect the pattern only in the Y axis and have almost no effect on the degree of similarity, provided a correlation coefficient is employed to perform the calculations.

The finer the resolution of the electrophoretic and densitometric systems, the less the probability that unrelated strains will show a high similarity by chance. High-resolution electrophoretic systems make strict demands on the reproducibility of the techniques, and preferably the resolution of the densitometer should be higher than the reproducibility of the electrophoretic system.

Some Applications of Numerical Analysis of Electrophoretic Protein Patterns in Microbial Systematics

A literature survey of the application of numerical analyses of polyacrylamide gel electrophoretic protein patterns to the classification and identification of different bacterial groups is given in Table 1. The majority of these studies were performed with homogeneous gel systems in non-denaturing conditions. Owen and Jackman (1982) and Jackman (1982) used gradient gels for taxonomic studies on *Pseudomonas paucimobilis,* and corynebacteria isolated from axillary

Table 1. *Contributions of computer-assisted comparisons of protein electrophoretic patterns to the classification of different bacterial groups*

Bacterial groups investigated	Approximate number of strains studied	Other techniques used in conjunction with PAGE[a]	References
PAGE in nondenaturing conditions			
Acetobacter	65	Phenotypic analysis	Gosselé *et al.* (1983b)
Agrobacterium	250	DNA–DNA hybridizations	Kersters and De Ley (1975)
Agrobacterium	90	Phenotypic analysis	du Plessis (1983)
Alcaligenes, Achromobacter, Bordetella, and allies	400	Phenotypic analysis DNA–rRNA hybridizations	Kersters (1978); Kersters and De Ley (1980); De Ley *et al.* (1983)
Corynebacterium	100	—	Jackman (1982)
Enterobacter agglomerans, Erwinia herbicola	200	—	Mergaert *et al.* (1983)
Enterobacteriaceae	40	Phenotypic analysis	Feltham and Sneath (1979)
Enterobacteriaceae from beer breweries	60	Phenotypic analysis	van Vuuren *et al.* (1981)
Frateuria	11	Phenotypic analysis, DNA–rRNA hybridizations	Swings *et al.* (1980)
Gluconobacter	100	Phenotypic analysis	Gosselé *et al.* (1983a)
Klebsiella	150	Phenotypic analysis, DNA–DNA hybridizations	Izard *et al.* (1981); Ferragut *et al.* (1983); Ferragut (1983)
Lactic streptococci	35	—	Jarvis and Wolff (1979)
Mycobacterium	28	—	Vanden Berghe and Pattyn (1979)
Pseudomonas paucimobilis	18	DNA–DNA hybridizations	Owen and Jackman (1982)
Xanthomonas campestris pv. *graminis*	70	Phenotypic analysis	K. Kersters *et al.*, unpublished data

Table 1 (*Continued*)

Bacterial groups investigated	Approximate number of strains studied	Other techniques used in conjunction with PAGE[a]	References
pv. *manihotis* and pv. *cassavae*	40	DNA–DNA hybridizations	K. Kersters *et al.*, unpublished data, and this chapter
pv. *oryzae* and pv. *oryzicola*	70	Phenotypic analysis	K. Kersters *et al.*, unpublished data
Yersinia pestis	160	Phenotypic analysis	Hudson *et al.* (1976)
Zymomonas	43	Phenotypic analysis DNA–DNA hybridizations	Swings *et al.* (1976)
PAGE in denaturing conditions			
Corynebacterium	100	—	P. J. H. Jackman, unpublished data
Oral streptococci	270	Phenotypic analysis	Whiley *et al.* (1982, 1983)

[a]PAGE, Polyacrylamide gel electrophoresis.

skin, respectively. It can be anticipated, however, that SDS-containing gel systems will be used increasingly in the future (Jackman, 1985a).

To demonstrate the possibilities and limitations of computer-assisted comparisons of protein patterns in the characterization of bacteria, some of our recent (mostly unpublished) results will be considered in some detail. Emphasis will be placed on taxonomic information derived from the combination of and the correlation between results obtained by protein electrophoresis and other approaches, notably DNA–DNA and DNA–rRNA hybridizations. The examples have been specially chosen from groups of bacteria isolated from different ecological niches: (i) *Klebsiella*-like strains isolated from the environment, (ii) plant-pathogenic bacteria of the genus *Xanthomonas,* and (iii) *Bordetella*-like bacteria isolated from turkeys suffering from rhinotracheitis. The first and the third examples are the results of collaboration with the research teams of, respectively, Professors H. Leclerc (INSERM, Unité 146, Villeneuve d'Ascq, France) and K.-H. Hinz (Klinik für Geflügel der Tierärztlichen Hochschule, Hannover, FRG).

Klebsiella

This example has been chosen to illustrate the overall agreement between the groupings obtained by numerical analysis of electrophoretic protein patterns and those resulting from DNA–DNA hybridization. At the same time it will demonstrate the role played by numerical analysis of protein electrophoregrams in the delineation of the new species *K. terrigena* (Izard *et al.*, 1981) and *K. trevisanii* (Ferragut *et al.*, 1983).

The following *Klebsiella* species are recognized on the *Approved Lists of Bacterial Names* (Skerman *et al.*, 1980): *K. mobilis, K. ozaenae, K. oxytoca, K. pneumoniae* (type species), and *K. rhinoscleromatis*. Gavini *et al.* (1977) and Naemura *et al.* (1979) investigated by numerical analysis of phenotypic features a large number of *Klebsiella* strains isolated from the medical environment as well as from sewage, surface water, drinking water, effluent water of paper mills, and unpolluted soils. Gavini *et al.* (1977) found that a great number of *Klebsiella*-like strains isolated from the environment clustered in two phena, K and L, which did not correspond to any known *Klebsiella* species. In 1979 C. Ferragut, in our laboratory, began an extensive, comparative protein electrophoresis study of 174 *Klebsiella* strains (Ferragut, 1983). She found that the majority of the phenon L strains (isolated from drinking and surface water and from unpolluted soils) displayed very similar protein patterns which differed from the electrophoregrams of *K. pneumoniae, K. ozaenae, K. rhinoscleromatis, K. oxytoca,* and phenon K strains. The protein patterns of representative strains [L84 (the phenotypic centrotype strain), L66, L81, L98, L67, L108, and L91] of phenon L (Gavini *et al.*, 1977) are depicted in Fig. 4. The patterns of these strains differed from those of *K. pneumoniae* ATCC 13882, strain K70 (the protein electrophoretic centrotype strain of phenon K), and from strains L28, L118, and L101, which also clustered in phenon L (Gavini *et al.*, 1977). Percentage DNA–DNA hybridization values of all these strains (Izard *et al.*, 1981; Ferragut *et al.*, 1983) with strains L84 and K70 are also shown in Fig. 4.

The *Klebsiella*-like strains L66, L67, L81, L91, L98, and L108 gave high levels (≥87%) of hybridization with labelled DNA of the phenotypic centrotype strain L84 (= CIP 80.07) of phenon L. The hybridization data clearly showed that the majority of the phenon L strains were genetically highly interrelated and that *K. pneumoniae, K. oxytoca, K. mobilis,* and the *Klebsiella*-like K strains displayed lower (≤70%) hybridization values with the L strains (Fig. 4; Izard *et al.*, 1981; Ferragut *et al.*, 1983). Although the protein electrophoregram of strain L108 showed an overall similarity to the patterns of other L strains, a heavy protein band in the lower part of the gel (Fig. 4) differentiated this strain from all the other L strains. This heavy band resulted in the removal of strain L108 from the core of group L in a numerical analysis of the patterns (Fig. 5). Yet, strain L108 hybridized at 94% with strain L84 (Fig. 4), indicating that for some strains the taxonomic interpretations of the comparisons of electrophoretic protein patterns should not rely solely on numerical analysis of the electrophoregrams. Visual inspection of the normalized photographs of the protein patterns is often helpful and very informative.

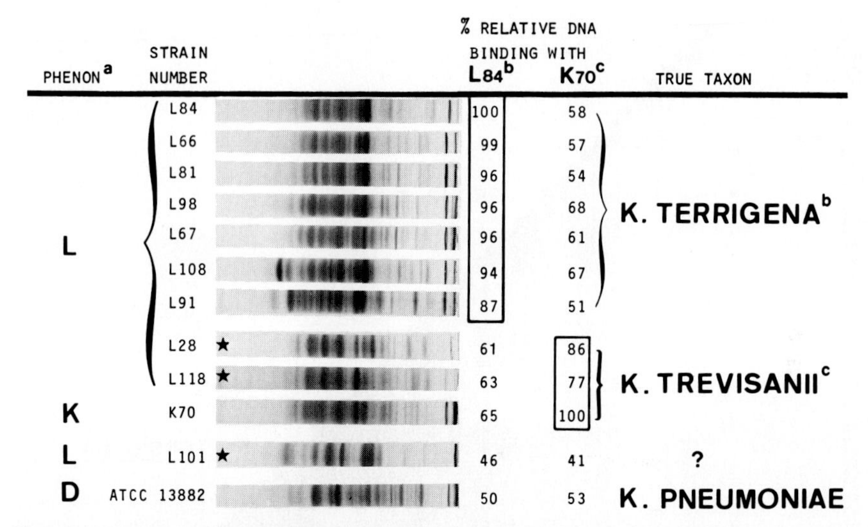

Fig. 4. Normalized electrophoregrams of soluble proteins of 12 *Klebsiella* strains and relative binding values of their DNAs to ^{3}H-labelled DNAs of the phenotypic centrotype strain of phenon L (L84) and the electrophoretic centrotype strain of phenon K (K70). The figure shows the protein electrophoretic and genotypic homogeneity of the majority of the phenon L strains (*K. terrigena*); the protein electrophoretic and genetic relationships of so-called L strains L28 and L118 (marked by a star) to *K. trevisanii* (phenon K), and the aberrant electrophoretic and genetic position of the so-called L strain L101 (marked by a star). [a]Phenon designation according to Gavini *et al.* (1977). [b]From Izard *et al.* (1981). [c] From Ferragut *et al.* (1983).

Polyacrylamide gel electrophoresis of these *Klebsiella* strains revealed also that three strains of phenon L (L28, L118, and L101, marked by a star in Figs. 4 and 5) displayed patterns which were atypical for L strains. The electrophoregrams of strains L28 and L118 resembled the pattern of the electrophoretic centrotype strain K70 of phenon K. DNA–DNA hybridizations confirmed that both strains did indeed belong to group K (Fig. 4). Strain L101 was not closely related to any of the strains studied (Figs. 4 and 5). The results of the numerical analysis of the protein patterns of all the L strains investigated together with representative strains of *K. pneumoniae, K. oxytoca,* and strains of phenon K are depicted in Fig. 5. This ordered matrix of correlation coefficients shows that the majority of the strains of phenon L constitute an electrophoretically fairly homogeneous group of bacteria, clearly differentiated from the other species and groups of *Klebsiella*. The genetic homogeneity of these L strains was proven by DNA–DNA hybridizations. The combination of phenotypic, electrophoretic, and genotypic data clearly indicated that this group of klebsiellae isolated from soil and water constituted a distinct new species for which the name *K. terrigena* was proposed (Izard *et al.*, 1981). The matrix of correlation coefficients (Fig. 5)

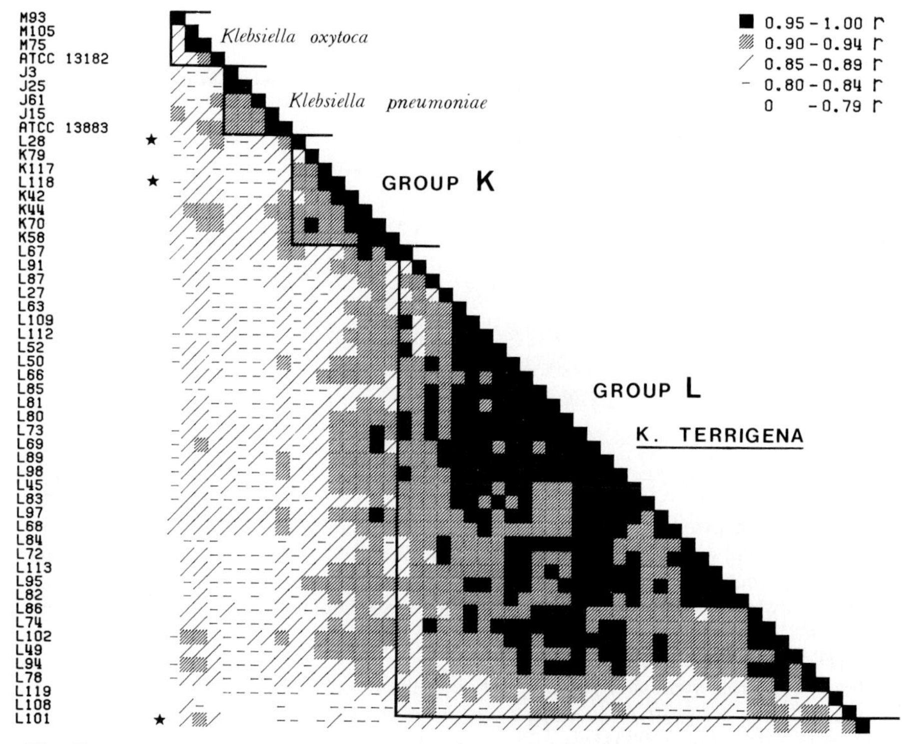

Fig. 5. Sorted and differentially shaded matrix of correlation coefficients calculated from the numerical analysis of protein patterns of all the investigated *Klebsiella* strains of phenon L and some representatives of other *Klebsiella* species. Aberrant L strains are marked by a star (see also Fig. 4). The range of correlation coefficients *r* is symbolized as indicated. Adapted from Izard *et al.* (1981).

shows also the borderline position of strain L108 and the separate position of strain L101. The correct taxonomic affiliation of the latter strain remains unknown. As expected, strains L28 and L118 (marked by a star in Figs. 4 and 5) clustered according to their protein patterns within phenon K. This phenon also forms a separate taxonomic entity (*K. trevisanii;* Ferragut *et al.*, 1983) in the genus *Klebsiella*.

Xanthomonads Pathogenic for Cassava

This example illustrates the usefulness of computer-assisted comparisons of protein electrophoregrams in studies on the classification and identification of *Xanthomonas* strains pathogenic for cassava. Again, there is an excellent correlation between the groupings obtained by numerical analysis of protein patterns and those revealed by DNA–DNA hybridizations.

Two pathovars (pv.) of *X. campestris* are known to be pathogenic for cassava: *X. campestris* pv. *manihotis* and *X. campestris* pv. *cassavae*. Formerly these bacteria were named '*X. manihotis*' and '*X. cassavae*', respectively. Whereas *X. campestris* pv. *manihotis* occurs in most cassava-growing areas of the world, the distribution of *X. campestris* pv. *cassavae* appears to be limited to the African highlands of Rwanda and Kenya at altitudes >1600 m (Maraite and Weyns, 1979). Phenotypically both pathogens are very similar. *Xanthomonas campestris* pv. *manihotis* forms whitish colonies and causes cassava bacterial blight, whereas *X. campestris* pv. *cassavae* forms yellow colonies and causes cassava bacterial leaf spot. The early leaf spot symptoms induced by both pathogens are so similar that plant pathologists frequently have difficulty distinguishing between the bacteria (Elango *et al.*, 1981). According to Robbs *et al.* (1972), the names '*X. cassavae*' and '*X. manihotis*' are synonymous, and Dye (1978) therefore considered '*X. cassavae*' to be a *nomen dubium*. However, Maraite and Weyns (1979) found a few, mainly quantitative, biochemical and physiological differences between the two pathogens. For this reason they are listed as separate pathovars (*X. campestris* pv. *manihotis* and *X. campestris* pv. *cassavae*) on the ISPP list of pathovar names (Dye *et al.*, 1980). Elango *et al.* (1981) found that the biochemical and physiological tests used by them did not differentiate between the two pathovars. The results of a numerical analysis of 296 phenotypic features of 268 *Xanthomonas* strains (M. Van den Mooter, unpublished) likewise indicated very few differences between pv. *manihotis* and pv. *cassavae*. Elango *et al.* (1981) found that serological techniques such as Ouchterlony double diffusion and direct immunofluorescence showed antigenic differences between the pathovars. Yellow *Xanthomonas* strains isolated from cassava in Colombia at sea level were found to be serologically similar to *X. campestris* pv. *cassavae* (Elango *et al.*, 1981).

To unravel the enigmatic taxonomic relationships of the *Xanthomonas* pathogens of cassava, the protein electrophoretic patterns of 30 *X. campestris* pv. *manihotis* strains, 6 *X. campestris* pv. *cassavae* strains (including the pathotype strains of both pathovars, NCPPB 1834 and NCPPB 101, respectively), 3 yellow strains isolated from cassava in Colombia, and the type strain of *X. campestris* (*X. campestris* pv. *campestris,* NCPPB 528) were compared (Meiresonne, 1983). The numerical analysis of the protein electrophoregrams of these strains (Fig. 6) showed that *X. campestris* pv. *manihotis* and *X. campestris* pv. *cassavae* could be differentiated easily from each other by protein electrophoresis and suggested that the two pathovars were probably not closely related genetically. The dendrogram (Fig. 6) shows that the yellow isolates from Colombia belong neither to the *X. campestris* pv. *cassavae* nor to the *X. campestris* pv. *manihotis* taxon. *Xanthomonas campestris* pv. *manihotis* strains isolated in South America could not be distinguished by electrophoresis from *X. campestris* pv. *manihotis* strains isolated in Africa, Mauritius, or Malaysia. No correlation was observed between the virulence of the *X. campestris* pv. *manihotis* strains investigated and

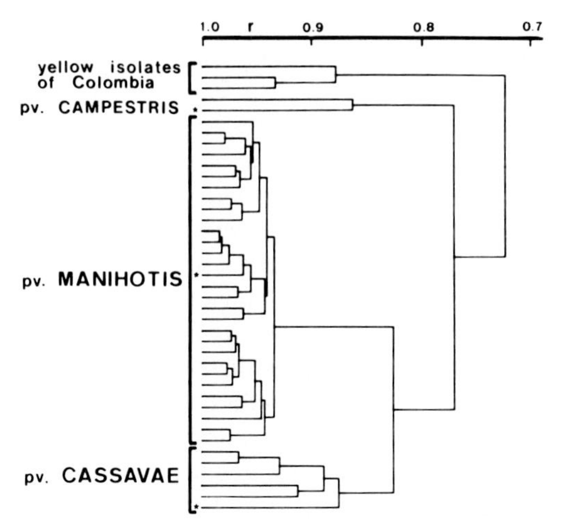

Fig. 6. Numerical analysis of electrophoregrams of soluble proteins of *Xanthomonas campestris* strains isolated from cassava. The figure shows the electrophoretic differentiation between strains of pathovar (pv.) *manihotis,* pv. *cassavae,* and the yellow isolates from Colombia, Pathotype strains are marked by an asterisk. The correlation coefficient *r* was used and clustering achieved by the UPGMA method. From Meiresonne (1983).

the electrophoretic groups. Strains isolated in 1946 were electrophoretically indistinguishable from recently isolated bacteria. Our results indicate that *X. campestris* pv. *manihotis* strains constitute an electrophoretic and hence probably homogeneous genetic group of bacteria, clearly differentiated from *X. campestris* pv. *cassavae* and *X. campestris* pv. *campestris* strains. These results are in agreement with the results of the protein electrophoretic investigations of El-Sharkawy and Huisingh (1971). These authors reported that electrophoretic protein patterns were generally identical for all isolates of the same species (most of which are currently considered as pathovars of a single species *X. campestris*), whereas strains of different species yielded different protein patterns.

Continuing our taxonomic investigations on the xanthomonads pathogenic for cassava, we determined the genetic relatedness between the protein electrophoretic clusters by DNA–DNA hybridizations. Because of the protein electrophoretic homogeneity of *X. campestris* pv. *manihotis* strains and strains of *X. campestris* pv. *cassavae* (Fig. 6), it was sufficient to investigate two strains from each pathovar group. The results of the DNA–DNA relatedness studies (Table 2) consolidated the protein electrophoretic groupings. Strains displaying almost identical protein patterns ($r \geq .90$) were found to be genetically highly related (e.g., the *X. campestris* pv. *manihotis* strains NCPPB 1834 and HMB 25; and the *X. campestris* pv. *cassavae* strains NCPPB 101 and HMB 38; Fig. 6 and Table 2). Conversely, strains with different protein patterns belonged to genetically

Table 2. *DNA characteristics and protein electrophoretic similarities of the type strain of* Xanthomonas campestris *and various* Xanthomonas *strains isolated from cassava*[a]

Strain	Where isolated	G + C content (mol %)	Percentage DNA–DNA hybridization[b] versus				Protein pattern similarity (%) *vs.* NCPPB 1834[e]
			NCPPB 1834[c]	NCPPB 101[c]	CIAT 1164	NCPPB 528[d]	
X. campestris pv. *manihotis*							
NCPPB 1834[c]	Brazil	65.9	100				95
HMB 25	Nigeria	—[f]	98				94
X. campestris pv. *cassavae*							
NCPPB 101[c]	Malawi	66.2	35	100			80
HMB 38	Rwanda	—	—	95			73
Yellow isolate							
CIAT 1164	Colombia	66.9	60	36	100		73
CIAT 1165	Colombia	—	—	—	98		72
X. campestris pv. *campestris*							
NCPPB 528[d]	Great Britain	65.2	18	34	37	100	79

[a]Data from Meiresonne (1983); L. Meiresonne and M. Gillis (unpublished results).
[b]Measured by the initial renaturation rate method (De Ley *et al.*, 1970). The values are averages of at least three independent hybridizations.
[c]Neopathotype strains according to Dye *et al.* (1980).
[d]Type strain of *X. campestris* (Skerman *et al.*, 1980).
[e]Mean value ($r \times 100$) of triplicate patterns relative to the pathotype strain of *X. campestris* pv. *manihotis* NCPPB 1834.
[f]Not determined.

different groups (e.g., *X. campestris* pv. *campestris* NCPPB 528, *X. campestris* pv. *manihotis* NCPPB 1834, *X. campestris* pv. *cassavae* NCPPB 101, and the yellow cassava isolate from Colombia CIAT 1164; Fig. 6 and Table 2).

It is difficult, often impossible, and always unwise to attempt deductions on the genetic relatedness of strains displaying differences in their electrophoretic protein patterns, without performing the actual DNA–DNA hybridizations. This is exemplified by comparison of data in Fig. 6 and Table 2. The results of the numerical analysis of the protein profiles of the investigated xanthomonads indicate that the patterns of *X. campestris* pv. *manihotis* are more similar to those of *X. campestris* pv. *cassavae* strains than to the patterns of the yellow isolates from Colombia (Fig. 6). However, the results in Table 2 indicate unambiguously that *X. campestris* pv. *manihotis* strains are more closely related to the isolates from Colombia (percentage of DNA homology of 60 ± 8) than to strains of *X. campestris* pv. *cassavae* (only 35 ± 9% DNA homology). Strains of both pathovars show approximately 30% of DNA homology with *X. campestris* pv. *campestris* strain NCPPB 528.

The overall conclusion of these taxonomic investigations on the xanthomonads pathogenic for cassava is that strains of each of the pathovars, *X. campestris* pv. *manihotis* and *X. campestris* pv. *cassavae,* constitute distinct protein electrophoretic and genetically homogeneous groups. They should be considered as different taxonomic entities as suggested by the phenotypic and phytopathological observations of Maraite and Weyns (1979). The rather low DNA–DNA relatedness between *X. campestris* pv. *campestris, X. campestris* pv. *manihotis,* and *X. campestris* pv. *cassavae* makes their present infra-subspecific nomenclatural status questionable. The taxonomic confusion which exists in the genus *Xanthomonas* has already been stressed by Starr (1981). In the context of the well-established DNA relationships of the Enterobacteriaceae, *X. campestris* pv. *manihotis* differs as much from *X. campestris* pv. *cassavae* as does, for example, the genus *Escherichia* from genera such as *Enterobacter, Klebsiella,* and *Salmonella* (Brenner *et al.*, 1972; Izard *et al.*, 1981).

Bordetella-*like Strains Isolated from Birds*

In 1977 K.-H. Hinz isolated, in Hannover, FRG, glucose-nonfermenting bacteria from the respiratory tract of turkeys suffering from a respiratory disease called turkey coryza or rhinotracheitis. The strains caused the same sickness upon reinoculation in turkey poults. In some instances the disease caused severe economic losses in turkey flocks. The causative organism was an oxidase-positive, peritrichously flagellated, urease-negative, glucose-nonfermenting, Gram-negative, small rod, resembling *Alcaligenes faecalis* and *Bordetella bronchiseptica*. Hinz named these organisms *Bordetella*-like bacteria (Hinz, 1981). Similar strains isolated in the United States were identified as *A. faecalis* (Simmons *et al.*, 1980). As we had built up a large data base of digitized electrophoretic protein patterns of more than 400 strains of *Alcaligenes, Achromobacter, B. bronchiseptica,* and allied groups (Kersters and De Ley, 1980), it was relatively easy to match the protein profiles of the *Bordetella*-like strains isolated from turkeys with the patterns of strains belonging to the above-mentioned groups.

It should be noted here that the genus *Alcaligenes* as defined in the eighth edition of *Bergey's Manual of Determinative Bacteriology* and in the *Approved Lists of Bacterial Names* (Skerman *et al.*, 1980) is extremely heterogeneous. A summary of the presently recognized species of the genera *Alcaligenes* and *Bordetella,* together with the reported mol % G + C span of their DNA is given in Fig. 7. The mol % G + C range of *Bordetella* is narrow, but that for *Alcaligenes* is broad. Johnson and Sneath (1973) found by numerical analysis of phenotypic features that representatives of the obligately parasitic *Bordetella* species shared many similarities with strains named *Alcaligenes faecalis* and *Alcaligenes denitrificans.* It is now established by DNA–rRNA hybridization techniques that the genus *Bordetella* belongs on the same rRNA branch as *A. faecalis* and *A. denitrificans* (De Ley *et al.*, 1983; Fig. 10). In this context it should be mentioned that *Achromobacter xylosoxidans* has been found to be highly related to *Alcaligenes denitrificans* (Kiredjian *et al.*, 1981; De Ley *et al.*, 1983). *Alcaligenes faecalis* and *A. denitrificans* are encountered in hospital environments and as free-living bacteria in, for example, soil. *Alcaligenes eutrophus, A. latus, A. paradoxus,* and *A. ruhlandii* occur in soil and are able to oxidize hydrogen gas. *Alcaligenes aestus, A. aquamarinus, A. cupidus, A. pacificus,* and *A. venustus* (Baumann *et al.*, 1972) have been isolated from the marine environment.

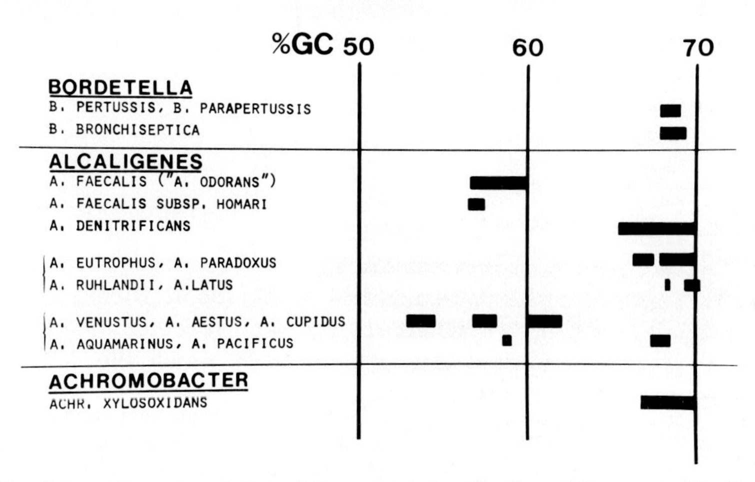

Fig. 7. Schematic representation of the present classification of the genera *Alcaligenes, Bordetella,* and *Achromobacter,* together with their mol % G + C span.

With the exception of *B. parapertussis* and *B. pertussis,* our protein electrophoretic data base contained patterns of a large number of strains of all the above-mentioned taxa, including those of their type strains, as well as protein profiles of unclassified groups of bacteria (e.g., CDC IVc-2 and IVe), phenotypically resembling *A. faecalis* and *B. bronchiseptica.* However, although an enormous variety of electrophoretic types was recorded (Kersters, 1978; Kersters and De Ley, 1980), the protein patterns of the *Bordetella*-like isolates could not be matched unambiguously with any of the profiles in the data base. They grouped approximately equidistant from both the *B. bronchiseptica* and the *Alcaligenes denitrificans–Achromobacter xylosoxidans* complexes. More than 25 avian *Bordetella*-like strains were compared by electrophoresis of their soluble proteins. The protein electrophoretic homogeneity of these organisms, and their differentiation from *B. bronchiseptica, Alcaligenes denitrificans, Achromobacter xylosoxidans, Alcaligenes faecalis,* and CDC group IVe are shown in Fig. 8. The protein patterns of strains isolated from turkeys in the Federal Republic of Germany were indistinguishable from those isolated in the United States, Spain, United Kingdom, Israel, and other countries. Also, protein electrophoregrams of strains isolated from a goose, a duck, and a chicken were very similar to those exhibited by the turkey isolates. The ordered and shadowed correlation coefficient matrix resulting from the computer-assisted comparisons of all the avian *Bordetella*-like strains, together with representative strains of *Alcaligenes, Bor-*

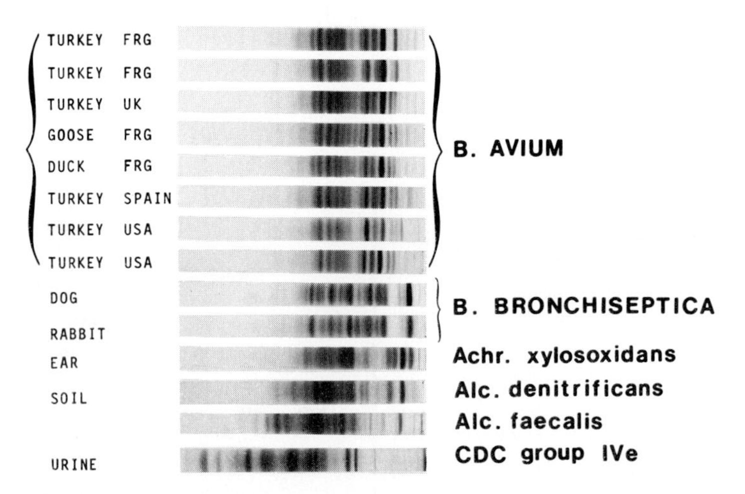

Fig. 8. Normalized electrophoregrams of soluble proteins from eight avian *Bordetella*-like strains (*B. avium*) and representative strains of various more or less allied bacteria. The origin of the strains is shown at the left side of the figure. The avian strains were isolated in the Federal Republic of Germany (FRG), United Kingdom (UK), Spain, and United States of America (USA).

detella, and *Achromobacter* is shown in Fig. 9. The avian strains form a tight electrophoretic cluster and also share similar biochemical, nutritional, physiological, and serological features.

It was impossible to deduce from the electrophoretic data alone, which taxa were genotypically the most closely related to the avian strains. It could only be concluded that the avian *Bordetella*-like strains were electrophoretically (and hence almost certainly genotypically) highly related to each other and different from all the other taxa investigated. The similarity matrix (Fig. 9) suggested that the avian strains might be more closely related to *B. bronchiseptica* or to the *Alcaligenes denitrificans–Achromobacter xylosoxidans* complex than to *Alcaligenes faecalis*. To detect the generic and suprageneric relationships of the avian isolates, DNA from representatives of the *Bordetella*-like strains was hybridized with labelled rRNA from the type strains of *B. bronchiseptica* (NCTC 452), *Alcaligenes faecalis* (NCIB 8156), and *Alcaligenes denitrificans* (ATCC 15173). DNA–rRNA hybridization techniques allow the determination of the generic and suprageneric relationships of bacteria because the cistrons coding for rRNA are conserved (Moore and McCarthy, 1967; De Smedt and De Ley, 1977; Fox *et al.*, 1980). These cistrons play a crucial role in the ubiquitous and very similar mechanism of protein synthesis present in all living beings. Some of the

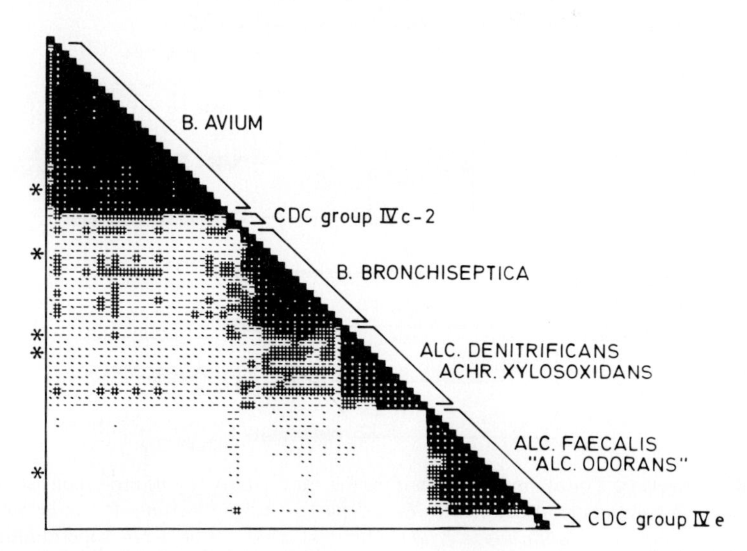

Fig. 9. Matrix of correlation coefficients calculated from the numerical analysis of protein patterns of 24 avian *Bordetella*-like strains (*B. avium*) and 46 strains belonging to various species and groups as shown. Type strains are marked by an asterisk and the range of correlation coefficients r is symbolized as follows: ■, 0.95–1.00; ◘, .90–.94; #, .85–.89; –, .80–.84; ·, .70–.79; □, 0–0.69. From Kersters *et al.* (1984), with permission.

results of the DNA–rRNA hybridizations in the so-called third rRNA superfamily (*sensu* De Ley, 1978) are summarized in Fig. 10.

The taxonomically most important parameter of each DNA–rRNA hybrid is the Tm(e) value. It is the midpoint temperature of the thermal elution curve of the hybrid and a measure of base similarity and mismatching. The higher the Tm(e) values, the more the organisms are genetically related. *Bordetella bronchiseptica, B. parapertussis,* and *B. pertussis* are genetically so closely related that they cannot be differentiated from each other by this technique (Fig. 10). The results are in perfect agreement with the high DNA–DNA hybridization values found by Kloos *et al.* (1981) among the three recognized *Bordetella* species. The following taxa are, in decreasing order, the closest neighbours of the genus *Bordetella:* (i) the avian *Bordetella*-like strains (mol % G + C: 61.6 to 62.6); (ii) the *Alcaligenes denitrificans–Achromobacter xylosoxidans* group (mol % G + C: 63.9 to 69.8), and (iii) *Alcaligenes faecalis* (*'A. odorans'*) (mol % G + C: 55.9 to 59.4). The last group is clearly different from both the genus *Bordetella* and the *Alcaligenes denitrificans–Achromobacter xylosoxidans* group (Fig. 10).

The results of the computer-assisted comparisons of protein electrophoregrams, phenotypic analysis, and DNA–rRNA hybridizations led us to conclude that the avian *Bordetella*-like strains were genetically highly interrelated, but

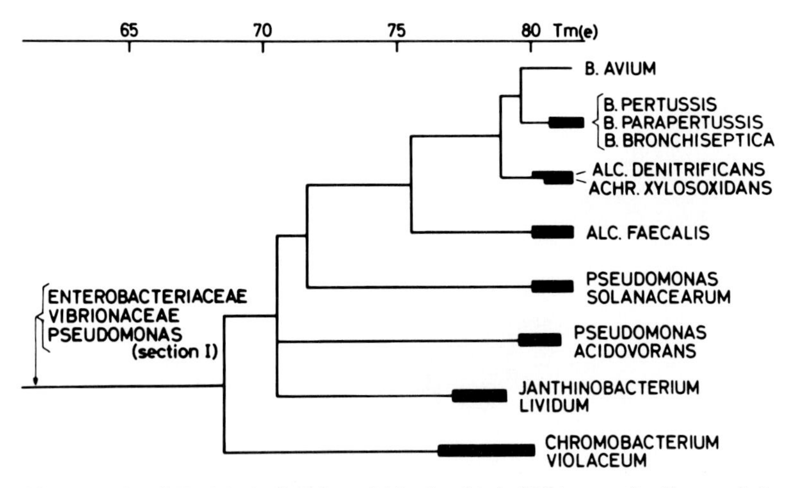

Fig. 10. Levels of Tm(e) similarities within the third rRNA superfamily consisting of *Bordetella,* the authentic *Alcaligenes, Pseudomonas solanacearum, P. acidovorans, Chromobacterium,* and *Janthinobacterium.* The next lower set of rRNA superfamilies is at an average Tm(e) of 62°C and consists of the Enterobacteriaceae, the Vibrionaceae, Pseudomonas (Section I), and a few other taxa. The data are from previous publications (De Ley *et al.* 1978, 1983; De Vos and De Ley, 1983) and from unpublished data of J. De Ley, P. De Vos, M. Gillis, W. Mannheim, and P. Segers. The black areas at the tips of most branches represent the range of Tm(e) values of the reference taxon. From Kersters *et al.* (1984), with permission.

distinct from all the other taxa investigated (Fig. 9). The parameters of the DNA–rRNA hybrids indicated that the closest relatives of the avian strains were the three recognized *Bordetella* species. A new species name, *B. avium,* has therefore been proposed for the homogeneous group of strains causing rhinotracheitis in turkey poults (De Ley *et al.*, 1983; Kersters *et al.*, 1984).

This chapter was presented at the symposium 'Twenty-five Years of Numerical Taxonomy'. It is therefore not only a fortunate coincidence but highly appropriate that Fig. 10 also shows the rRNA relationships of the chromobacteria. It was the computer-assisted analysis by Professor P. H. A. Sneath of his phenotypic data on these bacteria which resulted in the development of what is now termed numerical taxonomy. Professor Sneath was the first to apply computers to the study of bacterial taxonomy, and he unequivocally proved that *Chromobacterium violaceum* and *C. lividum* constituted two different taxa (Sneath, 1957). More than 20 years later molecular biological techniques showed that the results of those early computer analyses of phenotypic data reflected the profound genotypic differences between these two groups of chromobacteria. The results of DNA–rRNA hybridizations showed that the evolutionary split between *C. violaceum* and *C. lividum* is so deep that De Ley *et al.* (1978) proposed the elevation of *C. lividum* to genus rank and named the new genus, *Janthinobacterium*.

Conclusions

The application of computer-assisted comparisons of polyacrylamide gel electrophoresis of whole-cell proteins to the classification and identification of a wide variety of bacteria has been summarized (Table 1). Different electrophoretic techniques in nondenaturing (Kersters and De Ley, 1975, 1980) and denaturing (SDS) (Jackman, 1985a) conditions were briefly discussed. All of these studies have proved to be very successful in unravelling the similarities and differences between bacterial strains.

The commonest finding has been that the results of protein electrophoretic patterns are most discriminatory at the level of species, subspecies, or biotypes. The taxonomic level of discrimination depends on the delineation of the species, subspecies, and biotypes within the bacterial genera considered, and this varies from one bacterial group to another. With the exception of the genetically homogeneous genus *Zymomonas* (Swings *et al.*, 1976), no genus-specific protein patterns have been described hitherto. The most important reason for the application of these techniques in microbial systematics is that taxonomic relationships based on computer-assisted comparisons of protein electrophoregrams show a high correlation with those from DNA–DNA hybridizations: bacteria with high genome DNA similarity display very similar or almost identical protein patterns (see the examples on *Klebsiella* and *Xanthomonas*).

Jackman (1985a,b) has made proposals for a standard SDS polyacrylamide gel electrophoretic technique for microbial systematics, and programs for the numerical analysis of the patterns on a microcomputer are available (Jackman, 1983; Jackman *et al.*, 1983).

The advantages of the numerical analysis of electrophoretic protein patterns can be summarized as follows:

1. Large numbers of bacterial strains can be compared by the relatively simple and inexpensive technique of polyacrylamide gel electrophoresis.
2. The electrophoretic and hence genetic homogeneity or heterogeneity of bacterial groups can be determined.
3. Large numbers of processed electrophoretic traces can be stored on computer files and retrieved for the identification of unknown or new isolates, once electrophoretic groups of reliable taxonomic status have been delineated.
4. The electrophoretic techniques are applicable to a wide variety of bacteria from different ecological niches (see the examples on *Bordetella, Klebsiella,* and *Xanthomonas*).
5. Numerical analysis of protein electrophoregrams usually does not permit valid deductions on the genotypic relatedness between the electrophoretic clusters. Genetic inter-group relationships can be determined by DNA–DNA or DNA–rRNA hybridizations (see the examples on *Xanthomonas* and *Bordetella,* respectively). However, the number of DNAs which need to be prepared and the number of DNA–DNA or DNA–rRNA hybridizations can be drastically reduced, because only a few representative bacterial strains need to be examined genetically, once the electrophoretic groupings are known.

The use of numerical analysis of protein electrophoregrams has a very high potential in microbial systematics. The technique has played an important role in the delineation of new taxa (see the examples on *Bordetella* and *Klebsiella*) and in the detection and identification of misnamed strains. However, it does have some limitations:

1. It is not possible to identify a single colony in a few hours after isolation. The technique needs a larger amount of cells and is slower than, for example, pyrolysis gas–liquid chromatography or pyrolysis mass spectrometry.
2. Electrophoresis of bacterial proteins requires a relatively large number of experimental steps, each introducing its particular experimental error. All the experimental conditions should therefore be standardized and the reproducibility of the technique carefully controlled.

Despite these minor drawbacks, numerical analysis of protein electrophoregrams has already contributed towards major improvements and new insights into the systematics of different bacterial genera.

Acknowledgement

The author is indebted to the Nationaal Fonds voor Wetenschappelijk Onderzoek for research and travel grants.

References

Baptist, J. N., Shaw, C. R., and Mandel, M. (1969). Zone electrophoresis of enzymes in bacterial taxonomy. *Journal of Bacteriology* **99,** 180–188.

Baptist, J. N., Shaw, C. R., and Mandel, M. (1971). Comparative zone electrophoresis of enzymes of *Pseudomonas solanacearum* and *Pseudomonas cepacia. Journal of Bacteriology* **108,** 799–803.

Baumann, L., Baumann, P., Mandel, M., and Allen, R. D. (1972). Taxonomy of aerobic marine eubacteria. *Journal of Bacteriology* **110,** 402–429.

Biavati, B., Scardovi, V., and Moore W. E. C. (1982). Electrophoretic patterns of proteins in the genus *Bifidobacterium* and proposal of four new species. *International Journal of Systematic Bacteriology* **32,** 358–373.

Bousfield, I. J., Smith, G. L., Dando, T. R., and Hobbs, G. (1983). Numerical analysis of total fatty acid profiles in the identification of coryneform, nocardioform and some other bacteria. *Journal of General Microbiology* **129,** 375–394.

Brenner, D. J., Fanning, G. R., Skerman, F. J., and Falkow, S. (1972). Polynucleotide sequence divergence among strains of *Escherichia coli* and closely related organisms. *Journal of Bacteriology* **109,** 953–965.

Carlson, R. R., and Vidaver, A. K. (1982). Taxonomy of *Corynebacterium* plant pathogens, including a new pathogen of wheat, based on polyacrylamide gel electrophoresis of cellular proteins. *International Journal of Systematic Bacteriology* **32,** 315–326.

Cato, E. P., Holdeman, L. V., and Moore, W. E. C. (1982a). *Clostridium perenne* and *Clostridium paraperfringens:* later subjective synonyms of *Clostridium barati. International Journal of Systematic Bacteriology* **32,** 77–81.

Cato, E. P., Kelley, R. W., Moore, W. E., and Holdeman, L. V. (1982b). *Bacteroides zoogleoformans* (Weinberg, Nativelle, and Prévot 1937) corrig., comb. nov.: Emended Description. *International Journal of Systematic Bacteriology* **32,** 271–274.

Cato, E. P., Johnson, J. L., Hash, D. E., and Holdeman, L. V. (1983a). Synonymy of *Peptococcus glycinophilus* (Cardon and Barker 1946) Douglas 1957 with *Peptostreptococcus micros* (Prévot 1933) Smith 1957 and electrophoretic differentiation of *Peptostreptococcus micros* from *Peptococcus magnus* (Prévot 1933) Holdeman and Moore 1972. *International Journal of Systematic Bacteriology* **33,** 207–210.

Cato, E. P., Moore, W. E. C., and Johnson, J. L. (1983b). Synonymy of strains of *"Lactobacillus acidophilus"* group A2 (Johnson *et al.* 1980) with the type strain of *Lactobacillus crispatus* (Brygoo and Aladame 1953) Moore and Holdeman 1970. *International Journal of Systematic Bacteriology* **33,** 426–428.

Davis, B. J. (1964). Disc electrophoresis—II. Method and application to human serum proteins. *Annals of the New York Academy of Sciences* **121,** 404–427.

De Ley, J. (1978). Modern molecular methods in bacterial taxonomy: evaluation, application, prospects. *Proceedings of the 4th International Conference on Plant Pathogenic Bacteria, Angers* **1,** 347–357.

De Ley, J., Cattoir, H., and Reynaerts, A. (1970). The quantitative measurement of DNA hybridization from renaturation rates. *European Journal of Biochemistry* **12,** 133–142.

De Ley, J., Segers, P., and Gillis, M. (1978). Intra- and intergeneric similarities of *Chromobacterium* and *Janthinobacterium* ribosomal ribonucleic acid cistrons. *International Journal of Systematic Bacteriology* **28,** 154–168.

De Ley, J., Mannheim, W., Segers, P., Kersters, K., Hinz, K. H., and Lievens, A. (1983). Taxonomy of the genus *Bordetella. Colloque International de Bacteriologie, Lille.* Abstract, p. 30.

De Smedt, J., and De Ley, J. (1977). Intra- and intergeneric similarities of *Agrobacterium* ribosomal ribonucleic acid cistrons. *International Journal of Systematic Bacteriology* **27,** 222–240.

De Vos, P., and De Ley, J. (1983). The intra- and intergeneric similarities of *Pseudomonas* and *Xanthomonas* ribosomal ribonucleic acid cistrons. *International Journal of Systematic Bacteriology* **33,** 487–509.

du Plessis, H. J. (1983). A taxonomic study and biological control of *Agrobacterium* species in South Africa. Ph. D. thesis. The University of the Orange Free State, Bloemfontein, Republic of South Africa.

Dye, D. W. (1978). Genus IX. *Xanthomonas* Dowson 1939. *In* 'A Proposed Nomenclature and Classification for Plant Pathogenic Bacteria'. *New Zealand Journal of Agricultural Research* **21,** 153–177.

Dye, D. W., Bradbury, J. F., Goto, M., Hayward, A. C., Lelliott, R. A., and Schroth, M. N. (1980). International standards for naming pathovars of phytopathogenic bacteria and a list of pathovar names and pathotype strains. *Review of Plant Pathology* **59,** 153–168.

Elango, F. N., Lozano, J. C., and Peterson, J. F. (1981). Relationships between *Xanthomonas c.* pv. *manihotis, X. c.* pv. *cassavae* and Colombian yellowish isolates. *Proceedings of the 5th International Conference on Plant Pathogenic Bacteria, Cali,* pp. 96–104.

El-Sharkawy, T. A., and Huisingh, D. (1971). Electrophoretic analysis of esterases and other soluble proteins from representatives of phytopathogenic bacterial genera. *Journal of General Microbiology* **68,** 155–165.

Feltham, R. K. A., and Sneath, P. H. A. (1979). Quantitative comparison of electrophoretic traces of bacterial proteins. *Computers and Biomedical Research* **12,** 247–263.

Ferragut, C. (1983). Le genre *Klebsiella.* Contribution à l'étude taxonomique. Ph. D. thesis. Université des Sciences et Techniques de Lille, France.

Ferragut, C., Izard, D., Gavini, F., Kersters, K., De Ley, J., and Leclerc, H. (1983). *Klebsiella trevisanii:* a new species from water and soil. *International Journal of Systematic Bacteriology* **33,** 133–142.

Fox, G. E., Stackebrandt, E., Hespell, R. B., Gibson, J., Maniloff, J., Dyer, T. A., Wolfe, R. S., Balch, W. E., Tanner, R. S., Magrum, L. J., Zablen, L. B., Blakemore, R., Gupta, R., Bonen, L., Lewis, B. J., Stahl, D. A., Luehrsen, K. R., Chen, K. N., and Woese, C. R. (1980). The phylogeny of prokaryotes. *Science* **209,** 457–463.

Garber, E. D., and Rippon, J. W. (1968). Proteins and enzymes as taxonomic tools. *Advances in Applied Microbiology* **10,** 137–154.

Gavini, F., Leclerc, H., Lefebvre, B., Ferragut, C., and Izard, D. (1977). Etude taxonomique d'entérobactéries appartenant ou apparantées au genre *Klebsiella. Annales de Microbiologie (Paris)* **128B,** 45–59.

Gillis, M., Kersters, K., Gosselé, F., Swings, J., De Ley, J., MacKenzie, A. R., and Bousfield, I. J. (1983). Rediscovery of Bertrand's sorbose bacterium (*Acetobacter aceti* subsp. *xylinum*): Proposal to designate NCIB 11664 in place of NCIB 4112

(ATCC 23767) as the type strain of *Acetobacter aceti* subsp. *xylinum*. Request for an opinion. *International Journal of Systematic Bacteriology* **33,** 122–124.

Gosselé, F., Swings, J., Kersters, K., and De Ley, J. (1983a). Numerical analysis of phenotypic features and protein gel electropherograms of *Gluconobacter* Asai 1935 emend. mut. char. Asai, Iizuka, and Komagata 1964. *International Journal of Systematic Bacteriology* **33,** 65–81.

Gosselé, F., Swings, J., Kersters, K., Pauwels, P., and De Ley, J. (1983b). Numerical analysis of phenotypic features and protein gel electrophoregrams of a wide variety of *Acetobacter* strains. Proposal for the improvement of the taxonomy of the genus *Acetobacter* Beijerinck 1898, 215. *Systematic and Applied Microbiology* **4,** 338–368.

Gutteridge, C. S., and Norris, J. R. (1979). A review. The application of pyrolysis techniques to the identification of micro-organisms. *Journal of Applied Bacteriology* **47,** 5–43.

Hanna, J., Neill, S. D., O'Brien, J. J., and Ellis, W. A. (1983). Comparison of aerotolerant and reference strains of *Campylobacter* species by polyacrylamide gel electrophoresis. *International Journal of Systematic Bacteriology* **33,** 143–146.

Hill, A. C. (1983). *Mycoplasma cricetuli,* a new species from the conjunctivas of chinese hamsters. *International Journal of Systematic bacteriology* **33,** 113–117.

Hinz, K.-H. (1981). Some properties of the turkey coryza agent. *Abstracts of the 7th International Congress of the World Veterinary Poultry Association,* p. 79.

Holdeman, L. V., and Johnson, J. J. (1982). Description of *Bacteroides loescheii* sp. nov. and emendation of the descriptions of *Bacteroides melaninogenicus* (Oliver and Wherry) Roy and Kelly 1939 and *Bacteroides denticola* Shah and Collins 1981. *International Journal of Systematic Bacteriology* **32,** 399–409.

Holdeman, L. V., Moore, W. E. C., Churn, P. J., and Johnson, J. L. (1982). *Bacteroides oris* and *Bacteroides buccae,* new species from human periodontitis and other human infections. *International Journal of Systematic Bacteriology* **32,** 125–131.

Howard, C. J., and Gourlay, R. N. (1982). Proposal for a second species within the genus *Ureaplasma, Ureaplasma diversum* sp. nov. *International Journal of Systematic Bacteriology* **32,** 446–452.

Hudson, B. W., Quan, T. J., and Bailey, R. E. (1976). Electrophoretic studies of the geographic distribution of *Yersinia pestis* protein variants. *International Journal of Systematic Bacteriology* **26,** 1–16.

Izard, D., Ferragut, C., Gavini, F., Kersters, K., De Ley, J., and Leclerc, H. (1981). *Klebsiella terrigena:* a new species from soil and water. *International Journal of Systematic Bacteriology* **31,** 116–127.

Jackman, P. J. H. (1981). Taxonomy of aerobic axillary coryneforms based on electrophoretic protein patterns. Ph. D. thesis, Univ. of London.

Jackman, P. J. H. (1982). Classification of *Corynebacterium* species from axillary skin by numerical analysis of electrophoretic protein patterns. *Journal of Medical Microbiology* **15,** 485–492.

Jackman, P. J. H. (1983). A program in BASIC for numerical taxonomy of microorganisms based on electrophoretic band positions. *Microbios* **23,** 119–124.

Jackman, P. J. H. (1985a). Bacterial taxonomy based on electrophoretic whole-cell protein patterns. *In* 'Chemical Methods in Bacterial Systematics' (Eds. M. Goodfellow and D. E. Minnikin), pp. 115–128. Academic Press, London.

Jackman, P. J. H. (1985b). Characterisation of micro-organisms by electrophoretic protein patterns. *In* 'New Methods for the Detection and Characterisation of Micro-organisms'. (Ed. C. S. Gutteridge), in press. Wiley, Chichester.

Jackman, P. J. H., Feltham, R. K. A., and Sneath, P. H. A. (1983). A program in BASIC for numerical taxonomy of micro-organisms based on electrophoretic protein patterns. *Microbios Letters* **23,** 87–98.

Jarvis, A. W., and Wolff, J. M. (1979). Grouping of lactic streptococci by gel electrophoresis of soluble cell extracts. *Applied and Environmental Microbiology* **37,** 391–398.

Johnson, J. L., and Holdeman, L. V. (1983). *Bacteroides intermedius* comb. nov. and descriptions of *Bacteroides corporis* sp. nov. and *Bacteroides levii* sp. nov. *International Journal of Systematic Bacteriology* **33,** 15–25.

Johnson, R., and Sneath, P. H. A. (1973). Taxonomy of *Bordetella* and related organisms of the families Achromobacteriaceae, Brucellaceae, and Neisseriaceae. *International Journal of Systematic Bacteriology* **23,** 381–404.

Kersters, K. (1978). Taxonomy of *Alcaligenes* and *Achromobacter* by polyacrylamide gel electrophoresis of their soluble proteins. *Antonie van Leeuwenhoek* **44,** 116–117.

Kersters, K., and De Ley, J. (1975). Identification and grouping of bacteria by numerical analysis of their electrophoretic protein patterns. *Journal of General Microbiology* **87,** 333–342.

Kersters, K., and De Ley, J. (1980). Classification and identification of bacteria by electrophoresis of their proteins. *In* 'Microbiological Classification and Identification' (Eds. M. Goodfellow and R. G. Board), pp. 273–297. Academic Press, London.

Kersters, K., Hinz, K.-H., Hertle, A., Segers, P., Lievens, A., Siegmann, O., and De Ley, J. (1984). *Bordetella avium* sp. nov., isolated from the respiratory tracts of turkeys and other birds. *International Journal of Systematic Bacteriology* **34,** 56–70.

Kiredjian, M., Popoff, M., Coynault, C., Lefèvre, M., and Lemelin, M. (1981). Taxonomie du genre *Alcaligenes. Annales de Microbiologie (Paris)* **132B,** 337–374.

Kloos, W. E., Mohapatra, N., Dobrogosz, W. J., Ezzell, J. W., and Manclark, C. R. (1981). Deoxyribonucleotide sequence relationships among *Bordetella* species. *International Journal of Systematic Bacteriology* **31,** 173–176.

Laemmli, U. K. (1970). Cleavage of structural proteins during the assembly of the head of bacteriophage T4. *Nature (London)* **227,** 680–685.

Lambert, B., Kersters, K., Gosselé, F., Swings, J., and De Ley, J. (1981). Gluconobacters from honey bees. *Antonie van Leeuwenhoek* **47,** 147–157.

Leps, W. T., Roberts, G. P., and Brill, W. J. (1980). Use of two-dimensional polyacrylamide electrophoresis to demonstrate that putative *Rhizobium* cross-inoculation mutants actually are contaminants. *Applied and Environmental Microbiology* **39,** 460–462.

Maraite, H., and Weyns, J. (1979). Distinctive, physiological, biochemical and pathogenic characteristics of *Xanthomonas manihotis* and *X. cassavae. In* 'Diseases of Tropical Food Crops. Proceedings of an International Symposium' (Eds. H. Maraite and J. A. Meyer), pp. 103–117. Univezsite Catholique de Louvain, Louvain-la-Neuve, Belgium.

Matthew, M., and Harris, A. M. (1976). Identification of β-lactamases by analytical isoelectric focusing: correlation with bacterial taxonomy. *Journal of General Microbiology* **94,** 55–67.

Mays, T. D., Holdeman, L. V., Moore, W. E. C., Rogosa, M., and Johnson, J. L. (1982). Taxonomy of the genus *Veillonella* Prévot. *International Journal of Systematic Bacteriology* **32,** 28–36.

Meiresonne, L. (1983). Vergelijkende eiwit-electroforetische studie van *Xanthomonas campestris* pv. *manihotis* en pv. *cassavae* stammen. Tesis. Rijksuniversiteit, Gent.

Mergaert, J., Gavini, F., Kersters, K., Leclerc, H., and De Ley, J. (1983). Phenotypic and protein electrophoretic similarities between strains of *Enterobacter agglomerans,*

Erwinia herbicola and *Erwinia milletiae* from clinical and plant origin. *Current Microbiology* pp. 327–331.

Moore, R. L., and McCarthy, B. J. (1967). Comparative study of ribosomal ribonucleic acid cistrons in enterobacteria and myxobacteria. *Journal of Bacteriology* **94,** 1066–1074.

Mouches, C., Vignault, J. G., Tully, T. G., Whitcomb, R. F., and Bove, J. M. (1979). Characterization of *Spiroplasmas* by one and two-dimensional protein analysis on polyacrylamide slab gels. *Current Microbiology* **2,** 69–74.

Naemura, L. G., Bagley, S. T., Seidler, R. J., Kaper, J. B., and Colwell, R. R. (1979). Numerical taxonomy of *Klebsiella pneumoniae* strains isolated from clinical and non-clinical sources. *Current Microbiology* **2,** 175–180.

Nakajima, K., Muroga, K., and Hancock, R. E. W. (1983). Comparison of fatty acid, protein, and serological properties distinguishing outer membranes of *Pseudomonas anguilliseptica* strains from those of fish pathogens and other pseudomonads. *International Journal of Systematic Bacteriology* **33,** 1–8.

Norris, J. R. (1980). Introduction. *In* 'Microbiological Classification and Identification' (Eds. M. Goodfellow and R. G. Board), pp. 1–10. Academic Press, London.

O'Farrell, P. H. (1975). High resolution two-dimensional electrophoresis of proteins. *Journal of Biological Chemistry* **250,** 4007–4021.

Ornstein, L. (1964). Disc electrophoresis—I. Background and theory. *Annals of the New York Academy of Sciences* **121,** 321–349.

Owen, R. J., and Jackman, P. J. H. (1982). The similarities between *Pseudomonas paucimobilis* and allied bacteria derived from analysis of deoxyribonucleic acids and electrophoretic protein patterns. *Journal of General Microbiology* **128,** 2945–2954.

Philip, R. N., Casper, E. A., Anacker, R. L., Cory, J., Hayes, S. F., Burgdorfer, W., and Yunker, C. E. (1983). *Rickettsia bellii* sp. nov.: a tick-borne rickettsia, widely distributed in the United States, that is distinct from the spotted fever and typhus biogroups. *International Journal of Systematic Bacteriology* **33,** 94–106.

Razin, S., and Rottem, S. (1967). Identification of *Mycoplasma* and other microorganisms by polyacrylamide-gel electrophoresis of cell proteins. *Journal of Bacteriology* **94,** 1807–1810.

Robbs, C. F., de Ribeiro, R. L. D., Kimura, O., and Akiba, F. (1972). Variacoes en *Xanthomonas manihotis* (Arthaud Berthet) [sic] Starr. *Revisita da Sociedade Brasiliera de Fitopatologia* **5,** 67–75.

Roberts, G. P., Leps, W. T., Silver, L. E., and Brill, W. J. (1980). Use of two-dimensional polyacrylamide gel electrophoresis to identify and classify *Rhizobium* strains. *Applied and Environmental Microbiology* **39,** 414–422.

Rodwell, A. W., and Rodwell, E. S. (1978). Relationships between strains of *Mycoplasma mycoides* subspp. *mycoides* and *capri* studied by two-dimensional gel electrophoresis of cell proteins. *Journal of General Microbiology* **109,** 259–263.

Rosendal, S. (1973). Analysis of the electrophoretic pattern of *Mycoplasma* proteins for the identification of canine *Mycoplasma* species. *Acta Pathologica et Microbiologica Scandinavica, Section B* **81,** 273–281.

Seidler, R. J., Mandel, M., and Baptist, J. N. (1972). Molecular heterogeneity of the bdellovibrios: evidence of two new species. *Journal of Bacteriology* **109,** 209–217.

Simmons, D. G., Rose, L. P., and Gray, J. G. (1980). Some physical, biochemic and pathologic properties of *Alcaligenes faecalis,* the bacterium causing rhinotracheitis (coryza) in turkey poults. *Avian Diseases* **24,** 82–90.

Skerman, V. B. D., McGowan, V., and Sneath P. H. A. (1980). Approved lists of bacterial names. *International Journal of Systematic Bacteriology* **30,** 225–420.

Sneath, P. H. A. (1957). The application of computers to taxonomy. *Journal of General Microbiology* **15,** 70–98.

Starr, M. P. (1981). The genus *Xanthomonas. In* 'The Prokaryotes. A Handbook on Habitats, Isolation, and Identification of Bacteria' (Eds. M. P. Starr, H. Stolp, H. G. Trüper, A. Balows, and H. G. Schlegel), Vol. 1, pp. 742–763. Springer-Verlag, Berlin.

Swenson, C. E., Vanhamont J., and Dunbar, B. S. (1983). Specific protein differences among strains of *Ureaplasma urealyticum* as determined by two-dimensional gel electrophoresis and a sensitive silver stain. *International Journal of Systematic Bacteriology* **33,** 417–421.

Swings, J., Kersters, K., and De Ley, J. (1976). Numerical analysis of electrophoretic protein patterns of *Zymomonas* strains. *Journal of General Microbiology* **93,** 266–271.

Swings, J., Kersters, K., and De Ley, J. (1977). Taxonomic position of additional *Zymomonas mobilis* strains. *International Journal of Systematic Bacteriology* **27,** 271–273.

Swings, J., Gillis, M., Kersters, K., De Vos, P., Gosselé, F., and De Ley, J. (1980). *Frateuria,* a new genus for "*Acetobacter aurantius*". *International Journal of Systematic Bacteriology* **30,** 547–556.

Vanden Berghe, D. A., and Pattyn, S. R. (1979). Comparison of proteins from *Mycobacterium fortuitum, Mycobacterium nonchromogenicum* and *Mycobacterium terrae* using flat bed electrophoresis. *Journal of General Microbiology* **111,** 283–291.

Vantomme, R., Swings, J., Goor, M., Kersters, K., and De Ley, J. (1982). Phytopathological, serological, biochemical and protein electrophoretic characterization of *Erwinia amylovora* strains isolated in Belgium. *Phytopathologisches Zeitschrift* **103,** 349–360.

van Vuuren, H. J. J. (1978). Identification and physiology of Enterobacteriaceae isolated from South African lager beer breweries. Ph. D. thesis. Rijksuniversiteit, Gent, Belgium.

van Vuuren, H. J. J., Kersters, K., and De Ley, J. (1981). The identification of Enterobacteriaceae from breweries: combined use and comparison of API 20E system, gel electrophoresis of proteins and gas chromatography of volatile metabolites. *Journal of Applied Bacteriology* **51,** 51–65.

Weber, K., and Osborn, M. (1969). The reliability of molecular weight determinations by dodecylsulfate polyacrylamide gel electrophoresis. *Journal of Biological Chemistry* **244,** 4406–4412.

Weiss, E., and Dasch, G. A. (1982). Differential characteristics of strains of *Rochalimaea: Rochalimaea vinsonii* sp. nov., the canadian vole agent. *International Journal of Systematic Bacteriology* **32,** 305–314.

Whiley, R. A., Hardie, J. M., and Jackman, P. J. H. (1982). SDS–polyacrylamide gel electrophoresis of oral streptococci. *In* 'Basic Concepts of Streptococci and Streptococcal Diseases, Proceedings of the VIIIth International Symposium on Streptococci and Streptococcal Diseases' (Eds, S. E. Holm and P. Christensen), pp. 61–62. Reedbooks Ltd., Chertsey, England.

Whiley, R. A., Hardie, J. M., Sackin, M. J., and Jackman, P. J. H. (1983). A comparison of two numerical taxonomic studies on oral streptococci. *Society for General Microbiology Quarterly* **10,** M24.

Williams, R. A. D., and Shah, H. N. (1980). Enzyme patterns in bacterial classification and identification. *In* 'Microbiological Classification and Identification' (Eds. M. Goodfellow and R. G. Board), pp. 299–318. Academic Press, London.

14

Numerical Methods in the Classification of Micro-organisms by Pyrolysis Mass Spectrometry

C. S. GUTTERIDGE

Cadbury Schweppes PLC, Lord Zuckerman Research Centre, University of Reading, Reading, UK

L. VALLIS AND H. J. H. MACFIE

Agricultural and Food Research Council, Food Research Institute (Bristol), Bristol, UK

Introduction

If micro-organisms are subjected to a controlled thermal degradation process in an inert atmosphere (pyrolysis), the polymeric structures which make up the living cell fragment into a series of low molecular weight volatile compounds. In pyrolysis mass spectrometry (Py–ms) these volatile fragments are detected and counted by a mass analyser to produce a complex spectrum. In the related technique of pyrolysis gas chromatography (Py–gc) the volatiles are separated before detection (usually by flame ionisation) to produce a pyrogram (pyrolysis chromatogram).

The technology of Py–gc and Py–ms has been discussed in detail elsewhere (see Irwin, 1982, and Meuzelaar *et al.*, 1982, for extensive reviews). In microbiology, both techniques have been used principally for the characterisation of micro-organisms. For this application to be successful it is necessary to compare the quantitative aspects of the pyrolysis spectra or pyrograms. This article examines the techniques used to handle the data produced by analytical pyrolysis and reviews the impact of this methodology on microbial systematics. Most of the examples given relate to Py–ms, but the same techniques can be applied to Py–gc, and to other analytical data.

Pyrolysis Mass Spectra

To understand why data handling with Py–ms is such a complex problem it is necessary to consider the data structure and the factors which influence its

ISBN 0-12-289665-3

reproducibility. Figure 1 gives examples of pyrolysis mass spectra of two bacteria (*Citrobacter freundii* and *Serratia marcescens*) grown and handled under identical conditions. Examination of these spectra reveals that on pyrolysis, fragments are detected over the mass range m/z 16 to 180 with only very low-intensity ions above that. Observations on a variety of different micro-organisms suggest that spectra of different genera or species are qualitatively identical, and discrimination has to be based on differences in mass intensities. This is confirmed by a comparison of Fig. 1a and b.

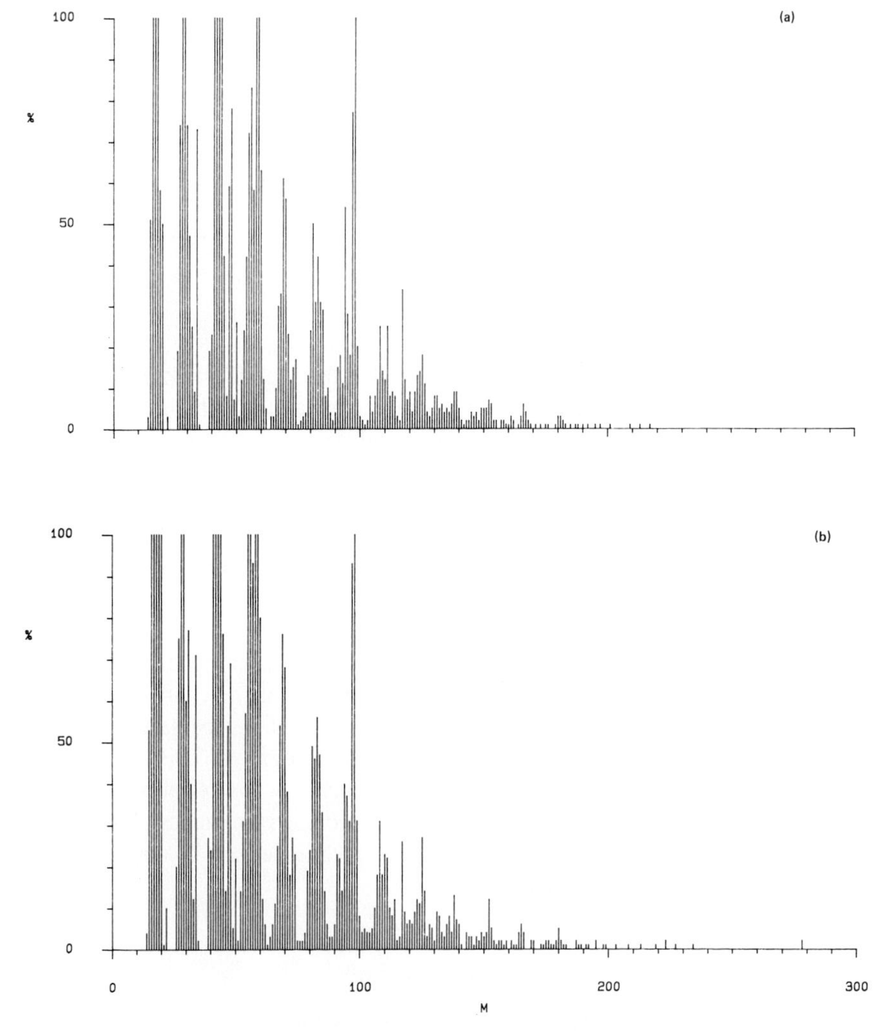

Fig. 1. Pyrolysis mass spectra of *Citrobacter freundii* (a) and *Serratia marcescens* (b).

The quantitative data are highly correlated both chemically and, after pre-processing, mathematically. Chemical correlations arise because each mass in the spectrum has a multiple origin during the thermal degradation process. This is illustrated by Fig. 2, which shows pyrolysis mass spectra for peptidoglycan and teichoic acid—both purified from *Bacillus subtilis* (Meuzelaar *et al.*, 1982). The overall fragmentation pattern of both polymers is clearly different, but they have many masses in common. Consequently if two organisms are to be differentiated by considering the relative amounts of the pyrolysis products of peptidoglycan and teichoic acid, it is unlikely to be achievable using a single mass, and usually a complex combination of several masses is required.

One of the advantages of Py–ms is that micro-organisms can be sampled directly from culture plates and analysed immediately with no further sample

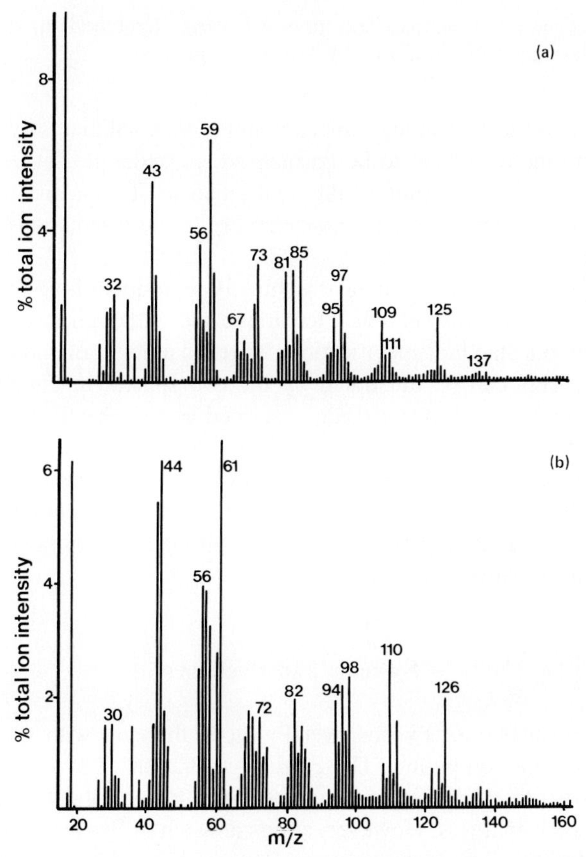

Fig. 2. Pyrolysis mass spectra of *Bacillus subtilis* peptidoglycan (a) and *Bacillus subtilis* teichoic acid (b) (Meuzelaar *et al.*, 1982).

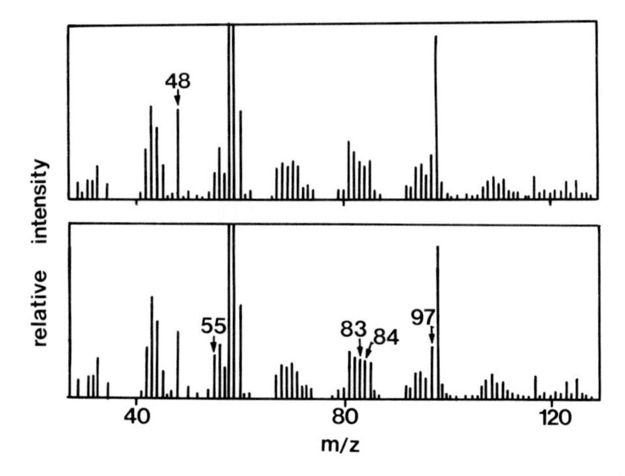

Fig. 3. Duplicate pyrolysis mass spectra of a *Listeria* (serotype IVb) strain (Eshuis *et al.*, 1977). Mass intensities differing by $>10\%$ are arrowed.

preparation. This means that the actual amount of material that is pyrolysed is not controlled and the data have to be normalised, for example, by expressing the mass intensities as a proportion of the total intensity. This is an important data-handling step but one which adds considerably to the correlations amongst the data.

The final influence on the data structure is reproducibility. Pyrolysis mass spectrometry is a complex analytical technique and, although instruments are set up to perform to a standard specification, a certain amount of 'noise' in the data is inevitable. This can be illustrated by the two spectra in Fig. 3, which are duplicate analyses of a *Listeria* strain described as serotype IVb (Eshuis *et al.*, 1977). Five masses have intensities that differ by more than 10%, a degree of difference between duplicates that is typical of Py–ms 'noise'.

The overall aim of the data analysis, therefore, is to amplify differences between spectra that are reproducible and can provide information useful for the discrimination of samples.

Systems and Packages

One of the advantages of Py–ms over Py–gc is the ease with which it can be integrated with the computing. The current generation of pyrolysis mass spectrometers have mini computer-based data systems to control data acquisition and to carry out some analyses. However, a typical batch of Py–ms data may contain more than 30,000 measurements (mass intensities), and most of the data handling is carried out off-line using main-frame computers. This situation will change over the next few years because of the advent of the affordable 16-bit microcomputer and prospects for the 32-bit microcomputer. Currently raw data are stored on magnetic media which are physically transferred to a main-frame or virtual-storage minicomputer.

A variety of packages have been used to analyse Py–ms data. Most are general-purpose statistical packages with a wide range of useful procedures, for example GENSTAT (Nelder, 1979), SPSS (Nie *et al.*, 1975), BMDP (Dixon, 1975), and ARTHUR (Kowalski, 1975). The ARTHUR package is particularly interesting because it is designed specifically for minicomputers rather than main-frame computers and it contains a version of the specialised program SIMCA (Wold, 1976, 1978). All these packages offer a comprehensive choice of routines such as data preprocessing, univariate and multivariate analyses, feature selection, factor and discriminant analysis, cluster analysis, and visualisation procedures.

In addition to these, the complexity of the Py–ms data-handling problem has stimulated the development of some purpose-written software including FOMPYR (Eshuis *et al.*, 1977), a package based around nonlinear mapping; NORMA, a program for interactive preprocessing developed by Eshuis (Huff *et al.*, 1981) which can be used to transform data into a format compatible with SPSS, ARTHUR, and FOMPYR; and HILDA (Gutteridge *et al.*, 1984), an interactive stepwise discriminant analysis program.

At the moment there is no single package or suite of programs that can satisfy all the possible requirements of the analyst faced with a large set of Py–ms data.

Experimental Design

We have already established that one feature of Py–ms data is an inherent amount of 'noise'. To monitor reproducibility, it is common practice (almost standard practice) with Py–ms to include in the data set replicate analyses of all the samples. This contrasts with conventional numerical taxonomy where usually a small percentage of the samples are repeated to get an assessment of test reproducibility. When Py–ms is applied to pure cultures, the usual approach is to grow each strain as two separate cultures and to analyse two samples from each culture. This procedure ensures that variability due to culturing and sample preparation methods is reflected fairly in the data set. Generally the replicates are analysed sequentially to avoid complex formatting problems when computing the data set.

To check on longer term reproducibility, it is advisable to analyse a few samples more than once. The rapid analysis speeds of Py–ms ($\leq$5 min per sample) allow this without adding significantly to the length of an experiment.

When a data set is analysed, replicates can be averaged to eliminate variability. However, with some data analysis techniques replicates can be included to contrast the degree of difference displayed between samples to that within samples (i.e., between the replicates).

In all other respects the design of Py–ms classification studies should adhere to the accepted practices of numerical taxonomy with regard to such factors as the numbers of strains compared and choice of type cultures. The use of discriminant analysis, however, may impose some constaints on the numbers of groups, the number of samples in a group, and the analysis of unrelated groups.

Review of Methods

A number of different data-handling techniques have been applied to pyrolysis data. The need for data handling was recognised in early reports of the application of Py–gc to the characterisation of micro-organisms (e.g., Sekhon and Carmichael, 1972; Carmichael *et al.*, 1973), and the momentum of technique development has continued unabated. Figure 4 summarises a cohesive approach to the handling of pyrolysis data that has been developed over a number of years (MacFie and Gutteridge, 1982). In some respects it is already outdated, but the major steps in a data analysis strategy remain valid and it serves as a suitable skeleton for describing the methods and their applications. It does not include methods aimed at the chemical interpretation of spectra which will be described separately.

Pre-processing

The detection of errors in data accumulation and transmission is not a problem with computer-controlled Py–ms systems but may be with Py–gc (Gutteridge *et al.*, 1979).

The major preprocessing operation is normalisation (pattern scaling), which is performed to compensate for variations in the overall ion intensity caused by factors unrelated to the analytical problem such as differences in sample size or changes in instrument sensitivity. The simplest way to remove these variations is to express mass intensity as a percentage of total ion intensity. For example, with 63 masses each individual mass m_{ij} of sample J is transformed to t_{ij}:

$$t_{ij} = m_{ij} \Big/ \sum_{i=1}^{63} m_{ij} 100$$

This procedure works better as the number of masses increases and as the variation in individual mass intensities decreases. The main problem in using this

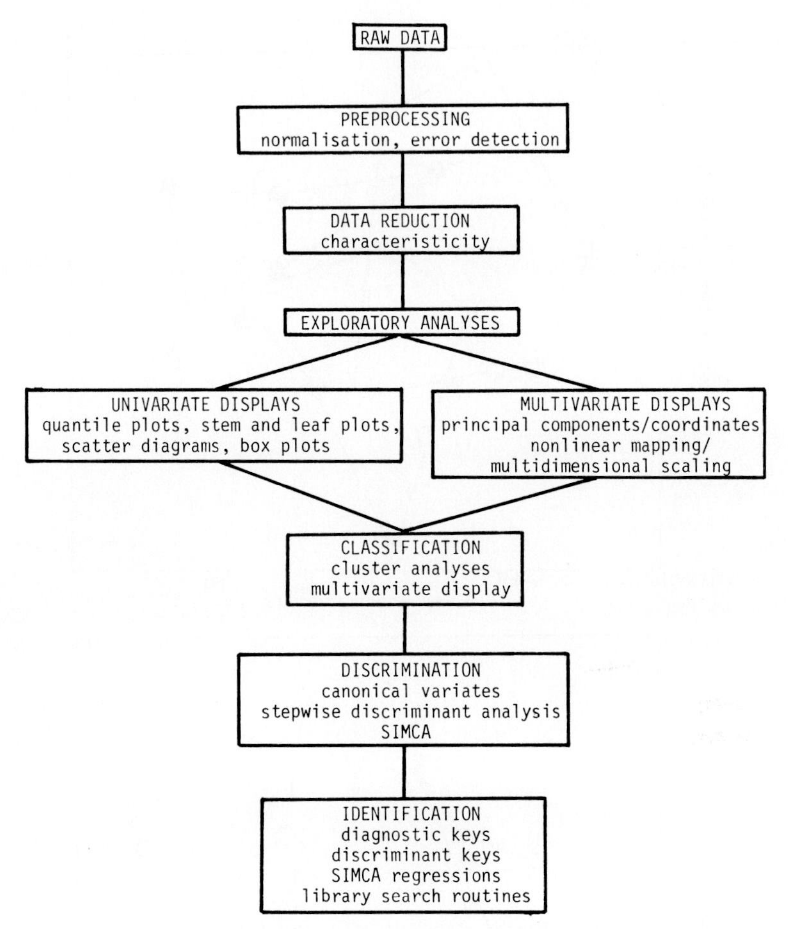

Fig. 4. Data-handling scheme (MacFie and Gutteridge, 1982).

procedure is the occurrence of very large masses, especially when these masses exhibit a high degree of intra- and/or inter-sample deviation. If such a mass happens to be unusually high in a given spectrum, then all other masses will be given low relative intensity values, which may confuse further quantitative and qualitative comparisons between the spectra. A simple solution to this problem is to exempt all masses larger than a certain percentage of total signal intensity, in one or more of the spectra compared, from the normalisation procedure.

Figures 5a and b are examples of the influence of large masses on a data base. Figure 5a represents an attempt to distinguish three heterogeneous octadecylsilyl polymer samples by Py–ms using 289 masses. The samples are not discriminated and there is considerable variation among the replicates. In Fig. 5b, the analysis

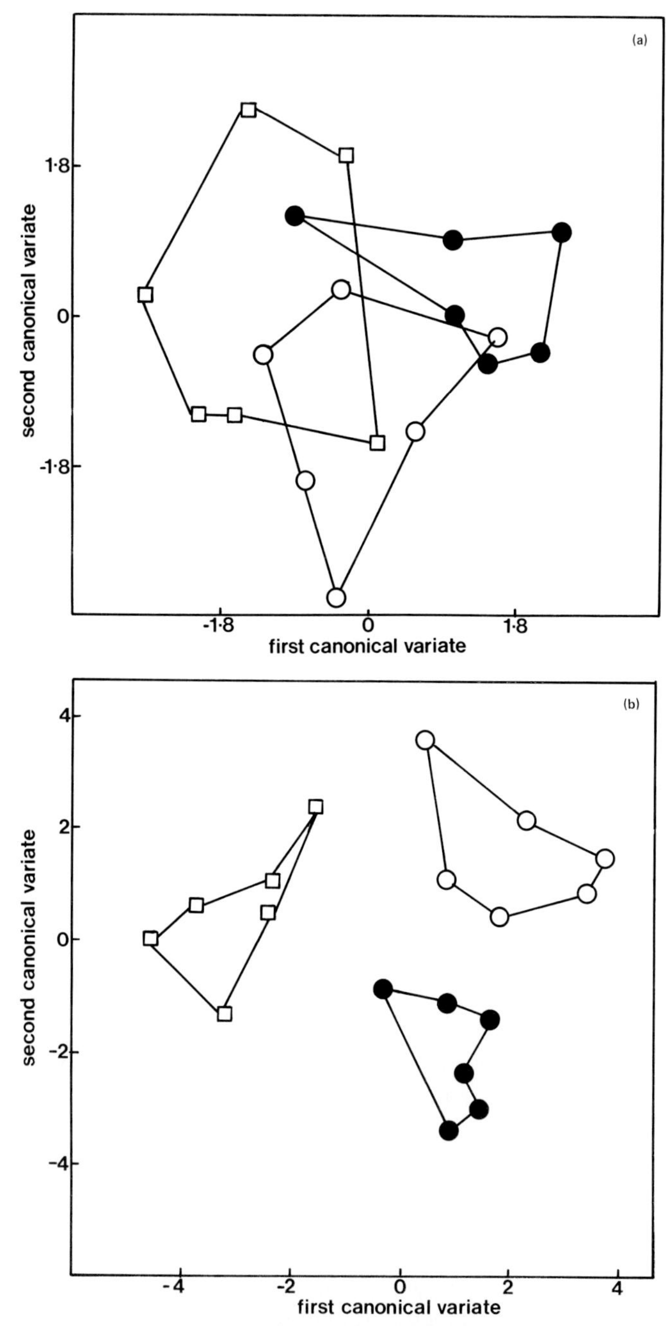

Fig. 5. Canonical variates analyses of three octadecylsilyl polymers, (a) using 289 masses with replicate analyses linked; (b) using 286 masses (three large masses removed); (c) using 289 masses and autoscaling.

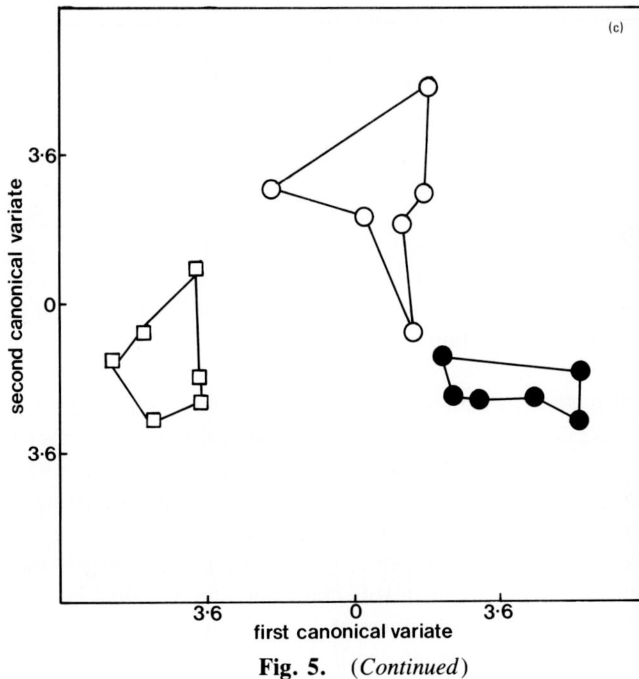

Fig. 5. (*Continued*)

is based on 286 masses; three large masses were removed, the polymers are discriminated, and the replicate variation is reduced.

The obvious shortcoming of this procedure is that elimination of the largest masses is, at best, a very rough way of eliminating potential sources of strong variation. An alternative would be to exempt all masses exhibiting more than a certain amount of intra- (between replicates) and/or inter- (between samples) deviation depending on the analytical problem. This requires a knowledge of the deviations which can only be calculated accurately after adequate normalisation. The NORMA programme (Huff *et al.*, 1981) was designed to overcome this problem and uses an iterative calculation of normalisation coefficients and variances while removing masses with high deviation values until no more can be found with deviation values above a certain level.

Another approach to solving the large mass problem is to use the so-called *auto-scaling procedure* (Harper *et al.*, 1977). This was applied by Blomquist *et al.* (1979a) to scale a set of *Penicillium* Py–gc data by subtracting the mean of each pyrogram peak and dividing by the standard deviation. The mean and standard deviation were calculated across all the samples. In Fig. 5c autoscaling has been applied to the polymers data base, producing a similar separation to Fig. 5b except that one of the replicates of group 2 is revealed as an outlier.

For classification studies on similar strains (e.g., within a genus or a species), normalisation does not usually present problems. In comparisons of widely differing samples (e.g., several genera or where the chemical interpretation of differences between spectra is required), normalisation is a critical step that has not received adequate attention (Klee *et al.*, 1981).

Data Reduction

Elimination of some data is usually necessary to reduce the amount of computer memory involved when dealing with large batches of Py–ms data. It may also be needed to comply with certain numerical constraints necessary for a statistically valid discriminant analysis (Dixon, 1975).

An elegant method of data reduction was designed by Eshuis *et al.* (1977) and involves the concept of *feature scaling*. Some features (i.e., mass intensities) may be constant across all the samples observed (i.e., features that represent some aspect of the instrument or a chemical compound occurring in constant proportions). Other features may show large variations between replicates of the same sample. This high inner variance may be due to variability in the sample or the experimental procedure and will contribute 'noise' that may obscure the detection of differences between samples, termed outer variance. Eshuis *et al.* (1977) therefore proposed mathematical expressions for inner variance (*reproducibility*) and outer variance (*specificity*) of each feature. The variability of each feature was scaled to unity and then weighted by these expressions. The most effective choice of weight was found to be the ratio of outer to inner variance (*characteristicity*). This characteristicity factor is closely related to the well-known Fisher ratio, and large data sets can be reduced by selecting only those mass intensities with high characteristicity values.

A classic illustration of the power of this data reduction/scaling technique is provided by the analysis of the spectra representing the two *Listeria* serotypes. Figure 6a is a nonlinear map of the raw data. The two serotypes are not discriminated and the replicate variation is large. In Fig. 6b the same data set is presented but with the mass intensities scaled for reproducibility. A small improvement is obvious but the two serotypes are still mixed. In Fig. 6c the data are weighted according to specificity and the two serotype groups are revealed. In Fig. 6d the data are weighted by characteristicity and the serotypes are completely discriminated, although the replication of some strains is still, comparatively, poor.

Figure 7 shows averaged spectra for these two serotypes, and the small differences in mass intensities responsible for the discrimination of the serotypes are marked. Figure 7 can be compared directly with Fig. 3 to illustrate that, in a complex data set, reproducible and discriminatory mass intensity values can be found by an appropriate data-handling technique. It should be emphasised that the differentiation of the serotypes is achieved without using prior information.

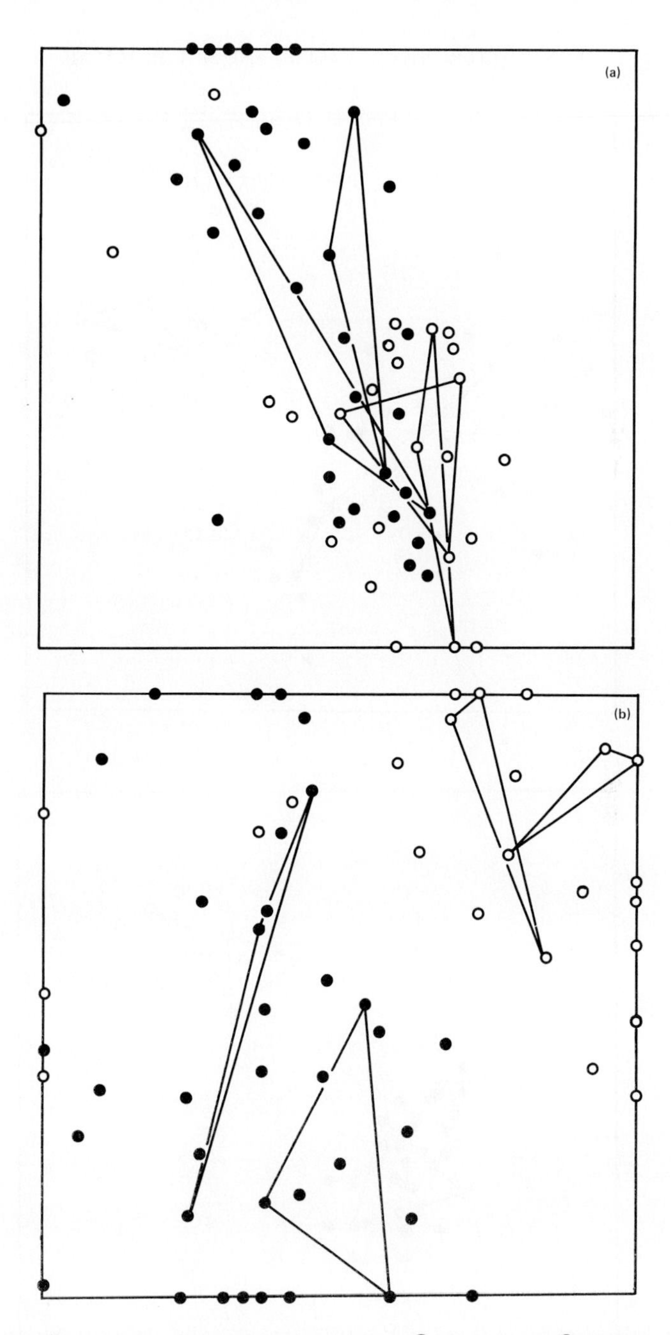

Fig. 6. Non-linear maps of two *Listeria* serotypes: ○, Serotype I; ●, Serotype IVb. (a) Analysis of raw data. Some replicate analyses are linked. (b) Analysis of data scaled by reproducibility. (c, next page) Analysis of data weighted by specificity. (d, next page) Analysis of data weighted by characteristicity (Eshuis *et al.*, 1977).

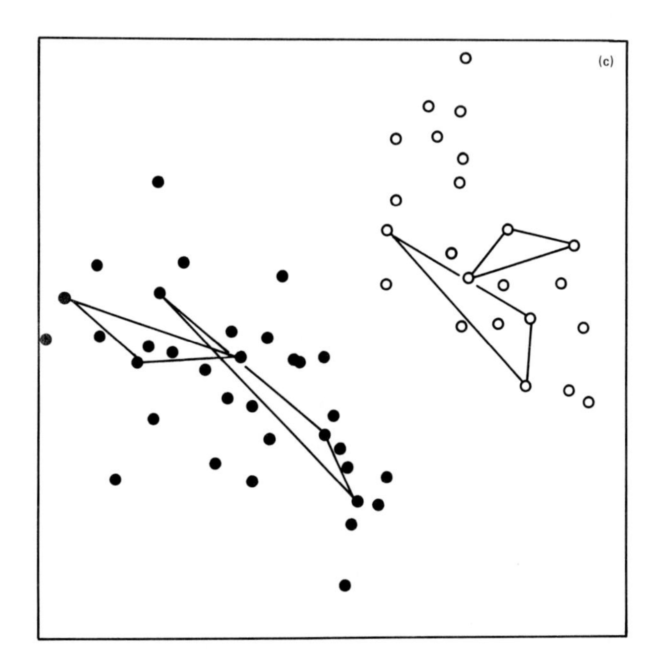

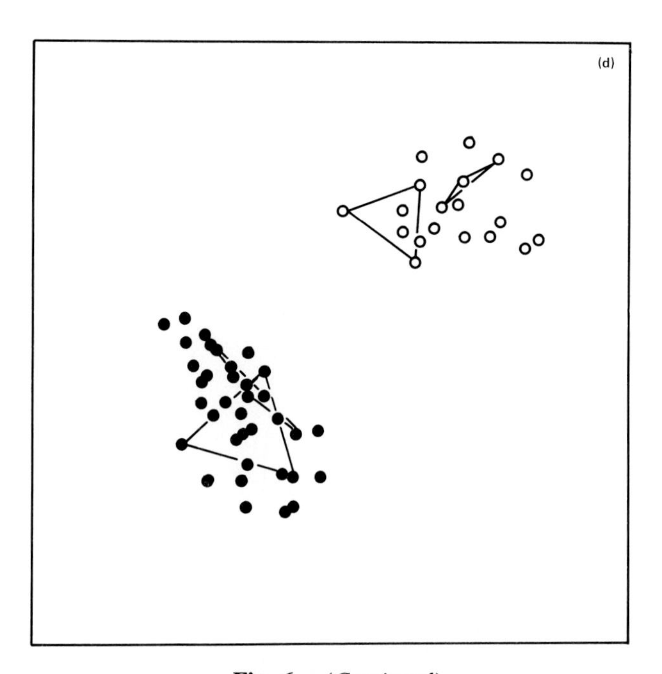

Fig. 6. (*Continued*)

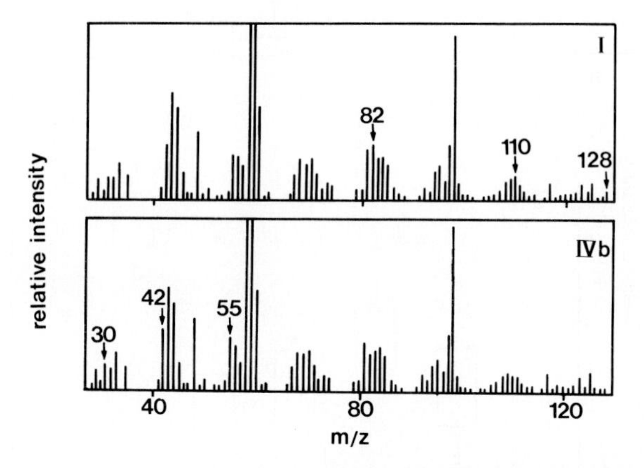

Fig. 7. Averaged pyrolysis mass spectra of two *Listeria* serotypes (I and IVb) (Eshuis *et al.*, 1977). Masses responsible for the discrimination seen in Fig. 6d are arrowed.

The success of the technique is based on the use of replicates to provide calculations of inner variance. This is a key feature of modern Py–ms data-handling strategies.

Exploratory Univariate Analyses

MacFie and Gutteridge (1982) have emphasised the value of data screening using univariate techniques. Although it rarely produces complete discrimination of groups, it is a valuable exercise for the detection of outliers, testing normality, and detecting natural grouping trends in the data. Plotting techniques such as *quantile plots, stem and leaf plots, scatter diagrams,* and *box plots* (MacFie and Gutteridge, 1982) are all capable of revealing useful information about a data set. The box plot technique of Tukey (1977) is particularly elegant, and an example is shown in Fig. 8, which shows known groups of Gram-negative bacteria (Gutteridge and Puckey, 1982).

The box plot is obtained by calculating the lower and upper quartiles and the median M of the data. These are the numbers below which one-fourth, three-fourths, and one-half of the observations fall. The inter-quartile range IQ is the range between the lower and upper quartiles. On the plot the median is denoted by an asterisk, and the IQ is contained in the box. The most extreme values outside the box but within the range M to IQ (vertical lines) are also plotted (+). All readings outside $M \pm$ IQ are plotted as circles; these are filled for points more extreme than $M \pm 1.5$ IQ. Careful inspection of Fig. 8 brings out many useful features of the box plot. The differences between groups 1, 6, 7 and 2, 3, 4, 5 are emphasised, as are the outliers in groups 4 and 5. In addition there is evidence of asymmetry in group 3 (median near top of box).

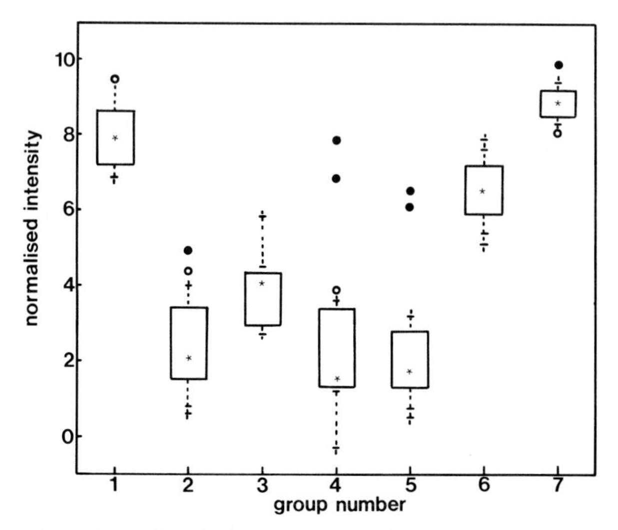

Fig. 8. Box plots of the distribution of seven Gram-negative bacterial groups for the normalised observations of m/z 185. (MacFie and Gutteridge, 1982).

Univariate techniques can also be of use when there is prior knowledge of the chemical differences between samples. For example, the spectrum shown in Fig. 9 is of colominic acid (Wieten, 1983), the purified form of the *Escherichia coli* K1 antigen. A scatter plot of the normalised intensities of two key masses (m/z 109 and 67) for a number of strains shows (Fig 10) that these two features can be used to discriminate $K1^+$ and $K1^-$ strains. It must be stressed that the ability to discriminate using two ions is unusual; the dissimilarity in this case reflects the presence or absence of the key polysaccharide in the bacterial cell.

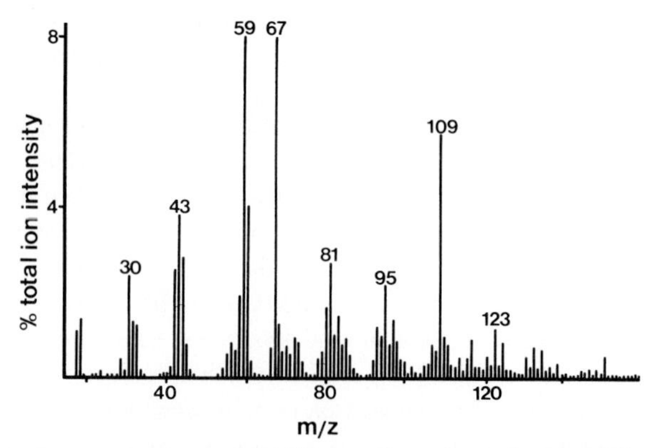

Fig. 9. Pyrolysis mass spectrum of the K1 capsular polysaccharide of *E. coli* (Wieten *et al.*, 1983).

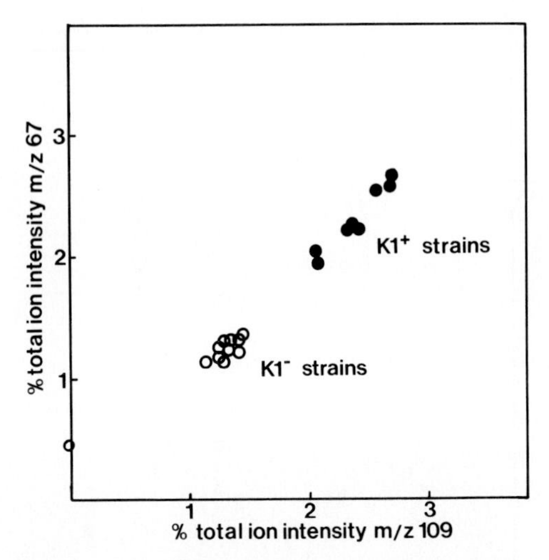

Fig. 10. Scatter diagram of ion intensities at *m/z* 67 and 109 in the pyrolysis mass spectra of a series of 18 *E. coli* strains (Wieten *et al.*, 1983).

Exploratory Multivariate Analysis

The rationale for using multivariate analysis is as follows: suppose M mass intensities are measured on each spectrum. If $M = 2$ then a simple way of representing the observed variation is to plot feature 1 against feature 2 for every sample. With $M > 3$ it is still conceptually convenient to consider each sample as a point in space with M dimensions (M space), but two- or three-dimensional scatter diagrams are selected that give the best approximations to the M-dimensional scatter.

Two methods have been used for obtaining these scatter diagrams. In principal components analysis (Blackith and Reyment, 1971; Marriott, 1974) new axes are found that approximate most closely the observed variation among the samples. The use of this technique for the analysis of bacterial Py–gc data was illustrated by Gutteridge *et al.* (1979). The second technique is multidimensional scaling, sometimes termed principal coordinates analysis (Gower, 1966), and operates on the matrix of inter-distances between the points. Nonlinear mapping (Kruskal, 1964a,b) is a related method which operates on the rankings of the inter-distances. In our experience these methods usually give very similar results, although Eshuis *et al.* (1977) have preferred nonlinear mapping.

The function of these techniques, as applied to Py–ms data, is to reveal natural grouping tendencies and to detect sample outliers. An example of this is shown in Figs. 11 and 12. Figure 11 is a principal components analysis of seven groups of Gram-negative bacteria (Gutteridge and Puckey, 1982). Four of the group 2

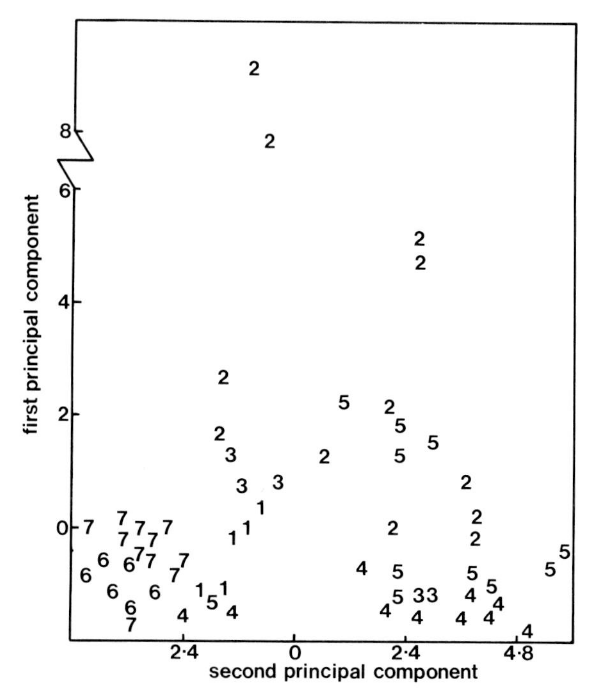

Fig. 11. Two-dimensional principal components plot of strains of seven Gram-negative groups (plotted using group numbers) (MacFie and Gutteridge, 1982).

strains are obvious outliers in the top half of the plot. If these outliers are removed and the analysis is recomputed (Fig. 12), the group 5 samples are separated. Clustering of other groups (e.g., 4) is now also apparent. This sequential approach to principal components analysis of multigroup data has been applied in numerical taxonomy by Bergan and Starr (1981).

The detection (and elimination) of outliers is an important operation in Py–ms data handling. Shute *et al.* (1984) have shown how outliers, presumed to be slow-growing strains, can affect the discrimination of four *Bacillus* species. Obviously care has to be taken not to eliminate outliers that are due to genuine taxonomic variation, but the analytical complexity of Py–ms is such that a small number of poorly reproduced analyses are present in most data sets (Gutteridge *et al.*, 1984).

Classification

Many authors have compared Py–gc traces using calculations of similarity (Stack *et al.*, 1977) or dissimilarity (Seviour *et al.*, 1974). Following the calculation of a similarity matrix, cluster analysis can be used to produce groupings. However, in our experience, cluster analysis works poorly with Py–ms data,

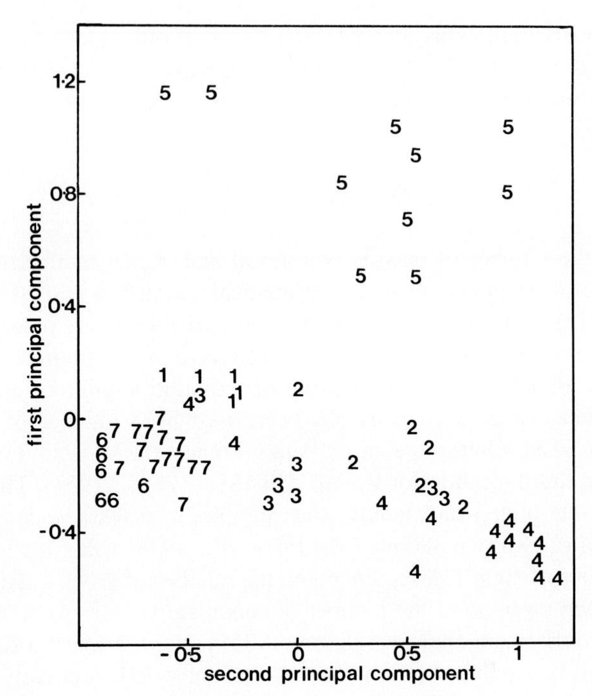

Fig. 12. Reanalysis of data used in Fig. 11 minus four group 2 outliers (Macfie and Gutteridge, 1982).

probably because of the complex correlations and inherent 'noise'. For example, when cluster analysis was applied to pyrolysis data for seven groups of Gram-negative bacteria (MacFie and Gutteridge, 1982), *Acinetobacter calcoaceticus* and *Serratia liquefaciens* were combined and certain replicates did not cluster satisfactorily. These problems were not apparent using discriminant analysis.

Other workers have had more success with cluster analysis. Magee *et al.* (1983) applied the technique successfully to Py–gc data of *Staphylococcus* species. However, considerable attention was paid to the selection of pyrogram peaks, and only those with good reproducibility characteristics were used. Thus, the Euclidean distance calculations were based on 'cleaned' data. The application of conventional numerical taxonomic methods to Py–ms data has not received much attention but, given a suitable detection system for outliers, there seems little doubt that a usable similarity coefficient could be developed.

Discrimination

Discriminant techniques have found wide and successful application to Py–gc and Py–ms data analysis. The basic linear discriminant analysis problem is to find a linear combination of the variables that best differentiates an established (*a priori*) group structure. This may be extended to K orthogonal linear combinations where $K < Ng - 1$ (Ng, number of groups). For a statistically valid analysis the following numerical constraint has to be complied with (Dixon, 1975):

$$N_v < N_s - N_g - 1$$

where N_v is the number of masses (variables) and N_s the number of samples.

One form of discriminant analysis is canonical variates analysis (CVA) (MacFie *et al.*, 1978), in which successive axes are derived that are independent and seek to maximise the ratio of variation between the *a priori* groups to that within these groups, assuming that the structure of variation within the groups is the same. Canonical variates analysis has been used by a number of authors to analyse Py–gc data (Gutteridge *et al.*, 1980; French *et al.*, 1980; O'Donnell *et al.*, 1980) and more recently for Py–ms data (Shute *et al.*, 1984). The results of CVA can be interpreted statistically, for example, to reach decisions about the discrimination of two populations (MacFie *et al.*, 1978) using a χ^2 procedure. The calculations of significance depend on the number of groups, the number of samples in each group, and the number of canonical variates to be used.

Stepwise discriminant analysis (Dixon, 1975) is used to select a subset of the masses that gives a stable solution. This procedure works iteratively, including masses in the subset one at a time on the basis of maximising the ratio of between-group to within-group variation. Instead of axes, discriminant (classification) functions are calculated which permit samples to be classified in one of the groups. Alternatively a generalised distance (Mahalanobis, 1936) of the samples from each of the group means can be obtained. Recording the number of misclassified samples gives the apparent error rate.

The stability of the solution is monitored at each stage by calculating the 'jacknifed' (leaving one out) error rate. This is accomplished by removing each sample in turn, recalculating the discriminant functions, and allocating the sample. Lachenbruch and Mickey (1968) show that, given normality and equality of within-group variation, the jacknifed error rate should be an unbiased estimate of the true error rate of classification. Minimisation of the error rates can be used as an alternative criterion for the selection of variables (Habbema and Hermans, 1977).

Stepwise discriminant analysis has been applied to Py–gc data (Emswiler and Kotula, 1978; O'Donnell *et al.*, 1980) and to Py–ms data (Gutteridge and Puckey, 1982; Shute *et al.*, 1984). It is a useful technique for evaluating the stability of a classification, but its uses are limited in multigroup situations by the problem of finding a small, stable subset of the masses that achieves discrimination. With most sets of Py–ms data this is difficult to satisfy. One advantage of the stepwise approach to data analysis is that most programs produce functions for the identification of unknowns at each step, so that unknowns can be analysed as a 'test set' alongside a 'training set' of recognised strains.

SIMCA (Wold and Sjostrom, 1977) is a specialised but powerful program and approach to the discrimination of *a priori* grouped data. A principal components analysis is applied to each group (Wold, 1976). By a process of removing and inserting elements (*cross-validation*) (Wold, 1978), the number of components that account for systematic, as opposed to random, variation within a group are determined. The class residual variances, found by summing the squares of residuals, are used to define the 'distances' of each sample from its 'class model' using the F distribution. SIMCA thus forms hypervolumes for each of the groups, and each sample is classified as belonging to its nearest group if its distance is within the normal range. The contribution of each variable to the definition of each class model (modelling power) can be assessed by calculating distances for each variable separately before and after fitting the class models. This should, in practice, enable 'good modelling' variates to be selected.

SIMCA has a number of theoretical advantages over the stepwise discriminant–canonical variates approach. As each class is modelled separately, there is no need to assume a common within-class variance–covariance matrix. The estimation of systematic components is an ingenious way of eliminating 'noise' which has no parallel in conventional discriminant analysis. A practical disadvantage is the large number of observations required to establish a stable class model.

As yet SIMCA has not found a wide application to the analysis of pyrolysis data. Blomquist *et al.* (1979a,b,c) applied SIMCA to the analysis of Py–gc data on *Penicillium* and showed how the technique could be used to filter out random variation between pyrograms. MacFie and Gutteridge (1982) and Meuzelaar (1982) applied SIMCA to Py–ms data with some success, but the method has not yet received the attention that it merits.

Identification and Operational Fingerprinting

The stated aim of most research workers applying pyrolysis techniques to the characterisation of micro-organisms is to develop systems for rapid identification. However, in practice, few studies have progressed to the stage where new isolates are compared, by whatever method, with an established library. MacFie and Gutteridge (1982) drew a distinction between 'confirmatory identification' and 'diagnostic identification'. The former is used to test the stability of discriminant functions, SIMCA class models, and so on, and is usually carried out either by jacknifing the samples in a data set (i.e., treating each one as an unknown using a leaving-one-out procedure) or by creating a training set and a test set at the time of establishing the data set.

Diagnostic identification is the process of identifying an unknown sample using an established data base. A very important distinction between this application and confirmatory identification is that the routine diagnostic user is not interested in which, or how many, of the components of a spectrum are used in the identification procedure. Thus, in addition to the discriminant approaches already described—all of which lead to simplified identification strategies—there is also a possibility of maintaining a library of spectra from each of the groups for matching with unknowns.

MacFie and Gutteridge (1982) compared four diagnostic identification methods—classification keys, classification functions, SIMCA regressions, and spectrum matching—for 28 unknown Gram-negative bacteria against a data set containing seven groups. Matching using the full spectrum was the most successful method, giving 27 of 28 and 25 of 28 correct identifications on two separate occasions. None of the other methods produced an acceptable number of successful matches.

In contrast, Magee *et al.* (1983) used discriminant functions of Py–gc data to identify 100 isolates of *Staphylococcus*. The method produced a 90% success rate when the identifications were compared to those obtained using conventional bacteriological methods.

The use of Py–ms for diagnostic identification over long periods of time obviously requires stability of the analytical instrumentation reflected by the

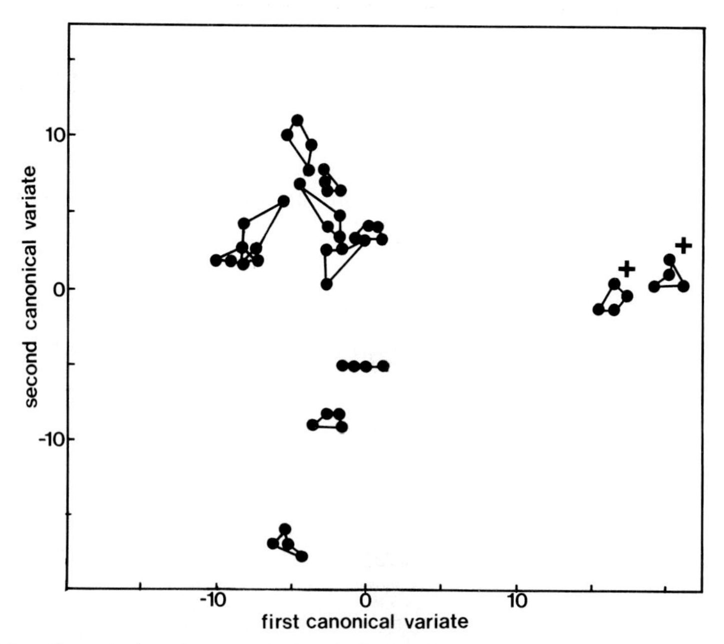

Fig. 13. Canonical variates analysis of 12 *Bacillus thuringiensis* strains showing the differentiation of two industrially important strains (+).

reproducibility of the spectra. Information on long-term effects is not yet available for microbiological data bases, although acceptable levels of reproducibility have been obtained for model compounds over periods of up to as long as 6 months (Meuzelaar, 1978; Windig *et al.*, 1979). An alternative approach to the identification of unknowns which eliminates any potential problems with long-term stability is known as *operational fingerprinting* (Meuzelaar *et al.*, 1982), that is, the analysis of batches of strains with the inclusion of selected reference strains. This approach was used by Wieten *et al.* (1981) for the analysis of data from 91 strains of mycobacteria. The purpose was to identify strains as either belonging to the 'tuberculosis complex' (*Mycobacterium tuberculosis, M. bovis, M. bovis* BCG) or not. Reference strains were included with each batch, and the identifications were based on 10 characteristic masses. A 92% positive correlation was obtained with conventional identification techniques (2.2% false negatives and 5.5% false positives).

An example of the potential for the operational fingerprinting approach to characterisation is shown in Figs. 13 and 14, which are canonical variates analyses of two separate Py–ms studies of 12 strains of the insect pathogen (and possible biological control agent) *Bacillus thuringiensis*. In both studies the production strains of industrial importance are differentiated (despite a change in

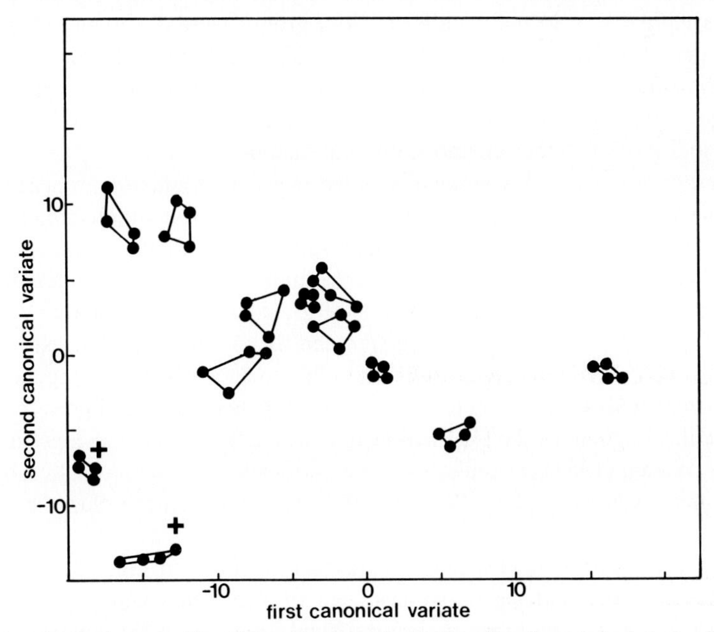

Fig. 14. Analysis as in 13, of data obtained 2 months after the original study. The two industrially important strains are still differentiated despite a change in the orientation of the plot.

orientation between the two figures). The significance of this example is that it shows the potential use of Py–ms as a rapid quality control method in cases where samples of whole micro-organisms are returned from the field after a negative or unusual result. The returned strains could be analysed rapidly in a batch with reference cultures of the production strains.

Taxonomic Considerations

Role of a Priori *Analysis*

All conventional discriminant analysis procedures require the establishment of *a priori* groups. Thus, in most applications of discriminant analysis to the characterisation of micro-organisms using pyrolysis data, prior knowledge of the identities of the strains under examination is assumed. Pyrolysis techniques have therefore been used mainly to test hypotheses about taxonomies and to examine the relationship of pyrolysis data to conventional microbiological data, and not for the establishment of novel taxonomies. For example, O'Donnell *et al.* (1980) and Shute *et al.* (1984) applied Py–gc and Py–ms, respectively, to the so-called *Bacillus subtilis* spectrum (i.e., *B. subtilis, B. pumilus, B. licheniformis,* and '*B. amyloliquefaciens*'). In both studies all four groups were distinguished, providing evidence for the existence of four species and for the distinction of '*B. amyloliquefaciens*' from *B. subtilis* in particular. These studies also showed up the high degree of correlation between the results obtained with pyrolysis techniques and those of other chemotaxonomic methods.

To use pyrolysis and discriminant analysis to establish a taxonomy, each strain would need to be treated as a separate group. One way of accomplishing this is to use the replicate analyses of each strain to form a group (usually of three or four). Since in a typical study this will increase the number of groups from less than 10 to more than 50, consideration has also to be given to the data reduction stage, (i) to satisfy the $N_v < N_s - N_g - 1$ equation and (ii) to reduce the intrinsic dimensionality of the data. An elegant solution is first to do a principal components (or coordinates) analysis and then to use the components (which are linear combinations of the original variables) in discriminant analysis. This concept has been used by Wieten (1983) in studies on the classification and identification by Py–ms of mycobacteria, and by Windig (1982) in studies on the classification of yeasts.

An example of this data-handling approach is shown in Fig. 15, which is a canonical variates analysis of Py–ms data of sporulated and non-sporulated strains of *B. licheniformis* using seven principal coordinates. The analysis, which uses only a knowledge of the replicates as *a priori* groups, distinguishes completely the two physiological states. With more complex examples several can-

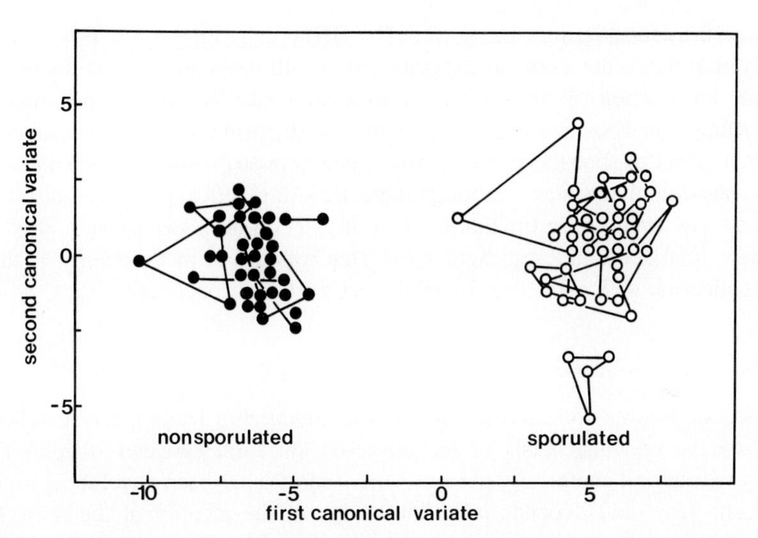

Fig. 15. Canonical variates analysis showing differentiation of sporulated and non-sporulated forms of *Bacillus licheniformis*.

onical variates would have to be examined to assess whether groups of strains have been discriminated. One way of simplifying this assessment is to transform the Mahalanobis distance matrix into a similarity matrix giving the similarity between the means of each of the groups of replicates. Cluster analysis can then be used to construct a dendrogram as in Fig. 16, which again shows the discrimination of the two physiological states.

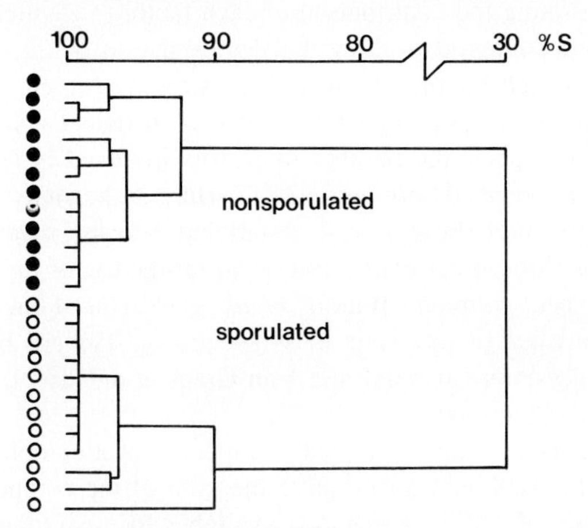

Fig. 16. Dendrogram obtained using average-linkage cluster analysis applied to similarities calculated from Mahalanobis distances and showing clustering of sporulated and nonsporulated forms of *Bacillus licheniformis*.

It is worth considering closely what Fig. 16 actually represents. Firstly, all of the available data are used for the principal co-ordinates analysis, so there is no *ad hoc* elimination of mass intensity measurements. Secondly, the power of discriminant analysis is used in combination with cluster analysis to produce the differentiation of sporulated and nonsporulated classes without prior knowledge of the physiological states. Although there are still practical problems to solve—such as how to optimise the number of principal co-ordinates to use—this data analysis approach could represent a real step forward in methodology as far as the application of Py–ms to microbial systematics is concerned.

Chemical Interpretation

The use of Py–ms for classification studies invariably leads the researcher to question the chemical basis of the observed variation between samples. The ability to interpret chemically pyrolysis mass spectra has been developing rapidly in recent years and depends on a knowledge of the identity of the masses as uncovered by conventional mass spectrometric techniques.

The simplest way of examining the chemical differences between two pyrolysis mass spectra is to use a spectral subtraction procedure (Meuzelaar *et al.*, 1982). A different approach is necessary when two classes of samples are to be compared and the differences between the classes are caused by multiple components. Using so-called factor analysis techniques, the contribution of each of the different components (factors) to each of the different sample classes can be determined (Burgard *et al.*, 1977a,b), provided the number of different spectra available is several times larger than the number of components involved. Apart from determining the contribution of each factor to each class of spectra, factor analysis procedures also allow the determination of the contribution of each mass peak to each factor. The resulting factor spectra can be regarded as characteristic of the corresponding components. If sufficiently large numbers of observations are available, the number of factors involved is fairly low (e.g., <10), and the components do not mutually interfere in the analytical procedure, the factor spectra should show a close resemblance to the spectra of the individual compounds (Meuzelaar *et al.*, 1982). So far the use of factor analysis for Py–ms data has been limited. Windig *et al.* (1980) used the technique for qualitative comparisons of pyrolysis mass spectra of standard biopolymers on changing the pyrolysis parameters, and Van Graas *et al.* (1979) have used the technique for studies on coals.

If a factor describes a single chemical component of a complex mixture, the factor score can be used as a quantitative measure of the compound, although there are a number of different methods available for quantitative analysis of mixtures (Vallis *et al.*, 1983).

The interpretation of the Py–ms spectra of whole micro-organisms can be considered to be a very complex mixture analysis. Because of the number of different biopolymers in the whole cell, factor spectra often represent correlated changes in a number of chemical components and the situation is complicated further by the normalisation of the spectra. Several specialised techniques are available to cope with this—for example, target transformation factor analysis (Malinowski and McCue, 1977)—but the special problems of Py–ms data analysis prompted the development by Windig *et al.* (1982a, 1983) of an empirical graphical rotation technique for interpreting pyrolysis mass spectra. An example of the use of this technique is shown in Fig. 17a–e (Meuzelaar *et al.*, 1982, based on Windig *et al.*, 1982b), which represents the analysis of a small set of pyrolysis mass spectra of yeast species.

The spectra of these yeasts (Fig. 17a) are composed mainly of series of fragment peaks indicative of neutral hexose and pentose-type carbohydrates, *N*-acetylamino sugars, and proteins, which are building blocks present in many different homo- and heterobiopolymers of the organisms. The first, unrotated

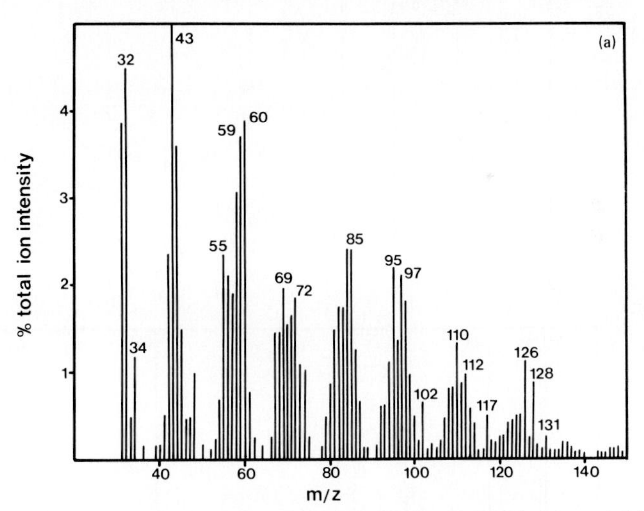

Fig. 17. (a) Pyrolysis mass spectrum of the yeast *Rhodosporidium toruloides* showing fragment peaks attributable to proteins (*m/z* 34, 48, 69, 83, 92, 94, 108, 117, 131), hexoses (*m/z* 55, 58, 68, 72, 74, 82, 84, 85, 96, 98, 102, 110, 112, 126, 144), pentoses (*m/z* 55, 58, 60, 68, 70, 72, 82, 84, 85, 86, 96, 98, 114), and *N*-acetylamino sugars (*m/z* 59, 73, 83, 97, 109, 123, 125, 137). (b, p. 394) First factor (unrotated); the positive part represents a protein subpattern and the negative part a mixed pattern of mainly carbohydrate fragment peaks. (c, p. 394) Negative part of the first factor after rotation of the feature space over 60°, showing a pentose subpattern. (d, p. 395) Positive part of the second factor in the 60° rotation configuration showing the strongly correlated hexose and *N*-acetylhexosamine subpatterns, which cannot be separated further. (e, p. 395) Plot showing the rotated factor scores: *Filobasidium capsuligenum* is relatively rich in pentose components, *Saccharomyces cerevisiae* in protein, and *Rhodosporidium toruloides* in hexoses and *N*-acetylamino sugars. From Meuzelaar *et al.* (1982).

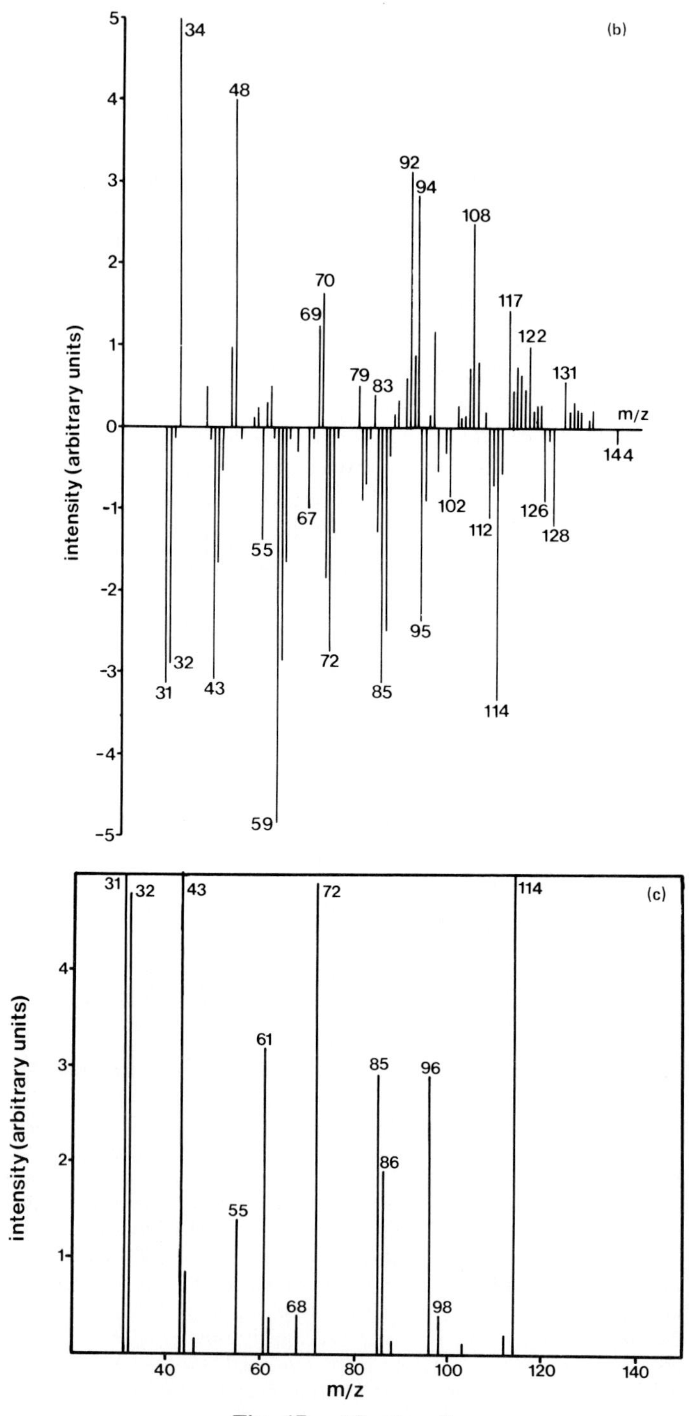

Fig. 17. (*Continued*)

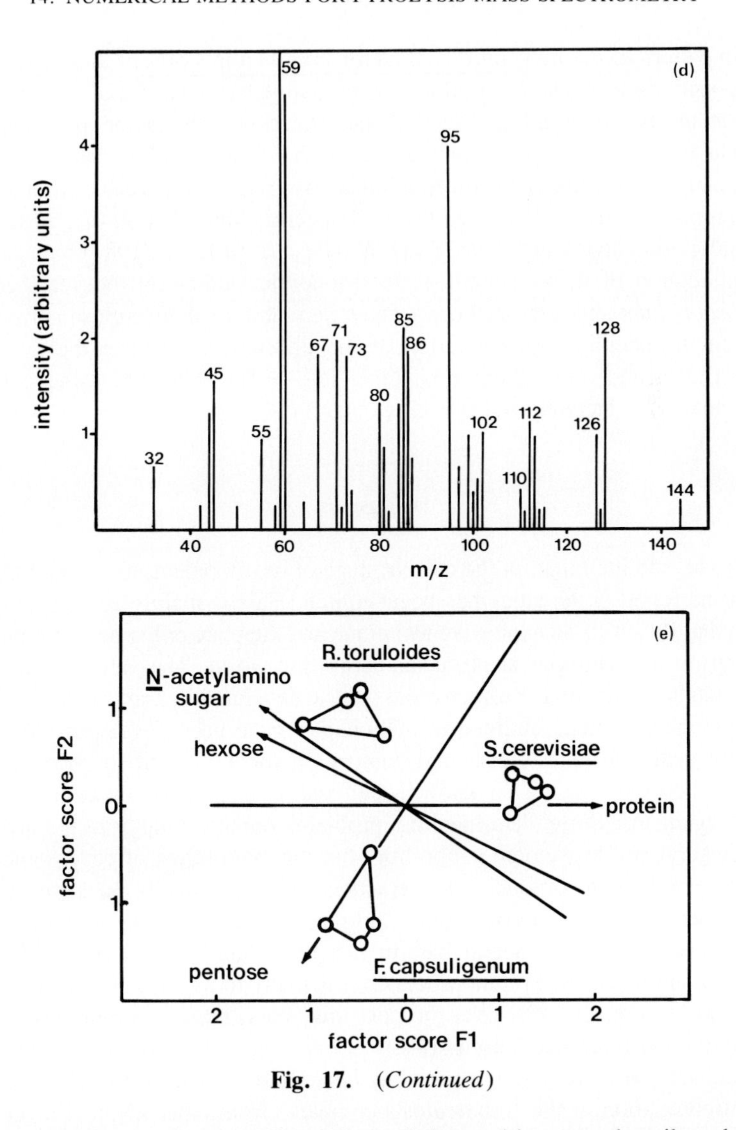

Fig. 17. (*Continued*)

factor calculated is given in Fig. 17b, and the positive part describes the differences in the overall protein subpatterns very well. The negative part of this factor represents a mixed pattern of fragment peaks attributable to a number of other component groups. On rotation in the plane through the two factor axes this pattern changes, until at 60° a set of peaks is observed (Fig. 17c) which represents optimally a fragment pattern of pentose-type carbohydrates. At the same rotation angle the positive part of factor 2 (Fig. 17d) represents optimally the highly correlated (unseparable) hexose and *N*-acetylhexosamine subpatterns, whereas the negative part of this second factor (not shown) now represents the protein pattern of the first, unrotated factor. The factor scores of these optimised 'component factors' can be used as a semiquantitative measure of the chemical components as shown in Fig. 17e. The interpretation of the factor spectra is only possible because of a prior knowledge of the fragmentation patterns of the individual components. This information is now generally available from a carefully prepared and standardised library of spectra (Meuzelaar *et al.*, 1982).

Windig and Haverkamp (1982) and Windig and de Hoog (1982) have shown the application of these techniques to taxonomic studies on the yeast genera *Rhodosporidium* and *Sporidiobolus*. They were able to show a close correlation between the chemical interpretation of the differences between the pyrolysis mass spectra and other chemotaxonomic information obtained using conventional analytical techniques.

Discussion

It has to be admitted that, at the current stage of development, the impact of Py–ms on microbial systematics has been limited. This is mainly for the practical reason that Py–ms is an expensive technique and there are only a few instruments working on microbiological problems across the world. The value of the technique stands or falls on the effectiveness of the data handling, and since 1980 the strategies and methods outlined in this chapter have been developed. Methods are now available to establish and compare taxonomies and to correlate differences between spectra to known chemical differences between classes of micro-organisms. Some data-handling problems remain, mainly in the areas of identification and instrument calibration, but the main thrust of development in the future will be to produce an interactive and user-friendly package of programs operating on a microcomputer, so that the data-handling operations become cheaper and more accessible. In fact the overall capital cost of Py–ms instrumentation, including data processing, is likely to fall from its current level (£80–100K) to under £40K over the next few years, making it more affordable for routine and reference laboratories.

Mycobacterial taxonomy has probably benefitted most from Py–ms research. The differentiation of the 'tuberculosis complex' from other atypical mycobacteria (Wieten *et al.*, 1981) remains a classic illustration of the ability to Py–ms to go rapidly from exploratory studies to a practical and pragmatic identification system. In fact, the data base used for this discrimination is said still to be stable after 3 years, which bodes well for the long-term reproducibility of Py–ms. The method has also been used to study the heterogeneity of *Mycobacterium kansasii* (Wieten *et al.*, 1984) and *Mycobacterium leprae* (Wieten *et al.*, 1982).

The key to the acceptance and use of Py–ms by microbiologists is that it should produce classifications that do not differ substantially from those obtained using other methods and that it should be stable enough to permit routine identification. Thus far the balance of the evidence suggests that the technique does produce recognisable classifications (Windig and Haverkamp, 1982; Windig and de Hoog, 1982). However, because Py–ms examines, albeit indirectly, the total cellular composition of micro-organisms, there may be situations where strains are differentiated irrespective of their taxonomic status. For example, a strain over-producing a secondary metabolite or depositing an intra-cellular storage compound may produce a spectrum dominated by the thermal degradation products of that compound. In fact, the ability of Py–ms to differentiate at the strain

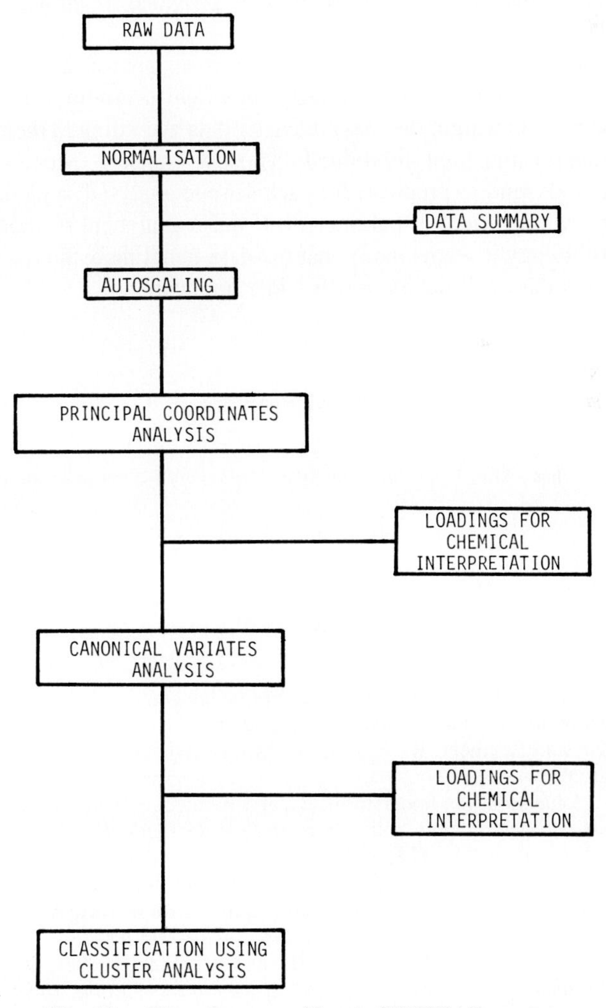

Fig. 18. Flow diagram of batch GENSTAT program.

level is particularly valuable, and one of the prime uses of the method may turn out to be in epidemiology as some initial studies on hospital isolates of *Klebsiella* (Meuzelaar *et al.*, 1982) have indicated. No work on the important problem of the transfer of data bases between instruments has yet been carried out.

The data-handling strategies outlined in this article may seem complex and long-winded, but most of the important computations can be carried out within a single programme. Figure 18 is a flow diagram of our current GENSTAT batch programme, which covers the preprocessing, data reduction, multivariate exploratory analyses, classification, and discrimination phases outlined in Fig. 4. In addition, 'factor analysis'-type information is provided as an aid to chemical interpretation.

In conclusion, handling Py–ms data requires an approach which is rather different from that used for conventional numerical taxonomy. The main difference is the need to weight the mass intensity data according to their usefulness for discrimination and their reproducibility (low 'noise'). Modern analytical chemical methods tend to produce, for each sample analysed, a plethora of data representing a wide variety of parameters. If these analytical methods are to be used in microbiology it seems likely that the data-handling techniques described here for Py–ms data will have a much wider application.

Acknowledgement

We would like to thank Miss L. A. Shute of Bristol University for the provision of the data on *Bacillus*.

References

Bergan, T., and Starr, M. P. (1981). Sequential principal components analysis, a tool for cluster detection in large bacteriophage-typing samples. *Current Microbiology* **6,** 1–6.

Blackith, R. E., and Reyment, R. A. (1971). 'Multivariate Morphometrics'. Academic Press, London.

Blomquist, G., Johansson, E., Soderstrom, B., and Wold, S. (1979a). Reproducibility of pyrolysis gas-chromatographic analyses of the mould *Penicillium brevi-compactum*. *Journal of Chromatography* **173,** 7–17.

Blomquist, G., Johansson, E., Soderstrom, B., and Wold, S. (1979b). Classification of fungi by means of pyrolysis–gas chromatography—pattern recognition. *Journal of Chromatography* **173,** 19–32.

Blomquist, G., Johansson, E., Soderstrom, B., and Wold, S. (1979c). Data analysis of pyrolysis-chromatograms by means of SIMCA pattern recognition. *Journal of Analytical and Applied Pyrolysis* **1,** 53–65.

Burgard, D. R., Perone, S. P., and Wiebers, J. L. (1977a). Sequence analysis of oligodeoxy-ribonucleotides by mass spectrometry 2. Application of computerised pattern recognition to sequence determination of di-, tri-, and tetranucleotides. *Biochemistry* **16,** 1051–1057.

Burgard, D. R., Perone, S. P., and Wiebers, J. L. (1977b). Factor analysis of the mass spectra of oligodeoxyribonucleotides. *Analytical Chemistry* **49,** 1444–1446.

Carmichael, J. W., Sekhon, A. S., and Sigler, L. (1973). Classification of some dermatophytes by pyrolysis gas–liquid chromatography. *Canadian Journal of Microbiology* **19,** 403–407.

Dixon, W. J. (1975). 'Biomedical Computer Programs'. Univ. of California Press, Los Angeles.

Emswiler, B. S., and Kotula, A. W. (1978). Differentiation of *Salmonella* serotypes by pyrolysis gas–liquid chromatography. *Applied and Environmental Microbiology* **35,** 97–104.

Eshuis, W., Kistemaker, P. G., and Meuzelaar, H. L. C. (1977). Some numerical aspects of reproducibility and specificity. *In* 'Analytical Pyrolysis', (Eds. C. E. R. Jones and C. A. Cramers), pp. 151–156. Elsevier, Amsterdam.

French, G. L., Gutteridge, C. S., and Phillips, I. (1980). Pyrolysis gas chromatography of *Pseudomonas* and *Acinetobacter* species. *Journal of Applied Bacteriology* **49,** 505–516.

Gower, J. C. (1966). Some distance properties of latent root and vector methods used in multivariate analysis. *Biometrika* **53,** 325–338.

Gutteridge, C. S., and Puckey, D. J. (1982). Discrimination of some Gram-negative bacteria by direct probe mass spectrometry. *Journal of General Microbiology* **128,** 721–730.

Gutteridge, C. S., MacFie, H. J. H., and Norris, J. R. (1979). Use of principal components analysis for displaying variation between pyrograms of micro-organisms. *Journal of Analytical and Applied Pyrolysis* **1,** 67–76.

Gutteridge, C. S., Mackey, B. M., and Norris, J. R. (1980). A pyrolysis gas–liquid chromatography study of *Clostridium botulinum* and related organisms. *Journal of Applied Bacteriology* **49,** 165–174.

Gutteridge, C. S., Sweatman, A. J., and Norris, J. R. (1984). Potential applications of Curie-point pyrolysis mass spectrometry with emphasis on food science. *In* 'Analytical Pyrolysis Techniques and Applications' (*Ed.* K. J. Voorhees), pp. 324–348. Butterworths, London.

Habbema, J. D. F., and Hermans, J. (1977). Selection of variables in discriminant analysis by F-statistic and error rate. *Technometrics* **19,** 487–493.

Harper, A. M., Duewer, D. L., and Kowalski, B. R. (1977). ARTHUR and experimental data analysis: the heuristic use of a polyalgorithm. *In* 'Chemometrics: Theory and Practice' (Ed. B. R. Kowalski), American Chemical Society Symposium Series No. 52, pp. 14–52. American Chemical Society, Washington, D.C.

Huff, S. M., Meuzelaar, H. L. C., Pope, D. L., and Kjeldsberg, C. R. (1981). Characterisation of leukemic and normal white blood cells by Curie-point pyrolysis mass spectrometry. 1. Numerical evaluation of the results of a pilot study. *Journal of Analytical and Applied Pyrolysis* **3,** 95–110.

Irwin, W. J. (1982). 'Analytical Pyrolysis; A Comprehensive Guide'. Dekker, New York.

Klee, M. S., Harper, A. M., and Rogers, L. B. (1981). Effects of normalisation on feature selection in pyrolysis gas chromatography of coal tar pitches. *Analytical Chemistry* **53,** 801–805.

Kowalski, B. R. (1975). Measurement analysis by pattern recognition. *Analytical Chemistry* **47,** 1152A–1162A.

Kruskal, J. B. (1964a). Multidimensional scaling by optimising goodness of fit to a nonmetric hypothesis. *Psychometrika* **29,** 1–27.

Kruskal, J. B. (1964b). A numerical method. *Psychometrika* **29,** 115–129.
Lachenbruch, P. A., and Mickey, M. R. (1968). Estimation of error rates in discriminant analysis. *Technometrics* **10,** 1–11.
MacFie, H. J. H., Gutteridge, C. S., and Norris, J. R. (1978). Use of canonical variates analysis in differentiation of bacteria by pyrolysis gas–liquid chromatography. *Journal of General Microbiology* **104,** 67–74.
MacFie, H. J. H., and Gutteridge, C. S. (1982). Comparative studies on some methods for handling quantitative data generated by analytical pyrolysis. *Journal of Analytical and Applied Pyrolysis* **4,** 175–204.
Magee, J. T., Hindmarch, J. M., and Meechan, D. F. (1983). Identification of staphylococci by pyrolysis gas–liquid chromatography. *Journal of Medical Microbiology* **16,** 483–495.
Mahalanobis, P. C. (1936). On the generalized distance in statistics. *Proceedings of the National Institute of Sciences of India* **2,** 49–55.
Malinowski, E. R., and McCue, M. (1977). Qualitative and quantitative determination of suspected components in mixtures by target transformation factor analysis of their mass spectra. *Analytical Chemistry* **49,** 284–287.
Marriott, F. H. C. (1974). 'The Interpretation of Multiple Observations'. Academic Press, London.
Meuzelaar, H. L. C. (1978). Pyrolysis mass specrometry; prospects for inter-laboratory standardisation. *In* 'Proceedings of the 26th American Society for Mass Spectrometry Conference on Mass Spectrometry and Allied Topics', pp. 29–41. American Society for Mass Spectrometry, St. Louis, Missouri.
Meuzelaar, H. L. C. (1982). 'Characterisation of Rocky Mountain Coals and Coal Liquids by Computerized Analytical Techniques', Report 30242-T4. U.S. Department of Energy, Washington, D.C.
Meuzelaar, H. L. C., Haverkamp, J., and Hileman, F. D. (1982). 'Pyrolysis Mass Spectrometry of Recent and Fossil Biomaterials; Compendium and Atlas'. Elsevier, Amsterdam.
Nelder, J. A. (1979). 'Genstat Reference Manual'. Scientific and Social Service Program Library, Univ. of Edinburgh, Edinburgh.
Nie, N. H., Hull, C. H., Jenkins, J. G., Steinbrenner, K., and Bent, D. H. (1975). 'Statistical Package for the Social Sciences (SPSS)', 2nd Edition. McGraw-Hill, New York.
O'Donnell, A. G., MacFie, H. J. H., and Norris, J. R. (1980). An investigation of the relationships between *B. cereus, B. thuringiensis* and *B. mycoides* using pyrolysis gas–liquid chromatography. *Journal of General Microbiology* **32,** 306–309.
Sekhon, A. S., and Carmichael, J. W. (1972). Pyrolysis gas–liquid chromatography of some dermatophytes. *Canadian Journal of Microbiology* **18,** 1593–1601.
Seviour, R. J., Chilvers, C. A., and Crow, W. D. (1974). Characterisation of eucalypt mycorrhizas by pyrolysis gas chromatography. *New Phytologist* **73,** 321–332.
Shute, L. A., Gutteridge, C. S., Norris, J. R., and Berkeley, R. C. W. (1984). Curie-point pyrolysis mass spectrometry applied to characterisation and identification of selected *Bacillus* species. *Journal of General Microbiology* **130,** 343–355.
Stack, M. V., Donoghue, H. D., Tyler, J. E., and Marshall, M. (1977). Comparison of oral streptococci by pyrolysis gas–liquid chromatography. *In* 'Analytical Pyrolysis' (Eds. C. E. R. Jones and C. A. Cramers), pp. 57–68. Elsevier, Amsterdam.
Tukey, J. W. (1977). 'Exploratory Data Analysis'. Addison-Wesley, Reading, Massachusetts.

Vallis, L. V., MacFie, H. J. H., and Gutteridge, C. S. (1983). Differentiation of simple biochemical mixtures by pyrolysis mass spectrometry: Some geometrical considerations. *Journal of Analytical and Applied Pyrolysis* **5,** 333–348.

Van Graas, G., de Leeuw, J. W., and Schenck, P. A. (1979). Analysis of coals of different rank by Curie-point pyrolysis–mass spectrometry and Curie-point pyrolysis–gas chromatography–mass spectrometry. *In* 'Advances in Organic Geochemistry 1979', (Eds. A. G. Douglas and J. R. Maxwell), pp. 485–494. Pergamon, Oxford.

Wieten, G. (1983). Studies on Classification and Identification of Mycobacteria by Pyrolysis Mass Spectrometry. Ph. D. Thesis, Univ. of Amsterdam, Amsterdam.

Wieten, G., Haverkamp, J., Meuzelaar, H. L. C., Engle, H. B. W., and Berwald, L. G. (1981). Pyrolysis mass spectrometry: a new method to differentiate between the mycobacteria of the 'Tuberculosis' complex and other mycobacteria. *Journal of General Microbiology* **122,** 109–118.

Wieten, G., Haverkamp, J., Berwald, L. G., Groothuis, D. G., and Draper, P. (1982). PMS: its application to mycobacteriology, including *Mycobacterium leprae*. *Annales de Microbiologie (Paris)* **133B,** 15–27.

Wieten, G., Meuzelaar, H. L. C., and Haverkamp, J. (1984). Analytical pyrolysis in clinical and pharmaceutical microbiology. *In* 'GC/MS Applications in Microbiology' (Eds. G. Odham, L. Larsson, and P. A. Mardh), pp. 335–380. Plenum, New York.

Windig, W. (1982). Pyrolysis Mass Spectrometry of Yeasts: A New Tool for Chemical Differentiation. Ph. D. Thesis, Univ. of Amsterdam, Amsterdam.

Windig, W., and Haverkamp, J. (1982). Pyrolysis mass spectrometry of selected yeast species: I. *Rhodosporidium*. *Studies in Mycology* **22,** 56–59, 69–74.

Windig, W., and de Hoog, G. S. (1982). Pyrolysis mass spectrometry of selected yeast species: II. *Sporidiobolus* and relationships. *Studies in Mycology* **22,** 60–64, 69–74.

Windig, W., Kistemaker, P. G., Haverkamp, J., and Meuzelaar, H. L. C. (1979). The effects of sample preparation, pyrolysis and pyrolysate transfer conditions on pyrolysis mass spectra. *Journal of Analytical and Applied Pyrolysis* **1,** 39–52.

Windig, W., Kistemaker, P. G., Haverkamp, J., and Meuzelaar, H. L. C. (1980). Factor analysis on the influence of changes in experimental conditions in pyrolysis mass spectrometry. *Journal of Analytical and Applied Pyrolysis* **12,** 7–18.

Windig, W., Kistemaker, P. G., and Haverkamp, J. (1982a). Chemical interpretation of differences in pyrolysis–mass spectra of simulated mixtures of biopolymers by factor analysis with graphical rotation. *Journal of Analytical and Applied Pyrolysis* **3,** 199–212.

Windig, W., de Hoog, G. S., and Haverkamp, J. (1982b). Chemical characterisation of yeasts and yeast-like fungi by factor analysis of their pyrolysis–mass spectra. *Journal of Analytical and Applied Pyrolysis* **3,** 213.

Windig, W., Haverkamp, J., and Kistemaker, P. G. (1983). Interpretation of sets of pyrolysis mass spectra by discriminant analysis and graphical rotation. *Analytical Chemistry* **55,** 81–88.

Wold, S. (1976). Pattern recognition by means of disjoint principal components models. *Pattern Recognition* **8,** 127–139.

Wold, S. (1978). Cross validatory estimation of the number of components in factor and principal components analysis. *Technometrics* **20,** 397–406.

Wold, S., and Sjostrom, M. (1977). SIMCA: A method for analysing chemical data in terms of similarity and analogy. *In* 'Chemometrics: Theory and Practice' (Ed. B. R. Kowalski), American Chemical Society Symposium Series No. 52, pp. 243–282. American Chemical Society, Washington, D.C.

15

Numerical Analysis of Chemotaxonomic Data

A. G. O'Donnell

Department of Soil Science, The University, Newcastle upon Tyne, UK

Introduction

Chemical methods such as the analysis of bacterial lipids (Minnikin *et al.*, 1979; Collins and Jones, 1982; Collins *et al.*, 1982; O'Donnell *et al.*, 1982a), peptidoglycans (Schleifer and Kandler, 1972; Keddie and Bousfield, 1980; Seidl *et al.*, 1980) and sugars (Lechevalier and Lechevalier, 1970; Aluyi and Drucker, 1983); fermentation end products (Drucker, 1981); polyacrylamide gel electrophoresis (Kersters and De Ley, 1980), and pyrolysis techniques (Gutteridge and Norris, 1979; O'Donnell and Norris, 1981) have contributed to an understanding of the interrelationships between bacteria, particularly in those areas of bacterial systematics where classical procedures have failed to provide reliable characters for the differentiation of taxa. In the majority of such chemotaxonomic studies, data have been interpreted by visual comparison of chemical profiles and this has usually limited interpretation of the results to definitions of taxa at the generic and suprageneric levels. In a number of studies using gas chromatography, however, a more objective, numerical approach has been adopted. The principal concern of this chapter is to review the numerical analysis of bacterial lipids and to outline some of the factors known to affect lipid composition. The numerical analysis of data produced using other chemotaxonomic methods such as polyacrylamide gel electrophoresis and pyrolysis techniques is considered elsewhere (see Chapters 13 and 14).

The volatility required in gas chromatography has meant that most numerical chemotaxonomic studies have used fatty acid profiles for classification. The grouping of organisms according to fatty acid composition has been carried out in several ways. In the majority of cases similarity values were derived using a variety of coefficients (Drucker, 1974; Ikemoto *et al.*, 1978; Sincoweay *et al.*, 1981; Bousfield *et al.*, 1983). In a study of the cellular fatty acids of a number of streptococci belonging to Lancefield groups A, B, C, D, N, and O, Drucker (1974) used three different correlation coefficients, Spearman, Fisher, and Bra-

ISBN 0-12-289665-3

vais-Pearson, to compare test strains with three reference strains representing Lancefield serological groups A, D, and O, and found that for all coefficients the highest similarity value was between duplicate analyses. He also noted that different coefficients gave different results and that of the three correlation measures tested, the Bravais-Pearson coefficient was the best because it enabled the Lancefield serological group to be predicted from the fatty acid profile. Drucker suggested that this was because the Bravais-Pearson coefficient placed greater weight on major than on minor peaks and that the areas of large peaks were more accurately measured than those of small peaks since the latter were subject to a greater proportion of error because of fluctuations in the analysis conditions. The modified Fisher procedure, which gives equal weight to all peaks regardless of size, was less successful as judged by the ability of the result to predict a serological group. This is interesting with regard to classical numerical taxonomy (Sneath, 1957), where equal weighting of biochemical characters is considered the best way of obtaining a 'natural classification'.

The effect of different correlation measures and transformation procedures has also been studied using the fatty acids of 23 strains of Enterobacteriaceae and Vibrionaceae (Bøe and Gjerde, 1980). In this study, the peak areas were standardised relative to hexadecanoate rather than to the total fatty acid composition. By doing so inter-class discrimination is favoured whilst the effect on intra-class relationships is minimal (Bøe and Gjerde, 1980). The reproducibility of the fatty acid patterns was examined using 10 subcultures of *Vibrio anguillarum* FSK14. Although the overall pattern was reproducible, results varied for individual fatty acid methyl esters (FAMEs) with relative peak areas less than 10%, a finding in agreement with previous studies (Drucker, 1974). For the purposes of classification, the fatty acid data were transformed using three different procedures which, together with the raw data (not transformed), gave four data sets. Each data set was then analysed using two similarity measures, the correlation coefficient (Sokal and Michener, 1958) and a coefficient based on Euclidean distance (Harper *et al.*, 1977), thereby giving a total of eight clustering algorithms. Bøe and Gjerde (1980) found that transformation procedures that gave equal weight to all peaks (autoscaling) had an adverse effect on the analysis and recommended a log transformation of the form $y = \ln(x + 1)$, where x is an element in the matrix and y is the corresponding element in the transformed matrix. Such a transformation affords greater weight to minor chemical constituents in the calculation of correlation coefficients (Jantzen *et al.*, 1974a,b; Drucker, 1981). In addition, Bøe and Gjerde proposed that for cluster analysis of bacterial FAMEs, the raw data should be screened prior to statistical analysis to remove peaks with values less than 10%.

A log transformation procedure and unweighted pair-group cluster analysis was used with success by Jantzen *et al.* (1974b) in studies on the fatty acids of representatives of the Micrococcaceae. Twenty-seven strains representing four species of *Micrococcus* and three species of *Staphylococcus* were examined together with two strains labelled *Micrococcus* sp. and *Staphylococcus* sp. The similarity in fatty acid composition between *Staphylococcus* and *Micrococcus* was less than 70% and both genera were readily separated. The inter-specific similarity amongst *S. aureus, S. epidermidis,* and *S. saprophyticus* was high (>85%S), thereby preventing species discrimination, although the coagulase-negative *S. epidermidis* strains formed two distinct clusters which were also distinguished using traditional biochemical tests. One of the *S. epidermidis*clusters comprised three strains which had a high similarity (>90%S) to *S. saprophyticus* and had been previously named '*S. lactis*' (Shaw *et al.*, 1951). The second cluster corresponded to *S. epidermidis sensu stricto* which was described by Schleifer and Kocur (1973) using biochemical and cultural characteristics, peptidoglycan composition, and teichoic acid type.

Studies on the classification of '*Acinetobacter*', *Moraxella,* and *Neisseria* (Jantzen *et al.*, 1974a, 1975) also demonstrated the value of numerical techniques in the analysis of bacterial fatty acid data. Using a combination of cluster analysis and principal-components analysis Jantzen *et al.* (1975) showed that groups based on fatty acid data were in general agreement with those found using DNA–DNA reassociation (Johnson *et al.*, 1970). Principal components analysis has also been used to evaluate the taxonomic potential of FAMEs in *Bacillus* systematics. A limited study, without replicate analyses (G. Dobson, A. G. O'Donnell, H. J. H. MacFie, D. E. Minnikin, and M. Goodfellow, unpublished data), suggested that several groups of aerobic endospore-forming bacilli could be distinguished (Fig. 1). The result of plotting the loadings on each peak for the first and second principal component is shown in Fig. 2. In this type of plot, the square of the distance between the origin (0,0) and the individual variables represents the amount of variation expressed by that variable in the original principal components plot (Fig. 1). In this study variables 6, 7, and 13, which correspond to 13-methyltetradecanoate (*iso*-15), 12-methyltetradecanoate (*anteiso*-15), and 14-methylhexadecanoate (*anteiso*-17), respectively, account for more of the between-strain variation than do the remaining variables which cluster around the origin. It is interesting that *iso*-15, *ai*-15, and *ai*-17 are major peaks in the *Bacillus* fatty acid profile and usually account for approximately 60 to 70% of the total fatty acid composition.

A new similarity coefficient, the 'overlap' coefficient, has been applied to the fatty acids of 'coryneform' bacteria (Bousfield *et al.*, 1983). This coefficient attempts to mimic the way in which fatty acid profiles might be compared visually. Chromatograms which could be superimposed exactly would be considered completely similar, whereas those showing no overlap would be considered completely dissimilar. The 'overlap' coefficient was compared with the correlation coefficient (Drucker, 1974; Jantzen *et al.*, 1974a,b), and the coefficient based on angular separation of vectors (Drucker, 1974; Ikemoto *et al.*, 1978) on

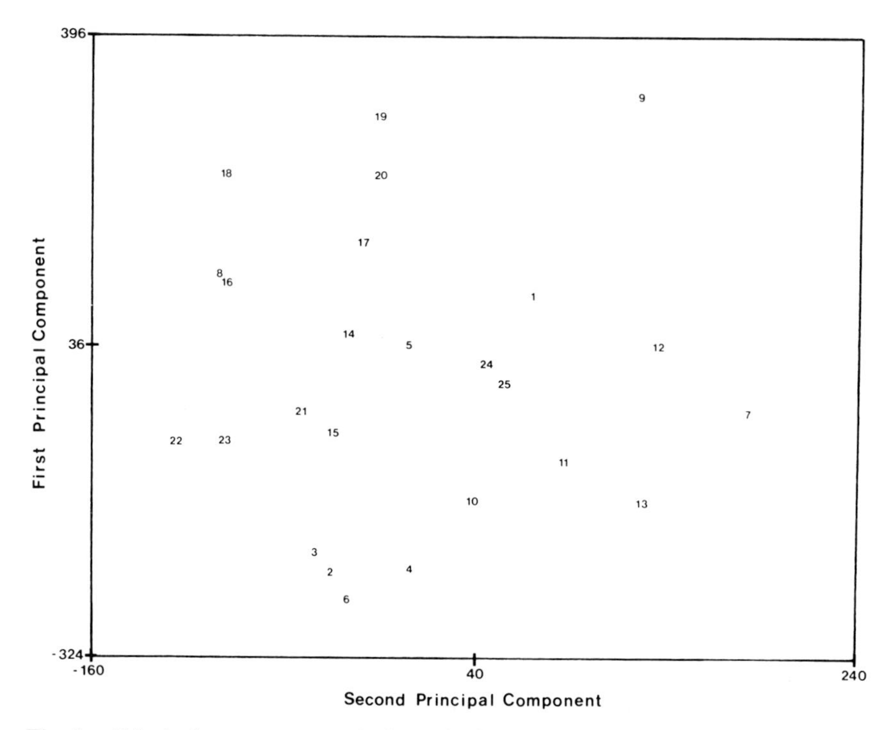

Fig. 1. Principal components analysis on the fatty acid methyl esters of representatives of the genus *Bacillus: B. firmus* (1), '*B. psychrophilus*' (2), *B. lentus* (3), *B. globisporus* (4), *B. insolitus* (5), *B. pantothenticus* (6), *B. laterosporus* (7), *B. badius* (8), *B. sphaericus* (9), *B. brevis* (10), *B. megaterium* (11–12), '*B. carotarum*' (13), *B. licheniformis* (14), *B. subtilis* (15), '*B. amyloliquefaciens*' (16), *B. pumilus* (17), *B. cereus* (18), *B. mycoides* (19), *B. thuringiensis* (20), *B. alvei* (21), *B. polymyxa* (22), *B. macerans* (23); *B. circulans* (24), '*B. macroides*' (25). Plot represents 78% of the between-strain variation.

a hypothetical set of data, which was chosen to test the robustness of the various similarity measures over a wide range of data and to remove the temptation to select a coefficient that would produce a classification predetermined by other criteria. Bousfield *et al.* (1983) found that, although all of the similarity measures tested gave 100%S for a pair of identical profiles, there was a lack of agreement over zero similarity. Of the three coefficients tested, the correlation coefficient was the least satisfactory. In some cases where the 'overlap' and angular separation coefficients gave values of 98%S, which was expected from the data, the correlation coefficient gave a value of 0%S. This occurred when a relatively high peak in one profile corresponded to a relatively low peak in the other. Conversely, when large peaks coincided in a pair of profiles the correlation coefficient returned a similarity value of 100%S, even though the relative

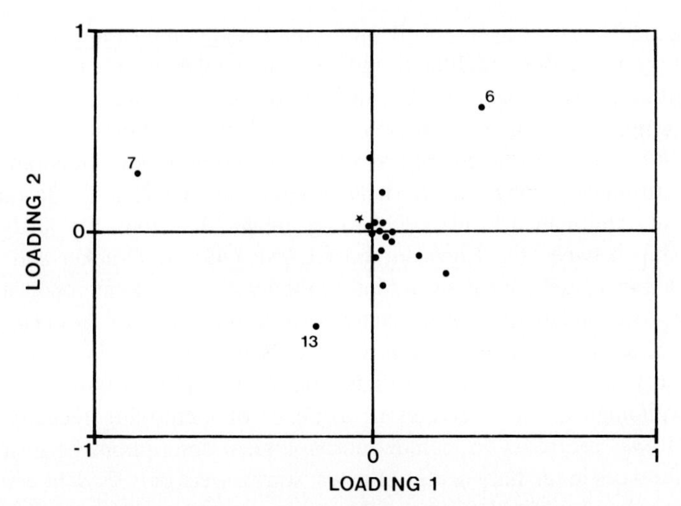

Fig. 2. A plot of the loadings on each variable used to define the first and second principal component axes in Fig. 1. The further a variable is from the origin (0,0), the greater is the effect of that variable on the analysis. This plot shows that peaks 6 (*iso*-15), 7 (*anteiso*-15), and 13 (*anteiso*-17) account for more of the between-strain variation shown in Fig. 1 than the remaining variables. ★, Co-incident points.

areas of the coincident peaks were vastly different and despite the lack of similarity between the two profiles in respect of other peaks. These workers also noted that, although log transformation of the raw data (Jantzen *et al.*, 1974a,b, 1975; Bøe and Gjerde, 1980) would reduce the distorting effect of large peaks, it would not remove it completely. Thus, the comparison of fatty acid profiles using the correlation coefficient remained unreliable. The 'overlap' and angular separation coefficients were in reasonably close agreement except that, with some combinations of test data, the angular separation coefficient gave high values where low values were expected. Nevertheless, despite these discrepancies, the angular separation coefficient and 'overlap' coefficient gave structurally similar clusters. When the three coefficients were tested, without log transformation, on the data of Bøe and Gjerde (1980), similar classifications were obtained but at differing similarity levels. The satisfactory performance of the correlation coefficient in this instance was ascribed to the lack of extreme comparisons in the data set. This implies that the robustness of a classification based on the correlation coefficient depends very much on the spread of the data and makes this coefficient of value only when comparing similar profiles.

Chemotaxonomic data were combined with the results of biochemical and morphological tests in a study by Drucker and Melville (1971, 1973). Two hundred fifty-two strains of the pyogenic, viridans, and enterococcus groups of streptococci were examined for colonial and cellular morphology, resistance to chemical and physical agents, cell wall amino acid composition, and fermentation end products. Both qualitative and quantitative data were analysed (Drucker and Melville, 1973) using the CLASP programme (Gower and Ross, 1969), which compared overall similarity values, and by the Harrison method (Harrison, 1968), which grouped the test strains according to a nonrandom distribution of attributes. Using the Harrison programme, the 252 strains formed 18 clusters at significance levels equal to or greater than four. By analysing the relationship between the Harrison and CLASP outputs, Drucker and Melville (1973) demonstrated a clear separation of the pyogenic and enterococcus groups of streptococci and distinguished groups corresponding to *Streptococcus mitis, S. mutans, S. salivarius,* and *S. sanguis.* A study by Romanovskaya *et al.* (1980) on the fatty acids of methane-oxidizing bacteria combined numerical analysis and morphological characteristics as a means of identifying recently isolated strains. It was necessary to include morphological descriptions because significant differences in the fatty acids of the test strains were only evident between the rod–coccoid and vibrioid groups and between the mesophilic and thermophilic forms. Nevertheless, combining the fatty acid and morphological data enabled Romanovskaya and co-workers to assign 10 recently isolated methylotrophs to six different species.

For chemical profiles to be of value in bacterial classification and identification, it is necessary to understand and control the environmental factors which have an influence on chemical composition. This is particularly important when using numerical procedures and quantitative data. The effect of various growth factors on fatty acid composition has been examined by several workers (see Drucker, 1976, 1981 for detailed discussion). Kaneda (1966), in studies on the genus *Bacillus,* demonstrated that branched-chain fatty acid synthesis is increased in media rich in branched-chain amino acids or in branched-chain acid precursors. The effect of substrate on the fatty acids of *Nocardia asteroides* has been reported by Farshtchi and McClung (1970), who found that different growth media gave fatty acid profiles that were both qualitatively and quantitatively different. They also noted that the amount of 10-methyloctadecanoic acid (tuberculostearic acid) was higher in methionine-rich media. This result supported the findings of Karlsson (1956) and of Lennarz *et al.* (1962), who found that the methyl group of tuberculostearic acid was derived from the amino acid methionine. An investigation into the effect of growth medium and culture age on the fatty acids of 41 strains of clostridia led Moss and Lewis (1967) to conclude that although culture age and growth medium influenced the relative proportions of certain acids, the ability to differentiate between species was not altered. Chromatograms were compared visually for qualitative and quantitative differences. In general, the largest differences were between young cultures (5 hr), which had higher amounts of hexadecenoate (16:1) and octadecenoate (18:1) and those harvested at 15 hr or later. There were only minor differences between cells harvested at 15 and 48 hr.

The effect of culture age on the fatty acid composition of *Rhizobium leguminosarum* TA101 and *Rhizobium japonicum* 61A76 was examined by MacKenzie *et al.* (1978) and shown to vary reproducibly. In the fast-growing *R. leguminosarum* TA101, the proportion of the main component, *cis*-vaccenic acid (Δ11-18:1) decreased from 76.5 to 25% of the total fatty acids over an 8-day period with the rate of decrease slowing after the fifth day. This decrease was accompanied by an increase in the amounts of lactobacillic acid (19:cyc), a finding consistent with the known biosynthetic pathway of lactobacillic acid (Gunstone, 1967). The variation in fatty acid composition of the slow-growing *R. japonicum* 61A76 with culture age differed considerably from that of *R. leguminosarum* TA101. The relative amounts of 16:0 and 18:0 + 11 Me-Δ11-18:1 (18:0 and 11 Me-Δ11- 18:1 were not fully resolved on the gas chromatographic system used) remained constant with only a gradual and less pronounced increase in lactobacillic acid. Having established that by careful standardisation fatty acid fingerprints were reproducible, MacKenzie *et al.* (1978) calculated similarity measures (Adams, 1975) on 42 isolates and, using single-linkage cluster analysis, showed that the rhizobial isolates, although constituting

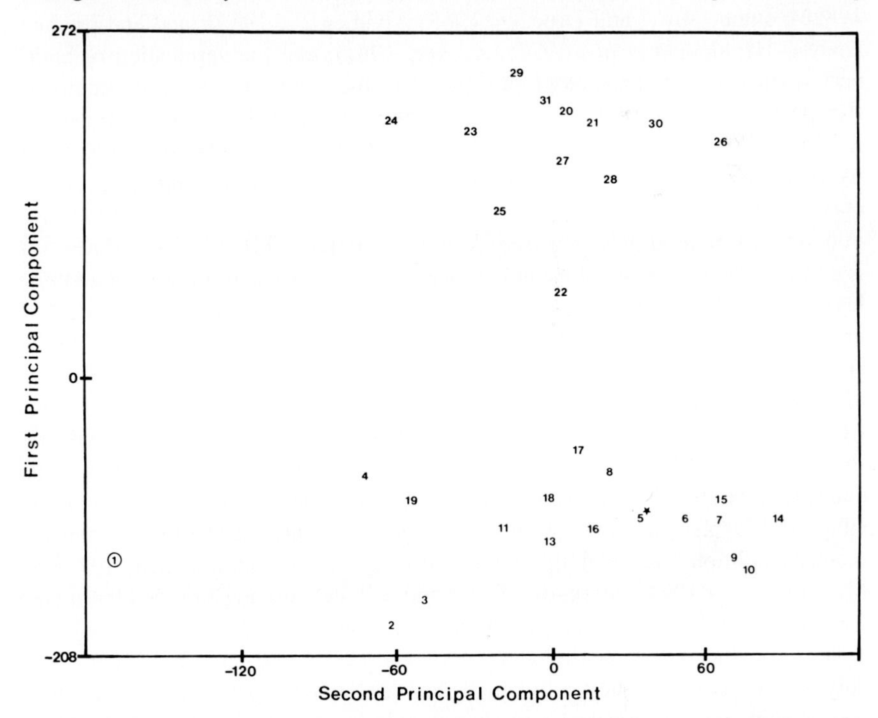

Fig. 3. Principal components analysis on the fatty acid methyl esters of *Staphylococcus aureus* (1–19; ★, strains 5 and 12 co-incident) and *S. intermedius* (20–31). Strain 1 was grown on a different medium from the others. Plot represents 92.7% of the between-strain variation.

a uniform group, could be subdivided into two major groups, the soybean–cowpea isolates and the pea–bean isolates. This result, although differing from the traditional plant-affinity grouping, was consistent with previous results (Graham, 1964; t'Mannetje, 1967; Moffett and Colwell, 1968).

Studies on the fatty acids of some coagulase-positive staphylococci (A. G. O'Donnell, M. Nahaie, M. Goodfellow, and D. E. Minnikin, unpublished data) with subsequent data analysis (Fig. 3) demonstrated the effect of different growth media and the value of numerical procedures such as principal components analysis in detecting aberrant samples. As shown in Fig. 3, *Staphylococcus aureus* (strains 1–19) can be distinguished from *S. intermedius* (strains 20–28). All of the strains except *S. aureus* 1, which is an outlier of the main cluster, were grown under identical conditions on sheep blood agar, whereas *S. aureus* 1 was grown in trypticase soy broth.

Future Trends

The use of gas chromatography in amino acid (Moss *et al.*, 1971; O'Donnell *et al.*, 1982b), sugar (Aluyi and Drucker, 1983; Alvin *et al.*, 1983), and end-product analysis (Holdeman *et al.*, 1977; Drucker, 1981), and the application of high-performance liquid chromatography (hplc) in the study of isoprenoid quinones (Kroppenstedt, 1982) and polar lipids (Batrakov and Bergelson, 1978) have made it easier to obtain quantitative reproducible chemical fingerprints and make it likely that a variety of chemical markers will be used in future numerical chemotaxonomic studies. The introduction of hplc to the analysis of the taxonomically important isoprenoid quinones (see Collins and Jones, 1981, for detailed review) is particularly promising, since unlike mass spectrometry and reverse-phase thin-layer chromatography, quantitative data are obtained. Analysing isoprenoid quinones on reverse-phase and on silver-loaded columns (Kroppenstedt, 1982) has overcome the limitations of reverse-phase hplc (Collins, 1982; Tamaoka *et al.*, 1983) by differentiating menaquinones with a partly saturated isoprenyl chain from those with a fully unsaturated side chain such as MK-9(H_6) and MK-11 (Kroppenstedt, 1982). Further exploitation of chemotaxonomic markers will depend on careful standardisation of growth conditions and the development of simple but sophisticated extraction and analysis procedures. The effect of environmental conditions on polar lipids (Minnikin and Abdolrahhimzadeh, 1974a,b; Minnikin *et al.*, 1974; Veerkamp, 1977) and cell wall amino acids (Schleifer and Kandler, 1972; Schleifer *et al.*, 1976) has been reported previously. Such investigations, together with the numerical analysis of quantitative data, should considerably increase the frequency with which chemical characters can be used to define taxa at subgeneric levels and provide valuable information on the relationships between structure and function in bacteria, thereby improving our understanding of the bacterial cell at the molecular level.

References

Adams, R. P. (1975). Statistical character weighting and similarity stability. *Brittonia* **27,** 305–316.

Aluyi, H. S., and Drucker, D. B. (1983). Trimethylsilyl-sugar profiles of *Streptococcus milleri* and *Streptococcus mitis. Journal of Applied Bacteriology* **54,** 391–397.

Alvin, C., Larsson, L., Magnusson, M., Mardh, P.-A., Odham, G., and Westerdahl, G. (1983). Determination of fatty acids and carbohydrate monomers in micro-organisms by means of glass capillary gas chromatography: Analysis of *Mycobacterium gordonae* and *Mycobacterium scrofulaceum. Journal of General Microbiology* **129,** 401–405.

Batrakov, S. G., and Bergelson, L. D. (1978). Lipids of the streptomycetes. Structural investigation and biological interrelation. *Chemistry and Physics of Lipids* **21,** 1–29.

Bøe, B., and Gjerde, J. (1980). Fatty acid patterns in the classification of some representatives of the families Enterobacteriaceae and Vibrionaceae. *Journal of General Microbiology* **116,** 41–49.

Bousfield, I. J., Smith, G. L., Dando, T. R., and Hobbs, G. (1983). Numerical analysis of total fatty acid profiles in the identification of coryneform, nocardioform and some other bacteria. *Journal of General Microbiology* **129,** 375–394.

Collins, M. D. (1982). A note on the separation of natural mixtures of bacterial menaquinones using reverse-phase high-performance liquid chromatography. *Journal of Applied Bacteriology* **52,** 457–460.

Collins, M. D., and Jones, D. (1981). The distribution of isoprenoid quinone structural types in bacteria and their taxonomic implications. *Bacteriological Reviews* **45,** 316–354.

Collins, M. D., and Jones, D. (1982). Reclassification of *Corynebacterium pyogenes* (Glage) in the genus *Actinomyces,* as *Actinomyces pyogenes* comb. nov. *Journal of General Microbiology* **128,** 901–903.

Collins, M. D., Goodfellow, M., and Minnikin, D. E. (1982). Polar lipid composition in the classification of *Arthrobacter* and *Microbacterium. FEMS Microbiology Letters* **15,** 299–302.

Drucker, D. B. (1974). Chemotaxonomic fatty acid fingerprints of some streptococci with subsequent statistical analysis. *Canadian Journal of Microbiology* **20,** 1723–1728.

Drucker, D. B. (1976). Gas–liquid chromatographic chemotaxonomy. *In* 'Methods in Microbiology' (Ed. J. R. Norris), Vol. 9, pp. 52–125. Academic Press, London.

Drucker, D. B. (1981). 'Microbiological Applications of Gas Chromatography'. Cambridge Univ. Press, Cambridge.

Drucker, D. B., and Melville, T. H. (1971). The classification of some oral streptococci of human or rat origin. *Archives of Oral Biology* **16,** 845–853.

Drucker, D. B., and Melville, T. H. (1973). Adansonian classification of *Streptococcus. Microbios* **7,** 117–130.

Farshtchi, D., and McClung, N. M. (1970). Effect of substrate on fatty acid production in *Nocardia asteroides. Canadian Journal of Microbiology* **16,** 213–217.

Gower, J. C., and Ross, G. J. S. (1969). Minimum spanning trees and single-linkage cluster analysis. *Applied Statistics* **18,** 54–56.

Graham, P. H. 1964. The application of computer techniques to the taxonomy of the root-nodule bacteria of legumes. *Journal of General Microbiology* **35,** 511–517.

Gunstone, F. D. 1967. 'An Introduction to the Chemistry and Biochemistry of Fatty Acids and their Glycerides', 2nd Edition. Chapman & Hall, London.

Gutteridge, C. S., and Norris, J. R. (1979). The application of pyrolysis techniques to the identification of microorganisms. *Journal of Applied Bacteriology* **47,** 5–43.

Harrison, P. J. (1968). Cluster analysis. *Applied Statistics* **17,** 226–236.

Harper, A. M., Duewer, D. L., Kowalski, B. R., and Fasching, J. L. (1977). *In* 'Chemometrics, Theory and Application' (Ed. B. R. Kowalski), pp. 14–52. American Chemical Society, Washington, D.C.

Holdeman, L. V., Cato, E. P., and Moore, W. E. C. (1977). Anaerobe Laboratory Manual 4th Edition. Anaerobe Laboratory, Virginia Polytechnic Institute and State University, Blacksburg.

Ikemoto, S., Kuraishi, H., Komagata, K., Azuma, R., Suto, T., and Murooka, H. (1978). Cellular fatty acid composition in *Pseudomonas* species. *Journal of General and Applied Microbiology* **24,** 199–213.

Jantzen, E., Bryn, K., Bergan, T., and Bøvre, K. (1974a). Gas chromatography of bacterial whole cell methanolysates. V. Fatty acid composition of neisseriae and moraxellae. *Acta Patholigica et Microbiologica Scandinavica, Section B* **82,** 767–779.

Jantzen, E., Bergan, T., and Bøvre, K. (1974b). Gas chromatography of bacterial whole cell methanolysates. VI. Fatty acid composition of strains within Micrococcaceae. *Acta Pathologica et Microbiologica Scandinavica Section B* **82,** 785–798.

Jantzen, E., Bryn, K., Bergan, T., and Bøvre, K. (1975). Gas chromatography of bacterial whole cell methanolysates. VII. Fatty acid composition of *Acinetobacter* in relation to the taxonomy of Neisseriaceae. *Acta Pathologica et Microbiologica Scandinavica Section B* **83,** 569–580.

Johnson, J. L., Anderson, R. S., and Ordal, E. J. (1970). Nucleic acid homologies among oxidase-negative *Moraxella* species. *Journal of Bacteriology* **101,** 568–573.

Kaneda, T. (1966). Biosynthesis of branched chain-fatty acids. IV. Factors affecting relative abundance of fatty acids produced by *Bacillus subtilis. Canadian Journal of Microbiology* **12,** 501–514.

Karlsson, J. L. (1956). Auxotrophic mutants of mycobacteria dependent on fatty acids derived from parent strain. *Journal of Bacteriology* **72,** 813–815.

Keddie, R. M., and Bousfield, I. J. (1980). Cell wall composition in the classification and identification of coryneform bacteria. *In* 'Microbiological Classification and Identification' (Eds. M. Goodfellow and R. G. Board), pp. 167–188. Academic Press, London.

Kersters, K., and De Ley, J. (1980). Classification and identification of bacteria by electrophoresis of their proteins. *In* 'Microbiological Classification and Identification' (Eds. M. Goodfellow and R. G. Board), pp. 273–297. Academic Press, London.

Kroppenstedt, R. M. (1982). Separation of bacterial menaquinones by HPLC using reverse phase (RP 18) and a silver loaded ion exchanger as stationary phases. *Journal of Liquid Chromatography* **5,** 2359–2367.

Lechevalier, M. P., and Lechevalier, H. (1970). Chemical composition as a criterion in the classification of aerobic actinomycetes. *International Journal of Systematic Bacteriology* **20,** 435–443.

Lennarz, W. J., Scheuerbrandt, G., and Bloch, K. (1962). The biosynthesis of oleic and 10-methylstearic acid in *Mycobacterium phlei. Journal of Biological Chemistry* **237,** 664–671.

MacKenzie, S. L., Lapp, M. S., and Child, J. J. (1978). Fatty acid composition of *Rhizobium* spp. *Canadian Journal of Microbiology* **25,** 68–74.

t'Mannetje, L. (1967). A re-examination of the taxonomy of the genus *Rhizobium* and related genera using numerical analysis. *Antonie van Leeuwenhoek* **33,** 477–491.

Minnikin, D. E., and Abdolrahhimzadeh, H. (1974a). The replacement of phosphatidylethanolamine and acidic phospholipids by an ornithine-amide lipid and a minor phosphorus-free lipid in *Pseudomonas fluorescens* NCMB 129. *FEBS Letters* **43,** 257–260.

Minnikin, D. E., and Abdolrahimzadeh, H. (1974b). Effect of pH on the properties of polar lipids, in chemostat cultures of *Bacillus subtilis. Journal of Bacteriology* **120,** 999–1003.

Minnikin, D. E., Abdolrahimzadeh, H., and Baddiley, J. (1974). Replacement of acidic phospholipids by acidic glycolipids in *Pseudomonas diminuta. Nature (London)* **249,** 268–269.

Minnikin, D. E., Collins, M. D., and Goodfellow, M. (1979). Fatty acid and polar lipid composition in the classification of *Cellulomonas, Oerskovia* and related taxa. *Journal of Applied Bacteriology* **47,** 87–95.

Moffett, M. L., and Colwell R. R. 1968. Adansonian analysis of the Rhizobiaceae. *Journal of General Microbiology* **51,** 245–266.

Moss, C. W., and Lewis, V. J. (1967). Characterization of clostridia by gas chromatography. 1. Differentiation of species by cellular fatty acids. *Applied Microbiology* **15,** 390–397.

Moss, C. W., Diaz, F. J., and Lambert, M. A. (1971). Determination of diaminopimelic acid, ornithine, and muramic acid by gas chromatography. *Analytical Biochemistry* **44,** 458–461.

O'Donnell, A. G., and Norris, J. R. (1981). Pyrolysis gas–liquid chromatographic studies in the genus *Bacillus. In* 'The Aerobic Endospore-forming Bacteria' (Eds. R. C. W. Berkeley and M. Goodfellow), pp. 141–179. Academic Press, London.

O'Donnell, A. G., Goodfellow, M., and Minnikin, D. E. (1982a). Lipids in the classification of *Nocardioides:* Reclassification of *Arthrobacter simplex* (Jensen) Lochhead in the genus *Nocardioides* (Prauser) emend. O'Donnell *et al.* as *Nocardioides simplex* comb. nov. *Archives of Microbiology* **133,** 323–329.

O'Donnell, A. G., Minnikin, D. E., Goodfellow, M., and Parlett, J. H. (1982b). The analysis of actinomycete wall amino acids by gas chromatography. *FEMS Microbiology Letters* **15,** 75(E)–78(E).

Romanovskaya, V. A., Malashenko, Y. R., and Grishchenko, N. I. (1980). Diagnosis of methane-oxidising bacteria by numerical methods based on cell fatty acid composition. *Mikrobiologiya* **49,** 969–975.

Schleifer, K. H., and Kandler, O. (1972). Peptidoglycan types of bacterial cell walls and their taxonomic implication. *Bacteriological Reviews* **36,** 407–477.

Schleifer, K. H., and Kocur, M. (1973). Classification of staphylococci based on chemical and biochemical properties. *Archiv für Mikrobiologie* **93,** 65–85.

Schleifer, K. H., Hammes, W. P., and Kandler, O. (1976). Effect of endogenous and exogenous factors on the primary structures of bacterial peptidoglycan. *Advances in Microbial Physiology* **13,** 246–292.

Seidl, P. H., Faller, A. H., and Schleifer, K. H. (1980). Peptidoglycan types and cytochrome patterns of strains of *Oerskovia turbata* and *O. xanthineolytica. Archives of Microbiology* **127,** 173–178.

Shaw, C., Stitt, J. M., and Cowan, S. T. (1951). Staphylococci and their classification. *Journal of General Microbiology* **3,** 1010–1023.

Sincoweay, H., Miyagawa, E., and Kume, T. (1981). Cellular fatty acid composition in staphylococci isolated from bovine milk. *National Institute of Animal Health Quarterly (Japan)* **21,** 14–20.

Sneath, P. H. A. (1957). Application of computers to taxonomy. *Journal of General Microbiology* **17,** 201–226.

Sokal, R. R., and Michener, C. C. (1958). A statistical method for evaluating systematic relationships. *University of Kansas Science Bulletin* **38,** 1409–1438.

Tamaoka, J., Katayama-Fujimura, Y., and Kuraishi, H. (1983). Analysis of bacterial menaquinone mixtures by high performance liquid chromatography. *Journal of Applied Bacteriology* **54,** 31–36.

Veerkamp, J. H. (1977). Effects of growth conditions on the lipid composition of *Bifidobacterium bifidum* subsp. *pennsylvanicum. Antonie van Leewenhoek* **43,** 101–110.

16

Future of Numerical Taxonomy

P. H. A. SNEATH

Department of Microbiology, University of Leicester, Leicester, UK

Introduction

It is now quite clear that the basic philosophy of numerical taxonomy is firmly established, that there is a consistent logic for quantification as applied to classification, and that this will remain secure even though the emphasis of future work may shift. It will be possible by and large to fit the new work into the present logical framework. Thus, if explicitly cladistic methods should become popular in microbiology they will be able to use (with appropriate modifications) schemes for coding and scaling characters that are available for phenetic work. Ideas of centre and variability of phenons can be extended to analogues, such as a common ancestor and evolutionary diversity. Again, extensive tables of DNA pairing values can be profitably treated as if they were similarity matrices of a particular kind, and appropriate methods of clustering and ordination can be applied to them.

Aims of Classification

The contributions of Sokal (Chapter 1) and J. Williams (Chapter 4) consider some of the current controversies that are exercising the attention of taxonomists in biology as a whole. It is difficult to know how far these will enter microbiology in the near future. At present, numerical taxonomy in microbiology remains in essence phenetic, but microbiologists are quite open-minded about applying methods of various kinds. Now that ribosomal RNA similarities and sequences are becoming used (particularly in the study of archaebacterial groups), and protein sequences are becoming more numerous, there will soon be more work with explicitly cladistic methods. So far there have been rather few such studies: Ambler (Chapter 12) mentions many of them, and Hill (1975) made a first attempt at a *character compatibility analysis* with micro-organisms. Many methods now used are in fact phenetic and not phylogenetic, despite their titles

ISBN 0-12-289665-3

(e.g., Stackebrandt *et al.*, 1980), because DNA sequences and the like are phenetic data and the methods of analysis give phenetic groupings, to which a phylogenetic explanation is then attached. The introduction of explicitly cladistic methods, such as minimum-length trees and character compatibility analyses, will be very exciting for exploring the higher groupings of bacteria. However, they are unlikely to be very useful in numerical taxonomies using individual strains of micro-organisms, for both conceptual and practical reasons.

Comparison of phenetic and cladistic analyses will no doubt be illuminating and lead to new biological insights. They will also focus attention on current techniques. There is already some concern as to the reliability of results from *oligonucleotide catalogues,* that is, some versions of catalogues—albeit with longer sub-sequences—have shown in the past some puzzling relations that seem to have little taxonomic relevance (e.g., see the doublet analysis of Subak-Sharpe *et al.*, 1974). The accuracy of some protein sequences is discussed by Ambler (Chapter 12), and the reliability of existing phylogenetic trees from protein sequences has also been called into question (Sneath, 1980). These points require more attention to statistical criteria if technical or chance effects are to be distinguished from those with biological meaning.

Whether Hennigian methods, as at present advocated, will be much used in microbiology is rather doubtful. One must first note that not all cladistic methods are Hennigian: the present use of the term cladistics as synonymous with Hennigian methods is incorrect, and a perversion of the original meaning that refers to relationships by ancestry as opposed to phenetic relationships on present-day properties. But the problems lie at a much deeper level than simply whether the term cladistics is to be restricted in its meaning to the concepts of a narrow school of thought. It is the adequacy of the thought that is in question. For example, J. Williams (Chapter 4) gives an illuminating discussion of 'transformed cladism' as propounded by Patterson (1982). He concludes that it is difficult to know whether Hennigian methods are now being directed towards phylogeny at all; instead, in the hands of some proponents, it is becoming a disguised form of phenetics.

At a yet deeper level it is evident that the attempts to evade the problems caused by our uncertainties over *evolutionary polarity* (i.e., which character states are ancestral or derived), and *homology* (i.e., which characters are in some sense 'the same' in different organisms) are misguided. These problems cannot be evaded. It is notable that those (such as protein analysts) who have to deal with real data are much more circumspect in their approach. Homology has not received much attention in microbiology: we tend to assume that common phenotypic properties (such as lactose fermentation) are homologous in some sense, without looking very deeply into this. Yet it remains an area where we are weak on theory, and where explanations of some puzzling observations may one day be found.

Congruence

The comprehensive review of Colwell (1973) showed that there was generally good agreement between numerical taxonomy based on phenotypic characters and other lines of evidence. Some incongruences were noted, however, and these foreshadow a larger number that are becoming more evident with time. Up to the present there have been rather few studies of congruence in microbiology (other than those that arose incidentally when combining data from different sources), and yet fewer have been studied numerically. Early on, Melville (1965) compared taxonomic relations based on characters from growth under aerobic conditions with those from anaerobic growth. A few studies (mentioned later) have compared results from different growth temperatures. It would be interesting to extend this more systematically.

Microbiologists will be among the first to have to face up to incongruence between different classes of information of an extent that cannot be brushed aside as biologically irrelevant or statistically nonsignificant. Some of these are unlikely to fit easily into current biological concepts. Izard *et al.* (1980) report differences in reciprocal pairing values that imply different relationships, and which seem well outside the expected errors due to technique. Goodfellow *et al.* (1980) found discrepancies in DNA pairing relationships under different technical conditions that cannot be dismissed by baptizing one technique as the only legitimate one. And Ambler has illustrated in Fig. 3 (Chapter 12) conflicting relations from molecular data that not long ago would all have been viewed as impeccably correct for taxonomy.

Similarly, discrepancies between DNA pairing and phenotypic resemblance will come in for close study. It is by no means clear why phenotypic resemblances, although generally very satisfactory up to the genus level, should be much less satisfactory at levels above this. If phenotypic data are poor guides to families and orders, it calls into question the criteria for determining higher groupings. Also, how do we decide whether other data are safe guides? In what way can we check whether our concepts, our data selection, or our techniques are wrong? Do we have explanations for why nucleic acid relationships, for example, are *not* reflected in phenotypic similarities? Is our selection of characters at fault, or is it that the redundancy of the genetic code, combined with differences in molar G + C content, can lead to nucleotide difference that is disproportionately greater than phenotypic difference? These and many other questions come to mind for the future.

Gene transfer is one explanation for incongruences in bacteria, even if it should prove to be not very common in nature at levels above the species (see Ambler, Chapter 12). It is already clear that plasmids can have a limited effect on diagnosis, because they can transfer genes for properties commonly used in identification (Jones, 1983). To what extent they influence overall phenetic relationships between species still awaits detailed study. There will also be a flood of molecular genetic information to assess. Instead of genetic data becoming a panacea, the complexity of the genome will initially make the task of the taxonomist more difficult.

Persistent incongruence may force microbiologists to be the first to develop explicitly alternative classifications for different purposes. Yet we should not expect to find congruence in every situation. Within a single homogeneous cluster of OTUs, with low correlations between characters, there will necessarily be poor congruence between similarities from different character subsets; this lack of congruence follows inevitably from the lack of character correlations. At higher taxonomic ranks there may be similar situations. For example, when there is no marked clustering of species into distinct genera, and the species are more or less equidistant from each other, small changes in the data may lead to instability of classifications, as noted by Sackin (Chapter 2). Different choices of OTUs may then produce major effects. This is illustrated by the study of Harris and Bisby (1980) on flowering plants; it is not clear whether the incongruence they found was mainly due to the pattern of sampling the OTUs. In such instances the use of several OTUs to represent each species might greatly increase the congruence values. This is an area where more work is required.

Theoretical and Statistical Advances

The complexities of the genome will lead to a more rigorous study of the theoretical basis of classification in relation to information theory. This development seems certain, because almost the only common factor in such complex data is 'information' (in a broad sense). One avenue will be along the lines of Gower (1974) and Barnett *et al.* (1975) on maximum predictivity. The properties of asymmetric matrices may also be useful in microbiology (Gower, 1980). We could do with more study on the relation between the choice of tests in bacteriology and the resulting numerical similarities, which Wayne (Chapter 5) has touched on. This will lead to renewed attention to the aims of classifications, and to what extent we can find universal parametric solutions to estimating 'overall' resemblance. A consistent theory of coding and scaling characters is much needed here. Bascomb (Chapter 3) compares the utility of quantitative and qualitative characters, and the handling of characters from some newer technical methods is mentioned later.

The complexities of modern taxonomy suggest that we may need new techniques such as nodal analysis (Lambert and Williams, 1962) to define joint groups of organisms and properties. We should perhaps be looking for altogether new types of taxonomic structure: instead of clustering OTUs around cluster centres—that is, around points in phenetic hyperspace—we could explore clustering them around other configurations, for example clustering to a line (Diday, 1974; Wold, 1975).

Particularly interesting is the suggestion of J. Williams (Chapter 4) that one might weight the dissimilarities in protein sequences by the functional resemblance of the amino acids. New methods of this kind may be useful in following up such points as the observation by Ambler (Chapter 12) that there may be something peculiar about the distribution of sequence resemblances in bacteria—with many very similar sequences but also many very dissimilar ones. Such concepts have been occasionally discussed, but the lack of significance tests has made them unattractive. There are now solutions to this, however, and these methods could be valuable for measuring relationships that are not amenable to counting simply the numbers of identical amino acids, nucleotides, or codons.

Consider, for example, the two small sequences *j* and *k*, where *j* is isoleucine–serine–arginine and *k* is valine–threonine–lysine. There are no matches, yet the sequences look intuitively similar because of the 'close resemblance' of isoleucine to valine, of serine to threonine, and of arginine to lysine. Thus, if we use the probabilities of mutation from one amino acid to another tabulated by Dayhoff (1972, p. 93; averaging symmetrical pairs), we can form a square matrix as follows (in which the usual one-letter abbreviations have been added):

			j		
			Isoleucine I	Serine S	Arginine R
k	Valine	V	.115	.06	.03
	Threonine	T	.055	.075	.03
	Lysine	K	.045	.04	.13

The resemblance between *j* and *k* can be taken as the product of the values for the pairs at the three sites, IV, ST, RK, that is, $.115 \times .075 \times .13 = 1.12 \times 10^{-3}$, whose logarithm is -2.95. We now wish to know if this value is significant.

In recent years significance tests based on combinatorial analysis have become better known (a good example is that of Mantel, 1967, much used in geographic analyses). If one asks what significance should be placed on the observation that *j* and *k* have the above resemblance, one answer is to judge this against all the possible resemblance values obtained by permuting the order of amino acids in *k* while keeping *j* fixed. For *n* sites there are *n*! permutations, here 6, and these, with the logarithmic resemblances, are shown here:

Serial no.	Permutation	Logarithm
1	V-T-K	−2.95
2	V-K-T	−3.86
3	T-K-V	−4.18
4	T-V-K	−3.37
5	K-T-V	−3.99
6	K-V-T	−4.09

The observed permutation has the smallest logarithmic value and shows the higher similarity. It lies at the $\frac{1}{6}$th quantile of the distribution, so it has a probability on this basis of about 1 in 6. It is possible also to calculate the mean (−3.74) and standard deviation (0.48), and when one has much larger examples than the one above, it is possible to utilize the observation that −2.95 lies 1.65 standard deviations from the mean. If the distribution were approximately normal, the observed permutation has a probability of about 1 in 20. It may not always be safe to assume normality even with large examples, but such problems can now be readily overcome by the ease with which a large random sample of the $n!$ values can be calculated by computer; this leads directly to good estimates of quantiles and probabilities.

Stability of Classifications

Stability of classifications receives a good deal of discussion (Sokal, Chapter 1; Sackin, Chapter 2). Various factors involved have been considered by Rohlf and Sokal (1980, 1981). Little has been done on treating the problem in terms of samples from populations (Milligan, 1979). Temple (1982) suggested a measure of robustness of a classification that examines the effect of omitting characters at random on the nearest-neighbour distances. The robustness of numerical taxonomies to small changes of technique is in urgent need of deeper study, as noted by Sackin (Chapter 2). Fortunately, however, this usually seems to affect only a minority of OTUs that happen to be almost equidistant from two or more clusters.

The proliferation of ingenious methods for comparing trees, stemming from the concepts of Farris (1973), is itself becoming a problem, and Sackin's evaluation is particularly useful here.

Statistical instability is paralleled by instability in the properties of micro-organisms over time, and under varied conditions. This is touched on in the contributions of MacDonnell and Colwell (Chapter 6), Priest and Barbour (Chapter 7), and Goodfellow and Dickinson (Chapter 8). It is difficult to foresee developments here. The effects of plasmids, of different environments, and of varied test conditions will need exploring. Very close standardization of test conditions, followed by statistical analyses such as analysis of variance, will be needed to distinguish the effects due to biological factors from those due to experimental error. It is welcome nowadays to see critical examination of experimental error (see Bascomb, Chapter 3; Gutteridge *et al.*, Chapter 14).

Tests for the significance of groups in numerical taxonomies will, it is hoped, be used more often than in the past. Examples are mentioned in Chapter 11 by Williams *et al.* Early studies sometimes accepted phenons on very slender grounds, but now that there are good bases for estimating the effects of test error, and for measuring cluster overlap, much more critical evaluation of phenons should be attempted.

Taxonomic Structure

We still know little about taxonomic structure in bacteria. It is usually believed, as the result of studies by numerical taxonomy and numerical identification, that the great majority of strains fall into distinct phenetic clusters separated by definite gaps. But the question, raised many years ago by Cowan (1955, 1962), of whether they may instead form continuous spectra, is never far from the surface. This view would imply that there are no phenetic gaps (and probably no genetic gaps either) between traditional species. It is difficult to believe that there are no gaps at all, because this would mean that every possible combination of properties would occur among micro-organisms. But one could readily envisage continuous variation in species which are fairly close to one another, arranged as chains or networks, as Cowan suggested.

We do have some statistical tests for continuous phenetic variation, although these have not yet been applied very widely. It is difficult, however, to envisage and test for complex patterns of intergrading variation in multidimensional spaces, so new statistical methods are needed here. Ordination may be useful (e.g., Darland, 1975), but it is probably not adequate. Further, such studies will demand very large and accurate data sets, with numerous strains and numerous properties, if one is to distinguish, for example, a network from a horseshoe pattern.

The concept of sharp limits to species variation was supported by early numerical taxonomic work when sharply distinct clusters were found in phenograms. It was given further emphasis by the advent of numerical identification, when it was found in studies like those of Lapage *et al.* (1970, 1973) and Bascomb *et al.* (1973) that the proportion of strains which did not fit into well-described species was low. There was, perhaps, a tendency to attribute less tidy findings to inadequacies of data or method.

Both types of work, however, dealt with selected material. The evidence for compact, well-separated clusters comes largely from studies on Gram-negative bacteria from rather restricted habitats such as bacteria of medical interest (see MacDonnell and Colwell, Chapter 6; Holmes and Hill, Chapter 10). The position is not so clear in Gram-positive groups. Wayne (Chapter 5) in his contribution describes the complex relations of *Mycobacterium avium, M. intracellulare,* and *M. scrofulaceum,* where clusters seem to overlap considerably (Hawkins, 1977). Williams *et al.* (Chapter 11) note similar phenomena in *Streptomyces,* and other reports (e.g., Orchard *et al.*, 1977; Orchard and Goodfellow, 1980) suggested that this is found in some other groups of soil bacteria. On the other hand, the overlap between species of *Bacillus,* although the strains came from a wide variety of habitats, including soil, does not appear to be pronounced (Logan and Berkeley, 1981; Bonde, 1981).

If considerable overlap does indeed occur (and is not an artefact of the selection of strains, of the testing methods, or of the classificatory algorithms), do most pairs of close taxa overlap mutually, or do only some adjoining taxa overlap to give long chains of variation? It is worth noting that such questions have been asked about other organisms, such as flowering plants; the variation in one section of *Plantago* seems almost completely continuous (Rahn, 1974), and it is difficult to believe that this is an artefact. These questions are particularly relevant to two areas of study that have recently come into prominence, DNA–DNA pairing and ecology.

There is evidence from a number of sources that phenotypic numerical taxonomy and DNA pairing are usually in good agreement. However, there are a number of reports of homogeneous phenotypic clusters which contain numerous 'DNA clusters', that is, clusters that are quite distinct on values from DNA–DNA pairing. This is rather different from just asking if DNA pairing and phenotypic similarities are concordant, because it is the *distinctness* of DNA clusters which is in question. Sometimes new phenotypic differentia are later discovered which show that the DNA clusters are also phenotypically different, even if not sharply distinct (e.g., Logan and Berkeley, 1981; Nakamura and Swezy, 1983).

Further, no one has seriously addressed the statistical evidence that the DNA clusters are actually distinct in the sense of being separated by significant gaps. This can be studied by tests of overlap when adequately large and complete matrices of DNA pairing values become available: until then the problem is intractable, because the great majority of studies provide only a very small proportion of the full matrix of DNA relationships that is required to test the problem. MacDonell and Colwell (Chapter 6) also note that the selection of strains for study may introduce bias, in that there is a risk of choosing from a cluster too many isolates for DNA examination that are phenotypically aberrant.

The difficulties of working with incomplete resemblance matrices (whether they be resemblances based on phenotypic similarities, DNA pairing, or serology) suggest that there are considerable dangers in relying on such incomplete matrices (Sneath, 1983). Apparent clusters from such matrices depend critically on the choice of reference strains, and one should preferably choose a reference strain from each cluster. This, however, begs the question because it presupposes one knows the clusters before commencing the analysis. It is evident that a good deal more work is needed here.

A related problem is the risk that ordination plots may cause clusters that are quite separate in the full phenetic space to overlap seriously in the ordination diagrams (Alderson, Chapter 9). The risk of this can be estimated (for a simple but probably widely applicable model) because of its relation to the χ^2 distribution (Sneath, 1983). In retrospect, unanticipated overlap may explain some of the rather few cases where an ordination diagram showed intermingling of bacteria that on other evidence would be expected to be well separated (e.g., Skyring and Quadling, 1969). I have noted that the extent of overlap in ordinations of *Bacillus* in the study of Logan and Berkeley (1981) is of the order of magnitude expected from the χ^2 distribution.

Taxonomy, Ecology, and Genetics

The relations of taxonomy to ecology and genetics are clearly very important. There are now many applications of numerical taxonomy to ecology, and taxonomy in turn is learning from ecology (e.g., MacDonell and Colwell, Chapter 6; Goodfellow and Dickinson, Chapter 8; Alderson, Chapter 9; Austin *et al.*, 1979; Bell *et al.*, 1982). Several contributors discuss one particular issue, that clear-cut phenetic groups are not always found in ecological surveys, and consequently a high proportion of strains may remain ungrouped (Priest and Barbour, Chapter 7; Goodfellow and Dickinson, Chapter 8; Williams *et al.*, Chapter 11). Reference strains from culture collections may be peripheral members of their clusters, for reasons that remain obscure (although they may have lost plasmid-borne genes during cultivation and preservation). Some wild strains may be particularly unstable genetically when brought into culture.

The pattern of taxonomic variation of bacteria from ecological surveys is thus presenting new challenges. Such surveys are being directed to wider ecological habitats than those studied in earlier numerical taxonomic work, and this may be a partial explanation for the new variation patterns that are being found. There is evidence that populations of Gram-negative bacteria in fresh water vary considerably at different hours of the day or at different seasons. This mainly involves changes in the relative numbers of different biovars, and these biovars do not necessarily form clear-cut groups (Holder-Franklin *et al.*, 1978, Bell *et al.*, 1980).

It may well be, therefore, that a continuum of variation (or a series of overlapping clusters) occurs in some habitats, because this permits bacterial populations to exploit minor fluctuations in the environment. The ability of many bacteria to persist with very little growth, and yet to multiply rapidly when conditions become favourable, may give them an advantage in fluctuating environments over organisms that persist less readily, multiply more slowly, and must generate phenotypic novelty largely by sexual reproduction. Bacterial populations may instead contain a wide range of phenotypic variation; although each variant may be present in very small numbers, those most suited to a changed environment may be able to exploit it quickly. The ability of bacteria to respond rapidly to environmental changes by swift multiplication of particular biovars may therefore be an important factor in the contribution of microbial ecology to ecology as a whole, as discussed in the contributions of MacDonell and Colwell (Chapter 6) and Goodfellow and Dickinson (Chapter 8).

Perhaps related to this is another phenomenon: certain phenons are only seen with tests carried out under particular conditions. Thus, certain phenons in *Yersinia enterocolitica* are only evident from tests performed at particular incubation temperatures (Stevens and Mair, 1973; Harvey and Pickett, 1980; Kapperud *et al.*, 1981; Kaneko and Hashimoto, 1982). The differences appear to turn on a minority of the tests, but the relevance to genetics and ecology remains to be explored, and the taxonomic significance is also uncertain.

Less can be said at present about the relation of numerical taxonomy to genetics, although some points have been mentioned. Better knowledge of bacterial variation is needed if we are to attempt the synthesis of taxonomy and genetics which has proved so interesting in higher organisms. Even the best-known bacterial genera are imperfectly known. We do not know how many new species of bacteria are still to be discovered, but there must be a great many (see Williams *et al.*, 1984). Also, there is difficulty in defining species, or indeed clusters, in nature. These difficulties go beyond the usual considerations of the taxonomist, which are to obtain representative samples of the variation in a habitat and to apply suitable statistical methods to the data from them. If a strain is very successful in a restricted habitat, then its descendants will become numerous there. One may then isolate many strains which show only minor differences from one another. These could form a tight cluster in a numerical taxonomy, and lead one to think it is a distinct taxon. Its status, however, would be more like a collection of replicate cultures. It is difficult to know in such cases what criteria should be used for formal recognition as a taxon. Nevertheless, phenomena like this will be of great significance in ecology and genetics, particularly because the acquisition of genes from other populations (through transmission of plasmids and the like) must play some part in bacterial evolution.

Numerical Identification

Numerical identification is now a rapidly advancing field. Identification matrices of the type discussed by Holmes and Hill (Chapter 10) and Williams *et al.* (Chapter 11) are becoming common, and are increasingly used for computer-assisted identification of unknown strains. Some of these methods are sophisticated; thus, Sielaff *et al.* (1982) have used discriminant functions with a separate covariance matrix for each taxon, including a two-stage procedure whereby closely similar taxa are reassessed with a second such matrix. This application uses quantitative variables based on growth rates in the presence of different inhibitors, and the degree of growth is adjusted for the amount of growth in the control. Bascomb (Chapter 3) also considers quantitative characters.

Even when identification matrices are not used directly to compare with unknowns, they are extensively employed as the data bases for constructing simplier systems of identification, such as profile indices and keys. It may be noted that numerical diagnostic systems depend critically on good numerical taxonomies and on careful test standardization. Also, they still require large resources. Wayne (Chapter 5) discusses how taxonomy, test standardization, and diagnostic systems can all be advanced through collaborative studies. It is to be hoped, too, that criteria for the quality of identification systems (discussed by Holmes and Hill, Chapter 10, and Williams *et al.*, Chapter 11) will be extended and used regularly.

New ways of handling very large matrices of comparative data, often in unfamiliar forms, will be needed. Ercolani (1978) discussed some of the problems, although taxon centroids are generally better than the HMOs that he used. Better ways will be needed to select characters for dividing large data bases into more convenient ones. Some proposals for this were made by Rypka and Babb (1970), Gower and Barnett (1971), and Willcox *et al.* (1980). It may be noted that numerical identification does not depend critically on sharply distinct groups. If the variation forms a linear continuum, it is possible to obtain identification of an unknown to a particular region of the continuum (Sneath, 1979), and this may be sufficient for certain applications.

Technical Methods

New technical methods offer much promise. The integration of automation with both classification and identification is described in this volume for the newer techniques of protein electrophoresis (Kersters, Chapter 13) and pyrolysis mass spectrometry (Gutteridge *et al.*, Chapter 14). Another example is the use of enzyme pattern (Bascomb, Chapter 3, and 1980). Chemotaxonomy and nucleic acid pairing are also developing rapidly. It is to be hoped that findings from these, at present treated as rather separate, will with the other new data be integrated into the numerical taxonomy of the future (see O'Donnell, Chapter 15).

There are, however, some logical and statistical problems to be overcome with these new methods, Appropriate coding and scaling of peaks may not be easy to achieve (Bousfield *et al.*, 1983), although progress is being made (Gutteridge *et al.*, Chapter 14). A deeper difficulty is how one should find discriminatory peaks against a background of nondiscriminatory peaks by methods that do not presuppose that one knows the groupings already. Gutteridge *et al.* (Chapter 14) presents some interesting ideas on this. The partitioning of sources of 'noise' in the way proposed by Eshuis *et al.* (1977) seems a promising approach. This will lead to quite elaborate preprocessing of data before classificatory or discriminatory methods are used.

Future Applications

New applications of numerical taxonomy will no doubt be numerous. One obvious field is to fungal classification, where numerical studies are only slowly being made. It is difficult to separate inherent difficulties with these organisms from historical dogmas on how fungi ought to be classified (e.g., see Kendrick and Weresub, 1966). Studies on yeasts have shown promise (Campbell, 1974, 1975; Barnett *et al.*, 1975), and work such as that of Vishniac and Hempfling (1979) suggested that numerical taxonomy could readily delineate clusters corresponding to yeast species. Progress is also being made with other fungi (e.g., Carmichael *et al.*, 1973; Whalley and Greenhalgh, 1973, 1975). The tradition that fungi imperfecti are somehow imperfect examples of living organisms and have to be treated differently from 'perfect' fungi is, it appears, now becoming a barrier to progress.

Another area that is now ripe for study is that of geographic variation in microorganisms. A few studies have already been made (e.g., Hudson *et al.*, 1976, on electrophoretic variants in relation to geography), but this will require the development of methods to obtain enough good data. Ecologists are now overcoming this, essentially with numerical methods, although they need not lead to traditional identifications. Thus, Griffiths and Lovitt (1980) obtained promising results from using the actual vectors of test results of isolates without attempting conventional identification of them. A related topic is the use of numerical taxonomy in epidemiology ('numerical epidemiology', e.g., Seal *et al.*, 1981), whereby different isolates or variants of an infecting strain may be distinguished numerically from adventitious strains, and traced through an epidemic.

Serology has not received much attention from numerical taxonomists. Studies such as that of Darbyshire *et al.* (1979) on viruses show that it can be usefully analysed by numerical methods (interestingly enough, they found pronounced asymmetry in the reciprocal cross-reactions). The large amount of work required to obtain complete matrices of cross-reactions, and certain technical problems with scoring antigens, has been a drawback. Monoclonal antibodies are now being extensively used, and their potential for numerical diagnostic work will soon be explored. Very soon, too, the development of 'molecular probes' in the form of DNA preparations that can recognize specific genes in lysates of microorganisms will add another weapon to the armory of diagnostic microbiology. Such probes will doubtless have many of the strengths and weaknesses of serology, in that it may be difficult to obtain the required degree of specificity that one would like. Very probably we will see (as with serology in the past) that some probes will be too specific to recognize all strains of a species, while others will give unwanted cross-reactions.

The trend to higher specificity may paradoxically be counterproductive in diagnostic or taxonomic applications, where a level of specificity that is neither too broad nor too narrow is often required.

A few novel applications deserve mention. Bergan (1972) used numerical taxonomy to find the most diverse sets, instead of the most homogeneous groupings, so as to choose a wide-ranging set of typing phages. Selwood and Hedges (1978) used similar methods to discover unusual antisera among antiserum panels with complex cross-reactions. One could envisage that numerical taxonomic studies of the type made by Green and Bousfield (1982) and Jenkins *et al.* (1984) could help the search for unusual organisms with novel properties. Williams *et al.* (1984) suggested that if one knows that a given species often produces a particular class of metabolite, and also that it often utilizes a certain substrate, one may then devise selective media containing that substrate to isolate further strains of the species. This strategy could prove very useful in the search for new microbial products. Methods for detecting hybrids, or strains with unusual plasmids, may also be promising in this regard. The topic has had little attention, and may draw usefully from work on plants (Baum and Lefkovitch, 1973; Schilling and Heiser, 1976).

Conclusion

Predictions are notoriously chancy; it does seem likely, however, that many of the lines of advance discussed here will turn out to be profitable. In exploring them we will no doubt observe phenomena that conflict with what we expected and with current views of taxonomic relationships. These may seem at the time to be awkward findings, but the seeds of new advances may lie in them. The enormous growth of information on micro-organisms will itself produce very large data bases. The techniques of numerical taxonomy will be one of the ways in which we will be able to marshal the new information, and will lead, we may hope, to new adventures of biological exploration.

References

Austin, B., Garges, S., Conrad, B., Harding, E. E., Colwell, R. R., Simidu, U., and Taga, N. (1979). Comparative study of the aerobic, heterotrophic bacteria flora of Chesapeake Bay and Tokyo Bay. *Applied and Environmental Microbiology* **37,** 704–714.

Barnett, J. A., Bascomb, S., and Gower, J. C. (1975). A maximal predictive classification of Klebsiellae and of the yeasts. *Journal of General Microbiology* **86,** 93–102.

Bascomb, S. (1980). Identification of bacteria by measurement of enzyme activities and its relevance to the clinical diagnostic laboratory. *In* 'Microbiological Classification and Identification (Eds. M. Goodfellow and R. G. Board), pp. 359–373. Academic Press, London.

Bascomb, S., Lapage, S. P., Curtis, M. A., and Willcox, W. R. (1973). Identification of bacteria by computer: identification of reference strains. *Journal of General Microbiology* **77,** 291–315.

Baum, B. R., and Lefkovitch, L. P. (1973). A numerical taxonomic study of phylogenetic and phenetic relationships in some cultivated oats, using known pedigrees. *Systematic Zoology* **22,** 118–131.

Bell, C. R., Holder-Franklin, M. A., and Franklin, M. (1980). Heterotrophic bacteria in two Canadian rivers. I. Seasonal variation in the predominant bacterial populations. *Water Research* **14,** 449–460.

Bell, C. R., Holder-Franklin, M. A., and Franklin, M. (1982). Correlations between predominantly heterotrophic bacteria and physicochemical water quality parameters in two Canadian rivers. *Applied and Environmental Microbiology* **43,** 269–283.

Bergan, T. (1972). A new bacteriophage typing set for *Pseudomonas aeruginosa.* 1. Selection procedure. *Acta Pathologica Microbiologia Scandinavica Section B* **80,** 177–188.

Bonde, G. J. (1981). *Bacillus* from marine habitats: allocation to phena established by numerical techniques. *In* 'The Aerobic Endospore-forming Bacteria' (Eds. R. C. W. Berkeley and M. Goodfellow), pp. 181–215. Academic Press, London.

Bousfield, I. J., Smith, G. L., Dando, T. R., and Hobbs, G. (1983). Numerical analysis of total fatty acid profiles in the identification of coryneform, nocardioform and some other bacteria. *Journal of General Microbiology* **129,** 375–394.

Campbell, I. (1974). Methods of numerical taxonomy for various genera of yeasts. *Advances in Applied Microbiology* **17,** 135–156.

Campbell, I. (1975). Numerical analysis and computerized identification of the yeast genera *Candida* and *Torulopsis. Journal of General Microbiology* **90,** 125–132.

Carmichael, J. W., Sekhon, A. S., and Sigler, L. (1973). Classification of some dermatophytes by pyrolysis–gas–liquid chromatography. *Canadian Journal of Microbiology* **19,** 403–407.

Colwell, R. R. (1973). Genetic and phenetic classification of bacteria. *Advances in Applied Microbiology* **16,** 137–175.

Cowan, S. T. (1955). The principles of microbial classification. Introduction: the philosophy of classification. *Journal of General Microbiology* **12,** 314–319.

Cowan, S. T. (1962). The microbial species—A macromyth? *Symposia of the Society for General Microbiology* **12,** 433–455.

Darbyshire, J. H., Rowell, J. G., Cook, J. K. A., and Peters, R. W. (1979). Taxonomic studies on strains of avian infectious bronchitis virus using neutralization tests in tracheal organ cultures. *Archives of Virology* **61,** 227–238.

Darland, G. (1975). Principal component analysis of intraspecific variation in bacteria. *Applied Microbiology* **30,** 282–289.

Dayhoff, M. O., Ed. (1972). 'Atlas of Protein Sequence and Structure 1972'. National Biomedical Research Foundation, Washington, D.C.

Diday, E. (1974). Optimization in non-hierarchical clustering. *Pattern Recognition* **6,** 17–33.

Ercolani, G. L. (1978). *Pseudomonas savastanoi* and other bacteria colonizing the surface of olive leaves in the field. *Journal of General Microbiology* **109,** 245–257.

Eshuis, W., Kistemaker, P. G., and Muezelaar, H. L. C. (1977). *In* 'Analytical Pyrolysis' (Eds. C. E. R. Jones and C. A. Cramers), pp. 151–156. Elsevier, Amsterdam.

Farris, J. S. (1973). On comparing the shapes of taxonomic trees. *Systematic Zoology* **22,** 50–54.

Goodfellow, M., Modarski, M., Tkacz, A., Syzba, L., and Pulverer, G. (1980). Polynucleotide sequence divergence among some coagulase-negative staphylococci. *Zentralblatt für Bakteriologie, Parasitenkunde, Infektionskrankheiten und Hygiene, Abteilung 1, Originale Reihe A* **246,** 10–22.

Gower, J. C. (1974). Maximal predictive classification. *Biometrics* **30,** 643–654.

Gower, J. C. (1980). Problems in interpreting asymmetrical chemical relationships. *In* 'Chemosystematics: Principles and Practice' (Eds. F. A. Bisby, J. G. Vaughan, and C. A. Wright), pp. 399–409. Academic Press, London.

Gower, J. C., and Barnett, J. A. (1971). Selecting tests in diagnostic keys with unknown responses. *Nature (London)* **232,** 491–493.

Green, P. N., and Bousfield, I. J. (1982). A taxonomic study of some Gram-negative facultatively methylotrophic bacteria. *Journal of General Microbiology* **128,** 623–638.

Griffiths, A. J., and Lovitt, R. (1980). Use of numerical profiles for studying bacterial diversity. *Microbial Ecology* **6,** 35–43.

Harris, J. A., and Bisby, F. A. (1980). Classification from chemical data. *In* 'Chemosystematics: Principles and Practice' (Eds. F. A. Bisby, J. G. Vaughan, and C. A. Wright), pp. 305–327. Academic Press, London.

Harvey, S., and Pickett, M. J. (1980). Comparison of Adansonian analysis and deoxyribonucleic acid hybridization results in the taxonomy of *Yersinia enterocolitica. International Journal of Systematic Bacteriology* **30,** 86–102.

Hawkins, J. (1977). Scotochromogenic mycobacteria which appear intermediate between *Mycobacterium avium/intracellulare* and *M. scrofulaceum. American Review of Respiratory Disease* **116,** 963–964.

Hill, L. R. (1975). Problems arising from some tests of Le Quesne's concept of uniquely derived characters. *In* 'Proceedings of the Eighth International Conference on Numerical Taxonomy' (Ed. G. F. Estabrook), pp. 375–398. Freeman, San Francisco.

Holder-Franklin, M. A., Franklin, M., Cashion, P., Cormier, C., and Wuest, L. (1978). Population shifts in heterotrophic bacteria in a tributory of the Saint John River as measured by taxometrics. *In* 'Microbial Ecology' (Eds. M. W. Loutit and J. A. R. Miles), pp. 44–50. Berlin, Springer-Verlag.

Hudson, B. W., Quan, T. J., and Bailey, R. E. (1976). Electrophoretic studies of the geographic distribution of *Yersinia pestis* protein variants. *International Journal of Systematic Bacteriology* **26,** 1–16.

Izard, D., Gavini, F., and Leclerc, H. (1980). Polynucleotide sequence relatedness and genome size among *Enterobacter intermedium* sp. nov. and the species *Enterobacter cloacae* and *Klebsiella pneumoniae. Zentralblatt für Bakteriologie, Mikrobiologie und Hygiene, Abteilung 1, Originale C* **1,** 51–60.

Jenkins, O., Byrom, D., and Jones, D. (1984). Taxonomic studies on some obligate methanol-utilizing bacteria. *In* 'Microbial Growth on C_1 Compounds' (Eds. R. L. Crawford and R. S. Hanson), pp. 255–261. American Society for Microbiology, Washington, D.C.

Jones, D. (1983). Impact of plasmids and transposons on microbial systematics. *In* 'Microbiology—1983' (Ed. D. Schlessinger), pp. 119–124. American Society for Microbiology, Washington, D.C.

Kaneko, K.-I., and Hashimoto, N. (1982). Five biovars of *Yersinia enterocolitica* delineated by numerical taxonomy. *International Journal of Systematic Bacteriology* **32,** 275–287.

Kapperud, G., Bergan, T., and Lassen, J. (1981). Numerical taxonomy of *Yersinia enterocolitica* and *Yersinia enterocolitica*-like bacteria. *International Journal of Systematic Bacteriology* **31,** 401–419.

Kendrick, W. B., and Weresub, L. K. (1966). Attempting neo-Adansonian computer taxonomy at the ordinal level in the basidiomycetes. *Systematic Zoology* **15,** 307–329.

Lambert, J. M., and Williams, W. T. (1962). Multivariate methods in plant ecology. IV. Nodal analysis. *Journal of Ecology* **50,** 775–802.

Lapage, S. P., Bascomb, S., Willcox, W. R., and Curtis, M. A. (1970). Computer identification of bacteria. *In* 'Automation Mechanization and Data Handling in Microbiology' (Eds. A. Baillie and R. J. Gilbert), pp. 1–22. Academic Press, London.

Lapage, S. P., Bascomb, S., Willcox, W. R., and Curtis, M. A. (1973). Identification of bacteria by computer: general aspects and perspectives. *Journal of General Microbiology* **77,** 273–290.

Logan, N. A., and Berkeley, R. C. W. (1981). Classification and identification of members of the genus *Bacillus* using API tests. *In* 'The Aerobic Endospore-forming Bacteria' (Eds. R. C. W. Berkeley and M. Goodfellow), pp. 105–140. Academic Press, London.

Mantel, N. (1967). The detection of disease clustering and a generalized regression approach. *Cancer Research* **27,** 209–220.

Melville, T. H. (1965). A study of the overall similarity of certain actinomycetes mainly of oral origin. *Journal of General Microbiology* **40,** 309–315.

Milligan, G. W. (1979). A note on the use of INDSCAL for the comparison of several classifications. *Systematic Zoology* **28,** 94–99.

Nakamura, L. K., and Swezy, J. (1983). Deoxyribonucleic acid relatedness of *Bacillus circulans* Jordan 1890 strains. *International Journal of Systematic Bacteriology* **33,** 703–708.

Orchard, V. A., and Goodfellow, M. (1980). Numerical classification of some named strains of *Nocardia asteroides* and related isolates from soil. *Journal of General Microbiology* **118,** 295–312.

Orchard, V. A., Goodfellow, M., and Williams, S. T. (1977). Selective isolation and occurrence of nocardiae in soil. *Soil Biology and Biochemistry* **9,** 233–238.

Patterson, C. (1982). Morphological characters and homology. *In* 'Problems of Phylogenetic Reconstruction' (Eds. K. A. Joysey and A. E. Friday), pp. 21–74. Academic Press, London.

Rahn, K. (1974). *Plantago* section *virginica.* A taxonomic revision of a group of American plantains, using experimental, taximetric and classical methods. *Dansk Botanisk Arkiv* **30,** 1–180.

Rohlf, F. J., and Sokal, R. R. (1980). Comments on taxonomic congruence. *Systematic Zoology* **29,** 97–101.

Rohlf, F. J., and Sokal, R. R. (1981). Comparing numerical taxonomic studies. *Systematic Zoology* **30,** 459–490.

Rypka, E. W., and Babb, R. (1970). Automatic construction and use of an identification scheme. *Medical Research Engineering* **9,** 9–19.

Schilling, E. E., Jr., and Heiser, C. B., Jr. (1976). Re-examination of a numerical taxonomic study of *Solanum* species and hybrids. *Taxon* **25,** 451–462.

Seal, D. V., McSwiggan, D. A., Datta, N., and Feltham, R. K. A. (1981). Characterization of an epidemic strain of *Klebsiella* and its variants by computer analysis. *Journal of Medical Microbiology* **14,** 295–305.

Selwood, N., and Hedges, A. (1978). 'Transplant Antigens—A Study in Serological Data Analysis'. Wiley, Chichester.

Sielaff, B. H., Matsen, J. M., and McKie, J. E. (1982). Novel approach to bacterial identification that uses the Autobac System. *Journal of Clinical Microbiology* **15,** 1103–1110.

Skyring, G. W., and Quadling, C. (1969). Soil bacteria: principal component analysis of descriptions of named clusters. *Canadian Journal of Microbiology* **15,** 141–158.
Sneath, P. H. A. (1979). BASIC program for identification of an unknown with presence–absence data against an identification matrix of percent positive characters. *Computers and Geosciences* **5,** 195–213.
Sneath, P. H. A. (1980). The estimation of differences in protein evolution rates. *Proceedings of the Geologists' Association* **91,** 71–79.
Sneath, P. H. A. (1983). Distortions of taxonomic structure from incomplete data on a restricted set of reference strains. *Journal of General Microbiology* **129,** 1045–1073.
Stackebrandt, E., Lewis, B. J., and Woese, C. (1980). The phylogenetic structure of the coryneform group of bacteria. *Zentralblatt für Bakteriologie, Mikrobiologie und Hygiene, Abteilung 1, Originale C* **1,** 137–149.
Stevens, M., and Mair, N. S. (1973). A numerical taxonomic study of *Yersinia enterocolitica* strains. *In* '*Yersinia, Pasteurella* and *Francisella*' (Ed. S. Winblad), pp. 17–22. Karger, Basel.
Subak-Sharpe, J. H., Elton, R. A., and Russell, G. J. (1974). Evolutionary implications of doublet analysis. *Symposia of the Society for General Microbiology* **24,** 131–150.
Temple, J. T. (1982). An empirical study of robustness of nearest-neighbor relations in numerical taxonomy. *Mathematical Geology* **14,** 675–678.
Vishniac, H. S., and Hempfling, W. P. (1979). *Cryptococcus vishniacii* sp. nov., an Antarctic yeast. *International Journal of Systematic Bacteriology* **29,** 153–158.
Whalley, A. J. S., and Greenhalgh, G. N. (1973). Numerical taxonomy of *Hypoxylon.* I. Comparison of classifications of the cultural and the perfect states. *Transactions of the British Mycological Society* **61,** 435–454.
Whalley, A. J. S., and Greenhalgh, G. N. (1975). Numerical taxonomy of *Hypoxylon.* III. Comparison of the cultural states of some *Hypoxylon* species with *Nodulisporium* species. *Transactions of the British Mycological Society* **64,** 229–233.
Willcox, W. R., Lapage, S. P., and Holmes, B. (1980). A review of numerical methods in bacterial identification. *Antonie van Leeuwenhoek* **46,** 233–299.
Williams, S. T., Goodfellow, M., and Vickers, J. C. (1984). New microbes from old habitats. *Symposium of the Society for General Microbiology* **36,** 219–256.
Wold, S. (1975). Analysis of similarities and dissimilarities between chromatographic liquid phases by means of pattern recognition. *Journal of Chromatographic Science* **13,** 525–532.

Index

M

N

O

Q

R

S

Z

Special Publications of the Society for General Microbiology

Publications Officer: Colin Ratledge, 62 London Road, Reading, UK

1. Coryneform Bacteria,
   eds. I. J. Bousfield and A. G. Callely
2. Adhesion of Microorganisms to Surfaces,
   eds. D. C. Ellwood, J. Melling and P. Rutter
3. Microbial Polysaccharides and Polysaccharases,
   eds. R. C. W. Berkeley, G. W. Gooday and D. C. Ellwood
4. The Aerobic Endospore-forming Bacteria: Classification and Identification,
   eds. R. C. W. Berkeley and M. Goodfellow
5. Mixed Culture Fermentations,
   eds. M. E. Bushell and J. H. Slater
6. Bioactive Microbial Products: Search and Discovery,
   eds. J. D. Bu'Lock, L. J. Nisbet and D. J. Winstanley
7. Sediment Microbiology,
   eds. D. B. Nedwell and C. M. Brown
8. Sourcebook of Experiments for the Teaching of Microbiology,
   eds. S. B. Primrose and A. C. Wardlaw
9. Microbial Diseases of Fish,
   ed. R. J. Roberts
10. Bioactive Microbial Products, Volume 2,
    eds. L. J. Nisbet and D. J. Winstanley
11. Aspects of Microbial Metabolism and Ecology,
    ed. G. A. Codd
12. Vectors in Virus Biology,
    eds. M. A. Mayo and K. A. Harrap
13. The Virulence of *Escherichia coli*,
    ed. M. Sussman
14. Microbial Gas Metabolism,
    eds. R. K. Poole and C. S. Dow
15. Computer-assisted Bacterial Systematics,
    eds. M. Goodfellow, D. Jones and F. G. Priest